T0334440

Unitary Reflection Groups

A unitary reflection is a linear transformation of a complex vector space which fixes each point in a hyperplane. Intuitively, it resembles the transformation an image undergoes when it is viewed through a kaleidoscope, or arrangement of mirrors.

This book gives a complete classification of all finite groups which are generated by unitary reflections, using the method of line systems. Irreducible groups are studied in detail, and are identified with finite linear groups. The new invariant theoretic proof of Steinberg's fixed point theorem is treated fully. The same approach is used to develop the theory of eigenspaces of elements of reflection groups and their twisted analogues. This includes an extension of Springer's theory of regular elements to reflection cosets. An appendix outlines links to representation theory, topology and mathematical physics.

Containing over 100 exercises ranging in difficulty from elementary to research level, this book is ideal for honours and graduate students, or for researchers in algebra, topology and mathematical physics.

 1 Introduction to Linear and Convex Programming, N. CAMERON
 2 Manifolds and Mechanics, A. JONES, A. GRAY & R. HUTTON
 3 Introduction to the Analysis of Metric Spaces, J. R. GILES
 4 An Introduction to Mathematical Physiology and Biology, J. MAZUMDAR
 5 2-Knots and their Groups, J. HILLMAN
 6 The Mathematics of Projectiles in Sport, N. DE MESTRE
 7 The Petersen Graph, D. A. HOLTON & J. SHEEHAN
 8 Low Rank Representations and Graphs for Sporadic Groups,
        C. E. PRAEGER & L. H. SOICHER
 9 Algebraic Groups and Lie Groups, G. I. LEHRER (ed.)
10 Modelling with Differential and Difference Equations,
        G. FULFORD, P. FORRESTER & A. JONES
11 Geometric Analysis and Lie Theory in Mathematics and Physics,
        A. L. CAREY & M. K. MURRAY (eds.)
12 Foundations of Convex Geometry, W. A. COPPEL
13 Introduction to the Analysis of Normed Linear Spaces, J. R. GILES
14 Integral: An Easy Approach after Kurzweil and Henstock, L. P. YEE & R. VYBORNY
15 Geometric Approaches to Differential Equations, P. J. VASSILIOU & I. G. LISLE (eds.)
16 Industrial Mathematics, G. R. FULFORD & P. BROADBRIDGE
17 A Course in Modern Analysis and its Applications, G. COHEN
18 Chaos: A Mathematical Introduction, J. BANKS, V. DRAGAN & A. JONES
19 Quantum Groups, R. STREET

Australian Mathematical Society Lecture Series: 20

Unitary Reflection Groups

GUSTAV I. LEHRER
DONALD E. TAYLOR
School of Mathematics and Statistics
University of Sydney

CAMBRIDGE
UNIVERSITY PRESS

CAMBRIDGE
UNIVERSITY PRESS

University Printing House, Cambridge CB2 8BS, United Kingdom

One Liberty Plaza, 20th Floor, New York, NY 10006, USA

477 Williamstown Road, Port Melbourne, VIC 3207, Australia

314-321, 3rd Floor, Plot 3, Splendor Forum, Jasola District Centre, New Delhi - 110025, India

103 Penang Road, #05-06/07, Visioncrest Commercial, Singapore 238467

Cambridge University Press is part of the University of Cambridge.

It furthers the University's mission by disseminating knowledge in the pursuit of education, learning and research at the highest international levels of excellence.

www.cambridge.org
Information on this title: www.cambridge.org/9780521749893

First published 2009

A catalogue record for this publication is available from the British Library

Library of Congress Cataloging in Publication data
Lehrer, Gus, 1947–
Unitary reflection groups / Gustav I. Lehrer, Donald E. Taylor.
p. cm. – (Australian Mathematical Society lecture series ; 20)
Includes bibliographical references and indexes.
ISBN 978-0-521-74989-3 (hardback : alk. paper)
1. Group theory – Reflections. I. Taylor, Donald E. (Donald Ewen), 1945– II. Title.
QA174.2.L44 2009
512′.2 – dc22 2008056031

ISBN 978-0-521-74989-3 Paperback

Contents

Introduction   *page* 1
   1.   Overview of this book   1
   2.   Some detail concerning the content   4
   3.   Acknowledgements   5
   4.   Leitfaden   5

Chapter 1.   Preliminaries   7
   1.   Hermitian forms   7
   2.   Reflections   9
   3.   Groups   12
   4.   Modules and representations   13
   5.   Irreducible unitary reflection groups   15
   6.   Cartan matrices   17
   7.   The field of definition   19
   Exercises   21

Chapter 2.   The groups $G(m, p, n)$ 23
   1.   Primitivity and imprimitivity   23
   2.   Wreath products and monomial representations   24
   3.   Properties of the groups $G(m, p, n)$ 25
   4.   The imprimitive unitary reflection groups   27
   5.   Imprimitive subgroups of primitive reflection groups   32
   6.   Root systems for $G(m, p, n)$ 34
   7.   Generators for $G(m, p, n)$ 35
   8.   Invariant polynomials for $G(m, p, n)$ 36
   Exercises   37

Chapter 3.   Polynomial invariants   39
   1.   Tensor and symmetric algebras   39
   2.   The algebra of invariants   41
   3.   Invariants of a finite group   42
   4.   The action of a reflection   46

5.   The Shephard–Todd–Chevalley Theorem   46
6.   The coinvariant algebra   51
Exercises   53

Chapter 4.   Poincaré series and characterisations of reflection groups   54
1.   Poincaré series   54
2.   Exterior and symmetric algebras and Molien's Theorem   56
3.   A characterisation of finite reflection groups   61
4.   Exponents   63
Exercises   65

Chapter 5.   Quaternions and the finite subgroups of $SU_2(\mathbb{C})$ 66
1.   The quaternions   67
2.   The groups $O_3(\mathbb{R})$ and $O_4(\mathbb{R})$ 69
3.   The groups $SU_2(\mathbb{C})$ and $U_2(\mathbb{C})$ 71
4.   The finite subgroups of the quaternions   72
5.   The finite subgroups of $SO_3(\mathbb{R})$ and $SU_2(\mathbb{C})$ 77
6.   Quaternions, reflections and root systems   79
Exercises   83

Chapter 6.   Finite unitary reflection groups of rank two   84
1.   The primitive reflection subgroups of $U_2(\mathbb{C})$ 84
2.   The reflection groups of type $\mathcal{T}$ 85
3.   The reflection groups of type $\mathcal{O}$ 87
4.   The reflection groups of type $\mathcal{I}$ 89
5.   Cartan matrices and the ring of definition   90
6.   Invariants   93
Exercises   98

Chapter 7.   Line systems   99
1.   Bounds on line systems   99
2.   Star-closed Euclidean line systems   100
3.   Reflections and star-closed line systems   101
4.   Extensions of line systems   103
5.   Line systems for imprimitive reflection groups   104
6.   Line systems for primitive reflection groups   105
7.   The Goethals–Seidel decomposition for 3-systems   111
8.   Extensions of $\mathcal{D}_n^{(2)}$ and $\mathcal{D}_n^{(3)}$ 115
9.   Further structure of line systems in $\mathbb{C}^n$ 119
10. Extensions of Euclidean line systems   120
11. Extensions of $\mathcal{A}_n$, $\mathcal{E}_n$ and $\mathcal{K}_n$ in $\mathbb{C}^n$ 125
12. Extensions of 4-systems   127
Exercises   133

Chapter 8.   The Shephard and Todd classification 137
   1.   Outline of the classification 137
   2.   Blichfeldt's Theorem 138
   3.   Consequences of Blichfeldt's Theorem 140
   4.   Extensions of 5-systems 142
   5.   Line systems and reflections of order three 146
   6.   Extensions of ternary 6-systems 149
   7.   The classification 151
   8.   Root systems and the ring of definition 153
   9.   Reduction modulo p 155
  10.   Identification of the primitive reflection groups .. 157
   Exercises 168

Chapter 9.   The orbit map, harmonic polynomials and semi-invariants 171
   1.   The orbit map 171
   2.   Skew invariants and the Jacobian 172
   3.   The rank of the Jacobian 174
   4.   Semi-invariants 176
   5.   Differential operators 179
   6.   The space of G-harmonic polynomials 183
   7.   Steinberg's fixed point theorem 186
   Exercises 189

Chapter 10.   Covariants and related polynomial identities 191
   1.   The space of covariants 191
   2.   Gutkin's Theorem 194
   3.   Differential invariants 198
   4.   Some special cases of covariants 199
   5.   Two-variable Poincaré series and specialisations . 201
   Exercises 206

Chapter 11.   Eigenspace theory and reflection subquotients ... 208
   1.   Basic affine algebraic geometry 208
   2.   Eigenspaces of elements of reflection groups ... 212
   3.   Reflection subquotients of unitary reflection groups 213
   4.   Regular elements 215
   5.   Properties of the reflection subquotients 218
   6.   Eigenvalues of pseudoregular elements 222

Chapter 12.   Reflection cosets and twisted invariant theory .. 228
   1.   Reflection cosets 228
   2.   Twisted invariant theory 229
   3.   Eigenspace theory for reflection cosets 231

4.   Subquotients and centralisers   237
5.   Parabolic subgroups and the coinvariant algebra   239
6.   Duality groups   242
Exercises   244

Appendix A.   Some background in commutative algebra   246

Appendix B.   Forms over finite fields   250
1.   Basic definitions   250
2.   Witt's Theorem   251
3.   The Wall form, the spinor norm and Dickson's invariant   251
4.   Order formulae   252
5.   Reflections in finite orthogonal groups   253

Appendix C.   Applications and further reading   255
1.   The space of regular elements   255
2.   Fundamental groups, braid groups, presentations   258
3.   Hecke algebras   261
4.   Reductive groups over finite fields   266

Appendix D.   Tables   271
1.   The primitive unitary reflection groups   272
2.   Degrees and codegrees   274
3.   Cartan matrices   276
4.   Maximal subsystems   277
5.   Reflection cosets   277

Bibliography   279

Index of notation   289

Index   291

Introduction

1. Overview of this book

In real Euclidean space a *reflection* is an orthogonal transformation which fixes every vector of some hyperplane, i.e. a subspace of codimension one. Thus a real reflection necessarily has order two. Finite groups generated by reflections in a real vector space have been studied in great depth and they play a central rôle in many branches of mathematics, particularly in the theory of Lie groups and Lie algebras, where many of them appear as 'Weyl groups'. They might be thought of as linking the discrete and continuous strands of Felix Klein's Erlangen programme, according to which geometry is studied through the group of symmetries of the space concerned. The standard work on these groups is Bourbaki's treatise [33] of 1968 and there is a more recent account in the monograph of Humphreys [119]. See [110] for a survey of the breadth of applications up to 1977.

In 1951 Shephard (see [191] and [192]) extended the concept of reflection to a complex vector space with an hermitian inner product. A reflection (sometimes called a pseudo-reflection) is a linear transformation of finite order, which fixes a hyperplane pointwise. Almost immediately, Shephard and Todd [193], building on the work of many authors over the preceding century, obtained the complete classification of finite groups generated by (unitary) reflections. These groups include the Euclidean reflection groups, and arise naturally from them when one considers certain subgroups and subquotients which act on subspaces of the complexification of the real space with which one begins. These more general 'unitary reflection groups' have a wide range of applications, including

(*i*)   the structure and representation theory of reductive algebraic groups;
(*ii*)   Hecke algebras;
(*iii*) knot theory;
(*iv*)   moduli spaces;
(*v*)   algebraic topology, particularly in low dimensions;
(*vi*)   invariant theory and algebraic geometry;
(*vii*) differential equations;
(*viii*) mathematical physics.

In writing this book we have had four principal objectives in mind. Firstly, although it is now more than half a century since the Shephard–Todd classification, there is still no complete and coherent account of this classification in book form in the literature, although there have been some research articles, e.g. [**54, 55**], which have addressed the subject. This is in sharp contrast to the situation for the real reflection groups, which are precisely the finite Coxeter groups (see [**209**, Appendix, Theorem 38] and [**33**, Chap. V, Th. 1 et 2]), and whose classification generally proceeds through the classification of root systems, which is readily available in the literature. Taking into account that the original Shephard–Todd classification itself depends on a significant body of earlier literature, and that the classification is much used and referred to, we thought it useful to provide a complete treatment of the classification of the unitary reflection groups.

For any unitary reflection group G, there is a corresponding collection of lines in the ambient complex space V, obtained by taking the lines orthogonal to the reflecting hyperplanes of G. We call this collection a '*line system*'; our treatment of the classification of the unitary reflection groups comes down to a classification of line systems. There are interrelationships among the various irreducible reflection groups, which may be studied through the relationships among their line systems. A consequence of our approach to the classification is that we are able to elucidate these systematically. In particular we indicate all the maximal reflection subgroups of any irreducible group, which of course essentially provides a complete list of reflection subgroups of each irreducible group. For analogous information concerning the real groups the classical references are [**31, 92, 90**]. More generally, we have sought to provide a good deal of detail concerning individual groups. We have also provided identifications of the irreducible groups with linear groups over finite fields where appropriate.

Our second objective relates to the invariant theory of unitary reflection groups. It is a beautiful result of Shephard and Todd that the unitary reflection groups are characterised among all complex linear groups as those whose algebra of invariants is free, or equivalently those which have a smooth variety of orbits on the vector space V in which they act. This is the merest hint that the invariant theory of unitary reflection groups is a rich vein for study. In this book we develop this theory in several directions. We give a complete treatment of the M-exponents of G, for any G-module M; this includes the usual exponents and the more recent 'coexponents', which are closely related to the topology of the complement M_G of the reflecting hyperplanes of G. These ideas are used to study parabolic subgroups, i.e. the stabilisers of points (or subspaces) of V; in particular, we give a simple proof of Steinberg's Theorem that the parabolic subgroups are reflection groups.

We give a comprehensive account of the application of invariant theoretic methods to the eigenspace theory of Springer and Lehrer–Springer. This has obviated the need for intersection theory, and requires only elementary concepts from affine algebraic geometry, which we provide in Appendix A. Our account includes material concerning centralisers of elements of reflection groups, which we regard as an integral part of the theory. In a related circle of ideas, we study harmonic polynomial functions on V through duality between the polynomial functions and differential operators. These two themes are united in applications to the structure of the coinvariant algebra. In particular, we prove results relating its module structure to that of parabolic subgroups of G.

Thirdly, in the study of reflection groups, it becomes apparent early, even if one confines attention to real groups, that it is important to consider situations where there is a linear transformation γ of the ambient space V which normalises the reflection group G. Examples include normalisers of parabolic subgroups, the ramification groups occurring in the representation theory of reductive groups, and many of the applications in the areas outlined in Appendix C. In view of this, we define 'reflection cosets' γG, and provide a chapter (12) on the 'twisted invariant theory' of such cosets. This theory is very close to the untwisted case, with only a certain set of roots of unity, the 'M-factors' for each $\langle \gamma, G \rangle$-module M entering the picture. The study of these fits well with the eigenspace theory alluded to above.

Finally, although the purpose of this book is to provide background in the core material on reflection groups, we are very conscious that current interest in this subject arises from its application to many and varied branches of mathematics, including those listed above. We have therefore provided, in Appendix C, a brief outline of how the subject matter in this book applies to various areas. We have attempted to write our development in such a way as to be accessible to people working in the diverse areas in which it may be applied. This appendix also contains a number of questions and open problems, which are suitable as research topics.

The reader is referred to the appendix for details, and we confine ourselves here to the following remark as to how these applications arise. A key observation which leads to links between the theory of reflection groups and other areas is that there is an important topological space associated with any unitary reflection group G, namely its associated hyperplane complement M_G, which is defined as the set of points of V which lie on no reflecting hyperplane of G. In the case where G is the symmetric group $\mathrm{Sym}(n)$, M_G is the space of ordered configurations $(z_1, z_2, \ldots, z_n)$ of distinct points $z \in \mathbb{C}$. Now M_G, and its quotient X_G by G, have the structure of complex analytic manifolds, but may also be regarded as the varieties of complex points of algebraic schemes over a number field. Moreover X_G has an interesting fundamental group, which in the example of $\mathrm{Sym}(n)$ is the classical Artin braid group. One may therefore consider differential equations for functions on X_G, or the geometry of its points over various rings; moreover the group algebra of its fundamental group has quotients which arise in various ways in the representation theory of reductive groups. It is these various ways of regarding M_G, X_G and associated spaces which lead to applications in many and varied areas of mathematics.

2. Some detail concerning the content

In this section we provide a brief description of the material in this book, chapter by chapter. In the next section we indicate the logical interdependencies among the chapters, and make some suggestions as to how the book may be used as a text for courses.

In Chapter 1, we introduce the elementary notions which underlie the subject, and define, for any unitary reflection group, the basic concepts of root, root system, and Cartan matrix. These are used later in the classification. In Chapter 2, we make a fairly detailed study of the imprimitive groups $G(m, p, n)$. The Shephard–Todd classification shows that any irreducible unitary reflection group is either one of these groups, or one of the 34 'exceptional groups', which were denoted in [193] as $G_4, G_5, \ldots, G_{37}$, a notation which is still commonly used today. Of these, 19 are two-dimensional, and Chapters 5 and 6 are devoted to their description and classification.

Chapters 3 and 4 provide the characterisation of reflection groups as precisely those groups with a free algebra of invariants. The former gives a general introduction to invariant theory and multilinear algebra, and introduces the coinvariant algebra for the first time; this is used to define the χ-exponents of G and the fake degree of any character χ of G. Chapter 4 uses Poincaré series to complete the proof of the Shephard–Todd characterisation.

In Chapters 7 and 8 the classification of the irreducible unitary reflection groups is completed. First, in Chapter 7, line systems are defined and studied in detail. It is explained how they may be extended, and what restrictions there are on them. In Chapter 8, a complete classification is given of all permissable line systems, and interrelationships among them. This is used to complete the classification. An interesting by-product of our development is the fact that any reflection group may be written over the ring of integers in the field generated by the character values of its defining representation. We call this the *ring of definition* of G, and show that it plays an important role in the description of reflection subgroups, and line subsystems.

The next two chapters, 9 and 10, provide a deeper study of the relationship between the structure and representations of G and its invariant theory. The orbit map $V \rightarrow V/G$ is studied in Chapter 9, and used to prove Steinberg's fixed point theorem and to study the semi-invariants of G. The space of G-harmonic functions is introduced here via duality and differential operators. The structure of the spaces of G-covariants for various representations M of G is studied in Chapter 10, and the M-exponents are defined. The usual exponents and coexponents are treated as special cases, and the structure theorems are translated here into statements concerning two-variable Poincaré series. In Chapter 11, all this is applied to give a complete treatment of the eigenspace theory of Springer and Lehrer–Springer, including related material on centralisers of elements of G.

Chapter 12 presents the twisted theory for reflection cosets which was mentioned above.

This book is intended to be suitable for a graduate student with a good background in undergraduate algebra. The books of Lang [142] and Atiyah–Macdonald [8] are more than adequate for our purpose, but we do not assume their content. On the few occasions where a little more background is required, we generally refer to these sources. The first two appendices, A and B contain background material necessary for some of the proofs in the text. The first contains some elementary affine algebraic geometry, which is needed in the exposition of the material on eigenspace theory. The second contains some material on the spinor norm which is used in the identification of some reflection groups as linear groups over finite fields.

Appendix C provides an introduction to some of the applications of the theory expounded in this book to various areas of algebra, topology and mathematical physics, and contains some suggestions for further reading. It also contains suggested research projects. Finally, Appendix D contains tables of various properties and invariants associated with irreducible finite unitary reflection groups.

3. Acknowledgements

This book is an outgrowth of several courses given over the years to honours and postgraduate students at the University of Sydney by the first author. The first of these was given in the spring of 1995, and we thank the students of those classes for their contribution to the shaping of the ideas presented here.

The authors also thank Michel Broué and Jean Michel for extensive discussions about the unitary reflection groups, particularly their applications to representation theory, Hecke algebras and related geometric themes, and Jean Michel for his valuable comments on the manuscript.

4. Leitfaden

The logical interdependencies among the chapters are indicated in the diagram below. From this diagram, it is clear that there are two main lines of development, one for the classification and specific properties of the various groups, and the other for the invariant theoretic ideas and their application to eigenspace theory. Either of these would be suitable for a one-semester course for graduate students; alternatively one could treat a subset of the chapters at the top of the diagram below, ensuring only that if a course includes a chapter, it should also include those above it in the diagram.

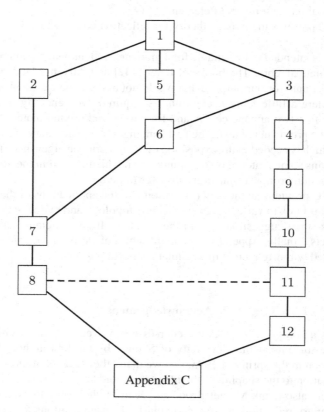

CHAPTER 1

Preliminaries

In this chapter we define unitary reflections and prove some elementary facts about them. We then introduce the important concepts of root, Cartan matrix and root system, which are used extensively in our development of the classification. The notion of the Weyl group of a Cartan matrix is discussed in the context of unitary reflection groups, and some elementary properties of root systems are pointed out. We also review the basic facts and terminology of group theory and representation theory needed throughout the book.

1. Hermitian forms

Definition 1.1. Given a vector space V of dimension n over the complex field $\mathbb{C}$, an *hermitian form* on V is a mapping

$$(-,-) : V \times V \to \mathbb{C}$$

such that

$$(v_1 + v_2, w) = (v_1, w) + (v_2, w)$$
$$(av, w) = a(v, w)$$
$$\overline{(v, w)} = (w, v)$$

for all v, w, v_1, $v_2 \in V$ and $a \in \mathbb{C}$. The hermitian form is *positive definite* if

$$(v, v) \geq 0 \quad \text{and}$$
$$(v, v) = 0 \quad \text{if and only if} \quad v = 0.$$

A positive definite hermitian form is also known as an *inner product*. For example, if V has a basis e_1, e_2, ..., e_n, we may define a positive definite hermitian form on V by

(1.2) $$(u, v) := a_1 \overline{b}_1 + a_2 \overline{b}_2 + \cdots + a_n \overline{b}_n,$$

where $u := a_1 e_1 + a_2 e_2 + \cdots + a_n e_n$ and $v := b_1 e_1 + b_2 e_2 + \cdots + b_n e_n$. It is an easy exercise to show that every positive definite hermitian form on V can be described in this fashion with respect to a suitable basis. In other words, if $(-, -)$ and $[-, -]$ are two positive definite hermitian forms on V, then they are *equivalent*

7

in the sense that there is an invertible linear transformation $\varphi : V \to V$ such that $(u, v) = [\varphi(u), \varphi(v)]$ for all $u, v \in V$.

Let $GL(V)$ be the group of all invertible linear transformations of V. A subgroup G of $GL(V)$ is said to leave the form $(-, -)$ *invariant* if

$$(gv, gw) = (v, w) \quad \text{for all } g \in G \text{ and for all } v, w \in V.$$

We also say that $(-, -)$ is a *G-invariant* form.

Lemma 1.3. *If G is a finite subgroup of $GL(V)$, there exists a G-invariant positive definite hermitian form on V.*

Proof. Choose a positive definite hermitian form $[-, -]$ on V and define a new form by

$$(v, w) := \sum_{g \in G} [gv, gw].$$

Then $(-, -)$ is easily seen to be hermitian and we have

$$(v, v) = \sum_{g \in G} [gv, gv] \geq 0.$$

This expression is 0 if and only if all $[gv, gv]$ are 0. Thus $(-, -)$ is positive definite. Finally, if $h \in G$, then as g runs through G, so does gh and therefore

$$(hv, hw) = \sum_{g \in G} [ghv, ghw] = \sum_{g \in G} [gv, gw] = (v, w).$$

Thus $(-, -)$ is G-invariant.   □

If $(-, -)$ is any positive definite hermitian form on V, we say that $x \in GL(V)$ is *unitary* (or an *isometry*) if $(xv, xw) = (v, w)$ for all $v, w \in V$; that is, $(-, -)$ is $\langle x \rangle$-invariant, where $\langle x \rangle$ is the cyclic group generated by x.

A basis $e_1, e_2, \ldots, e_n$ for V is *orthogonal* if $(e_i, e_j) = 0$ for all $i \neq j$; it is *orthonormal* if in addition $(e_i, e_i) = 1$ for all i.

Let M be the matrix of $x \in GL(V)$ with respect to an orthonormal basis of V. Then x is unitary if and only if M is a *unitary matrix*; i.e. $M\overline{M}^t = I$, where $\overline{M}^t$ denotes the transpose of the complex conjugate of M and I is the identity matrix.

The group of all isometries of V is denoted by $U(V)$ and called the *unitary group* of the form. Its subgroup of transformations of determinant 1 is called the *special unitary group*. The corresponding groups of unitary matrices will be denoted by $U_n(\mathbb{C})$ and $SU_n(\mathbb{C})$, where $n := \dim V$. The group $U(V)$ depends on the form but as any two positive definite hermitian forms on V are equivalent, $U(V)$ is unique up to conjugacy in $GL(V)$. With this notation Lemma 1.3 says that any finite subgroup of $GL(V)$ is a subgroup of $U(V)$ for an appropriate hermitian form.

2. Reflections

Throughout this section V denotes a vector space of dimension n with a positive definite hermitian form $(-, -)$.

Definition 1.4. If U is a subset of V we define the *orthogonal complement* of U to be the subspace $U^\perp := \{ v \in V \mid (u, v) = 0 \text{ for all } u \in U \}$.

If U and W are subspaces of V, we write $V = U \perp W$ to indicate that $V = U \oplus W$ and $(u, w) = 0$ for all $u \in U$ and $w \in W$. It is an easy exercise to check that $V = U \perp W$ if and only if $W = U^\perp$. Further, $U^{\perp\perp} = U$ and $\dim U + \dim U^\perp = \dim V$ for any subspace $U \subseteq V$.

Definition 1.5. Let 1 be the identity element of $GL(V)$. For $g \in GL(V)$ and $H \subseteq GL(V)$, put

(i) $\operatorname{Fix} g := \operatorname{Ker}(1 - g) = \{ v \in V \mid gv = v \}$,

(ii) $V^H := \operatorname{Fix}_V(H) := \{ v \in V \mid hv = v \text{ for all } h \in H \}$, and

(iii) $[V, g] := \operatorname{Im}(1 - g)$.

Lemma 1.6. *If $g \in U(V)$, then $[V, g] = (\operatorname{Fix} g)^\perp$.*

Proof. Suppose that $u := (1 - g)w$ and that $v \in \operatorname{Fix} g$. Then

$$(u, v) = (w - gw, v) = (w, v) - (gw, v)$$
$$= (gw, gv) - (gw, v) = (gw, gv - v) = 0.$$

Thus $[V, g] \subseteq (\operatorname{Fix} g)^\perp$ and on comparing dimensions we see that equality holds. □

Definition 1.7. A linear transformation g is a *reflection* if the order of g is finite and if $\dim[V, g] = 1$. (In some references, such as Bourbaki [**33**], such a transformation is called a pseudo-reflection.)

If g is a reflection, the subspace $\operatorname{Fix} g$ is a hyperplane, called the *reflecting hyperplane* of g.

If a spans $[V, g]$, then for all $v \in V$, there exists $\varphi(v) \in \mathbb{C}$ such that $v - gv = \varphi(v)a$. It is clear that $\varphi : V \to C$ is a linear functional such that $\operatorname{Fix} g = \operatorname{Ker} \varphi$.

We call g a *unitary* reflection if it preserves the hermitian form $(-, -)$. In this case $\operatorname{Fix} g$ is orthogonal to $[V, g]$ and $V = [V, g] \perp \operatorname{Fix} g$.

Suppose $g \in GL(V)$ is a reflection of order m. Then the cyclic group $\langle g \rangle$ has order m and hence, by Lemma 1.3, it leaves invariant a positive definite hermitian form. Thus every reflection g is a unitary reflection with respect to some form. If $H = \operatorname{Fix} g$, then g leaves invariant the line (one-dimensional subspace) $H^\perp$. Hence with respect to a basis adapted to the decomposition $V = H^\perp \perp H$, g has matrix $\operatorname{diag}[\zeta, 1, \dots, 1]$, where ζ is a primitive m^{th} root of unity.

Definition 1.8. A *root* of a line ℓ of V is any non-zero vector of ℓ. If g is a unitary reflection, a *root* of g is a root of the line $[V, g]$. A root a is *short*, *long* or *tall* if (a, a) is 1, 2 or 3, respectively. For the most part we consider only short roots. However, in Chapters 2, 7 and 8 it will be useful to use roots of other lengths.

Any line in $\mathbb{C}^n$ contains long, short and tall roots, each of which is unique up to multiplication by an element of $S^1 := \{\, z \in \mathbb{C} \mid |z| = 1 \,\}$.

Lemma 1.9. *If $g, h \in GL(V)$, then $\mathrm{Fix}(ghg^{-1}) = g\,\mathrm{Fix}\,h$. In particular, if r is a reflection with reflecting hyperplane $H := \mathrm{Fix}\,r$, then grg^{-1} is a reflection with reflecting hyperplane $gH = \mathrm{Fix}(grg^{-1})$.*

Definition 1.10. A *unitary reflection group* is a finite subgroup of $U(V)$ that is generated by reflections. These groups are also referred to by several authors as *complex* reflection groups.

Because of Lemma 1.3, every finite subgroup of $GL(V)$ that is generated by reflections is a unitary reflection group with respect to some positive definite hermitian form on V.

It is important to note that the concept 'unitary reflection group' includes the representation as well as the group. A given group may act as a reflection group or otherwise. For example, for $\zeta := \exp(2\pi i/m)$, the element $\mathrm{diag}[\zeta, 1, \ldots, 1]$ generates a cyclic reflection group of order m, but the (isomorphic) group generated by $\mathrm{diag}[\zeta, \zeta, 1, \ldots, 1]$ is not a reflection group.

Remark 1.11. The sentence 'G is a unitary reflection group in V' will indicate that G is a finite group, generated by reflections in V. By Lemma 1.3, there is then a positive definite G-invariant hermitian form on V, and by Corollary 1.26, this form is unique up to a non-zero positive multiple if G acts irreducibly on V.

Example 1.12. If ω is a cube root of unity, the matrices

$$r := \begin{bmatrix} \omega & 0 \\ -\omega^2 & 1 \end{bmatrix} \quad \text{and} \quad s := \begin{bmatrix} 1 & \omega^2 \\ 0 & \omega \end{bmatrix}$$

are reflections of order 3 and they generate a group of order 24. This is the group G_4 in the list of Shephard and Todd [**193**]. See the exercises at the end of the chapter for further details.

Definition 1.13. The *dual space* of V is the vector space V^* of all linear maps $\varphi : V \to \mathbb{C}$ with addition and multiplication by scalars given by

$$(\varphi + \psi)(v) := \varphi(v) + \psi(v)$$
$$(\alpha\varphi)(v) := \alpha\varphi(v).$$

The elements of V^* are sometimes referred to as (linear) functionals.

Example 1.14. Another example of a unitary reflection group is the *symmetric group* $\mathrm{Sym}(n)$ of all permutations of $\{1, 2, \ldots, n\}$. In order to represent $\mathrm{Sym}(n)$ as a reflection group we first choose an orthonormal basis e_1, e_2, $\ldots$, e_n of V and let $X_1, X_2, \ldots, X_n$ be the dual basis for V^*; that is, $X_i(e_j) = \delta_{ij}$ (Kronecker delta) for all i, j such that $1 \leq i, j \leq n$.

To each permutation $\pi \in \mathrm{Sym}(n)$ we associate the linear transformation of V that sends e_i to $e_{\pi(i)}$; its matrix is the *permutation matrix* associated to π. In particular, the transposition (i, j) corresponds to the reflection that interchanges e_i and e_j and fixes the other basis elements; its hyperplane of fixed points is $\mathrm{Ker}(X_i - X_j)$. The group $\mathrm{Sym}(n)$ is generated by its transpositions (i, j) and therefore it is a unitary reflection group.

Given a hyperplane H in V, let $L_H : V \to \mathbb{C}$ be a linear map such that $H = \mathrm{Ker}\, L_H$. The element $L_H \in V^*$ is determined by H up to a non-zero scalar multiple.

Lemma 1.15. *Suppose that r is a reflection of order m in $GL(V)$, let $H := \mathrm{Fix}\, r$ and suppose that a spans $[V, r]$. Then there exists a primitive m^{th} root of unity α such that*

$$rv = v - (1 - \alpha)\frac{L_H(v)}{L_H(a)}a.$$

Proof. From the definition of a reflection we have $rv = v - \varphi(v)a$ for some linear functional φ such that $H = \mathrm{Ker}\, \varphi$ and $a \notin H$. Thus $ra = \alpha a$ for some $\alpha \in \mathbb{C}$ of order m and hence $\varphi(a) = 1 - \alpha$. Finally, we have $\varphi = \lambda L_H$ for some $\lambda \neq 0$ and we see immediately that $\lambda = (1 - \alpha)/L_H(a)$, and the proof is complete. $\quad\square$

Corollary 1.16. *Suppose that r is a unitary reflection of order m in $U(V)$ and that a is a root of r of length 1. Then there exists a primitive m^{th} root of unity α such that for all $v \in V$ we have*

$$rv = v - (1 - \alpha)(v, a)a.$$

Proof. In this case we may take $L_H(v) = (v, a)$. $\quad\square$

Definition 1.17. Given a non-zero vector $a \in V$ and an mth root of unity $\alpha \neq 1$, define the reflection $r_{a,\alpha}$ by

(1.18) $$r_{a,\alpha}(v) := v - (1 - \alpha)\frac{(v, a)}{(a, a)}a.$$

It follows immediately from this definition that $r_{a,\alpha}$ is a unitary transformation of order m.

Proposition 1.19.

(i) $\quad r_{a,\alpha} r_{a,\beta} = r_{a,\alpha\beta}$.

(ii) $\quad$ *For $g \in U(V)$, $g r_{a,\alpha} g^{-1} = r_{ga,\alpha}$.*

(iii) $\quad$ *For $\lambda \in \mathbb{C}$ such that $\lambda \neq 0$, $r_{\lambda a,\alpha} = r_{a,\alpha}$.*

Proof. All parts follow directly from the definition.   □

Proposition 1.20. *If $g \in U(V)$ and $g r_{a,\alpha} g^{-1} = r_{a,\alpha}^k$ for some k, then $k = 1$. In other words, a reflection is not conjugate to any proper power.*

Proof. From Proposition 1.19 *(ii)* we have $ga = \theta a$ for some θ and hence g leaves both $\mathrm{Fix}(r_{a,\alpha})$ and $\mathbb{C}a$ invariant. It follows that g commutes with $r_{a,\alpha}$.   □

Proposition 1.21. *The unitary reflections $r_{a,\alpha}$ and $r_{b,\beta}$ commute if and only if $\mathbb{C}a = \mathbb{C}b$ or $(a, b) = 0$.*

Proof. A direct calculation yields

$$r_{a,\alpha} r_{b,\beta}(v) = v - (1 - \alpha)\frac{(v, a)}{(a, a)}a - (1 - \beta)\frac{(v, b)}{(b, b)}b + (1 - \alpha)(1 - \beta)\frac{(b, a)}{(a, a)}\frac{(v, b)}{(b, b)}a$$

and by symmetry a similar expression holds for $r_{b,\beta} r_{a,\alpha}(v)$. Thus when a and b are linearly independent, $r_{a,\alpha}$ and $r_{b,\beta}$ commute if and only if the last terms of both expressions vanish; that is, if and only if $(a, b) = 0$.   □

Proposition 1.22. *A subspace W of V is invariant with respect to the reflection r if and only if $W \subseteq \mathrm{Fix}\, r$ or $[V, r] \subseteq W$.*

Proof. If $W \subseteq \mathrm{Fix}\, r$, then W is certainly r-invariant. Also, if $[V, r] \subseteq W$, then the equation (1.18) for r shows that W is r-invariant. Conversely, suppose that W is r-invariant but that $W \nsubseteq \mathrm{Fix}\, r$. Then $[W, r] \neq \{0\}$ and consequently $[V, r] = [W, r] \subseteq W$.   □

Corollary 1.23. *If r is a unitary reflection with root a, then the subspace W is invariant with respect to r if and only if $a \in W$ or $a \in W^{\perp}$.*

3. Groups

The previous two sections introduced some notation associated with groups of linear transformations. In this section we record some standard notions of elementary group theory needed later. As a general reference we use Lang's *Algebra* [**142**, Chapter I].

As in Lang and as is evident from the previous sections we write group actions on the left. That is, if G is a group and X is a set on which G acts, gx denotes the element of X obtained by applying $g \in G$ to $x \in X$. The *G-orbit* of x is the set $\{gx \mid g \in G\}$ and the set of all G-orbits on X is denoted by X/G. For any subset A of X, the *pointwise stabiliser* of A is the subgroup

$$G_A := \{g \in G \mid ga = a \text{ for all } a \in A\}.$$

Given a group G we write $H \leq G$ to indicate that H is a subgroup of G and we write $H \trianglelefteq G$ to indicate that H is a *normal* subgroup of G; that is, $gHg^{-1} = H$ for all $g \in G$.

The *trivial* group, whose only element is the identity, will be denoted by 1. A group G is *simple* if 1 and G are its only normal subgroups.

If X is a subset of G, the *normaliser* $N_G(X)$ and *centraliser* $C_G(X)$ are the subgroups

$$N_G(X) := \{\, g \in G \mid gXg^{-1} = X \,\} \quad \text{and}$$
$$C_G(X) := \{\, g \in G \mid gx = xg \text{ for all } x \in X \,\}.$$

The *centre* of G is the subgroup $Z(G) := C_G(G)$.

If p is a prime, a group whose order is a power of p is called a *p-group*. If p^n is the highest power of p dividing the order of G, then a subgroup of G of order p^n is called a *Sylow p-subgroup* of G. The intersection of the Sylow p-subgroups of G is $O_p(G)$, the largest normal p-subgroup of G.

The cyclic group of order n is denoted by C_n. For a prime p, the direct product of n copies of C_p is known as the *elementary abelian* group of order p^n.

The *commutator* of $x, y \in G$ is $[x, y] := xyx^{-1}y^{-1}$. If X and Y are subsets of G, the subgroup generated by the commutators $[x, y]$ is

$$[X, Y] := \langle\, [x, y] \mid x \in X, \, y \in Y \,\rangle.$$

The *commutator subgroup* (also known as the *derived* group) of G is $[G, G]$ and we sometimes write G' to denote $[G, G]$.

A group G is the *central product* of subgroups H and K (written $G = H \circ K$) if $G = HK$ and $[H, K] = 1$. In this case H and K are normal subgroups of G and $H \cap K \subseteq Z(G)$.

The *normal closure* in G of a subset X is the smallest normal subgroup of G that contains X. In particular, the normal closure of an element $g \in G$ is the subgroup $\langle g^G \rangle$ generated by the set g^G of all conjugates of g.

A group G is an *extension* of a group N by a group H is $N \trianglelefteq G$ and $G/N \simeq H$. The extension is *split* if there is a subgroup K of G such that $K \cap N = 1$ and $G = NK$; otherwise it is *non-split*. A split extension NK is also called the *semidirect product* of N by K.

The *Frattini* subgroup $\Phi(P)$ of a group P is the intersection of the maximal subgroups of P. An *extraspecial p-group* is a p-group P such that $|P'| = p$ and $P' = Z(P) = \Phi(P)$.

4. Modules and representations

In this section we review some of the basic facts about modules needed in later chapters. Once again, a general reference for proofs and for further information is Lang's *Algebra* [**142**].

As in the previous sections V is a vector space of dimension n over the field $\mathbb{C}$ and $GL(V)$ is the group of all invertible linear transformations of V. A *linear representation* of a group G with *representation space* V is a homomorphism ρ : $G \rightarrow GL(V)$. The representation is *faithful* if ρ is injective.

If $\rho : G \rightarrow GL(V)$ is a representation, we say that G *acts* on V and we call V a G-*module*. The action of $g \in G$ on $v \in V$ is defined by $gv := \rho(g)v$. The representation ρ may be extended by linearity to the *group algebra* $\mathbb{C}G$ of G ([**142**, Chapter XVIII]). Thus V becomes a $\mathbb{C}G$-module and if $\xi = \sum_{g \in G} a_g g$ and $v \in V$, then $\xi v = \sum_{g \in G} a_g \rho(g)v$.

The *contragredient* representation of ρ is the homomorphism $\rho^* : G \rightarrow GL(V^*)$ defined by $(\rho^*(g)\varphi)(v) = \varphi(\rho(g)^{-1}v)$ for all $g \in G, \varphi \in V^*$ and $v \in V$. In particular, V^* becomes a G-module with action $(g\varphi)(v) = \varphi(g^{-1}v)$.

A G-*submodule* of V is a vector subspace U such that $gu \in U$ for all $g \in G$ and all $u \in U$. The G-module V is *irreducible* if 0 and V are its only G-submodules.

If U is a G-submodule of V, then V/U is a G-module with action $g(v + U) := gv + U$. We say that V is an *extension* of U by V/U. This extension is said to *split* if there is a G-submodule W of V such that $V = U \oplus W$.

If V and W are G-modules, a G-*homomorphism* is a linear transformation α : $V \rightarrow W$ such that $\alpha(gv) = g\alpha(v)$ for all $g \in G$ and $v \in V$. We denote the set of all G-homomorphisms from V to W by $\operatorname{Hom}_G(V, W)$; it is a vector space over $\mathbb{C}$. A linear transformation $\alpha : V \rightarrow V$ is called an *endomorphism* and the set $\operatorname{End}_G(V) := \operatorname{Hom}_G(V, V)$ of all endomorphisms is a ring in which multiplication is defined to be composition of functions.

If $\alpha : V \rightarrow V$ is a linear transformation, the matrix of α with respect to a basis e_1, $e_2, \ldots, e_n$ of V is $A = (a_{ij})$, where

$$\alpha(e_j) = \sum_{i=1}^{n} a_{ij} e_i.$$

If B is the matrix of $\beta : V \rightarrow V$, then AB is the matrix of $\alpha\beta$. Therefore, from a matrix point of view, a representation of G can be thought of as a homomorphism from G to the group $GL_n(\mathbb{C})$ of invertible matrices over $\mathbb{C}$.

If F is a subring of $\mathbb{C}$, we say that the representation is *definable over* F if there is a basis for V such that the entries of the matrices representing the elements of G belong to F.

Theorem 1.24 (Maschke). *If G is a finite group and if U is a submodule of the G-module V, then the extension V of U by V/U splits. In other words, every submodule of V is a direct summand.*

Proof. By Lemma 1.3 we may suppose that G preserves a positive definite hermitian form on V. Then $U^{\perp}$ is G-invariant and $V = U \oplus U^{\perp}$. For another proof see [**142**, p. 666]. □

As a consequence of Maschke's Theorem, it is always possible to write V as a direct sum of irreducible G-submodules.

Theorem 1.25 (Schur's Lemma). *If V and W are irreducible G-modules, every non-zero homomorphism from V to W is an isomorphism. In addition, $\mathrm{End}_G(V) \simeq \mathbb{C}$.*

Proof. (Lang [**142**, p. 643].) Let $f : V \to W$ be a non-zero homomorphism. Then $\mathrm{Ker}\, f$ and $\mathrm{Im}\, f$ are G-submodules of V and W respectively, hence $\mathrm{Ker}\, f = 0$ and $\mathrm{Im}\, f = W$. If $f \in \mathrm{End}_G(V)$ and $f \neq 0$, let λ be an eigenvalue of f. Then $f - \lambda 1$ is not an isomorphism and hence it is 0. That is, $f = \lambda 1$ and the map $\mathrm{End}_G(V) \to \mathbb{C} : f \mapsto \lambda$ is an isomorphism. $\qquad\square$

Corollary 1.26. *If V is an irreducible G-module and if $(-, -)$ and $[-, -]$ are positive definite G-invariant hermitian forms on V, then for some real number $\lambda > 0$ we have $(u, v) = \lambda[u, v]$ for all $u, v \in V$.*

Proof. Define $f : V \to V^*$ by $f(v)w := (w, v)$ and define $g : V \to V^*$ by $g(v)w := [w, v]$. Then f and g are bijections and $g^{-1}f \in \mathrm{End}_G(V)$. It follows from Schur's Lemma that $g^{-1}f = \lambda 1$ for some $\lambda \in \mathbb{C}^\times$, hence $(u, v) = \lambda[u, v]$ for all $u, v \in V$. Taking $u = v$ we see that $\lambda > 0$. $\qquad\square$

5. Irreducible unitary reflection groups

If G is a unitary reflection subgroup of $U(V)$, the G-module V is called the *natural* (or *reflection*) representation of G. If V is an irreducible G-module, we say that G is an *irreducible unitary reflection group.*

As an application of the results of the previous sections we shall show that, for many purposes, the study of unitary reflection groups reduces to the irreducible case.

Theorem 1.27. *Suppose that G is a finite group generated by reflections on V, which leaves the positive definite hermitian form $(-, -)$ invariant. Then V is the direct sum of pairwise orthogonal subspaces $V_1, V_2, \ldots, V_m$ such that the restriction G_i of G to V_i acts irreducibly on V_i and $G \simeq G_1 \times G_2 \times \cdots \times G_m$. If W is an irreducible subspace of V that is not fixed pointwise by every element of G, then $W = V_i$ for some i.*

Proof. The proof of Maschke's Theorem shows that V is the direct sum of mutually orthogonal irreducible subspaces $V_1, V_2, \ldots, V_m$. By Corollary 1.23, if a is a root of a reflection in G, then $a \in V_i$ for some i. Let G_i be the subgroup of G generated by the reflections whose roots belong to V_i. Then for $i \neq j$, G_i fixes every vector in V_j and by Proposition 1.21, the elements of G_i commute with the elements of G_j. It follows that G is the direct product of the groups G_i and that G_i may be identified with the restriction of G to V_i.

If $W \not\subseteq \operatorname{Fix} r$ for some reflection r, then a root a of r belongs to W. But we also have $a \in V_i$ for some i and therefore $W \cap V_i \neq 0$, whence $W = V_i$. $\square$

It follows from this theorem and its proof that

$$V = V^G \perp V_1 \perp \cdots \perp V_k,$$

where the V_i are non-trivial irreducible G-modules.

Definition 1.28. The *support* of a unitary reflection group $G \subseteq U(V)$ is the subspace M of V spanned by the roots of the reflections in G. Equivalently, the support of G is the orthogonal complement of the subspace V^G of fixed points of G. If the subspace V_i of the previous theorem is not contained in the support of G, then the corresponding group G_i is trivial.

Definition 1.29. The *rank* of a reflection group G is the dimension of its support. A *reflection subgroup* of G is a subgroup H that is generated by reflections.

If G is generated by the reflections $r_1, r_2, \ldots, r_k$ with roots $a_1, a_2, \ldots, a_k$, define a graph Γ with vertex set $R := \{a_1, a_2, \ldots, a_k\}$ by joining a_i to a_j whenever a_i and a_j are neither equal nor orthogonal. Note that in this definition, we may have $|R| < k$.

Proposition 1.30. *If G is a unitary reflection subgroup of $U(V)$ with support V and graph Γ, then V is an irreducible G-module if and only if Γ is connected.*

Proof. If G is reducible, it follows from Theorem 1.27 that V is the direct sum of non-trivial pairwise orthogonal subspaces $V_1, V_2, \ldots, V_k$ with $k > 1$, and that each element of Γ belongs to one of the V_i. Furthermore, since $V^G = 0$, every subspace V_i contains at least one element of Γ. In particular, Γ is not connected.

Conversely, suppose that Γ is not connected. Let V_1 be the subspace spanned by the elements in one of the connected components of Γ and let V_2 be the subspace spanned by the remaining elements. Then V_1 and V_2 are orthogonal and fixed by all the reflections whose roots are in Γ. But these reflections generate G and so V_1 and V_2 are G-invariant. Thus G is reducible. $\square$

Corollary 1.31. *Suppose that G is a unitary reflection group and that H is a reflection subgroup of G acting irreducibly on its support W. If $r \in G$ is a reflection with root a such that $a \notin W \cup W^\perp$, then $\langle H, r \rangle$ acts irreducibly on $W \oplus \mathbb{C}a$.*

Proof. Observe first that since $\langle H, r \rangle$ is generated by reflections with roots in $W \oplus \mathbb{C}a$, $\langle H, r \rangle$ preserves the subspace $W \oplus \mathbb{C}a$. If Γ is the graph corresponding to reflection generators of H, then the graph obtained by adjoining a to Γ remains connected. $\square$

Corollary 1.32. *Suppose that G is a unitary reflection group of rank n that acts irreducibly on its support. Then there is a chain of subgroups*

$$1 = G_0 \subset G_1 \subset G_2 \subset \cdots \subset G_n \subseteq G_\ell = G$$

where $n \leq \ell$, such that for $1 \leq i \leq \ell$, there are reflections r_i such that $G_i = \langle G_{i-1}, r_i \rangle$ and for $1 \leq i \leq n$, G_i is generated by i reflections, has rank i, and acts irreducibly on its support.

Proof. We construct the G_i inductively, beginning with $G_0 = 1$. Suppose that G_i is a reflection subgroup of rank i that acts irreducibly on its support W. If $i < n$, since G has rank n there are reflections of G whose roots are not in W. Since G is irreducible, these roots are not all contained in $W^\perp$ and so there is a reflection $r \in G$ with root $a \notin W \cup W^\perp$. By the previous corollary, $G_{i+1} := \langle G_i, r \rangle$ acts irreducibly on its support $W \oplus \mathbb{C}a$. By construction, the rank of G_n is n and we may successively adjoin reflection generators of G to obtain the groups $G_{n+1}, \ldots, G_\ell = G$. $\qquad\square$

Note that it is a consequence of Corollary 1.32 that any irreducible reflection group of rank n has an irreducible reflection subgroup of rank n which has n generators.

Theorem 1.33. *If G_1 and G_2 are finite irreducible unitary reflection subgroups of $U(V)$, then G_1 and G_2 are conjugate in $GL(V)$ if and only if they are conjugate in $U(V)$.*

Proof. We may represent the elements of G_1 and G_2 by (unitary) matrices with respect to an orthonormal basis. Suppose that for some matrix M we have $g' := MgM^{-1} \in G_2$ for all $g \in G_1$. Then, on taking the conjugate transpose of this equation and using the fact that $\overline{g}^t = g^{-1}$, we find that $\overline{M}^t M$ commutes with every element of G_1. It follows from Schur's Lemma that $\overline{M}^t M$ is a scalar matrix. In fact, its diagonal entries must be positive real numbers and hence it is equal to $\overline{c}cI$ for some $c \neq 0$. But now $c^{-1}M$ is a unitary matrix that conjugates G_1 to G_2. The converse is obvious. $\qquad\square$

6. Cartan matrices

Suppose that G is a unitary reflection subgroup of $U(V)$ and that $\dim V = n$. Let $r_1, r_2, \ldots, r_\ell$ be reflections in G with roots $a_1, a_2, \ldots, a_\ell$. From Definition 1.7, for $1 \leq i \leq \ell$ there exist linear maps $\varphi_i \in V^*$ such that

$$r_i(v) = v - \varphi_i(v)a_i$$

for all $v \in V$. The *Cartan coefficient* of the pair a_i, a_j of roots is defined as $\langle a_i \,|\, a_j \rangle := \varphi_j(a_i)$. Hence $r_i(a_i) = (1 - \langle a_i \,|\, a_i \rangle)a_i$ and therefore $1 - \langle a_i \,|\, a_i \rangle$ is a primitive m_i^{th} root of unity, where m_i is the order of r_i.

Definition 1.34. The *Cartan matrix* of the reflections $r_1, r_2, \ldots, r_\ell$ with respect to the roots a_i is the $\ell \times \ell$ matrix $C := (\langle a_i \mid a_j \rangle)$.

If G is a Euclidean reflection group and if the a_i are the simple roots (see, for example, Humphreys [**119**, p. 39]), then C is the usual Cartan matrix of G.

The following construction reverses the above process, by associating to an appropriate matrix C, a set of reflections, together with one root for each reflection in a vector space V.

Construction 1.35. Suppose that C is an $\ell \times \ell$ matrix such that the diagonal entries α_i of $I - C$ are roots of unity distinct from 1. Define $\ell \times \ell$ matrices $\hat{r}_1, \hat{r}_2, \ldots, \hat{r}_\ell$ as follows: $\hat{r}_i$ is obtained by replacing the ith row of the identity matrix by the ith row of $I - C^t$. The eigenvalues of $\hat{r}_i$ are 1 and α_i with multiplicities $\ell - 1$ and 1, respectively.

Regard $\widetilde{V} \simeq \mathbb{C}^\ell$ as the space of column vectors of length ℓ. Matrices then define linear transformations of $\widetilde{V}$ by left multiplication. Let $K = \operatorname{Ker}_{\widetilde{V}}(C^t)$. The ith row of C^t defines a linear map $\widetilde{\varphi}_i : \widetilde{V} \to \mathbb{C}$ by left multiplication and we have $\widetilde{\varphi}_i(v) = 0$ for all $v \in K$. Then, writing $V = \widetilde{V}/K$, $\widetilde{\varphi}_i$ induces a linear map $\varphi_i : V \to \mathbb{C}$. If a_i is the image in V of the ith standard basis vector of $\widetilde{V}$, then $\hat{r}_i$ defines a reflection r_i on V given by

$$r_i(v) = v - \varphi_i(v)a_i,$$

and C is the Cartan matrix of the reflections $r_1, r_2, \ldots, r_\ell$ with respect to the roots $a_1, a_2, \ldots, a_\ell$.

Definition 1.36. By analogy with the Euclidean case, the group $W(C)$ generated by the reflections $r_1, r_2, \ldots, r_\ell$ is called the *Weyl group* of C. The construction just given shows that $W(C)$ is completely determined by the Cartan matrix C. In general $W(C)$ is not finite.

If $g \in GL(V)$, the reflection $r_i' = g r_i g^{-1}$ corresponds to the root $g(a_i)$ and the linear functional $\varphi_i g^{-1}$. Thus the Cartan matrices of r_i (with respect to the roots a_i) and r_i' (with respect to the roots $g(a_i)$) are the same. On the other hand, if we replace a_i by $\mu_i a_i$ and replace φ_i by $\mu_i^{-1} \varphi_i$, then the matrix C becomes DCD^{-1}, where D is the diagonal matrix with entries $\mu_1, \mu_2, \ldots, \mu_\ell$.

Now suppose that the reflections r_i preserve the hermitian form $(-, -)$. This is the case if and only if a_i is orthogonal to $\operatorname{Ker} \varphi_i$ for all i. Then, in the notation of equation (1.18), we have $r_i = r_{a_i, \alpha_i}$ and

$$\langle a_i \mid a_j \rangle = (1 - \alpha_j) \frac{(a_i, a_j)}{(a_j, a_j)}.$$

Readers familiar with the theory of root systems will note the resemblance of the above to the usual formula involving coroots. In particular, the function $\langle a_i \mid a_j \rangle$ is linear in the first variable, but not the second.

7. The field of definition

Suppose that G is a finite unitary reflection group generated by reflections r_1, $r_2, \ldots, r_\ell$, with Cartan matrix C corresponding to a certain set of corresponding roots. If the entries of C belong to a field F, then G is definable over F. In this section we shall prove that a finite unitary reflection group G has a unique minimal field F over which it is defined and that G has a Cartan matrix whose entries belong to F.

Definition 1.37. If G is a subgroup of $GL(V)$ and if $g \in G$, let $\chi(g)$ denote the trace of the linear transformation g. The function $\chi : G \to \mathbb{C}$ is called the *character* of the representation of G in V, and $\mathbb{Q}(\chi)$ is the field generated by the values $\chi(g)$, for all $g \in G$.

Definition 1.38. The *field of definition* $\mathbb{Q}(G)$ of a unitary reflection group G is the field $\mathbb{Q}(\chi)$, where χ is the character of the natural representation.

As a consequence of a theorem of Clark and Ewing [52], the natural representation of every finite unitary reflection group is definable over its field of definition. More generally, this result extends to any irreducible group that contains a reflection. (See Benson [12, Proposition 7.1.1].)

Theorem 1.39. *Let V be an irreducible module for a finite group G. If G contains a reflection, then the representation of G on V is definable over $\mathbb{Q}(G)$.*

Proof. Let χ be the character of V and let $F := \mathbb{Q}(\chi)$. By Wedderburn's Theorem (Lang [142, Chapter XVII]), the group algebra $F[G]$ of G is semisimple and a direct sum of matrix algebras over division rings. Let A be the component corresponding to the given (reflection) representation of G; this is the unique component on which χ does not vanish. Let K be a maximal subfield of the corresponding division ring D and let V_0 be an irreducible A-module. If $r \in G$ is a reflection, then the set of fixed points of r is a D-subspace of V_0. Hence the dimension of the fixed-point space of r on $K \otimes_F V_0$ is $|D : K|$ times the dimension of the fixed-point space on V_0. It follows that $K = F$ and that G can be represented by matrices with entries in F.

Sketch of an alternative proof. (Clark and Ewing [52], Feit [96]) A reflection r has a unique eigenvalue other than 1. Therefore, by Frobenius reciprocity (Lang [142, p. 688]), χ occurs with multiplicity 1 in the representation induced from a faithful representation of $\langle r \rangle$. It follows that the Schur index of χ is 1. $\square$

Corollary 1.40. *If G is a finite unitary reflection group, then the action of G on its natural representation is definable over $\mathbb{Q}(G)$.*

In fact much more is known. In 1976 Benard [11] proved that *every* representation of a unitary reflection group is definable over its field of definition. See Bessis [13] for a shorter proof.

Corollary 1.41. *If G is a finite unitary reflection group generated by reflections r_1, $r_2, \ldots, r_\ell$, then there is a Cartan matrix of these reflections with entries in $\mathbb{Q}(G)$.*

Proof. We may suppose that the rank of G is n and that G acts on a vector space V of dimension n over the field $\mathbb{Q}(G)$. If $\lambda_i \neq 1$ is an eigenvalue of r_i, then $\lambda_i \in \mathbb{Q}(G)$. Thus r_i has an eigenvector $a_i \in V$ with eigenvalue λ_i and we may choose the notation so that $a_1, a_2, \ldots, a_n$ is a basis for V, where $n \leq \ell$. Then the reflections $r_1, r_2, \ldots, r_\ell$ are represented by matrices with entries in $\mathbb{Q}(G)$. It follows that the entries of the Cartan matrix belong to $\mathbb{Q}(G)$.   □

For real reflection groups that are also Weyl groups (see [**33**, Chap. VI]) it is always possible to choose roots for the reflections so that the entries of the Cartan matrix are rational integers. The best that can be hoped for in the case of unitary reflection groups is that the entries of a Cartan matrix belong to the algebraic integers of the field $\mathbb{Q}(G)$. Accordingly, we make the following definition.

Definition 1.42. The *ring of definition* $\mathbb{Z}(G)$ of G is the ring of algebraic integers in $\mathbb{Q}(G)$.

In general, it is not the case that if a finite group G can be represented by matrices over a number field F, then G can also be represented by matrices over the integers of F; see [**53**] for an example. However, for irreducible unitary reflection groups G, we shall see (in later chapters) that it always possible to represent G by matrices over $\mathbb{Z}(G)$. The key idea is to choose suitable roots for the reflections and then represent G on the $\mathbb{Z}(G)$-module that they generate. We conclude this chapter by setting up the necessary machinery and notation (see also Nebe [**171**]).

By a theorem of Brauer, every representation of a finite group G over the complex numbers is definable over the field $\mathbb{Q}(\zeta_m)$, where ζ_m is a primitive m^{th} root of unity such that $g^m = 1$ for all $g \in G$ (Lang [**142**, XIII, §11]). In later chapters we shall prove this for unitary reflection groups without recourse to Brauer's theorem.

Suppose that F is a finite abelian extension of $\mathbb{Q}$, let A be the ring of integers of F and let $\mu(A)$ be the (finite cyclic) group of roots of unity in A. The field F is fixed by complex conjugation under any embedding of F in $\mathbb{C}$ and since F is abelian there is a well-defined operation of complex conjugation on F which inverts every element of $\mu(A)$. Consequently, if V is a vector space over F, Definition 1.1 applies and we may consider hermitian forms on V. In particular, if $e_1, e_2, \ldots, e_n$ is a basis for V, there is a well-defined hermitian form (u, v) on V given by equation (1.2). However, given an hermitian form $(-, -)$, it is not always possible to choose an orthonormal basis for V with respect to $(-, -)$. (See Feit [**95**] or Nebe [**171**] for more information about this situation.)

Suppose that G is a finite unitary reflection group preserving a free A-submodule L of V. If $v \in L$, $g \in G$ and $g(v) = \theta v$, then $\theta^k = 1$, where k is the order of g; that is, $\theta \in \mu(A)$. It follows that we may choose roots for the reflections of G that satisfy the following refinement of Definition 4.9 of Cohen [55]. See [**1, 47, 117, 118, 195**] for additional developments based on Cohen's definition.

Definition 1.43. An *A-root system* in a vector space V over F with hermitian inner product $(-, -)$ is a pair (Σ, f) where Σ is a finite subset of V and $f : \Sigma \to \mu(A)$ is a function such that

(*i*) Σ spans V and $0 \notin \Sigma$;

(*ii*)   for all $a \in \Sigma$ and $\lambda \in F$, we have $\lambda a \in \Sigma$ if and only if $\lambda \in \mu(A)$;

(*iii*)   for all $a \in \Sigma$ and $\lambda \in \mu(A)$, we have $f(\lambda a) = f(a) \neq 1$;

(*iv*)   for all $a, b \in \Sigma$, the Cartan coefficient

$$\langle a \,|\, b \rangle = (1 - f(b)) \frac{(a, b)}{(b, b)}$$

belongs to A;

(*v*)   for all $a, b \in \Sigma$, we have $r_{a, f(a)}(b) \in \Sigma$ and $f(r_{a, f(a)}(b)) = f(b)$.

We say that (Σ, f) is a root system *defined over* A. If $W := W(\Sigma, f)$ is the group generated by the reflections $r_{a, f(a)}$, then W is a finite unitary reflection group, called the *Weyl group* of the root system. It follows from (*ii*) and (*v*) that $\mu(A)w\Sigma = \Sigma$ and W is a group of permutations of the finite set Σ. From (*i*) only the identity element fixes every element of Σ and therefore W is finite.

The order of $r_{a, f(a)}$ is two if and only if $f(a) = -1$ and therefore every reflection of order two is uniquely determined by the line spanned by its root a. In this case we denote $r_{a, f(a)}$ by r_a. Furthermore, if the order of $r_{a, f(a)}$ is two for all $a \in \Sigma$, then the root system (Σ, f) may be abbreviated to Σ, without ambiguity.

In the following chapters, for each irreducible unitary reflection group G, we construct a root system for G defined over $\mathbb{Z}(G)$.

Exercises

1. Prove that if $(-, -)$ and $[-, -]$ are two positive definite hermitian forms on V there exists an invertible linear transformation $\varphi : V \to V$ such that $(u, v) = [\varphi(u), \varphi(v)]$ for all $u, v \in V$.

2. Show that when U is a subspace of the finite dimensional vector space V, $\dim U + \dim U^\perp = \dim V$ and $V = U \oplus U^\perp$.

3. Suppose that $r \in GL(V)$ is a reflection and that a is a non-zero element of $[V, r]$. Show that $[V^*, r]$ is the hyperplane $\{\, \varphi \in V^* \mid \varphi(a) = 0 \,\}$ and hence r acts on V^* as a reflection.

4. Let G be the group generated by the matrices r and s of Example 1.12.

   (*i*)   Check directly that r and s are reflections of order 3.

   (*ii*) Show that $a := r^{-1}s^{-1}rs$ is an element of order 4 and the conjugates of a under the action of G generate the quaternion group of order 8.

(*iii*)   Let L_1, L_2, L_3 and L_4 be the 1-dimensional subspaces spanned by the vectors $\begin{bmatrix} 1 \\ 0 \end{bmatrix}$, $\begin{bmatrix} 0 \\ 1 \end{bmatrix}$, $\begin{bmatrix} 1 \\ -\omega \end{bmatrix}$ and $\begin{bmatrix} 1 \\ \omega^2 \end{bmatrix}$. Show that G acts on the set $\{L_1, L_2, L_3, L_4\}$ as the alternating group Alt(4).

(*iv*)   Deduce that the order of G is 24.

5. If ω is a cube root of unity, show that the matrices

$$\frac{\omega(i-1)}{2} \begin{bmatrix} 1 & 1 \\ -i & i \end{bmatrix} \quad \text{and} \quad \frac{\omega(i-1)}{2} \begin{bmatrix} 1 & -1 \\ i & i \end{bmatrix}$$

are reflections and that the group they generate is conjugate in $GL_2(\mathbb{C})$ to the group G of the previous exercise.

6. Let G be a reflection group of rank two generated by reflections r and s of orders p and q, respectively. Show that it is possible to choose roots a and b for r and s, respectively, so that their Cartan matrix has the form $\begin{bmatrix} 1-\alpha & \theta \\ 1 & 1-\beta \end{bmatrix}$, where α is a primitive p^{th} root of unity, β is a primitive q^{th} root of unity and θ is an algebraic integer such that $\alpha, \beta, \theta \in \mathbb{Z}(G)$.

CHAPTER 2

The groups $G(m, p, n)$

This chapter introduces the three-parameter family of unitary reflection groups known as $G(m, p, n)$ in the notation of Shephard and Todd [**193**]. These groups are relatively easy to describe in terms of monomial matrices and they serve as a useful source of examples for the material that follows.

1. Primitivity and imprimitivity

Definition 2.1. The G-module V is *imprimitive* if for some $m > 1$ it is a direct sum $V = V_1 \oplus V_2 \oplus \cdots \oplus V_m$ of non-zero subspaces V_i ($1 \le i \le m$) such that the action of G on V permutes the subspaces $V_1, V_2, \ldots, V_m$ among themselves; otherwise V is *primitive*. The set $\{V_1, V_2, \ldots, V_m\}$ is called a *system of imprimitivity* for V. Note that according to this definition, if G is primitive group, then V is necessarily an irreducible G-module.

If V is irreducible, then G acts transitively on any system of imprimitivity. Moreover, if G acts transitively on the set of imprimitivity $\{V_1, V_2, \ldots, V_m\}$ and if H is the stabiliser of V_1 in G, then V_1 is an H-module and V is isomorphic to the module obtained by inducing the representation of H on V_1 to G. (See Lang [**142**, p. 688 ff] for a description of induced representations.)

In the next section we define the groups $G(m, p, n)$ and show that every irreducible imprimitive unitary reflection group is conjugate in $U(V)$ to some $G(m, p, n)$.

Definition 2.2. Given an irreducible G-module I, the *isotypic component* V_I of V corresponding to I is the sum of all G-submodules of V isomorphic to I.

Every irreducible G-submodule of the isotypic component V_I is isomorphic to I and V is the direct sum of its distinct isotypic components (Exercise 2 at the end of this chapter).

Theorem 2.3. *If G is a finite primitive subgroup of $GL(V)$ and if A is an abelian normal subgroup of G, then A is cyclic and contained in the centre $Z(G)$ of G.*

Proof. Since A is abelian, its elements have a common eigenvector $w \in V$. That is, for all $a \in A$ we have $aw = \chi(a)w$ for some $\chi(a) \in \mathbb{C}^\times$. Then $\chi : A \to \mathbb{C}^\times$ is a homomorphism (in other words, a *character* of A) and the isotypic component of V corresponding to χ is

$$V_\chi := \{\, v \in V \mid av = \chi(a)v \text{ for all } a \in A \,\}.$$

For $g \in G$, the normality of A implies that gV_χ is the isotypic component corresponding to the character $g\chi$ of A, where $(g\chi)(a) := \chi(g^{-1}ag)$. But V is the direct sum of its isotypic components for A and we have just shown that G permutes them among themselves. Thus the assumption that G is primitive means that there is only one isotypic component, namely $V = V_\chi$. This shows that A acts on V by scalar transformations, whence $A \subseteq Z(G)$. In addition, $\chi : A \to \mathbb{C}^\times$ is one-to-one and hence A is cyclic since, *a fortiori*, every finite subgroup of $\mathbb{C}^\times$ is cyclic (see Lang [**142**, p. 177]). $\square$

In essence we have just proved a special case of Clifford's Theorem (see Dornhoff [**86**, Lemma 29.6]). For finite unitary reflection groups there is a converse to this result, namely an imprimitive irreducible unitary reflection group has a non-central abelian normal subgroup. This is a consequence of Theorem 2.14 and the structure of the groups $G(m, p, n)$, described in the next two sections.

In the case of a group which in addition to being primitive is also generated by reflections we can say a little more about its normal subgroups.

Theorem 2.4. *If $G \subseteq U(V)$ is a finite primitive unitary reflection group and if N is a normal subgroup of G, then either V is an irreducible N-module or $N \subseteq Z(G)$.*

Proof. Suppose that V is reducible as an N-module. Then $V = V_1 \oplus \cdots \oplus V_k$, where $k > 1$ and the V_i are proper irreducible N-submodules. For each i there exists a reflection $r \in G$ with root a such that $rV_i \neq V_i$. Since N is normal in G, rV_i is an irreducible N-submodule and hence $rV_i \cap V_i = \{0\}$. On the other hand if $a^\perp \cap V_i \neq \{0\}$, then $rV_i = V_i$ and to avoid a contradiction we must have $a^\perp \cap V_i = \{0\}$. Since $a^\perp$ is a hyperplane in V it follows that $\dim V_i = 1$ for all i. But now N is abelian and by the previous theorem we have $N \subseteq Z(G)$. $\square$

2. Wreath products and monomial representations

Definition 2.5. Given a group G acting on the set $\Omega := \{1, 2, \ldots, n\}$ and any group H, we can form the direct product $B := H \times H \times \cdots \times H$ of n copies of H and then define an action of G on B as follows. For $g \in G$ and $h := (h_1, h_2, \ldots, h_n) \in B$ we put

$$g.h := (h_{g(1)}, h_{g(2)}, \ldots, h_{g(n)}).$$

The *wreath product* $H \wr G$ of H by G is the semidirect product of B by G. Its elements will be written $(h; g)$ and its multiplication is given by

$$(h; g)(h'; g') = (h\, g.h'; gg').$$

If H and G are finite, the order of $H \wr G$ is $|H|^n |G|$.

For the remainder of this section we shall assume that H is a finite subgroup of the multiplicative group $\mathbb{C}^\times$ of $\mathbb{C}$. That is, for some m, H is the (cyclic) group $\boldsymbol{\mu}_m$ of mth roots of unity.

If V is a vector space of dimension n over $\mathbb{C}$ with a positive definite hermitian form $(-, -)$ and an orthonormal basis $e_1, e_2, \ldots, e_n$, then the elements of B can be represented by diagonal transformations of V. That is, with h as above, we put

$$he_i = h_i e_i \quad \text{for } 1 \leq i \leq n.$$

Because $h_i \bar{h}_i = 1$ for all $h_i \in H$, B is a subgroup of $U(V)$.

The group G acts on V by permuting the basis vectors:

$$ge_i := e_{g(i)} \quad \text{for } 1 \leq i \leq n$$

and thus it too may be regarded as a subgroup of $U(V)$. The actions of $B = \boldsymbol{\mu}_m^n$ and G on V are compatible and lead to an action of $\boldsymbol{\mu}_m \wr G$ on V given by

$$(h; g)e_i := h_{g(i)} e_{g(i)} \quad \text{for } 1 \leq i \leq n.$$

The matrices of these linear transformations (with respect to $e_1, e_2, \ldots, e_n$) have exactly one non-zero entry in each row and column and are known as *monomial matrices*. Consequently we call this representation of $\boldsymbol{\mu}_m \wr G$ the *standard monomial representation*. Thus $\boldsymbol{\mu}_m \wr G$ is represented as a subgroup of $U(V)$ and the subspaces $\mathbb{C}e_1, \mathbb{C}e_2, \ldots, \mathbb{C}e_n$ form a system of imprimitivity for $\boldsymbol{\mu}_m \wr G$.

Definition 2.6. With $B := \boldsymbol{\mu}_m^n$ as above and for each divisor p of m let

$$A(m, p, n) := \{\, (\theta_1, \theta_2, \ldots, \theta_n) \in B \mid (\theta_1 \theta_2 \cdots \theta_n)^{m/p} = 1 \,\}.$$

It is immediate that $A(m, p, n)$ is a subgroup of index p in B that is invariant under the action of $\mathrm{Sym}(n)$.

The group $G(m, p, n)$ is defined to be the semidirect product of $A(m, p, n)$ by the symmetric group $\mathrm{Sym}(n)$. This notation was introduced by Shephard and Todd.

The group $G(m, p, n)$ is a normal subgroup of index p in $\boldsymbol{\mu}_m \wr \mathrm{Sym}(n)$ and consequently we may represent it as a group of linear transformations in the standard monomial representation. Its order is $m^n \, n!/p$.

3. Properties of the groups $G(m, p, n)$

As a step towards showing that $G(m, p, n)$ is generated by reflections we first determine how a reflection can act non-trivially on a system of imprimitivity and then apply this to the reflections of $G(m, p, n)$.

Lemma 2.7. *Suppose that V is a vector space of dimension n over $\mathbb{C}$ with positive definite hermitian form $(-, -)$ and that $V_1, V_2, \ldots, V_k$ are 1-dimensional subspaces* *spanned by linearly independent vectors $e_1, e_2, \ldots, e_k$. If r is a reflection that permutes the V_i among themselves and if $rV_i = V_j$ for some $i \neq j$, then the order of r is two and $V_h \subseteq \mathrm{Fix}\, r$ for all $h \neq i, j$.*

Proof. We have $re_i = \theta e_j$ for some θ. If a is a short root of r, then

$$\theta e_j = e_i - (1 - \alpha)(e_i, a)a$$

for some α and hence $a \in V_i + V_j$. It follows from Proposition 1.22 that $V_i + V_j$ is r-invariant. Furthermore, the same argument shows that r cannot interchange more than one pair of the subspaces V_h and hence $V_h \subseteq \mathrm{Fix}\, r$ for all $h \neq i, j$. In particular, we have $re_j = \theta' e_i$ for some θ'. But now r^2 has $\theta\theta'$ as an eigenvalue with multiplicity at least two and hence $\theta\theta' = 1$. This shows that the order of r is two and r acts on the set $\{V_1, V_2, \ldots, V_k\}$ as a transposition. $\qquad\square$

We shall always regard $G(m, p, n)$ as being defined with respect to the orthonormal basis $e_1, e_2, \ldots, e_n$ for V introduced in the previous section.

Lemma 2.8. *The element $r \in G(m, p, n)$ is a reflection if and only if r has one of the following forms:*

(i)   *For some i and for some $(m/p)^{\mathrm{th}}$ root of unity $\theta \neq 1$, $r = r_{e_i, \theta}$.*
(ii)   *For some $i \neq j$ and for some m^{th} root of unity θ, $r = r_{e_i - \theta e_j, -1}$. In this case the order of r is two, $re_i = \theta e_j$, $re_j = \theta^{-1} e_i$ and $re_k = e_k$ for all $k \neq i, j$.*

Proof. It is clear from its definition that $G(m, p, n)$ contains the reflections described in (i) and (ii).

Conversely, suppose that $r \in G(m, p, n)$ is a reflection. If $r \in A(m, p, n)$, then we have case (i). If $r \notin A(m, p, n)$, then for some $i \neq j$ we have $re_i = \theta e_j$, where θ is an mth root of unity. It follows from the previous lemma that the order of r is two. Finally, it is clear that $e_i - \theta e_j \in [V, r]$ and hence (ii) holds. $\qquad\square$

Proposition 2.9. *The group $G(m, m, n)$ contains $m\binom{n}{2}$ reflections and the order of every reflection is two. If $n > 2$ or if m is odd, the reflections form a single conjugacy class. If m is even, $G(m, m, 2)$ contains two conjugacy classes of reflections.*

Proof. It follows from Lemma 2.8 that the subgroup of $G(m, p, n)$ generated by the reflections of type (ii) is the group $G(m, m, n)$ and this group does not contain any reflections of type (i). If $n \geq 3$ the group $G(m, m, n)$ has a single orbit on the lines spanned by the vectors $e_i - \theta e_j$, where θ is an mth root of unity. Therefore all reflections in $G(m, m, n)$ are conjugate. The group $G(m, m, 2)$ is a dihedral group of order $2m$ and it can be seen directly that this group has a single conjugacy class of reflections when m is odd and two classes otherwise. $\qquad\square$

Proposition 2.10. *If $m > 1$, then $G(m, p, n)$ is an imprimitive unitary reflection group. It is irreducible except when $(m, p, n) = (2, 2, 2)$.*

Proof. The reflections $r_{e_i-e_j,-1}$ generate the symmetric group $\mathrm{Sym}(n)$ and the reflections $r_{e_i,\eta}$ together with the products $r_{e_i-\theta e_j,-1}r_{e_i-e_j,-1}$ generate $A(m,p,n)$, where η is an $(m/p)^{\mathrm{th}}$ root of unity and θ is an mth root of unity. Thus $G(m,p,n)$ is generated by reflections.

We shall prove that $G(m,p,n)$ is irreducible except for the groups $G(1,1,n)$ and $G(2,2,2)$. To this end, suppose that W is a proper invariant subspace. In particular, W is fixed by each reflection $r_{e_i-e_j,-1}$ and, by Corollary 1.23, we have $e_i-e_j\in W$ or $W^\perp$. If, for distinct indices i, j and k, we have $e_i-e_j\in W$ and $e_j-e_k\in W^\perp$, then e_i-e_k cannot belong to W or to $W^\perp$, which is a contradiction. Thus, without loss of generality, we may suppose that $e_i-e_j\in W^\perp$ for all i, j and therefore $W=\langle e_1+\cdots+e_n\rangle$. If $h:=(\theta_1,\ldots,\theta_n)\in A(m,p,n)$, then h preserves W and so $\theta_1=\theta_2=\cdots=\theta_n$. From this it is clear that either $m=1$ or $m=p=n=2$. $\square$

Example 2.11. When $m\le 2$, the matrices of the standard monomial representation for the elements of $G(m,p,n)$ have real entries and hence the group may be regarded as a Euclidean reflection group. There are several other cases where the groups $G(m,p,n)$ are familiar.

(*i*) $G(m,p,1)$ is the cyclic group of order m/p.

(*ii*) $G(1,1,n)$ is the symmetric group $\mathrm{Sym}(n)$. Its action on the hyperplane $\langle e_1+e_2+\cdots+e_n\rangle^\perp$ is primitive whenever $n\ge 5$. This group is known as the Coxeter group of type A_{n-1} and the hyperplane is its reflection module.

(*iii*) $G(2,1,n)\simeq\boldsymbol{\mu}_2\wr\mathrm{Sym}(n)$ is known as the Coxeter group of type B_n.

(*iv*) $G(2,2,n)$ is known as the Coxeter group of type D_n.

(*v*) $G(m,m,2)$ is the dihedral group of order $2m$. It is also known as the Coxeter group of type $I_2(m)$ but the standard monomial representation is not the Euclidean reflection module for $I_2(m)$.

(*vi*) $G(3,3,2)$ is the symmetric group $\mathrm{Sym}(3)$ and $G(2,2,3)$ is the symmetric group $\mathrm{Sym}(4)$.

(*vii*)   The group $G(4,4,2)$ is a dihedral group of order 8, generated by $\left[\begin{smallmatrix} i & 0 \\ 0 & -i \end{smallmatrix}\right]$ and $\left[\begin{smallmatrix} 0 & 1 \\ 1 & 0 \end{smallmatrix}\right]$. On the other hand, $G(2,1,2)$, with generators $\left[\begin{smallmatrix} -1 & 0 \\ 0 & 1 \end{smallmatrix}\right]$ and $\left[\begin{smallmatrix} 0 & 1 \\ 1 & 0 \end{smallmatrix}\right]$ is also a dihedral group of order 8. These groups are conjugate in $U_2(\mathbb{C})$.

(*viii*)   The group $G(4,2,2)$ can be written in four ways as the central product $\mathcal{C}_4\circ D$ of the cyclic group $\mathcal{C}_4$ generated by $\left[\begin{smallmatrix} i & 0 \\ 0 & i \end{smallmatrix}\right]$ and a subgroup D of order 8. The group D is either the quaternion group Q_8 of order 8 generated by $\left[\begin{smallmatrix} 0 & -1 \\ 1 & 0 \end{smallmatrix}\right]$ and $\left[\begin{smallmatrix} 0 & i \\ i & 0 \end{smallmatrix}\right]$ or one of three dihedral subgroups of order 8: namely $G(4,4,2)$, $G(2,1,2)$ or the subgroup generated by $\left[\begin{smallmatrix} 0 & i \\ i & 0 \end{smallmatrix}\right]$ and $\left[\begin{smallmatrix} 1 & 0 \\ 0 & -1 \end{smallmatrix}\right]$.

4. The imprimitive unitary reflection groups

In this section we prove a theorem of Shephard and Todd [193], which determines all imprimitive unitary reflection groups. The proof is essentially that of Cohen [54].

Lemma 2.12. *If G is an irreducible imprimitive unitary reflection group and if $\{V_1, V_2, \ldots, V_k\}$ is a system of imprimitivity for G, then $\dim V_i = 1$ for all i.*

Proof. Since G is irreducible it is transitive on $\{V_1, V_2, \ldots, V_k\}$ and so for each i there is a reflection r such that $rV_i \neq V_i$. In this case $V_i \not\subseteq \mathrm{Fix}\, r$, hence $V_i \cap \mathrm{Fix}\, r = \{0\}$ and therefore $\dim V_i = 1$.   □

Lemma 2.13. *If G is a transitive group of permutations of the finite set Ω, and if G is generated by transpositions, then $G = \mathrm{Sym}(\Omega)$.*

Proof. Let T be a set of transpositions and define a graph on Ω by joining i to j whenever (i, j) is a transposition in T. The connected components of this graph are the orbits of the group generated by T. In particular, if T generates G, the graph is connected. The lemma is clearly true when $|\Omega| = 1$, thus we may suppose that $|\Omega| > 1$ and choose $i \in \Omega$ so that the induced graph on $\Omega \setminus \{i\}$ remains connected. By induction, the elements of T that do not involve i generate $\mathrm{Sym}(\Omega \setminus \{i\})$. Since G is transitive on Ω, this group has index $|\Omega|$ in G and therefore $|G| = |\mathrm{Sym}(\Omega)|$. Thus $G = \mathrm{Sym}(\Omega)$, as required.   □

Theorem 2.14. *If V is a vector space of dimension n over $\mathbb{C}$ with positive definite hermitian form $(-, -)$ and if G is an irreducible imprimitive finite subgroup of $U(V)$, which is generated by reflections, then $n > 1$ and G is conjugate to $G(m, p, n)$ for some $m > 1$ and some divisor p of m.*

Proof. Let $\Omega := \{V_1, \ldots, V_k\}$ be a system of imprimitivity for G. By Lemma 2.12 we have $\dim V_i = 1$ for all i and so $k = n$ and we may write $V_i = \mathbb{C}e_i$, where $(e_i, e_i) = 1$.

If the reflection r fixes every subspace V_i, then it has the form (i) of Lemma 2.8. On the other hand, if $rV_i = V_j$ for some $j \neq i$, then by Lemma 2.7 the order of r is two and acts on Ω as a transposition. Thus the group G^Ω of permutations of Ω induced by the action of G is transitive and generated by transpositions. Now Lemma 2.13 shows that $G^\Omega = \mathrm{Sym}(\Omega)$.

For $i := 2, 3, \ldots, n$ choose a reflection $r_i \in G$ that interchanges V_1 and V_i and then choose the notation so that $e_i = r_i e_1$. The subgroup $\Sigma := \langle r_2, r_3, \ldots, r_n \rangle$ is isomorphic to $\mathrm{Sym}(n)$.

With respect to the basis $e_1, e_2, \ldots, e_n$, every element $g \in G$ has the form

$$g\, e_i := \theta_i e_{\pi(i)} \quad \text{for } 1 \leq i \leq n$$

where π is a permutation of $\{1, 2, \ldots, n\}$ and $\theta_1, \theta_2, \ldots, \theta_n \in \mathbb{C}$.

If $ge_i = \theta e_j$ for some θ, then $1 = (e_i, e_i) = (ge_i, ge_i) = \theta\bar{\theta}(e_j, e_j)$ and hence $\theta\bar{\theta} = 1$. Therefore, if $[-, -]$ is the hermitian form on V such that $[e_i, e_j] = \delta_{ij}$ for all i, j, then $[-, -]$ is G-invariant. By Corollary 1.26 to Schur's Lemma $[-, -]$ is a scalar multiple of $(-, -)$ and thus $(e_i, e_j) = 0$ for all $i \neq j$.

Let $\Theta := \{\, \theta \in \mathbb{C} \mid re_i = \theta e_j \text{ for some reflection } r \in G \text{ and some } i \neq j \,\}$. If $\theta \in \Theta$, we may suppose that there is a reflection $r \in G$ such that $re_1 = \theta e_2$ because the subgroup Σ acts doubly transitively on the basis $e_1, e_2, \ldots, e_n$. If s is a reflection such that $se_1 = \eta e_2$, then $rr_2 s$ is a reflection and $rr_2 se_1 = \theta \eta e_2$. It follows that $\theta \eta \in \Theta$ and hence Θ is a finite subgroup of $\mathbb{C}$. Thus for some m, Θ is the group of mth roots of unity.

If $r \in G$ is a reflection such that $re_1 = \theta e_1$, where $\theta \neq 1$, then $r^{-1}r_2 r$ is a reflection such that $r^{-1}r_2 re_1 = \theta e_2$. Therefore the set

$$\{\, \theta \in \mathbb{C} \mid ge_1 = \theta e_1 \text{ for some } g \in G \text{ such that } V_1^{\perp} = \text{Fix}\, g \,\}$$

is a subgroup of Θ and hence the group of qth roots of unity for some divisor q of m. We have $G(q, 1, n) \subseteq G$ and $m > 1$, otherwise $G = \Sigma$ would be reducible.

If $\theta \in \Theta$ and r is a reflection such that $re_1 = \theta e_2$, then $r_2 re_1 = \theta e_1$, $r_2 re_2 = \theta^{-1} e_2$ and $r_2 r \in A(m, m, n)$. It follows that $A(m, m, n)$ is a subgroup of G and therefore $G(m, p, n) \subseteq G$, where $p = m/q$. On the other hand, $G(m, p, n)$ contains all reflections of G and so $G = G(m, p, n)$. $\qquad\square$

Lemma 2.7 of the preceding section can be used to prove a little more. In order to state the result we introduce the notation $W(A_m)$ to denote the symmetric group $\text{Sym}(m + 1)$ in its guise as $G(1, 1, m + 1)$ acting on the hyperplane $\langle e_1 + e_2 + \cdots + e_{m+1}\rangle^{\perp}$ described in Example 2.11 (i).

Theorem 2.15. *Suppose that V is a vector space of dimension n over $\mathbb{C}$ with positive definite hermitian form $(-, -)$ and that G is an irreducible imprimitive finite subgroup of $U(V)$, which is generated by reflections. Suppose that H is a subgroup of G generated by reflections with roots in a subspace M of V. If $m := \dim M > 1$ and if the action of H on M is primitive, then H is conjugate to $W(A_m)$.*

Proof. Suppose that $\Omega := \{V_1, \ldots, V_n\}$ is a system of imprimitivity for G. If for some i we have $V_i \subseteq M$, then the images of V_i under H form a system of imprimitivity for H, contrary to assumption. Thus $V_i \not\subseteq M$ for all i.

Suppose that $r \in H$ is a reflection with root a. If r fixes every V_k then for some i, the subspace V_i is spanned by a and hence $V_i \subseteq M$, contrary to what was proved in the previous paragraph. Thus for some $i \neq j$ we have $rV_i = V_j$ and Lemma 2.7 shows that r has order two and acts on Ω as a transposition. Moreover, $(V_i + V_j) \cap M$ contains the root a of r and consequently $M + V_i = M + V_j$.

If Γ is an orbit of H on Ω we may choose the notation so that $\Gamma = \{V_1, \ldots, V_k\}$. Put $N := V_1 + \cdots + V_k$ and observe that the subspace $M \cap N$ is H-invariant and non-zero, hence coincides with M. It is now clear that $N = M + V_1$ and that M is a subspace of codimension 1 in N. Therefore $k = m + 1$. If $i, j > k$ and if $r \in H$ is a reflection that interchanges V_i and V_j, then $(V_i + V_j) \cap M \neq \{0\}$, whence $(V_i + V_j) \cap N \neq \{0\}$, which is a contradiction. It follows that H fixes $N^{\perp}$ and acts on N as $G(1, 1, k)$. Thus it acts on M as $W(A_m)$, as required. $\qquad\square$

We are now in a position to determine which irreducible groups $G(m, p, n)$ have more than one system of imprimitivity. This is Lemma (2.7) of [**54**] but note that the list of groups in the lemma and the list of systems of imprimitivity in Remark (2.8) are incomplete.

We retain the notation of the previous sections. That is, $e_1, e_2, \ldots, e_n$ is an orthonormal basis for V and the matrices of $G(m, p, n)$ with respect to this basis are monomial. In addition, we note that when $G(m, p, n)$ is irreducible, the system of imprimitivity $\{\mathbb{C}e_1, \mathbb{C}e_2, \ldots, \mathbb{C}e_n\}$ consists of the distinct isotypic components of the abelian normal subgroup $A(m, p, n)$.

Theorem 2.16. *Suppose that $G := G(m, p, n)$ is irreducible and has more than one system of imprimitivity. Then G is one of the following groups.*

(i) $G(2, 1, 2) \simeq G(4, 4, 2)$ or $G(4, 2, 2)$, *each of which has the same three systems of imprimitivity:* $\{\mathbb{C}e_1, \mathbb{C}e_2\}$, $\{\mathbb{C}(e_1 + e_2), \mathbb{C}(e_1 - e_2)\}$ *and* $\{\mathbb{C}(e_1 + ie_2), \mathbb{C}(e_1 - ie_2)\}$.

(ii) $G(3, 3, 3)$ *with four systems of imprimitivity:* $\Pi_0 := \{\mathbb{C}e_1, \mathbb{C}e_2, \mathbb{C}e_3\}$ *and, for a fixed primitive cube root of unity ω,*

$$\Pi_i := \{\mathbb{C}(e_1 + \omega_2 e_2 + \omega_3 e_3) \mid \omega_2, \omega_3 \in \{1, \omega, \omega^2\} \text{ and } \omega_2 \omega_3 = \omega^i\},$$

for $i := 1, 2$ and 3.

(iii) $G(2, 2, 4)$ *with three systems of imprimitivity:*

$$\Lambda_0 := \{\mathbb{C}e_1, \mathbb{C}e_2, \mathbb{C}e_3, \mathbb{C}e_4\} \text{ and}$$

$$\Lambda_i := \{\mathbb{C}(e_1 + \varepsilon_2 e_2 + \varepsilon_3 e_3 + \varepsilon_4 e_4) \mid \varepsilon_j = \pm 1 \text{ and } \varepsilon_2 \varepsilon_3 \varepsilon_4 = (-1)^i\},$$

for $i := 1$ and 2.

Proof. From Lemma 2.12 the subspaces that occur in a system of imprimitivity all have dimension 1. In addition, if $\{\mathbb{C}u_1, \mathbb{C}u_2, \ldots, \mathbb{C}u_n\}$ is a system of imprimitivity, and if B is the subgroup of G that leaves each $\mathbb{C}u_i$ fixed, then B is an abelian normal subgroup of G and the $\mathbb{C}u_i$ are its isotypic components.

The system of imprimitivity $\{\mathbb{C}e_1, \mathbb{C}e_2, \ldots, \mathbb{C}e_n\}$ is the set of isotypic components of $A := A(m, p, n)$ and, by assumption, we may choose the u_i so that $B \neq A$. But then AB/A is an abelian normal subgroup of $G/A \simeq \text{Sym}(n)$, hence n is 2, 3 or 4 and $|AB/A| = n$. In each case, B acts transitively on $\{\mathbb{C}e_1, \mathbb{C}e_2, \ldots, \mathbb{C}e_n\}$ and since $A \cap B$ consists of diagonal matrices, which commute with all elements of B, it follows that $A \cap B$ is the group of scalar matrices θI, where $\theta^{nm/p} = 1$. In particular, $|A \cap B| = \gcd(n, p)\, m/p$.

We also have $G/B \simeq \text{Sym}(n)$ and so $|B| = |A| = m^n/p$. Thus from $n = |AB/A| = |B/A \cap B|$ we derive the condition

$$m^{n-1} = n \gcd(n, p),$$

where p divides m and n is 2, 3 or 4. It is immediate that the only solutions for (m, p, n) are $(2, 1, 2)$, $(4, 2, 2)$, $(4, 4, 2)$, $(3, 3, 3)$ and $(2, 2, 4)$.

Suppose that $n = 2$ and that $u_1 := e_1 + \eta e_2$, where $\eta \neq 0$. If $\mathbb{C}u_1$ is fixed by $x := \begin{bmatrix} 0 & 1 \\ 1 & 0 \end{bmatrix}$, then $\eta = \pm 1$ and we have the system of imprimitivity $\{\, \mathbb{C}(e_1 + e_2),$ $\mathbb{C}(e_1 - e_2) \,\}$. If $\mathbb{C}u_1$ and $\mathbb{C}u_2$ are interchanged by x, we may take $u_1 := e_1 + \eta e_2$ and $u_2 := e_1 + \eta^{-1} e_2$, where $\eta \neq 0, \pm 1$. If G is $G(2, 1, 2)$, or $G(4, 2, 2)$, then $y := \begin{bmatrix} -1 & 0 \\ 0 & 1 \end{bmatrix} \in G$ and y interchanges $\mathbb{C}u_1$ and $\mathbb{C}u_2$ and hence $\eta = \pm i$. In the remaining case, of $G(4, 4, 2)$, we have $\begin{bmatrix} i & 0 \\ 0 & -i \end{bmatrix} \in G$ and again $\eta = \pm i$. Thus the groups $G(2, 1, 2)$, $G(4, 2, 2)$ and $G(4, 4, 2)$ have the same three systems of imprimitivity and this completes the proof of (i).

Suppose that $n = 3$. In this case $G(3, 3, 3)$ has order 54 and if ω is a primitive cube root of unity, the subgroup E generated by $\begin{bmatrix} 1 & 0 & 0 \\ 0 & \omega & 0 \\ 0 & 0 & \omega^2 \end{bmatrix}$ and $\begin{bmatrix} 0 & 0 & 1 \\ 1 & 0 & 0 \\ 0 & 1 & 0 \end{bmatrix}$ has order 27. The centre of E has order 3 and every non-identity element of E has order 3. Thus E has four (abelian) subgroups of order 9. On the other hand, one can check directly that the sets of subspaces listed in part (ii) of the statement of the theorem do provide systems of imprimitivity for $G(3, 3, 3)$ and therefore they must be the isotypic components of the subgroups of order 9 in E. This also shows that these four subgroups are normal in $G(3, 3, 3)$ and thus conjugation by $\begin{bmatrix} 0 & 1 & 0 \\ 1 & 0 & 0 \\ 0 & 0 & 1 \end{bmatrix}$ takes each element of E to its inverse (modulo the centre).

Finally, suppose that $n = 4$. Then the order of $G := G(2, 2, 4)$ is 192. The order of the subgroup E generated by

$$\begin{bmatrix} -1 & 0 & 0 & 0 \\ 0 & -1 & 0 & 0 \\ 0 & 0 & 1 & 0 \\ 0 & 0 & 0 & 1 \end{bmatrix}, \quad \begin{bmatrix} -1 & 0 & 0 & 0 \\ 0 & 1 & 0 & 0 \\ 0 & 0 & -1 & 0 \\ 0 & 0 & 0 & 1 \end{bmatrix}, \quad \begin{bmatrix} 0 & 1 & 0 & 0 \\ 1 & 0 & 0 & 0 \\ 0 & 0 & 0 & 1 \\ 0 & 0 & 1 & 0 \end{bmatrix} \quad \text{and} \quad \begin{bmatrix} 0 & 0 & 1 & 0 \\ 0 & 0 & 0 & 1 \\ 1 & 0 & 0 & 0 \\ 0 & 1 & 0 & 0 \end{bmatrix}$$

is 32 and $E = O_2(G)$, the largest normal 2-subgroup of G. There are six normal abelian subgroups of E of order 8 but only three of these are normal in G; they correspond to the three systems of imprimitivity in part (iii) of the theorem. $\square$

Remark 2.17. The groups E that occur in the proof of this theorem are examples of *extraspecial p-groups* for the primes 3 and 2. That is, the derived group E' has order p and coincides with the centre of E. In addition, every non-identity element of E/E' has order p.

We shall see in later chapters, and in the exercises at the end of this chapter, that the groups $G(4, 2, 2)$, $G(3, 3, 3)$ and $G(2, 2, 4)$ occur as normal subgroups of primitive reflection groups. On the other hand, for all divisors p of m we have $G(m, p, n) \trianglelefteq G(m, 1, n)$ and when $n = 2$ we also have $G(m, p, 2) \trianglelefteq G(2m, 2, 2)$. From the theorem just proved we have the following partial converse.

Corollary 2.18. *Suppose that $m \geq 2$ and that G is a finite reflection subgroup of $U_n(\mathbb{C})$ which contains $G(m, p, n)$ as a normal subgroup.*

(i)   If $n \geq 3$ and (m, p, n) is neither $(3, 3, 3)$ nor $(2, 2, 4)$, then for some divisor q
       of p we have $G = G(m, q, n)$.

(ii)   If $n = 2$ and $(m, p, 2)$ is not $(4, 2, 2)$, then G is contained in $G(2m, 2, 2)$.

Proof. Suppose first that $n \geq 3$. From the theorem, $G(m, p, n)$ has a unique system of imprimitivity $\{ \mathbb{C}e_1, \dots, \mathbb{C}e_n \}$ which must be preserved by G. It follows from Theorem 2.14 that $G = G(k, q, n)$ for some k and q, where q divides k. If θ is a kth root of unity, then $h := \text{diag}(\theta^{-1}, \theta, 1, \dots, 1) \in G$ and we may choose $r \in G(m, p, n)$ to be the reflection that interchanges e_1 and e_3. Then $hrh^{-1} \in G(m, p, n)$ and $hrh^{-1}(e_1) = \theta e_3$. It follows that θ is an mth root of unity and hence $k = m$. To complete the proof, note that $A(m, p, n) \subseteq A(m, q, n)$ if and only if q divides p.

The groups $G(2, 1, 2)$ and $G(4, 4, 2)$ are dihedral of order 8 and they cannot be normalised by elements of order 3 that permute their 3 systems of imprimitivity. Furthermore, if θ is a $2m^{\text{th}}$ root of unity, then $G(2m, 2, 2)$ is generated by its subgroup $G(m, 1, 2)$ and the central element $\left[\begin{smallmatrix} \theta & 0 \\ 0 & \theta \end{smallmatrix} \right]$. Thus when $n = 2$, essentially the same proof as above shows that, except for $G(4, 2, 2)$, the largest reflection subgroup of $U_2(\mathbb{C})$ containing $G(m, p, 2)$ as a normal subgroup is $G(2m, 2, 2)$. $\qquad \square$

The reflection $r := \frac{\omega}{2} \left[\begin{smallmatrix} i-1 & i-1 \\ i+1 & -i-1 \end{smallmatrix} \right]$ has order 3, normalises $G(4, 2, 2)$ and permutes its three systems of imprimitivity. It follows that $G(4, 2, 2) \langle r \rangle$ is primitive. The reflection r together with $G(8, 2, 2)$ generates a group of order 576 which is the largest reflection subgroup of $U_2(\mathbb{C})$ containing $G(4, 2, 2)$ as a normal subgroup. (This is the group G_{11} in the Shephard and Todd [**193**] notation. See Chapter 6 for more details.)

5. Imprimitive subgroups of primitive reflection groups

As an application of the determination of the imprimitive reflection groups with a unique system of imprimitivity we prove the following version of Proposition (2.9) of Cohen [**54**].

Theorem 2.19. *Suppose that G is a primitive unitary reflection group in $U(V)$ and that W is a proper subspace of V of dimension at least two. Suppose that H is a subgroup of G that fixes $W^{\perp}$ pointwise and which acts on W as an irreducible imprimitive reflection group not conjugate to $G(2, 1, 2)$, $G(3, 3, 2)$, $G(4, 4, 2)$, $G(4, 2, 2)$, $G(2, 2, 3)$, $G(3, 3, 3)$ nor $G(2, 2, 4)$. Then there is a reflection $r \in G$ with root $a \notin W$ such that the action of $\langle H, r \rangle$ on $W \oplus \mathbb{C}a$ is primitive.*

Proof. Let $m := \dim W$. By Theorem 2.16 and the restrictions on H there is a unique system of imprimitivity $W_1, W_2, \dots, W_m$ for H acting on W.

Suppose that $r \in G$ is a reflection, with root a, which does not fix W and suppose that the action of $\langle H, r \rangle$ on $W \oplus \mathbb{C}a$ is imprimitive. Let $\Omega := \{ V_1, V_2, \dots, V_{m+1} \}$ be a system of imprimitivity for $\langle H, r \rangle$ and suppose at first that $V_i \not\subseteq W$ for all i. Then by Lemma 2.7 the reflections in H act on Ω as transpositions and the argument of Theorem 2.14 shows that H acts on W as $W(A_m)$. We are assuming that the action of H on W is imprimitive and therefore $m \leq 3$. But then H is conjugate to $G(3,3,2)$ or $G(2,2,3)$, contrary to our initial choice of H. It follows that $V_i \subseteq W$ for some i. Since W is an irreducible H-module, the images of V_i under H span W and form a system of imprimitivity for H which, by uniqueness, must be $\{W_1, \ldots, W_m\}$. Thus $\Omega = \{W_1, \ldots, W_m, W_{m+1}\}$, where $W_{m+1} := W^{\perp} \cap rW$ and we may choose the notation so that $rW_m = W_{m+1}$. In particular, $\langle H, r \rangle$ acts irreducibly on $W \oplus \mathbb{C}a$ and we note that by Lemma 2.7, $r^2 = 1$. If $W_m = \mathbb{C}w_m$, where $(w_m, w_m) = 1$, then $(w_m, rw_m) = 0$ and therefore $|(w_m, a)|^2 = \frac{1}{2}$.

If $\dim V = m+1$, then the primitivity of G ensures that there is a reflection r that does not fix the set $\{W_1, W_2, \ldots, W_m, W^{\perp}\}$. The considerations of the previous paragraph show that $\langle H, r \rangle$ must be primitive in this case.

Thus from now on we may suppose that $\dim V > m+1$. By induction on $\dim V - m$ we may choose a reflection s with root b such that $\langle H, r, s \rangle$ is primitive and, in particular, such that s does not fix $W \oplus \mathbb{C}a$. The group $\langle H, r \rangle$ acts transitively on $\{W_1, \ldots, W_{m+1}\}$ and so, on replacing s by a suitable conjugate, we may suppose that $sW_m \not\subseteq W \oplus \mathbb{C}a$. If $\langle H, s \rangle$ is not primitive on $W \oplus \mathbb{C}b$, then the results obtained above for r and a imply that $|(w_m, b)|^2 = \frac{1}{2}$. Similarly, if $\langle H, rsr \rangle$ is not primitive on $W \oplus \mathbb{C}rb$ and if $rsrW_m \neq W_m$, then $|(w_m, rb)|^2 = \frac{1}{2}$. On the other hand, if $rsrw_m = w_m$, then $(w_m, rb) = 0$. Putting $w_{m+1} := rw_m$ we may write $b := \kappa w_m + \lambda w_{m+1} + \mu v$, where $(v, v) = 1$ and v is orthogonal to w_i for $1 \leq i \leq m+1$. Then

$$1 = (b, b) = |\kappa|^2 + |\lambda|^2 + |\mu|^2.$$

Now $\kappa = (b, w_m)$ and $\lambda = (b, w_{m+1}) = (rb, w_m)$. Thus $|(w_m, rb)|^2 = \frac{1}{2}$ implies $\mu = 0$ and hence $b \in W \oplus \mathbb{C}a$, which is a contradiction. It follows that $(w_m, rb) = 0$. But now $sw_m = v$ and hence $\langle H, r, s \rangle$ is imprimitive, contradicting the choice of s. Thus there is a reflection r such that $\langle H, r \rangle$ is primitive.   □

The groups excluded as possibilities for H are genuine exceptions to this theorem. For example, $G(2,1,2) \simeq G(4,4,2)$, $G(3,3,2)$ and $G(4,2,2)$ are all subgroups of the primitive group G_{31} (in the Shephard and Todd notation) and from Theorem 8.30 the ring of definition of G_{31} is $\mathbb{Z}[i]$ whereas the primitive reflection groups of rank 3 cannot be written over $\mathbb{Q}[i]$. Therefore G_{31} does not contain primitive reflection subgroups of rank 3.

Similarly, we find that $G(2,2,3)$ is contained in G_{28}, $G(3,3,3)$ is contained in G_{33} and $G(2,2,4)$ is contained in G_{35}. In all cases there is no primitive subgroup that is generated by the given group and another reflection. (See Theorems 7.21 and 7.22.)

6. Root systems for $G(m, p, n)$

The root systems for the groups $G(m, p, n)$ can be obtained directly from the description of the reflections given in Lemma 2.8. To define them, let ζ_m be a primitive m^{th} root of unity (in $\mathbb{C}$), let $\boldsymbol{\mu}_m$ be the group of all m^{th} roots of unity, and let e_1, $e_2, \ldots, e_n$ be an orthonormal basis for the vector space V. We begin with the set

$$\Sigma(m, m, n) := \{ \xi e_i - \eta e_j \mid \xi, \eta \in \boldsymbol{\mu}_m \text{ and } i \neq j \}$$

and define $f : \Sigma(m, m, n) \to \boldsymbol{\mu}_2$ by $f(a) = -1$ for all $a \in \Sigma(m, m, n)$. The elements of $\Sigma(m, m, n)$ are long roots for the reflections in $G(m, m, n)$ and the order of every reflection is two. It can be checked directly that $\Sigma(m, m, n)$ is a $\mathbb{Z}[\zeta_m]$-root system for $G(m, m, n)$, according to Definition 1.43.

Let $\Delta := \{ \pm \xi e_i \mid 1 \leq i \leq n, \ \xi \in \boldsymbol{\mu}_m \}$ and define $f : \Delta \to \boldsymbol{\mu}_m$ by $f(a) = \zeta_m^p$ for all $a \in \Delta$. If $p \neq m$, put $\Sigma(m, p, n) = \Sigma(m, m, n) \cup \Delta$.

The pair $(\Sigma(m, p, n), f)$ is a $\mathbb{Z}[\zeta_m]$-root system for $G(m, p, n)$. The next theorem shows that, except for the dihedral groups $G(m, m, 2)$, these root systems are defined over the ring of definition of $G(m, p, n)$.

Theorem 2.20 (Clark–Ewing [52]).

(i)   *Except for the dihedral groups $G(m, m, 2)$, the ring of definition of $G(m, p, n)$ is $\mathbb{Z}[\zeta_m]$.*

(ii)   *The ring of definition of $G(m, m, 2)$ is $\mathbb{Z}[\zeta_m + \zeta_m^{-1}]$.*

Proof. Suppose that $n \geq 3$ and let $G := G(m, p, n)$. Then G contains the matrices $\operatorname{diag}(\zeta_m, \zeta_m^a, \zeta_m^{-1-a}, 1, \ldots, 1)$, for all a. On summing over a we see that $\zeta_m \in \mathbb{Q}(G)$. Conversely, it is clear from its definition that $\mathbb{Q}(G) \subset \mathbb{Q}[\zeta_m]$, hence equality holds. Then $\mathbb{Z}(G) = \mathbb{Z}[\zeta_m]$ because $\mathbb{Z}[\zeta_m]$ is the ring of integers of $\mathbb{Q}(\zeta_m)$ (see Washington [**223**, Theorem 2.6]).

Suppose that $n = 2$. Then $\mathbb{Q}[\zeta_m + \zeta_m^{-1}] \subseteq \mathbb{Q}(G) \subseteq \mathbb{Q}[\zeta_m]$ and $\zeta_m^p \in \mathbb{Q}(G)$. Complex conjugation is an automorphism of $\mathbb{Q}[\zeta_m]$ that inverts every root of unity and its fixed field is $\mathbb{Q}[\zeta_m + \zeta_m^{-1}]$. Therefore, if $\zeta_m^p \in \mathbb{Q}[\zeta_m + \zeta_m^{-1}]$, then $\zeta_m^p = \zeta_m^{-p}$ and hence $m = p$ or $m = 2p$. If $m = 2p$, then $\zeta_m^p = -1$ and therefore $\operatorname{diag}(\zeta_m, \pm\zeta_m^{-1}, 1, \ldots, 1) \in G$. Thus $\mathbb{Q}(G) = \mathbb{Q}[\zeta_m]$ when $p < m$. If $m = p$, direct inspection of the matrices shows that $\mathbb{Q}(G) = \mathbb{Q}[\zeta_m + \zeta_m^{-1}]$. The ring of integers of $\mathbb{Q}[\zeta_m + \zeta_m^{-1}]$ is $\mathbb{Z}[\zeta_m + \zeta_m^{-1}]$ (Washington [**223**, Theorem 2.16]) and therefore $\mathbb{Z}(G) = \mathbb{Z}[\zeta_m + \zeta_m^{-1}]$. $\square$

When $n = 2$ we modify the definition of $\Sigma(m, m, 2)$ given above to obtain a root system defined over $A := \mathbb{Z}[\zeta_m + \zeta_m^{-1}]$. In this case $\mu(A) = \{\pm 1\}$, since $\mathbb{Q}[\zeta_m + \zeta_m^{-1}]$ is real.

If m is odd, the group $G(m, m, 2)$ has a single conjugacy class of reflections and we take $\Sigma(m, m, 2) := \{ \pm(\xi e_1 - \xi^{-1} e_2) \mid \xi \in \boldsymbol{\mu}_m \}$. This is an A-root system for $G(m, m, 2)$. The reflections (of order two) with roots $a := e_1 - e_2$ and $b := \zeta_m e_1 - \zeta_m^{-1} e_2$ generate $G(m, m, 2)$. The Cartan matrix of r_a and r_b with respect to a and b is

$$\begin{bmatrix} 2 & \zeta_m + \zeta_m^{-1} \\ \zeta_m + \zeta_m^{-1} & 2 \end{bmatrix}.$$

If m is even, $G(m, m, 2)$ has two conjugacy classes of reflections and in order to represent the group by matrices over $A := \mathbb{Z}[\zeta_m + \zeta_m^{-1}]$ we use roots of two different lengths. For the Cartan matrix we take

$$\begin{bmatrix} 2 & 2 + \zeta_m + \zeta_m^{-1} \\ 1 & 2 \end{bmatrix}.$$

If V is a vector space of dimension two over the field $F := \mathbb{Q}[\zeta_m + \zeta_m^{-1}]$, this choice of Cartan matrix corresponds to choosing a basis a, b for V and defining an orthogonal inner product on V by $(a, a) = 2$, $(a, b) = (b, a) = 2 + \zeta_m + \zeta_m^{-1}$ and $(b, b) = 2(a, b)$. The matrices of the reflections r_a and r_b are

$$\begin{bmatrix} -1 & -1 \\ 0 & 1 \end{bmatrix} \quad \text{and} \quad \begin{bmatrix} 1 & 0 \\ -2 - \zeta_m - \zeta_m^{-1} & -1 \end{bmatrix}.$$

Under the embedding of F in $\mathbb{R}$ that sends ζ_m to $\exp(2\pi i/m)$ we have

$$\frac{(a, b)}{\sqrt{(a, a)}\sqrt{(b, b)}} = \cos\frac{\pi}{m}$$

and therefore $G := \langle r_a, r_b \rangle$ is a dihedral group of order $2m$, namely $G(m, m, 2)$.

The root system $\Sigma(m, m, 2)$ for G is the union of the m images of a with the m images of b under the action of G. By construction, the roots are A-linear combinations of a and b.

Theorem 2.21. *Every irreducible imprimitive reflection group $G(m, p, n)$ can be represented by $n \times n$ matrices over its ring of definition.*

Proof. This is clear when $n > 2$ and when $m \neq p$. The description just given for the root system of $G(m, m, 2)$ completes the proof. $\quad\square$

7. Generators for $G(m, p, n)$

Minimal sets of generating reflections for the groups $G(m, p, n)$ are easy to determine. For $i := 1, \ldots, n - 1$ let $r_i := r_{e_i - e_{i+1}, -1}$ be the reflection of order 2 that interchanges the basis vectors e_i and e_{i+1} and fixes e_j for $j \neq i, i + 1$. Let $t := r_{e_1, \zeta_m}$ be the reflection of order m that fixes $e_2, \ldots, e_n$ and sends e_1 to $\zeta_m e_1$ and let $s := t^{-1} r_1 t$ be the reflection of order 2 that interchanges e_1 and $\zeta_m e_2$. For $m > 1$, the groups $G(m, m, n)$ and $G(m, 1, n)$ can be generated by n reflections;

however, the minimum number of reflections required to generate $G(m, p, n)$, for $p \neq 1, m$ is $n + 1$. In particular, we have

$$G(1, 1, n) = \langle r_1, r_2, \ldots, r_{n-1} \rangle \simeq \text{Sym}(n),$$
$$G(m, m, n) = \langle s, r_1, r_2, \ldots, r_{n-1} \rangle,$$
$$G(m, 1, n) = \langle t, r_1, r_2, \ldots, r_{n-1} \rangle, \quad \text{and}$$
$$G(m, p, n) = \langle s, t^p, r_1, r_2, \ldots, r_{n-1} \rangle \quad \text{for } p \neq 1, m.$$

8. Invariant polynomials for $G(m, p, n)$

In the next chapter we consider group actions on polynomials in considerable detail and so for now we restrict our attention to $G(m, p, n)$ and describe its action on polynomials $P(X_1, \ldots, X_n)$ in the n variables $X_1, \ldots, X_n$ in a very explicit form.

Given $h := (\theta_1, \ldots, \theta_n) \in A(m, p, n)$ and $\pi \in \text{Sym}(n)$ we define

$$(h\, P)(X_1, \ldots, X_n) := P(\theta_1^{-1} X_1, \ldots, \theta_n^{-1} X_n) \quad \text{and}$$
$$(\pi\, P)(X_1, \ldots, X_n) := P(X_{\pi(1)}, \ldots, X_{\pi(n)}).$$

These actions combine to give an action of $G(m, p, n)$ on polynomials.

A polynomial $P \in \mathbb{C}[X_1, \ldots, X_n]$ is an *invariant* of $G(m, p, n)$ if $gP = P$ for all $g \in G(m, p, n)$. For $1 \leq r \leq n$, the rth elementary symmetric polynomial in the variables $X_1, X_2, \ldots, X_n$ is

$$e_r(X_1, X_2, \ldots, X_n) := \sum_{i_1 < i_2 < \cdots < i_r} X_{i_1} X_{i_2} \cdots X_{i_r}.$$

From the given action of $G(m, p, n)$ on polynomials we see immediately that for $1 \leq r < n$ the polynomials $\sigma_r := e_r(X_1^m, X_2^m, \ldots, X_n^m)$ are invariants of $G(m, p, n)$ and so is the polynomial $\sigma_n := e_n^{m/p} = (X_1 X_2 \cdots X_n)^{m/p}$.

Proposition 2.22. *The polynomials $\sigma_1, \sigma_2, \ldots, \sigma_n$ are algebraically independent.*

Proof. The proof is by induction on n. If $n = 1$, then $\sigma_1 = X_1^{m/p}$ and the result is clear. So suppose that $n > 1$ and that the result holds for all groups $G(m', p', n')$ with $n' < n$. By way of contradiction, suppose that $H \in \mathbb{C}[Y_1, \ldots, Y_n]$ is a non-zero polynomial of smallest degree in Y_n such that $H(\sigma_1, \sigma_2, \ldots, \sigma_n) = 0$. We may write

$$H = H_0 + H_1 Y_n + \cdots + H_m Y_n^m$$

where $H_i \in \mathbb{C}[Y_1, \ldots, Y_{n-1}]$ and where by minimality of m we have $H_0 \neq 0$. On setting $X_n = 0$, the polynomials $\sigma_1, \ldots, \sigma_{n-1}$ reduce to the corresponding invariants $\sigma_1', \ldots, \sigma_{n-1}'$ for $G(m, 1, n - 1)$ and we have a non-trivial relation

$$H_0(\sigma_1', \sigma_2', \ldots, \sigma_{n-1}') = 0.$$

But this contradicts the induction hypothesis and completes the proof. $\qquad \square$

The degrees of the polynomials $\sigma_1, \sigma_2, \ldots, \sigma_n$ are

$$m, \; 2m, \; \ldots, \; (n-1)m, \; nm/p$$

and the product of their degrees is $n! \, m^n / p$, which is the order of $G(m, p, n)$. In the next chapter we shall prove a general result that shows that they form a basis for the algebra of all invariant polynomials for $G(m, p, n)$.

From Lemma 2.8, the number of reflections in $G(m, p, n)$ is $n(m/p - 1) + m\binom{n}{2}$. If we add n to this quantity we get the sum of the degrees. Once again this is a special case of a general result that will be proved in a later chapter.

Exercises

1. Verify all the statements of Example 2.11

2. If V is a G-module and if I is an irreducible G-module, show that every irreducible G-submodule of the isotypic component V_I is isomorphic to I.

3. Show that the groups $G(4, 4, 2)$ and $G(2, 1, 2)$ are conjugate in $U_2(\mathbb{C})$ and that both are subgroups of index 2 in $G(4, 2, 2)$.

4. Show that $G(4, 2, 2)$ contains a unique subgroup of order 8 all of whose non-central elements have order 4. (This is the quaternion group Q_8.)

5. Prove that for all divisors p of m we have $G(m, p, n) \trianglelefteq G(m, 1, n)$.

6. Using the fact that $G(3, 3, 3)$ is a normal subgroup of $G(3, 1, 3)$, show that $G(3, 3, 3)$ is a normal subgroup of a primitive reflection group of order 1296 (the group G_{26} in the Shephard and Todd notation) and determine its structure.

7. If Z is the centre of $G(2m, 2, 2)$, show that the order of Z is $2m$ and that $G(2m, 2, 2) = Z \, G(m, 1, 2)$. Deduce that for all divisors p of m we have $G(m, p, 2) \trianglelefteq G(2m, 2, 2)$.

8. Show that $G(4, 2, 2)$ is a normal subgroup of a primitive reflection group of order 576 (the group G_{11} in the Shephard and Todd notation).

9. Show that $G(2, 2, 4)$ is a normal subgroup of a primitive reflection group of order 1152 (the group G_{28} in the Shephard and Todd notation).

10. Show that the elementary symmetric functions $e_r(X_1, X_2, \ldots, X_n)$ for $1 \leq r \leq n$ (defined in §8) are algebraically independent and invariant under the action of $\mathrm{Sym}(n)$.

11. Find $n - 1$ algebraically independent invariant polynomials for the group $G(1, 1, n) \simeq \mathrm{Sym}(n)$ acting on $\langle e_1 + \cdots + e_n \rangle^\perp$.

12. Let s and r_1 be the reflections in $G(m, m, n)$ defined in §7. Show that the order of sr_1 is m.

13. Let s, t^p and r_1 be the reflections in $G(m, p, n)$ defined in §7. Show that $(x - 1)^{n-2}(x^2 - \zeta_m^p)$ is the minimal polynomial of both st^p and $r_1 t^p$.

14. Show that in $G(m, m, n)$ the order of the product of two distinct reflections is 2, 3 or a divisor of m. If $n \geq 3$, show that the reflections in $G(m, m, n)$ form a single conjugacy class.

15. Show that every reflection in $G(m, p, n)$ has order two if and only if $m = p$ or $m = 2p$.

16. Let G be an irreducible imprimitive group generated by reflections of order two.
    (i)   If the order of the product of two reflections is at most 3, show that G is $G(2, 2, n)$ or $G(3, 3, n)$, for some n.
    (ii) If the order of the product of two reflections is at most 4, show that G is either one of the groups listed in (i) or one of $G(2, 1, n)$, $G(4, 2, n)$ or $G(4, 4, n)$, for some n.

17. In §6 the Cartan matrix
$$\begin{bmatrix} 2 & 2 + \zeta_m + \zeta_m^{-1} \\ 1 & 2 \end{bmatrix}.$$

    was used to define the reflection group $G(m, m, 2)$ when m is even. Show that the construction also applies when m is odd. Explain how the root system derived from this construction relates to the root system
$$\{ \pm(\xi e_1 - \xi^{-1} e_2) \mid \xi \in \boldsymbol{\mu}_m \}.$$

18. Suppose that e_1, e_2 is an orthonormal basis for the Euclidean space $\mathbb{R}^2$. Given $m \geq 2$, show that the reflection group defined by the $\mathbb{Z}$-root system
$$\{ \cos \frac{2k\pi}{m} e_1 + \sin \frac{2k\pi}{m} e_2 \mid 1 \leq k \leq 2m \}$$
    is conjugate to $G(m, m, 2)$. Explain how this relates to the root system of the previous question.

CHAPTER 3

Polynomial invariants

In this chapter we present the basic results on polynomial invariants of finite reflection groups. There are other expositions of this material in Bourbaki [**33**], Flatto [**98**], Hiller [**113**], Humphreys [**119**] and Springer [**202**].

1. Tensor and symmetric algebras

Throughout this section V will be a vector space of dimension n over $\mathbb{C}$. In preparation for our study of the invariants of finite reflection groups we review the construction of the tensor and symmetric algebras of V. As a general reference for the this material see Lang [**142**, Chapter XVI].

Definition 3.1. For each $r := 0, 1, \ldots$ we let $T^r(V)$ denote the r-fold tensor power of V. That is, $T^0(V) := \mathbb{C}$, $T^1(V) := V$, $T^2(V) := V \otimes V$, and so on. The direct sum

$$T(V) := \bigoplus_{r=0}^{\infty} T^r(V)$$

is the called the *tensor algebra* of V. Multiplication is defined on $T(V)$ by declaring the product of $v_{i_1} \otimes v_{i_2} \cdots \otimes v_{i_k} \in T^k(V)$ with $v_{j_1} \otimes v_{j_2} \cdots \otimes v_{j_\ell} \in T^\ell(V)$ to be $v_{i_1} \otimes v_{i_2} \cdots \otimes v_{i_k} \otimes v_{j_1} \otimes v_{j_2} \cdots \otimes v_{j_\ell} \in T^{k+\ell}(V)$, then extending this to all of $T(V)$ by linearity.

There is a natural embedding of V in $T(V)$ that identifies $v \in V$ with the element $(0, v, 0, \ldots) \in T(V)$. This makes $T(V)$ the free associative algebra on V in the sense that if A is any associative $\mathbb{C}$-algebra and if $\varphi : V \to A$ is a vector space homomorphism, there is a unique algebra homomorphism $\Phi : T(V) \to A$ such that $\varphi(v) = \Phi(v)$ for all $v \in V$.

Given a linear transformation $f : V \to W$ there is a unique algebra homomorphism $T(f) : T(V) \to T(W)$ such that $T(f)v = f(v)$ for all $v \in V$. Thus $T(f)(v_{i_1} \otimes v_{i_2} \cdots \otimes v_{i_k}) = f(v_{i_1}) \otimes f(v_{i_2}) \cdots \otimes f(v_{i_k})$. Moreover, $T(1_V) = 1_{T(V)}$ and if $g : W \to U$ is another linear transformation, then $T(gf) = T(g)T(f)$. These properties are usually expressed by saying that T is a *functor* from vector spaces to associative algebras.

Definition 3.2. Let I be the two-sided ideal of $T(V)$ generated by the elements $v \otimes w - w \otimes v$ for all $v, w \in V$. Define the *symmetric algebra* of V to be the quotient $S(V) := T(V)/I$. It is the free commutative algebra on V. The product of $\xi := u+I$ and $\eta := v + I \in S(V)$ is $\xi\eta := u \otimes v + I$, where $u \otimes v$ is the product of u and v in $T(V)$.

We have $V \cap I = \{0\}$ and therefore V may be identified with its image in $S(V)$. If $v_1, v_2, \ldots, v_n$ is a basis for V, then $S(V)$ is isomorphic to the algebra of polynomials in the symbols $v_1, v_2, \ldots, v_n$. That is, $S(V) \simeq \mathbb{C}[v_1, v_2, \ldots, v_n]$.

As in Chapter 1, $V^* := \mathrm{Hom}(V, \mathbb{C})$ denotes the dual space of V and we let $X_1, X_2, \ldots, X_n$ be the basis of V^* dual to the basis $v_1, v_2, \ldots, v_n$ of V. The linear maps $X_1, X_2, \ldots, X_n$ are the *coordinate functions* of V and $S := S(V^*) \simeq \mathbb{C}[X_1, X_2, \ldots, X_n]$ is the *coordinate ring* of V. Thus S may be identified with the ring of polynomial functions on V.

The one variable case of the next result is centuries old, and is often referred to as 'Lagrange interpolation'.

Lemma 3.3. *Let $w_1, w_2, \ldots, w_s$ be distinct elements of V, and let $a_1, a_2, \ldots, a_s$ be arbitrary elements of $\mathbb{C}$. Then there is a polynomial $P \in S$ such that $P(w_i) = a_i$ for $i = 1, 2, \ldots, s$.*

Proof. It clearly suffices to prove that there exist polynomials $P_i \in S$ such that $P_i(w_j) = \delta_{ij}$, $i, j = 1, \ldots, s$. Moreover by symmetry it suffices to prove the existence of P_1. We do this by induction on s. When $s = 1$ the result is trivial and so we may suppose $s > 1$.

By induction, there is a polynomial $Q \in S$ such that $Q(w_1) = 1$ and $Q(w_i) = 0$ for $i = 2, \ldots, s-1$. Since $w_1 \neq w_s$, there is a linear function $L \in S$ such that $L(w_1) \neq L(w_s)$. Define $P_1 := Q \dfrac{L - L(w_s)}{L(w_1) - L(w_s)}$. Then P_1 has the required properties. $\qquad\square$

The *degree* of the monomial $X_1^{m_1} X_2^{m_2} \cdots X_n^{m_n}$ is $m_1 + m_2 + \cdots + m_n$. A polynomial P is *homogeneous* of degree r, and we write $\deg P := r$, if P is a linear combination of monomials of degree r. The ring S has a natural grading by degree. That is, if S_r denotes the subspace of homogeneous polynomials of degree r, then

$$S = \bigoplus_{r=0}^{\infty} S_r$$

and $S_r S_t \subseteq S_{r+t}$. Note that there are $\binom{n+r-1}{r}$ distinct monomials of degree r and therefore $\dim S_r = \binom{n+r-1}{r}$.

The above remarks about grading apply equally to the non-commutative tensor algebra $T(V) = \bigoplus_{r=0}^{\infty} T^r(V)$, and we clearly have $\dim T^r(V) = n^r$. If I is any homogeneous ideal of $T(V)$, i.e., if I is an ideal which contains the homogeneous components of all its elements, then the quotient algebra $T^r(V)/I$ inherits a grading, and the example $S(V)$ above is a special case of this construction. Others will occur later in this book.

2. The algebra of invariants

We retain the notation of the previous section. The group $GL(V)$ acts on V^* and hence on S. That is, for $g \in GL(V)$ and for a polynomial function $P \in S$ we define gP by

$$(gP)(v) := P(g^{-1}v) \quad \text{for all } v \in V.$$

This is a linear action that preserves both the degree and algebra structure of S, i.e. $gS_r = S_r$ and $g(PQ) = (gP)(gQ)$.

If the matrix of g with respect to the basis $v_1, v_2, \ldots, v_n$ is $A := (a_{ij})$, where $gv_j = \sum_i a_{ij} v_i$, then the matrix of the action of g on V^* with respect to the basis $X_1, X_2, \ldots, X_n$ is A^{-t}, the inverse transpose of A. Thus for a polynomial $P(X_1, \ldots, X_n)$ we have $(gP)(X_1, \ldots, X_n) = P(X_1', \ldots, X_n')$, where

$$\begin{bmatrix} X_1' \\ \vdots \\ X_n' \end{bmatrix} = A^{-1} \begin{bmatrix} X_1 \\ \vdots \\ X_n \end{bmatrix}.$$

For example, if $n = 2$ and the matrix of g is $\left(\begin{smallmatrix} 1 & 1 \\ 0 & 1 \end{smallmatrix}\right)$, then the action of g on S is given by

$$g(X_1^i X_2^j) = (X_1 - X_2)^i X_2^j.$$

From now on, let G be a subgroup of $GL(V)$. We say that $P \in S$ is G-*invariant* if $gP = P$ for all $g \in G$.

If M is a G-module, then M^G denotes the isotypic component corresponding to the trivial representation; in other words

$$M^G := \{ m \in M \mid gm = m \text{ for all } g \in G \}$$

is the submodule of G-invariant elements, also known as the space of fixed points of G.

Definition 3.4. The *algebra of invariants* of G is the algebra of G-invariant polynomial functions $J := S^G$.

If $P \in J$, then P is constant on each G-orbit of V and therefore J may be interpreted as an algebra of polynomial functions on the set V/G of all G-orbits on V. A useful direct connection between the action of G and the algebra of invariants is given in the next result, which is due to E. Noether.

Theorem 3.5. *Suppose that $v, w \in V$. Then there exists $g \in G$ such that $g(v) = w$ if and only if $P(v) = P(w)$ for all $P \in J$.*

Proof. If $g(v) = w$, then certainly $P(w) = P(g(v)) = P(v)$. Conversely, if $w \in V$ is not in the orbit Gv of $v \in V$, by Lemma 3.3 there exists $Q \in S$ such that $Q(gv) = 1$ for each element $g \in G$ while $Q(w) = 0$. Let $P = \prod_{g \in G} gQ$. Then $P \in J = S^G$, and $P(v) = \prod_{g \in G} Q(g^{-1}v) = 1$, while $P(w) = 0$. Thus distinct G-orbits are separated by J. $\qquad\square$

3. Invariants of a finite group

Hilbert's Fourteenth Problem (see Mumford [168]) asks whether the algebra of invariants J of an arbitrary group G is finitely generated as an algebra. This was answered in the negative by Nagata [169] in 1958. However, when G is finite and the field is $\mathbb{C}$, Hilbert [112] had shown in 1890, as an application of his celebrated 'Basis Theorem', that J is a finitely generated $\mathbb{C}$-algebra. According to Dieudonné (see [79]): 'this may be considered to be the first paper in "modern algebra" ... but Hilbert's success also spelled the doom of XIXth Century invariant theory.'

In 1914 Noether [173] showed that if G is finite, then J is finitely generated even when $\mathbb{C}$ is replaced by an arbitrary field (see Flatto [98]). It is also the case that J is finitely generated when G is a reductive algebraic group; for more details see the introduction and Chapter 1 of the book by Benson [12] and the references given there.

A fundamental concept in this circle of ideas is that of a *Noetherian ring*, which is a commutative ring R all of whose ideals are finitely generated (as R-modules). Hilbert's basis theorem (Lang [142, Chapter IV, Theorem 4.1]) asserts that if R is a Noetherian ring then the polynomial ring $R[X]$ is again Noetherian, and therefore so is $R[X_1, \ldots, X_n]$. We shall also use the related concept of Noetherian module: an A-module M, where A is a commutative ring, is said to be *Noetherian* if all submodules of M are finitely generated. It is an easy consequence of Hilbert's basis theorem that any finitely generated module over a Noetherian ring is a Noetherian module.

Now suppose that V is a finite dimensional vector space over $\mathbb{C}$, and that G is a finite subgroup of $GL(V)$. Continuing with the notation of the previous section, let $S := S(V^*)$ be the algebra of polynomial functions on V and let $J := S^G$. Furthermore, let J^+ be the (maximal) ideal of J consisting of polynomials with zero constant term and let F be the ideal of S generated by J^+. That is,

$$J^+ := \{ P \in J \mid P(0) = 0 \} \quad \text{and} \quad F := SJ^+.$$

The ring S is isomorphic to the ring $\mathbb{C}[X_1, \ldots, X_n]$ of polynomials in n variables and so by Hilbert's basis theorem, the ideal F is finitely generated. Indeed, we may write $F = \langle I_1, I_2, \ldots, I_r \rangle$, where $I_1, I_2, \ldots, I_r$ are G-invariant and homogeneous.

Definition 3.6. For use in the next proposition (and later) we define a linear 'averaging operator' $\mathrm{Av} : S \to J$ (also known as the *Reynolds operator*). Specifically, for

$P \in S$ we put

$$\mathrm{Av}(P) := |G|^{-1} \sum_{g \in G} gP.$$

It is clear from the definition that $\mathrm{Av}(P) \in J$ and that $\mathrm{Av}(P)$ is either 0 or has the same degree as P. Moreover, for $P \in J$ we have $\mathrm{Av}(P) = P$ and therefore $\mathrm{Av}^2 = \mathrm{Av}$. Thus Av is a projection of S onto J. In fact, a somewhat stronger statement is true, namely that for $P \in J$ and $Q \in S$ we have $\mathrm{Av}(PQ) = P\,\mathrm{Av}(Q)$ so that Av is a J-module homomorphism.

Proposition 3.7. *Suppose that G is finite and that $F = \langle I_1, I_2, \ldots, I_r \rangle$, where I_1, $I_2, \ldots, I_r$ are G-invariant and homogeneous. Then J is generated as an algebra by $I_1, I_2, \ldots, I_r$.*

Proof. Without loss of generality we may suppose that $I_i \neq 0$ for all i. We shall show by induction on the degree that every homogeneous polynomial $P \in J$ is a (homogeneous) polynomial in $I_1, I_2, \ldots, I_r$. This is certainly true for polynomials of degree 0 and so we shall suppose that the degree of P is at least 1. In this case we have $P \in J^+ \subseteq F$ and therefore we may write

$$P = P_1 I_1 + P_2 I_2 + \cdots + P_r I_r$$

for some $P_1, P_2, \ldots, P_r \in S$. In addition, for all i, we may choose P_i to be homogeneous of degree $\deg P - \deg I_i$. It follows that

$$P = \mathrm{Av}(P) = \mathrm{Av}(P_1)I_1 + \mathrm{Av}(P_2)I_2 + \cdots + \mathrm{Av}(P_r)I_r.$$

If $\mathrm{Av}(P_i) \neq 0$, then $\deg \mathrm{Av}(P_i) = \deg P_i < \deg P$ and by induction $\mathrm{Av}(P_i)$ is a polynomial in $I_1, I_2, \ldots, I_r$. This completes the proof. $\square$

Corollary 3.8 (Hilbert). *If G is finite, then J is finitely generated as an algebra.*

See Flatto [**98**] for an account of Noether's generalisation of this result. In the remainder of this section we present another approach based on knowing that S is an *integral* extension of J.

Let A and B be commutative rings and suppose that A is a subring of B.

Definition 3.9. The extension $A \subseteq B$ is of *finite type* if B is finitely generated as an A-algebra; it is *finite* if B is finitely generated as an A-module.

Definition 3.10. An element $b \in B$ is *integral* over A if b is a root of a monic polynomial with coefficients in A. We say that B is *integral over A* if every element of B is integral over A.

We refer to Lang [**142**, Chapter VII] for elementary properties of integral dependence (see also Atiyah and Macdonald [**8**, Chapter 5], especially the exercises). A fundamental result is that the extension $A \subseteq B$ is finite if and only if it is integral and of finite type (Lang [**142**, Chapter VII, Proposition 1.2]).

Lemma 3.11. *If G is a finite group of automorphisms of the commutative ring B then the extension $B^G \subseteq B$ is integral.*

Proof. Given $b \in B$, the polynomial $\prod_{g \in G} (x - g(b))$ is monic and has b as a root. The coefficients of this polynomial are symmetric functions of $\{\, g(b) \mid g \in G \,\}$ and therefore they are G-invariant. That is, the coefficients lie in the ring of invariants B^G, as required. $\square$

Suppose that k is a field and that B is a finitely generated k-algebra. We may refine the proof just given to extract information about the ring of invariants B^G (see Benson [**12**, p. 5] and also Kane [**124**, p. 174]).

Theorem 3.12. *Suppose that G is a finite group of automorphisms of the finitely generated commutative k-algebra B. Then B^G is a finitely generated k-algebra and the extension $B^G \subseteq B$ is finite.*

Proof. Let $A := B^G$. Let $b_1, b_2, \ldots, b_n$ be generators for B as a k-algebra. The previous lemma shows that each b_i satisfies a monic polynomial with coefficients in A. Let C be the k-subalgebra of A generated by these coefficients. Then for all i, b_i is integral over C. The set of elements which are integral over C form a subring and therefore B is integral over C. It follows that B is a finitely generated C-module. But C is a finitely generated k-algebra and hence Noetherian. Hence B is a finitely generated module over the Noetherian ring C, and by the remarks above, it follows that B is a Noetherian C-module, so that all its submodules are finitely generated. In particular A is finitely generated as a C-module, which in turn is finitely generated as a k-algebra. It follows that A is finitely generated as a k-algebra.

Finally, the extension $A \subseteq B$ is both integral and of finite type, hence finite. $\square$

Corollary 3.13 (Hilbert–Noether). *Suppose that V is a finite dimensional vector space over a field k and that G is a finite subgroup of $GL(V)$. Then the algebra $S(V^*)^G$ of invariants is finitely generated.*

This approach has other useful consequences (see Springer [**201**]). These arise from the fact that points of V correspond bijectively to maximal ideals of $S(V^*)$. Underlying all this are algebraic geometric ideas whose general context is beyond the scope of this work, but which we shall explore a little in the ensuing paragraphs, and later in Chapter 11 and Appendix A.

We therefore return to the case where V is a finite dimensional vector space over $\mathbb{C}$ and G is a finite subgroup of $GL(V)$. As before, let $S := S(V^*)$ and $J := S^G$. Given $v \in V$, the set

$$\mathfrak{m}_v := \{\, P \in S \mid P(v) = 0 \,\}$$

is the kernel of the homomorphism $S \to \mathbb{C} : P \mapsto P(v)$ and hence a maximal ideal of S. Conversely, since $S \simeq \mathbb{C}[X_1, X_2, \ldots, X_n]$, it is a consequence of Hilbert's

Nullstellensatz [**142**, p. 378 *et seq.*] that every maximal ideal of S has this form. Furthermore, for $g \in G$ we have $g(\mathfrak{m}_v) = \mathfrak{m}_{g(v)}$.

Let $F_1, \ldots, F_\ell$ be elements of S; i.e. polynomial functions on V. We wish to consider properties of the map $\varphi : V \to \mathbb{C}^\ell$ defined by $\varphi(v) = (F_1(v), \ldots, F_\ell(v))$. There is a unique algebra homomorphism $\varphi^* : \mathbb{C}[X_1, \ldots, X_\ell] \to S$ such that $\varphi^*(X_i) = F_i$ for $i = 1, \ldots, \ell$ and for $P \in \mathbb{C}[X_1, \ldots, X_\ell]$ we have $\varphi^*(P) = P\varphi$, whence $\varphi^{*-1}(\mathfrak{m}_v) = \mathfrak{m}_{\varphi(v)}$ for all $v \in V$. It follows that φ^* is injective if and only if there is no polynomial $P \neq 0$ such that $P(F_1, \ldots, F_\ell) = 0$; i.e. if and only if the F_i are algebraically independent. In order to explore the surjectivity of φ, we require the following consequence of integral dependence.

Lemma 3.14. *Suppose B is an integral domain and A is a subring such that the extension $A \subseteq B$ is finite. If $\mathfrak{m}$ is any maximal ideal of A, there exists a maximal ideal of $\mathfrak{n}$ of B such that $\mathfrak{m} = \mathfrak{n} \cap A$.*

Proof. By definition B is finitely generated as an A-module. Therefore, if $B = \mathfrak{m}B$, then by Corollary A.4 (*i*) there exists $a \in \mathfrak{m}$ such that $(1+a)B = 0$. But this is false since B is an integral domain and $1 \notin \mathfrak{m}$. Consequently, $B \neq \mathfrak{m}B$ and taking $\mathfrak{n}$ to be any maximal ideal of B such that $\mathfrak{m}B \subseteq \mathfrak{n}$ we have $\mathfrak{m} = \mathfrak{n} \cap A$. $\square$

As an application of this result we have the following theorem.

Theorem 3.15. *Let G be a finite group acting on $V = \mathbb{C}^n$ and let $S = S(V^*)$. Let $\{F_1, \ldots, F_n\}$ be a set of algebraically independent elements of S^G such that S^G is integral over $\mathbb{C}[F_1, \ldots, F_n]$. Then the map $\varphi : V \to \mathbb{C}^n$ given by $v \mapsto (F_1(v), \ldots, F_n(v))$ is surjective.*

Proof. Since $F_1, \ldots, F_n$ are algebraically independent, the homomorphism $\varphi^* : \mathbb{C}[X_1, \ldots, X_n] \to S$ defined above is injective, so we may identify $\mathbb{C}[X_1, \ldots, X_n]$ with $\mathbb{C}[F_1, \ldots, F_n]$. By Lemma 3.11 S is integral over S^G and by assumption S^G is integral over $S' := \mathbb{C}[F_1, \ldots, F_n]$, hence S is integral over S'. Since S is finitely generated (e.g. by $X_1, \ldots, X_n$), the extension $S' \subseteq S$ is finite.

The point $v \in V$ corresponds to the maximal ideal $\mathfrak{m}_v$ of S and under the identification of $\mathbb{C}[X_1, \ldots, X_n]$ with S', its image $\varphi(v)$ corresponds to the maximal ideal $\mathfrak{m}_{\varphi(v)} = \varphi^{*-1}(\mathfrak{m}_v) = \mathfrak{m}_v \cap S'$ of S'. That is, if we identify points of $\mathbb{C}^n$ with maximal ideals of S, as above, the map φ may be thought of as taking maximal ideals of S to their intersection with S'. But by Lemma 3.14, the latter map is surjective, whence so is φ. $\square$

Remarks 3.16.

1. For any finite group G acting on $V = \mathbb{C}^n$, it is a consequence of Noether's normalisation lemma that there is always a set $\{F_1, \ldots, F_n\}$ of homogeneous invariants such that S^G is integral over $\mathbb{C}[F_1, \ldots, F_n]$ (see Theorem A.12).

2. The examples $\varphi(x,y) = (x^2, xy)$ (not surjective) and $\varphi(x,y) = (x^2 + y^2, xy)$ (surjective) show that the question of surjectivity is quite delicate. All polynomials in these formulae are invariant under the cyclic group of order 2, acting via scalar multiplication on $\mathbb{C}^2$.

3. Although the map φ of Theorem 3.15 is surjective, in practice it may be very difficult to find a point mapping to a given point $(a_1, \ldots, a_n) \in \mathbb{C}^n$. For example, if $F_i(X_1, \ldots, X_n)$ is the ith elementary symmetric function of the X_i $(i = 1, \ldots, n)$ this question reduces to finding the roots $t = x_j$ of the polynomial equation $t^n + \sum_{i=1}^n (-1)^i a_i t^{n-i} = 0$.

4. The action of a reflection

Up until now we have not assumed that the group G contains reflections. It turns out that the presence of reflections places considerable restrictions on the invariants of G. Indeed, if G is generated by reflections, a theorem of Chevalley asserts that the ring of invariants is a polynomial ring. In preparation for Chevalley's Theorem we investigate next the effect of a reflection on a polynomial.

Recall from Chapter 1 that if H is a hyperplane in V, then $L_H \in V^*$ denotes a linear map such that $H = \operatorname{Ker} L_H$. We regard L_H as an element of S.

Lemma 3.17. *If r is a reflection in $GL(V)$ and if $H := \operatorname{Fix} r$ is its reflecting hyperplane, then for all $P \in S$ there exists $Q \in S$ such that*

$$rP = P + L_H Q.$$

Proof. Let a be a root of r. It follows from Lemma 1.15 that there exists $\alpha \in \mathbb{C}$ such that $r^{-1}v = v + \alpha L_H(v)a$ for all $v \in V$. Thus for all $\varphi \in V^*$ we have $r\varphi(v) = \varphi(v) + \alpha L_H(v)\varphi(a)$; that is, $r\varphi - \varphi = \beta L_H$ for some $\beta \in \mathbb{C}$.

Now suppose that for polynomials $P_1, P_2 \in S$ that there exist $Q_1, Q_2 \in S$ such that $rP_1 = P_1 + L_H Q_1$ and $rP_2 = P_2 + L_H Q_2$. Then

$$r(P_1 P_2) - P_1 P_2 = r(P_1)(r(P_2) - P_2) + (r(P_1) - P_1)P_2$$
$$= r(P_1)L_H Q_2 + L_H Q_1 P_2 = L_H(r(P_1)Q_2 + Q_1 P_2)$$

and the result holds for $P_1 P_2$. Since S is generated by V^* the result holds for all $P \in S$. $\qquad\square$

5. The Shephard–Todd–Chevalley Theorem

The rest of this chapter is devoted to the study of J when G is a finite unitary reflection group. In particular, we shall show that the ring J is itself a polynomial algebra and in the next chapter we shall complete the proof that unitary reflection groups are characterised by this property (Shephard and Todd [193], Chevalley [51]). There are proofs of these results in the context of reflection groups defined over $\mathbb{R}$ in Humphreys [119]. See also Bourbaki [33], Flatto [98] and Hiller [113].

The next lemma is the key technical tool needed in Chevalley's proof of the main theorem.

Lemma 3.18 (Chevalley). *Suppose that G is a finite unitary reflection group and that $U_1, U_2, \ldots, U_r$ are homogeneous elements of J such that U_1 is not in the ideal of J generated by $U_2, U_3, \ldots, U_r$. If $P_1, P_2, \ldots, P_r$ are homogeneous elements of S such that*

$$(3.19) \qquad P_1 U_1 + P_2 U_2 + \cdots + P_r U_r = 0,$$

then $P_1 \in F$.

Proof. The proof is by induction on the degree of P_1. Applying the projection operator $\mathrm{Av} : S \to J$ to (3.19) we find that

$$\mathrm{Av}(P_1)U_1 + \mathrm{Av}(P_2)U_2 + \cdots + \mathrm{Av}(P_r)U_r = 0.$$

If P_1 is a non-zero constant, then $\mathrm{Av}(P_1) = P_1$ and we see that U_1 is in the ideal of J generated by $U_2, U_3, \ldots, U_r$, contrary to assumption. Thus we may suppose that $d := \deg P_1 > 0$.

If $G = 1$, then $F = J^+$ and so $P_1 \in F$. Thus we may suppose that G contains a reflection r_H with hyperplane $H := \mathrm{Ker}\, L_H$. On applying r_H to (3.19) and subtracting the result from (3.19), we find that

$$(r_H P_1 - P_1)U_1 + (r_H P_2 - P_2)U_2 + \cdots + (r_H P_r - P_r)U_r = 0.$$

But, by the previous lemma, for each i there exists a homogeneous polynomial $Q_i \in S$ of degree less than $\deg P_i$ such that

$$r_H P_i - P_i = L_H Q_i.$$

After cancelling L_H we find that

$$Q_1 U_1 + Q_2 U_2 + \cdots + Q_r U_r = 0.$$

It follows by induction that $Q_1 \in F$ and therefore $r_H P_1 - P_1 = L_H Q_1 \in F$. That is, $r_H P_1 \equiv P_1 \pmod{F}$. On the other hand, G is generated by reflections of the form r_H and therefore $gP_1 \equiv P_1 \pmod{F}$ for all $g \in G$. But now $\mathrm{Av}(P_1) \equiv P_1 \pmod{F}$ and hence $P_1 \in F$, as required. $\qquad\square$

The next theorem shows that for a unitary reflection group, the algebra of invariants is itself a polynomial algebra. The theorem was proved by Shephard and Todd [**193**] by providing, for each irreducible unitary reflection group, an explicit set of algebraically independent invariants, the product of whose degrees is equal to the order of the group. Chevalley [**51**] gave a uniform proof for real reflection groups and it was pointed out by Serre [**189**] that Chevalley's proof remains valid for unitary reflection groups. Chevalley stated the result for the real case in his paper [**50**] for the 1950 International Congress of Mathematicians. Borel has given a more detailed

account of the provenance of this theorem in his book [**30**, Chapter VII] and in his
earlier article [**29**].

For the basic facts about algebraic dependence used in this and subsequent proofs,
see Lang [**142**].

Theorem 3.20 (Shephard–Todd). *If G is a finite unitary reflection group on the
vector space V of dimension n, then the ring J of G-invariant polynomials is a
polynomial algebra, i.e. it is generated by a collection of algebraically independent
homogeneous polynomials.*

Proof. (Chevalley [**51**]) Let $I_1, I_2, \ldots, I_r$ be a minimal set of homogeneous invariant
polynomials that generate the ideal $F := SJ^+$ of S. Then, by Proposition 3.7, J is
generated by $I_1, I_2, \ldots, I_r$ as an algebra and we may suppose, by way of contradiction, that these polynomials are algebraically dependent. That is, there is a non-trivial
polynomial $H(Y_1, Y_2, \ldots, Y_r)$ such that $H(I_1, I_2, \ldots, I_r) = 0$. If $d_i := \deg I_i$, for
$1 \le i \le r$, we may assume that there exists an integer h such that for each monomial
$Y_1^{k_1} Y_2^{k_2} \cdots Y_r^{k_r}$ of H we have

$$k_1 d_1 + k_2 d_2 + \cdots + k_r d_r = h.$$

Moreover, we choose H so that h is as small as possible.

For $1 \le i \le r$ set $H_i := \dfrac{\partial H}{\partial Y_i}$ and $U_i := H_i(I_1, I_2, \ldots, I_r)$. Then $U_1, U_2, \ldots, U_r$
are homogeneous polynomials in the ring J. Not all the partial derivatives of H can
be zero, and if H_i is not the zero polynomial, then by the minimal choice of h, we
have $U_i = H_i(I_1, I_2, \ldots, I_r) \ne 0$. Thus we may choose the notation so that U_1,
$U_2, \ldots, U_s$ is a set of generators for the ideal of J generated by all the U_i and so that
s is minimal. In particular we can find homogeneous G-invariant polynomials V_{kj} of
degree $d_j - d_k$ such that for $k > s$

$$U_k = \sum_{j=1}^{s} V_{kj} U_j.$$

Since $H(I_1, I_2, \ldots, I_r) = 0$, for $1 \le i \le n$, we have

$$0 = \frac{\partial H}{\partial X_i} = \sum_{j=1}^{r} U_j \frac{\partial I_j}{\partial X_i}$$

$$= \sum_{j=1}^{s} U_j \frac{\partial I_j}{\partial X_i} + \sum_{k=s+1}^{r} \sum_{j=1}^{s} V_{kj} U_j \frac{\partial I_k}{\partial X_i}$$

$$= \sum_{j=1}^{s} U_j \left(\frac{\partial I_j}{\partial X_i} + \sum_{k=s+1}^{r} V_{kj} \frac{\partial I_k}{\partial X_i} \right).$$

For $i \leq s$ we have chosen the notation so that U_i is not in the ideal generated by $\{U_1, \ldots, U_s\} \setminus \{U_i\}$. Thus by Lemma 3.18, for $1 \leq i \leq n$ and $1 \leq j \leq s$, we have

$$\frac{\partial I_j}{\partial X_i} + \sum_{k=s+1}^{r} V_{kj} \frac{\partial I_k}{\partial X_i} \in F.$$

This polynomial has degree $d_j - 1$ and so

$$(3.21) \qquad \frac{\partial I_j}{\partial X_i} + \sum_{k=s+1}^{r} V_{kj} \frac{\partial I_k}{\partial X_i} = \sum_{\ell=1}^{r} B_{ij\ell} I_\ell,$$

where the $B_{ij\ell}$ are homogeneous polynomials such that $B_{ij\ell} = 0$ for all ℓ such that $d_\ell \geq d_j$. From Euler's formula we have

$$\sum_{i=1}^{n} X_i \frac{\partial I_j}{\partial X_i} = d_j I_j$$

and so, on multiplying equation (3.21) by X_i and summing over i, we obtain

$$d_j I_j + \sum_{k=s+1}^{r} V_{kj} d_k I_k = \sum_{\ell=1}^{r} A_\ell I_\ell,$$

where $A_j = 0$. In particular, I_1 is in the ideal of S generated by $I_2, \ldots, I_r$, which contradicts the minimality of r. $\qquad\square$

A set of algebraically independent homogeneous polynomials that generate the algebra J of G-invariant polynomials is called a set of *basic invariants* for G. Such a set is not unique but we shall soon see that their number and their degrees are uniquely determined by G.

Let $\mathbb{C}(V) := \mathbb{C}(X_1, X_2, \ldots, X_n)$ denote the field of fractions of S, i.e. the field of rational functions P/Q, where $P, Q \in S$. Then the action of G on S extends to an action on $\mathbb{C}(V)$ given by $g(P/Q) := gP/gQ$.

Lemma 3.22. *The fixed field $\mathbb{C}(V)^G$ of G is the field of fractions*

$$\mathbb{C}(J) := \mathbb{C}(I_1, I_2, \ldots, I_r)$$

of J. In particular, $\mathbb{C}(V)$ is a Galois extension of $\mathbb{C}(J)$ of degree $|G|$ with Galois group G.

Proof. If $P/Q \in \mathbb{C}(V)$ is fixed by g, then multiplying numerator and denominator by $\prod_{g \neq 1} gQ$, if necessary, we may suppose that $gQ = Q$ and hence $gP = P$. That is, $P/Q \in \mathbb{C}(J)$. Now Artin's Theorem (Lang [142, Chapter VI, Theorem 1.8]) says that $\mathbb{C}(V)$ is a Galois extension of $\mathbb{C}(J)$ of degree $|G|$ with Galois group G. $\qquad\square$

Corollary 3.23. *The action of G on $\mathbb{C}(V)$, regarded as a vector space over the field $\mathbb{C}(J)$, is the regular representation.*

Proof. By the Normal Basis Theorem (Lang [**142**, Chapter VI, Theorem 13.1]), there is an element $\theta \in \mathbb{C}(V)$ such that $\{\, g\theta \mid g \in G \,\}$ is a $\mathbb{C}(J)$-basis for $\mathbb{C}(V)$. From this we see directly that $\mathbb{C}(V)$ affords the regular representation of G.   □

Corollary 3.24. *If the unitary reflection group G is irreducible and if d_1, d_2, ..., d_r are the degrees of a set of basic invariants of G, then the centre $Z(G)$ of G is a cyclic group of order $\gcd(d_1, d_2, \ldots, d_r)$.*

Proof. Since G is irreducible, $Z(G)$ is cyclic and the action of $z \in Z(G)$ on $\mathbb{C}(V)$ is simply multiplication by a scalar λ. If I_j is a homogeneous invariant of degree d_j, then $I_j = zI_j = \lambda^{-d_j} I_j$ and hence $\lambda^{d_j} = 1$. This proves that $|Z(G)|$ divides $\gcd(d_1, d_2, \ldots, d_r)$.

Conversely, suppose that λ is a dth root of unity, where d is a divisor of d_1, $d_2, \ldots, d_r$. Then multiplication by λ defines a linear transformation z of $\mathbb{C}(V)$ that fixes $\mathbb{C}(J)$. That is, z belongs to the Galois group of $\mathbb{C}(V)/\mathbb{C}(J)$ which, by Lemma 3.22, is the group G. Combining this with the previous paragraph proves that $|Z(G)| = \gcd(d_1, d_2, \ldots, d_r)$.   □

Proposition 3.25. *If $\{I_1, I_2, \ldots, I_r\}$ is a set of basic invariants for G, then $r = n$ and the degrees $\deg I_i$ (including multiplicities) are uniquely determined by G.*

Proof. Since $\mathbb{C}(V)$ is a Galois extension of $\mathbb{C}(J)$, the transcendence degrees of $\mathbb{C}(V)$ and $\mathbb{C}(J)$ coincide (Lang [**142**, p. 356]). However, the transcendence degree of $\mathbb{C}(V)$ is n and the transcendence degree of $\mathbb{C}(J)$ is r, and therefore $r = n$.

Now suppose that $I_1', I_2', \ldots, I_n'$ is another set of basic invariants for G and that $d_i' := \deg I_i'$ for $i := 1, 2, \ldots, n$. Then there are polynomials $f_1, f_2, \ldots, f_n$ and polynomials $g_1, g_2, \ldots, g_n$ such that $I_i' = f_i(I_1, \ldots, I_n)$ and $I_i = g_i(I_1', \ldots, I_n')$. On differentiating with respect to I_j' and using the chain rule we find that

$$\sum_{k=1}^{n} \frac{\partial f_i}{\partial I_k} \frac{\partial g_k}{\partial I_j'} = \delta_{ij}.$$

In particular, $\det\left(\dfrac{\partial f_i}{\partial I_k}\right)$ is not zero and so there is a permutation π of $\{1, 2, \ldots, n\}$ such that

$$\prod_{i=1}^{n} \frac{\partial f_i}{\partial I_{\pi(i)}} \neq 0.$$

We may choose the numbering so that $\pi(i) = i$ for all i and then it follows that I_i occurs in the expression for I_i' as a polynomial in the I_j. In particular, $d_i' \geq d_i$. But then $\sum_{i=1}^{n} d_i' \geq \sum_{i=1}^{n} d_i$ and by symmetry the reverse inequality holds. Thus $d_i' = d_i$ for all i.   □

The integers $d_1, d_2, \ldots, d_n$ are called the *degrees* of G and the integers $m_i :=$ $d_i - 1$ are called the *exponents* of G. Another proof that the exponents depend only on G and not on the choice of basic invariants can be deduced from the form of the Poincaré series of J given in Chapter 4.

6. The coinvariant algebra

Recall that F is the ideal of S generated by the elements of J with 0 constant term. The quotient S/F is called the *coinvariant algebra* of the reflection group G. The goal of this section is to show that, as a $\mathbb{C}G$-module, S is the tensor product $S/F \otimes_{\mathbb{C}} J$ and that S/F affords the regular representation of G. Later we shall show that S/F may be identified with the space of harmonic polynomials on V.

Lemma 3.26 (Chevalley [51]). *Let G be a finite unitary reflection group on the vector space V of dimension n and let $P_1, P_2, \ldots, P_n$ be homogeneous elements of S whose residues modulo F are linearly independent over $\mathbb{C}$. Then $P_1, P_2, \ldots, P_n$ are linearly independent as elements of the vector space $\mathbb{C}(V)$ over the field $\mathbb{C}(J)$.*

Proof. Suppose that we have a non-trivial linear relation
$$(A_1/B_1)P_1 + \cdots + (A_s/B_s)P_s = 0$$
for some $A_i, B_i \in J$. By clearing denominators we may suppose that we have a non-trivial relation $U_1 P_1 + \cdots + U_s P_s = 0$, where the U_i are homogeneous elements of J such that for some h and for all i, $\deg U_i + \deg P_i = h$. We may also suppose that the U_i are ordered so that
$$\deg U_1 = \cdots = \deg U_t < \deg U_{t+1} \leq \cdots \leq \deg U_s.$$
By Lemma 3.18 if U_i is not in the ideal of J generated by the U_j with $j \neq i$, then $P_i \in F$, contrary to hypothesis. It follows that there are non-zero elements $a_1, a_2, \ldots, a_t \in \mathbb{C}$ such that
$$a_1 U_1 + a_2 U_2 + \cdots + a_t U_t = 0.$$
But now we may eliminate U_1 from the given relation between the P_i to obtain
$$U_2(P_2 - \frac{a_2}{a_1}P_1) + \cdots + U_t(P_t - \frac{a_t}{a_1}P_1) + U_{t+1}P_{t+1} + \cdots + U_s P_s = 0.$$
On repeating this argument $t-1$ times we find that there are elements $b_1, \ldots, b_t \in \mathbb{C}$, not all 0, such that
$$U_t \sum_{i=1}^{t} b_i P_i + U_{t+1}P_{t+1} + \cdots + U_s P_s = 0.$$
Since U_t cannot be in the ideal of J generated by $U_{t+1}, \ldots, U_s$ it follows from Lemma 3.18 that $\sum_{i=1}^{t} b_i P_i \in F$, which is a contradiction. $\square$

Corollary 3.27. $\dim_{\mathbb{C}(J)} \mathbb{C}(V) = |G| \geq \dim_{\mathbb{C}} S/F.$

Proof. The ideal F is generated by homogeneous polynomials and therefore it is the direct sum of its homogeneous components $S_i \cap F$. Thus we can find homogeneous polynomials $P_1, P_2, \ldots, P_s$ in S whose residues modulo F form a basis for S/F. Then, by the lemma, $P_1, P_2, \ldots, P_s$ can be extended to a basis for $\mathbb{C}(V)$. $\qquad\square$

Lemma 3.28. *Let $A_1, A_2, \ldots, A_s$ be homogeneous polynomials whose residues modulo F form a basis for S/F. Then every polynomial $P \in S$ has a unique expression of the form $P := U_1 A_1 + U_2 A_2 + \cdots + U_s A_s$, where $U_i \in J$ for all i.*

Proof. (Chevalley [**51**]) It is sufficient to prove that every homogeneous polynomial P has an expression of the given form. Then uniqueness follows from the previous lemma. The result is clearly true when $\deg P = 0$, so suppose that $\deg P > 0$. By hypothesis there are elements $a_1, a_2, \ldots, a_s \in \mathbb{C}$ such that

$$P = a_1 A_1 + \cdots + a_s A_s + Q$$

for some $Q \in F$. The polynomials $I_1, I_2, \ldots, I_n$ generate J as an algebra and F as an ideal. Thus $Q = \sum P_j I_j$ for some homogeneous polynomials P_j and as $\deg P_j < \deg P$, the result follows by induction. $\qquad\square$

Corollary 3.29. $\dim_{\mathbb{C}} S/F = |G|$.

Proof. Suppose that $A_1, A_2, \ldots, A_s$ are homogeneous polynomials of S whose residues modulo F form a basis for S/F. We shall show that these polynomials span $\mathbb{C}(V)$ as a vector space over $\mathbb{C}(J)$. To this end, suppose that $P/Q \in \mathbb{C}(V)$. Multiplying numerator and denominator by $\prod_{g \neq 1} gQ$, if necessary, we may assume that $Q \in J$. By the lemma we have $P = U_1 A_1 + \cdots + U_s A_s$ for some $U_i \in J$ so that $P/Q = (U_1/Q) A_1 + \cdots + (U_s/Q) A_s$, as claimed. Thus $\dim_{\mathbb{C}} S/F \geq |G|$ and from Corollary 3.27 we see that equality holds. $\qquad\square$

Definition 3.30. An $\mathbb{N}$-*graded G-module* is a vector space M with a direct sum decomposition

$$M = \bigoplus_{i \geq 0} M_i$$

where each subspace M_i is a finite dimensional G-module. Usually we omit reference to $\mathbb{N}$ and call M a *graded G-module*.

Given the graded G-modules $M := \bigoplus_{i \geq 0} M_i$ and $N := \bigoplus_{i \geq 0} N_i$, the tensor product $M \otimes_{\mathbb{C}} N$ is a graded G-module with G-action given by $g(m \otimes n) := gm \otimes gn$ and whose k^{th} graded component is defined to be

$$(M \otimes N)_k := \bigoplus_{i=1}^{k} M_i \otimes N_{k-i}.$$

Corollary 3.31. $S/F \otimes_{\mathbb{C}} J \simeq S$ *as graded G-modules.*

Proof. Choose homogeneous polynomials A_1, A_2, ..., A_s in S whose reductions $\overline{A}_1, \overline{A}_2, \ldots, \overline{A}_s$ modulo F form a basis for S/F. Define a bilinear map

$$S/F \times J \to S$$

that sends $(\overline{A}_i, U)$ to $A_i U$ for all $U \in J$. This induces an isomorphism $\varphi : S/F \otimes J \to S$ of graded G-modules such that $\varphi(\overline{A}_i \otimes U) = A_i U$. $\square$

Proposition 3.32. *The representation of G on S/F is the regular representation.*

Proof. Choose homogeneous polynomials A_1, A_2, ..., A_s in S whose reductions modulo F form a basis for S/F. It follows from the previous lemma that for every $g \in G$ there are homogeneous polynomials $U_{ij}(g) \in J$ such that

$$g A_j = \sum_i U_{ij}(g) A_i.$$

Thus the trace of g as a linear transformation of the vector space $\mathbb{C}(V)$ over $C(J)$ is $\sum_{i=1}^{s} U_{ii}(g)$ which, by Corollary 3.23, is 0 whenever $g \neq 1$. But now reduction modulo F shows that the trace of g acting on S/F is 0 whenever $g \neq 1$. Thus the action of G on S/F is the regular representation. $\square$

Exercises

1. Prove Lemma 3.3 directly by showing that for $i \neq j$, there is a linear map $L_{ij} \in S$ such that $L_{ij}(w_i) \neq L_{ij}(w_j)$ and the polynomial function

$$f_i = \prod_{\substack{j=1 \\ j \neq i}}^{s} \frac{L_{ij} - L_{ij}(w_j)}{L_{ij}(w_i) - L_{ij}(w_j)}$$

   has the property that $f_i(w_j) = \delta_{ij}$ for $i, j = 1, \ldots, s$.

2. Show that if $\dim V = n$, the dimension of the rth homogeneous component of the coordinate ring of V is $\binom{n+r-1}{r}$.

3. Let G_1 and G_2 be finite unitary reflection groups acting on V_1 and V_2 respectively. If $d_1, d_2, \ldots, d_m$ are the degrees of G_1 and if $e_1, e_2, \ldots, e_n$ are the degrees of G_2, show that $G_1 \times G_2$ is a reflection group in $V_1 \oplus V_2$ with degrees $d_1, d_2, \ldots, d_m, e_1, e_2, \ldots, e_n$.

4. Deduce Lemma 1.15 from Hilbert's Nullstellensatz (Lang [**142**, Chapter IX, Theorem 1.5]) using the fact that $rP - P$ vanishes on H.

CHAPTER 4

Poincaré series and characterisations of reflection groups

In order to study the invariants and degrees of a finite reflection group we introduce a tool of great general utility: the Poincaré series of a graded module. The first two sections are mainly concerned with general results which, in the third section, are used to show that finite reflection groups are precisely those finite groups whose ring of invariants is a polynomial algebra.

1. Poincaré series

Let G be a finite group. In preparation for the definition of equivariant Poincaré series we first define the Euler–Grothendieck ring of finite dimensional $\mathbb{C}G$-modules (see Lang [142, Chapter XX, §3]).

Definition 4.1. For each finite dimensional G-module M, let $[M]$ denote the isomorphism class of M. We make the set $\mathcal{M}$ of all $[M]$ into a monoid by defining the sum of $[M]$ and $[N]$ to be $[M \oplus N]$. By Maschke's Theorem (Chapter 1 1.24), every module M is a direct sum of irreducible submodules and the isomorphism classes of these submodules are uniquely determined by M. Hence $\mathcal{M}$ embeds in the free abelian group $R(G)$ with basis $\{ [N] \mid N$ is irreducible $\}$. We make $R(G)$ into a ring by defining the product of $[M]$ and $[N]$ to be $[M \otimes N]$ and extending this definition to all of $R(G)$ by bilinearity; this is the *Euler–Grothendieck ring* of G. It is isomorphic to the ring of complex valued class functions on G.

Definition 4.2. If A is a commutative ring we let $A[\![t]\!]$ denote the ring of *formal power series* in t with coefficients in A (Lang [142, Chapter IV, §9]). The field of fractions of $A[\![t]\!]$ is denoted by $A(\!(t)\!)$.

Recall from Chapter 3 that a *graded vector space* is a vector space M with a direct sum decomposition

$$M = \bigoplus_{i \geq 0} M_i$$

where each subspace M_i is finite dimensional. It is a *graded G-module* if each M_i is a G-module.

Definition 4.3. Suppose that M is a graded module.

(i)   The *Poincaré series* of M is the formal power series

$$P_M(t) := \sum_{i \geq 0} \dim M_i \, t^i \in \mathbb{Z}[\![t]\!].$$

(ii) If M is a graded G-module, the *equivariant Poincaré series* of M is the formal power series

$$P_M^G(t) := \sum_{i \geq 0} [M_i] \, t^i \in R(G)[\![t]\!].$$

(iii) For $g \in G$ the *relative Poincaré series* of the graded G-module M with respect to g is the formal power series

$$P_M(g, t) := \sum_{i \geq 0} \operatorname{trace}(g, M_i) \, t^i \in \mathbb{C}[\![t]\!]$$

where $\operatorname{trace}(g, M_i)$ denotes the trace of the action of g on M_i. When $g = 1$ we recover the usual Poincaré series; i.e. $P_M^G(1, t) = P_M(t)$.

(iv) If χ is an irreducible character of G and if V_χ is a G-module that affords χ, then the *relative Poincaré series* of the graded G-module M with respect to χ is the formal power series

$$P_{M,\chi}(t) := \sum_{i \geq 0} (V_\chi, M_i) \, t^i \in \mathbb{Z}[\![t]\!]$$

where (V_χ, M_i) denotes the multiplicity with which V_χ occurs in M_i; namely $\dim \operatorname{Hom}(V_\chi, M_i)$.

For $g \in G$ we have a ring homomorphism $\rho_g : R(G) \to \mathbb{C}$ such that $\rho_g([M]) = \operatorname{trace}(g, M)$. It is a standard fact of representation theory (Lang [**142**, Chapter XVIII, Theorem 2.3]) that $[M] = [N]$ in $R(G)$ if and only if $\rho_g([M]) = \rho_g([N])$ for all $g \in G$. Thus identities satisfied by $P_M^G(t)$ correspond to identities satisfied by all $P_M(g, t)$, and conversely.

When χ is an irreducible character we see that $P_{M,\chi}(t)[V_\chi]$ is the equivariant Poincaré series $P_{M_\chi}^G(t)$ of the χ-isotypic component M_χ of M and $P_{M,\chi}(t) = \chi(1)^{-1} P_{M_\chi}(t)$.

Lemma 4.4. *For all $g \in G$, we have $P_{M \oplus N}(g, t) = P_M(g, t) + P_N(g, t)$ and $P_{M \otimes N}(g, t) = P_M(g, t) P_N(g, t)$.*

Proof. The kth graded component of $M \oplus N$ is $M_k \oplus N_k$ and so

$$\operatorname{trace}(g, (M \oplus N)_k) = \operatorname{trace}(g, M_k) + \operatorname{trace}(g, N_k),$$

whence $P_{M \oplus N}(g, t) = P_M(g, t) + P_N(g, t)$.

The kth graded component of $M \otimes N$ is $\bigoplus_{i=1}^{k} M_i \otimes N_{k-i}$ and so the coefficient of t^k in $P_{M \otimes N}(g, t)$ is $\sum_{i=1}^{k} \operatorname{trace}(g, M_i) \operatorname{trace}(g, N_{k-i})$. The result now follows from the definition of the product of two formal power series together with the fact that $\text{trace}(g, M_i)\,\text{trace}(g, N_{k-i}) = \text{trace}(g, M_i \otimes N_{k-i})$. $\hspace{1cm}\square$

Corollary 4.5. $P_{M \oplus N}(t) = P_M(t) + P_N(t)$ and $P_{M \otimes N}(t) = P_M(t)P_N(t)$.

1.1. Bigraded modules. In some of the later chapters of this work, we shall encounter vector spaces M which are graded by $\mathbb{N} \times \mathbb{N}$, i.e.

$$M = \bigoplus_{i,j \geq 0} M_{i,j}.$$

Such modules are called *bigraded*, and have Poincaré series

$$P_M(t, u) = \sum_{i,j \geq 0} \dim M_{ij} t^i u^j \in \mathbb{Z}[\![t, u]\!],$$

whose properties are analogous to those of the single variable Poincaré series which we discuss in detail here.

2. Exterior and symmetric algebras and Molien's Theorem

The *symmetric algebra* of V was defined in Chapter 3 as the quotient $T(V)/I$ where $T(V)$ is the tensor algebra of V and I is the two-sided ideal generated by elements of the form $v \otimes w - w \otimes v$ for all $v, w \in V$. If $v_1, v_2, \ldots, v_n$ is a basis for V, then $S(V)$ is isomorphic to the polynomial ring $\mathbb{C}[v_1, \ldots, v_n]$.

Lemma 4.6. *If G is a finite group acting on the vector space V of dimension n over $\mathbb{C}$, then for all $g \in G$,*

$$P_{S(V)}(g, t) = 1/\det(1 - gt).$$

Proof. Since G is finite, for every element $g \in G$ we may choose a basis for V consisting of eigenvectors $v_1, v_2, \ldots, v_n$ of g with the eigenvalues $\lambda_1, \lambda_2, \ldots, \lambda_n$.

The monomials in the v_i form a basis for $S(V)$ and, in particular, the kth homogeneous component $S_k(V)$ of $S(V)$ has a basis consisting of the monomials $v_1^{m_1} \cdots v_n^{m_n}$ where $m_1 + \cdots + m_n = k$. These monomials are eigenvectors for the action of g on $S_k(V)$ and the corresponding eigenvalues are $\lambda_1^{m_1} \cdots \lambda_n^{m_n}$. Therefore the trace of g on $S_k(V)$ is

$$\text{trace}(g, S_k(V)) = \sum \lambda_1^{m_1} \cdots \lambda_n^{m_n},$$

where the sum is taken over all n-tuples $(m_1, \ldots, m_n)$ such that $\sum_{i=1}^{n} m_i = k$. This is the coefficient of t^k in the product

$$\prod_{i=1}^{n} \frac{1}{1 - \lambda_i t} = \frac{1}{\det(1 - gt)}$$

and so $P_{S(V)}(g, t) = \sum \text{trace}(g, S_k(V)) t^k = 1/\det(1 - gt)$. $\hspace{1cm}\square$

The *exterior algebra* of the vector space V is the quotient $\Lambda(V) := T(V)/I$ where $T(V)$ is the tensor algebra of V and I is the two-sided ideal generated by the elements $v \otimes v$ for all $v \in V$. The product of $\xi = u + I$ and $\eta = v + I \in \Lambda(V)$ is $\xi \wedge \eta = u \otimes v + I$, where $u \otimes v$ denotes the product of u and v in $T(V)$. The ideal I is generated by homogeneous elements and therefore $\Lambda(V)$ inherits the grading of $T(V)$. That is,

$$\Lambda(V) = \bigoplus_{r \geq 0} \Lambda^r(V),$$

where $\Lambda^r(V) := T^r(V)/(I \cap T^r(V))$.

We have $V \cap I = \{0\}$ and therefore V may be identified with its image in $\Lambda(V)$. If $v_1, v_2, \ldots, v_n$ is a basis for V, then $\Lambda^r(V)$ has a basis consisting of the products $v_{i_1} \wedge v_{i_2} \wedge \cdots \wedge v_{i_r}$ where $1 \leq i_1 < i_2 < \cdots < i_r \leq n$. In particular, the dimension of $\Lambda^r(V)$ is $\binom{n}{r}$ and consequently,

$$P_{\Lambda(V)}(t) = (1 + t)^n.$$

More generally, we have the following result which permits the easy computation of the Poincaré series of symmetric and exterior algebras (see Lang [**142**, Chap. XVI Prop. 8.2; Chap. XIX Prop. 1.2]).

Lemma 4.7.

(i) *If W_1 and W_2 are vector spaces and $W = W_1 \oplus W_2$, then we have vector space isomorphisms*

$$S(W) \simeq S(W_1) \otimes S(W_2)$$
$$\Lambda(W) \simeq \Lambda(W_1) \otimes \Lambda(W_2),$$

where $S(W)$ and $\Lambda(W)$ are respectively the symmetric and exterior algebras on W.

(ii) *Let $V = \bigoplus_{i \geq 1} V_i$ be a finite dimensional graded vector space, with degree i component V_i. Then*

$$P_{S(V)}(t) = \prod_{i \geq 1} \left(\frac{1}{1 - t^i}\right)^{\dim V_i}, \quad and$$

$$P_{\Lambda(V)}(t) = \prod_{i \geq 1} (1 + t^i)^{\dim V_i}.$$

Proof. The statement (i) is standard linear algebra, and (ii) follows easily, given that for a homogeneous one dimensional space U of degree d, $P_{S(U)}(t) = \dfrac{1}{1 - t^d}$ and $P_{\Lambda(U)}(t) = 1 + t^d$. $\qquad \square$

Example 4.8. For the coordinate ring S of the vector space V of dimension n, we have

$$S = \mathbb{C}[X_1, \ldots, X_n] \simeq \mathbb{C}[X_1] \otimes \cdots \otimes \mathbb{C}[X_n]$$

and, since $P_{\mathbb{C}[X]}(t) = 1 + t + t^2 + \cdots = (1-t)^{-1}$, it follows that

$$P_S(t) = (1-t)^{-n}.$$

Example 4.9. If G is a unitary reflection group in V and if $J := S^G$ is its ring of invariants then we know from Chapter 3 that J is a polynomial ring in the basic invariants $I_1, I_2, \ldots, I_n$, of degrees $d_1, d_2, \ldots, d_n$. The ring J inherits the grading of S and therefore $P_{\mathbb{C}[I_j]}(t) = 1 + t^{d_j} + t^{2d_j} + \cdots = (1 - t^{d_j})^{-1}$, whence

$$P_J(t) = \prod_{j=1}^{n}(1 - t^{d_j})^{-1}.$$

This provides another proof that the degrees d_i are uniquely determined by G: if $\zeta \in \mathbb{C}$ is a kth root of unity, where k is a positive integer, then the multiplicity of ζ as a root of $1/P_J(t)$ is equal to the number of d_i that are multiples of k; this determines the d_i by induction, beginning with the largest d_i.

Lemma 4.10. *If G is a finite group acting on the vector space V of dimension n over $\mathbb{C}$, then for all $g \in G$,*

$$P_{\Lambda(V)}(g, t) = \det(1 + gt).$$

Proof. The proof proceeds as before except that the kth homogeneous component $\Lambda^k := \Lambda^k(V)$ of $\Lambda(V)$ has basis $v_{i_1} \wedge v_{i_2} \wedge \cdots \wedge v_{i_k}$ where $1 \leq i_1 < i_2 < \cdots < i_k \leq n$. Once again these elements are eigenvectors for g and in this case the corresponding eigenvalues are $\lambda_{i_1} \lambda_{i_2} \cdots \lambda_{i_k}$. Thus the trace of g on Λ^k is

$$\text{trace}(g, \Lambda^k) = \sum_{i_1 < i_2 < \cdots < i_k} \lambda_{i_1} \lambda_{i_2} \cdots \lambda_{i_k}.$$

This is the kth elementary symmetric function in the λ_i and also the coefficient of t^k in the product

$$\prod_{i=1}^{n}(1 + \lambda_i t) = \det(1 + gt).$$

Hence $P_{\Lambda(V)}(g, t) = \sum \text{trace}(g, \Lambda^k)t^k = \det(1 + gt)$. □

Corollary 4.11. *If $S := S(V^*)$ and $L := \Lambda(V^*)$, then $P_S^G(-t)P_L^G(t) = 1$ in $R(G)[\![t]\!]$.*

Proof. On applying the previous two lemmas to V^* we have $P_S(g, -t)P_L(g, t) = 1$ for all $g \in G$. In addition, $[M] = [N]$ in $R(G)$ if and only if $\text{trace}(g, M) = \text{trace}(g, N)$ for all $g \in G$ and therefore $P_S^G(-t)P_L^G(t) = 1$. □

Lemma 4.12. *If χ is an irreducible character of the finite group G and if M is a graded G-module with χ-isotypic component M_χ, then*

$$P_{M_\chi}(t) = \chi(1) P_{M,\chi}(t) = \frac{\chi(1)}{|G|} \sum_{g \in G} \overline{\chi(g)} P_M(g,t).$$

Proof. The multiplicity (V_χ, M_i) in M_i of the G-module V_χ affording χ is

$$|G|^{-1} \sum_{g \in G} \overline{\chi(g)} \operatorname{trace}(g, M_i)$$

(see Lang [**142**, Chapter XVIII]). The dimension of V_χ is $\chi(1)$ and so the dimension of the ith graded component of M_χ is $\chi(1)(V_\chi, M_i)$. Thus the coefficient of t^i in $P_{M_\chi}(t)$ is

$$\frac{\chi(1)}{|G|} \sum_{g \in G} \overline{\chi(g)} \operatorname{trace}(g, M_i)$$

and the result follows from the definition $P_M(g,t)$. $\qquad\square$

Theorem 4.13 (Molien [**167**]). *If G is a finite group acting on the vector space V and if J is the ring of invariants of G, then*

$$P_J(t) = \frac{1}{|G|} \sum_{g \in G} \frac{1}{\det(1 - gt)}.$$

Proof. The ring J is the isotypic component of S corresponding to the trivial character and therefore, by the previous lemma, we have

$$P_J(t) = \frac{1}{|G|} \sum_{g \in G} P_S(g,t).$$

On the other hand, by Lemma 4.6, we have $P_S(g,t) = 1/\det(1 - g^{-1}t)$ and on replacing g by g^{-1} in the summation we see that

$$P_J(t) = \frac{1}{|G|} \sum_{g \in G} \frac{1}{\det(1 - gt)}. \qquad\square$$

Theorem 4.14 (Shephard and Todd [**193**]). *If G is a finite unitary reflection group and if $d_1, d_2, \ldots, d_n$ are the degrees of G, then*

(*i*)   *the order of G is $d_1 d_2 \cdots d_n$,*

(*ii*)   *the number N of reflections in G is $\sum_{i=1}^{n}(d_i - 1)$.*

Proof. If $\lambda_1, \lambda_2, \ldots, \lambda_n$ are the eigenvalues of $g \in G$, then

$$\det(1 - gt)^{-1} = \frac{1}{(1 - \lambda_1 t)(1 - \lambda_2 t) \cdots (1 - \lambda_n t)}.$$

On multiplying this expression by $(1 - t)^n$ and setting $t = 1$ we get 0 unless all eigenvalues are 1, in which case g is the identity element and we get 1. Now

$$(1 - t)/(1 - t^{d_i}) = 1/(1 + t + t^2 + \cdots + t^{d_i - 1})$$

and on setting $t = 1$ we get $1/d_i$. By Molien's Theorem and the expression given in Example 4.9 for $P_J(t)$ we have

(4.15)
$$|G| \prod_{i=1}^{n} (1 - t^{d_i})^{-1} = \sum_{g \in G} \det(1 - gt)^{-1}.$$

Multiplying this by $(1 - t)^n$, setting $t = 1$ and using the values obtained above produces $|G|/d_1 d_2 \cdots d_n = 1$, which proves (i).

An element of $g \in G$ is a reflection if it has exactly one eigenvalue $\lambda(g)$ that is not equal to 1; in this case we have $\det(1 - gt) = (1 - t)^{n-1}(1 - \lambda(g)t)$. Thus if R denotes the set of all reflections in G then, after subtracting $(1 - t)^{-n}$ from both sides of (4.15) and multiplying by $(1 - t)^{n-1}$, we obtain

(4.16)
$$\frac{|G| - \prod_{i=1}^{n}(1 + t + \cdots + t^{d_i - 1})}{(1 - t)\prod_{i=1}^{n}(1 + t + \cdots + t^{d_i - 1})} = \sum_{g \in R} \frac{1}{1 - \lambda(g)t} + (1 - t)h(t),$$

where $h(t)$ is a rational function whose denominator is not divisible by $1 - t$. We wish to evaluate this equality when $t = 1$ and therefore we put $H(t) := |G| - \prod_{i=1}^{n}(1 + t + \cdots + t^{d_i - 1})$ and use the equality $|G| = d_1 \cdots d_n$ from (i) to obtain $H(1) = 0$. Hence $H(t) = (t - 1)H'(t) + \frac{1}{2}(t - 1)^2 H''(t) + \cdots$ and then setting $t = 1$ in (4.16) leads to the equation

$$-\frac{H'(1)}{|G|} = \sum_{g \in R} \frac{1}{1 - \lambda(g)}.$$

Now

$$H'(t) = -\sum_{i=1}^{n}(1 + 2t + 3t^2 + \cdots + (d_i - 1)t^{d_i - 2}) \prod_{j \neq i}(1 + t + \cdots + t^{d_j - 1})$$

whence $H'(1) = -\sum_{i=1}^{n} \frac{1}{2}d_i(d_i - 1)\prod_{j \neq i} d_j = -\frac{1}{2}|G|\sum_{i=1}^{n}(d_i - 1)$ and consequently

(4.17)
$$\frac{1}{2}\sum_{i=1}^{n}(d_i - 1) = \sum_{g \in R} \frac{1}{1 - \lambda(g)}.$$

In particular, if $n = 1$ and $G = \langle g \rangle$, where g is a reflection of order d with non-identity eigenvalue λ, then $d_1 = d$ and G contains $d - 1$ reflections. Therefore

$$\tfrac{1}{2}(d - 1) = \sum_{i=1}^{d-1} \frac{1}{1 - \lambda^i}.$$

Returning to the general case, the set R is the disjoint union of the sets of reflections corresponding to (maximal) cyclic subgroups of reflections. Now each such cyclic subgroup of order d contributes $\frac{1}{2}(d-1)$ to the right hand side of (4.17) and so

$$\frac{1}{2}\sum_{i=1}^{n}(d_i - 1) = \frac{1}{2}|R|,$$

which proves (ii).   □

3. A characterisation of finite reflection groups

In the previous chapter we proved that the ring of invariants of a finite unitary reflection group G in $U_n(\mathbb{C})$ is a polynomial algebra generated by n invariants with uniquely determined degrees $d_1, d_2, \ldots, d_n$. Moreover, we have just seen that the order of G is the product of the degrees. We now show that these conditions characterise unitary reflection groups.

Lemma 4.18 (Springer [201]). *Suppose that V is a vector space of dimension n over $\mathbb{C}$ and that $S := S(V^*)$. Let R be a subalgebra of S generated by n algebraically independent homogeneous polynomials $f_1, f_2, \ldots, f_n$ of degrees $d_1, d_2, \ldots, d_n$, respectively, where $d_1 \leq d_2 \leq \cdots \leq d_n$. Then*

(i)   *for all i, d_i is the smallest degree of a homogeneous polynomial of R not in the subalgebra generated by $f_1; f_2, \ldots, f_{i-1}$;*

(ii)   *if $g_1, g_2, \ldots, g_n$ are n algebraically independent homogeneous elements of R with degrees $e_1, e_2, \ldots e_n$, where $e_1 \leq e_2 \leq \cdots \leq e_n$, then $d_i \leq e_i$ for all i.*

Proof. Let $f \in R$ be a homogeneous polynomial of degree d not in the subalgebra generated by $f_1, \ldots, f_{i-1}$. Then there exists a polynomial $P \in \mathbb{C}[X_1, \ldots, X_n]$ such that $P(f_1, \ldots, f_n) = f$.

For any monomial $X_1^{m_1} \cdots X_n^{m_n}$ that occurs in P the algebraic independence of the f_j implies that $d_1 m_1 + \cdots + d_n m_n = d$ and by the choice of f, we have $m_j \neq 0$ for some $j \geq i$. Thus $d \geq d_j \geq d_i$ and this proves (i).

To prove (ii) we first choose polynomials $G_j \in \mathbb{C}[X_1, \ldots, X_n]$ such that $g_j = G_j(f_1, \ldots, f_n)$. Given i, the algebraic independence of the g_j means that not all polynomials $G_1, \ldots, G_i$ involve only $X_1, \ldots, X_{i-1}$. Thus for some $j \leq i$ and for some $k \geq i$, the polynomial G_j involves X_k. But then $e_i \geq e_j \geq d_k \geq d_i$, as required.   □

The next theorem is essentially due to Shephard and Todd [193] but the version given here is Theorem 2.4 of Springer [201].

Theorem 4.19. *Let G be a finite group acting on the vector space V of dimension n over $\mathbb{C}$. Suppose that $f_1, f_2, \ldots, f_n$ are homogeneous algebraically independent elements of $J := S(V^*)^G$ and set $d_i := \deg f_i$ for $i := 1, 2, \ldots, n$. Then*

(i) $|G| \leq d_1 d_2 \cdots d_n$;

(ii)   if $|G| = d_1 d_2 \cdots d_n$, then G is a reflection group in V and J is generated by $f_1, f_2, \ldots, f_n$ as an algebra;

(iii)   if J is generated by $f_1, f_2, \ldots, f_n$ as an algebra, then equality holds in (i) and G is a reflection group.

Proof. (i) Let S be the symmetric algebra of V^*, let J_i be the ith homogeneous component of $J := S^G$ and let a_i be the dimension of J_i. By Molien's Theorem we have

$$\sum_{i \geq 0} a_i t^i = \frac{1}{|G|} \sum_{g \in G} \det(1 - gt)^{-1}.$$

For all i we have $0 \leq a_i \leq \dim S_i$ and $P_S(t) = (1-t)^{-n}$ whence the series $\sum a_i t^i$ is absolutely convergent for $|t| < 1$.

Define B to be the subalgebra of J generated by the elements $f_1, \ldots, f_n$ and let b_i be the dimension of its ith homogeneous component $B \cap S_i$. Since $B \cap S_i \subseteq J_i$ we have $b_i \leq a_i$ and so the series $\sum_{i \geq 0} b_i t^i = \prod_{j=1}^n (1 - t^{d_j})^{-1}$ is absolutely convergent for $|t| < 1$.

Hence for $0 \leq t < 1$ we have

(4.20) $$\prod_{j=1}^n \frac{1}{1 - t^{d_j}} \leq \frac{1}{|G|} \sum_{g \in G} \frac{1}{\det(1 - gt)},$$

and on multiplying by $(1 - t)^n$ it follows that

$$\prod_{j=1}^n (1 + t + t^2 + \cdots + t^{d_j - 1})^{-1} \leq |G|^{-1} + (1 - t)h(t)$$

where $h(t)$ is a rational function whose denominator is not divisible by $1 - t$. In the limit, as t approaches 1, we have $1/d_1 \cdots d_n \leq 1/|G|$; i.e. $|G| \leq d_1 \cdots d_n$, proving (i).

(ii) Now suppose equality holds, i.e. $|G| = d_1 \cdots d_n$. Then from (4.20), after subtracting $(1 - t)^{-n}$ from both sides and multiplying by $(1 - t)^{n-1}$, we find that

$$\frac{|G| - \prod_{i=1}^n (1 + t + \cdots + t^{d_i - 1})}{(1 - t) \prod_{i=1}^n (1 + t + \cdots + t^{d_i - 1})} \leq \sum_{g \in R} \frac{1}{1 - \lambda(g)t} + (1 - t)h(t),$$

where R is the set of reflections in G and $h(t)$ is a rational function whose denominator is not divisible by $1 - t$. Taking the limit as $t \to 1$ the argument of Theorem 4.14 shows that

$$\tfrac{1}{2} \sum_{i=1}^n (d_i - 1) \leq \tfrac{1}{2}|R|.$$

Let G_1 be the subgroup of G generated by the reflections in G and let $e_1, \ldots, e_n$ be the degrees of G_1. The polynomials f_i are invariants for G and hence for G_1 and so by Lemma 4.18 we have $d_i \geq e_i$ for all i. On the other hand, from Theorem 4.14 we have $|R| = \sum_i (e_i - 1)$ and so

$$\sum_i (d_i - 1) \leq |R| = \sum_i (e_i - 1) \leq \sum_i (d_i - 1).$$

Thus $d_i = e_i$ for all i and therefore $|G_1| = e_1 \cdots e_n = d_1 \cdots d_n = |G|$. It follows that $G_1 = G$ and this proves (ii).

(iii) Suppose that J is generated by $f_1, \ldots, f_n$ as an algebra. Then $a_i = b_i$ for all i and so (4.20) becomes

$$\prod_{j=1}^{n} \frac{1}{1 - t^{d_j}} = \frac{1}{|G|} \sum_{g \in G} \frac{1}{\det(1 - gt)}.$$

Multiplying by $(1 - t)^n$ and setting $t = 1$ leads to the equality $|G| = d_1 \cdots d_n$ and then (iii) follows from (ii).   □

4. Exponents

Throughout this section G will denote a unitary reflection group acting on V and S will be the coordinate ring of V. In addition, F will be the ideal of S generated by the G-invariant polynomials with non-zero constant term and so S/F will be the coinvariant algebra of G. The results of this section come from §2.5 of Springer [201].

If χ is any character of G, define $f_\chi(t)$ to be the Poincaré polynomial $P_{S/F,\chi}(t) = \sum_{i \geq 0} (M_\chi, (S/F)_i) t^i$, where $(M_\chi, (S/F)_i)$ is the multiplicity with which the representation M_χ affording χ occurs in the ith homogeneous component of S/F. The formal power series $f_\chi(t)$ is actually a polynomial in t and since S/F affords the regular representation of G we have $f_\chi(1) = \chi(1)$. Because of their connections with the representation theory of reductive groups over finite fields when G is a Weyl group (see Appendix C), the polynomials $f_\chi(t)$ are referred to as the *fake degrees* of G.

Since the coefficients of $f_\chi(t)$ are integers we may write

$$f_\chi(t) = \sum_{j=0}^{\chi(1)} t^{q_j(\chi)},$$

where $q_0(\chi) \leq q_1(\chi) \leq \cdots$

The integers $q_j(\chi)$ are called the χ-*exponents* of G.

Lemma 4.21. *If $d_1, d_2, \ldots, d_n$ are the degrees of G, then*

$$f_\chi(t) = |G|^{-1} \prod_{i=1}^{n} (1 - t^{d_i}) \sum_{g \in G} \chi(g) \det(1 - gt)^{-1}.$$

Proof. From Lemma 4.12 we have

$$f_\chi(t) = \frac{1}{|G|} \sum_{g \in G} \overline{\chi(g)} P_{S/F}(g,t)$$

and from Lemma 4.4 and Corollary 3.31 we have $P_S(g,t) = P_J(t)P_{S/F}(g,t)$. We know that $P_J(t) = \prod_{i=1}^n (1 - t^{d_i})^{-1}$ and from Lemma 4.6 we have $P_S(g,t) = \det(1 - g^{-1}t)^{-1}$. The present lemma follows on substituting these values in the above expression for $f_\chi(t)$. □

Define the character $\delta : G \to \mathbb{C}$ by $\delta(g) := \det(g)$.

Lemma 4.22. *If N is the number of reflections in G and χ is any character, then*

$$t^N f_\chi(t^{-1}) = f_{\delta\overline{\chi}}(t).$$

Proof. We have $N = \sum_{i=1}^n (d_i - 1)$ and from the previous lemma,

$$f_\chi(t^{-1}) = |G|^{-1} \prod_{i=1}^n (1 - t^{-d_i}) \sum_{g \in G} \chi(g) \det(1 - gt^{-1})^{-1}$$

$$= |G|^{-1} t^{-\sum d_i} \prod_{i=1}^n (t^{d_i} - 1) \sum_{g \in G} \chi(g) \det(t - g)^{-1} t^n$$

$$= |G|^{-1}(-1)^n t^{-N} \prod_{i=1}^n (1 - t^{d_i}) \sum_{g \in G} \chi(g)\delta(g)^{-1} \det(1 - g^{-1}t)^{-1}(-1)^n$$

$$= t^{-N} f_{\delta\overline{\chi}}(t).$$ □

Corollary 4.23. *(cf. Corollary 9.12 and Theorem 9.38).*

(i) $(S/F)_i = 0$ *for* $i > N$.
(ii) $(S/F)_0 = 1_G$.
(iii) $(S/F)_N = M_\delta$.

Proof. From the lemma, no polynomial $f_\chi(t)$ contains t^i with $i > N$, hence (i) holds. The component $(S/F)_0$ of S/F is $(\mathbb{C} + F)/F$ and this is the isotypic component of S/F corresponding to the trivial representation of G, hence $f_1(t) = 1$, proving (ii).

Taking $\chi = 1$ in the lemma we have $f_\delta(t) = t^N$. On the other hand, if χ is any irreducible character, the lemma shows that the multiplicity of M_χ in $(S/F)_N$ equals the multiplicity of $M_{\delta\overline{\chi}}$ in $(S/F)_0$, and this is 0 except when $\chi = \delta$. Thus $(S/F)_N = M_\delta$, completing the proof. □

If M is a G-module with character χ, we write $q_j(M) := q_j(\chi)$. We shall return to the theme of the 'M-exponents' $q_j(M)$, for M any finite dimensional G-module, in Chapter 10.

Exercises

In the following exercises, G is a unitary reflection group acting on a vector space V of dimension n and containing N reflections.

1. Use Lemma 4.12 to show that $P_J(t^{-1}) = (-t)^n P_{S,\delta}(t)$, where $\delta(g) := \det(g)$ for $g \in G$. Deduce that $P_{S,\delta}(t) = t^N P_J(t)$ and hence every polynomial $f \in S_\delta$ can be written in the form ΠI, where $I \in J$ and where Π is an element of S_δ of degree N.

2. Use Lemma 4.21 to prove Proposition 3.32 by showing that the multiplicity of M_χ in S/F is $\chi(1)$.

3. Using Corollary 3.24 and Theorem 4.14, or otherwise, show that the degrees of the group G_4 defined in Example 1.12 are 4 and 6.

4. Use the Poincaré series to show that the degrees of the basic invariants of a finite reflection group are uniquely determined.

5. Show that if G is a finite reflection subgroup of $U(V)$ and if g is an element of the normaliser of G in $U(V)$, then the basic invariants of G can be chosen to be eigenfunctions of g.

6. Suppose that G is a group generated by reflections represented by matrices all of whose entries are real; that is, G is a *Euclidean reflection group*. Show that G has a basic invariant of degree 2.

7. Suppose that G is a finite irreducible unitary reflection group, which has a basic invariant of degree 2. Prove that G can be generated by reflections represented by matrices with real entries. [Hint. Use the invariant of degree 2 to construct a non-degenerate G-invariant bilinear form. From the bilinear form and the hermitian inner product, construct a semi-linear map $f : V \to V$ and prove that there is a positive real number λ such that $f^2(v) = \lambda v$ for all $v \in V$. Show that the set V_0 of fixed points of a suitable multiple of f is a real vector space such that $V = \mathbb{C} \otimes V_0$.]

CHAPTER 5

Quaternions and the finite subgroups of $SU_2(\mathbb{C})$

The main result of this chapter is the classification of the finite subgroups of the special unitary group $SU_2(\mathbb{C})$. This is the key to the determination of all finite unitary reflection groups of rank two, which is carried out in the next chapter. The complete list of the rank two reflection groups first appeared in Shephard and Todd [193] as part of their classification of *all* unitary reflection groups and it has been derived anew by several other authors (for example, Crowe [65, 66], Cohen [55], Coxeter [62]).

As a starting point for this work most authors refer to earlier classifications of the finite collineation groups of the complex projective line. It has been known since the nineteenth century (essentially due to Klein [131], but anticipated by Hessel [111]) that the only finite primitive collineation groups of the complex projective line are the so-called tetrahedral, octahedral and icosahedral groups: namely the groups of rotations of a tetrahedron, an octahedron and an icosahedron, respectively. Shephard and Todd begin with these groups and, using Klein's generators, they determine which of their central extensions in $U_2(\mathbb{C})$ are reflections groups.

On the other hand, the group $U_2(\mathbb{C})$ can be interpreted as a subgroup of the group $O_4(\mathbb{R})$ of orthogonal transformations of Euclidean 4-space and the group $SO_3(\mathbb{R})$ of rotations of Euclidean 3-space is isomorphic to $SU_2(\mathbb{C})$ modulo its centre. The enumeration of all finite subgroups of these groups can be found in many places (for example, Blichfeldt [26, 163], Coxeter [62], Crowe [65], Dornhoff [86], Du Val [88], Gordon [102], Goursat [103], Hessel [111], Threlfall and Seifert [217]). The methods vary, but the theme emerges, particularly from the works of Crowe, Coxeter and Du Val, that the division algebra of quaternions is an effective way to unify the various interpretations of the groups and determine their finite subgroups. A recent account, from a geometric point of view, can be found in Conway and Smith [59], where it is pointed out that all of the previous enumerations are to some degree incomplete.

Building on the insights contained in the books and papers just mentioned, it is possible to give a reasonably compact, self-contained account of the necessary results. We begin by describing the division algebra of quaternions and go on to describe its finite subgroups, a result that was obtained by Stringham [214] in 1881.

66

Along the way we show that $SU_2(\mathbb{C})$ is isomorphic to the group of quaternions of norm 1 and that the groups $O_3(\mathbb{R})$, $O_4(\mathbb{R})$ and $U_2(\mathbb{C})$ are closely related.

In the next chapter we apply these results to the determination of the finite unitary reflection subgroups of $U_2(\mathbb{C})$.

1. The quaternions

The algebra $\mathbb{H}$ of *quaternions* was discovered by Hamilton in 1843 and it can be defined as the two-dimensional vector space over $\mathbb{C}$ with basis 1 and j, and with a multiplication where 1 is the identity element and j satisfies $j^2 = -1$ and $ij = -ji$. (It is straightforward to check that this extends to an associative multiplication for $\mathbb{H}$.)

We regard $\mathbb{C}$ (and hence $\mathbb{R}$) as a subalgebra of $\mathbb{H}$ by identifying $\alpha \in \mathbb{C}$ with $\alpha \cdot 1$. Then for $\alpha \in \mathbb{C}$ we have $j\alpha = \overline{\alpha}j$, where $\overline{\alpha}$ denotes the usual complex conjugate of α. For $\alpha, \beta \in \mathbb{C}$, the quaternion $\alpha + \beta j$ commutes with i if and only if $\beta = 0$ and it commutes with j if and only if both α and β are real. In particular, the only elements of $\mathbb{H}$ that commute with both i and j are the real multiples of 1; that is, the centre of $\mathbb{H}$ is $\mathbb{R}$.

If we restrict the scalars to $\mathbb{R}$, then $\mathbb{H}$ becomes a four-dimensional algebra with basis 1, i, j, k, where $k = ij$. This is the usual description of $\mathbb{H}$ and the generators satisfy the symmetrical relations

$$i^2 = j^2 = k^2 = ijk = -1.$$

Given $q = a + bi + cj + dk$, where $a, b, c, d \in \mathbb{R}$, define the *conjugate* of q to be $\overline{q} = a - bi - cj - dk$. (Note that for $q \in \mathbb{C}$, the conjugate of q coincides with the usual complex conjugate.) Next define the *norm* of q to be $\mathrm{N}(q) = q\overline{q}$ and the *trace* of q to be $\mathrm{Tr}(q) = q + \overline{q}$. Then $\mathrm{N}(q) = a^2 + b^2 + c^2 + d^2 \in \mathbb{R}$ and $\mathrm{N}(q) = 0$ if and only if $q = 0$. Thus, for $q \neq 0$, the inverse of q is $\mathrm{N}(q)^{-1}\overline{q}$ and this proves that $\mathbb{H}$ is a *division algebra*, namely an algebra in which every non-zero element has an inverse.

Proposition 5.1.

(i)   If $q = \alpha + \beta j$ with $\alpha, \beta \in \mathbb{C}$, then $\overline{q} = \overline{\alpha} - \beta j$, $\mathrm{N}(q) = \alpha\overline{\alpha} + \beta\overline{\beta}$ and $\mathrm{Tr}(q) = \alpha + \overline{\alpha}$.

(ii)   Conjugation is an anti-automorphism of order 2 : $\overline{q + r} = \overline{q} + \overline{r}$, $\overline{qr} = \overline{r}\,\overline{q}$, and $\overline{\overline{q}} = q$.

(iii) $\mathrm{N}(qr) = \mathrm{N}(q)\,\mathrm{N}(r)$.

(iv) $\mathrm{Tr}(q + r) = \mathrm{Tr}(q) + \mathrm{Tr}(r)$.

(v) $\mathrm{Tr}(qr) = \mathrm{Tr}(rq)$.

(vi) $q^2 - \mathrm{Tr}(q)q + \mathrm{N}(q) = 0$.

(vii) $\mathrm{Tr}(q\overline{r}) = \mathrm{N}(q + r) - \mathrm{N}(q) - \mathrm{N}(r)$.

Proof. We may verify (*ii*) by direct calculation and then (*iii*) follows because

$$N(qr) = qr\overline{r}\,\overline{q} = q\,N(r)\overline{q} = N(q)\,N(r).$$

The other properties of conjugation, norm and trace are even easier to establish. □

As shown by Proposition 5.1 (*vi*), every quaternion satisfies a quadratic equation with real coefficients. In particular, if $q \notin \mathbb{R}$, the field $K := \mathbb{R}[q]$ is a quadratic extension of $\mathbb{R}$ and hence isomorphic to $\mathbb{C}$. Multiplication on the left by the elements of K turns $\mathbb{H}$ into a vector space of dimension two over K. Then multiplication on the *right* by any element $h \in \mathbb{H}$ effects a K-linear transformation of this vector space and its eigenvalues will be the roots in K of the equation $x^2 - \mathrm{Tr}(h)x + N(h) = 0$.

For example, if $q^2 = -1$, then multiplication on the right by q is an $\mathbb{R}[q]$-linear transformation with eigenvalues q and $-q$. If r is an eigenvector of norm 1 corresponding to the eigenvalue $-q$, then $rq = -qr$. Since r^2 commutes with r and q it commutes with all of $\mathbb{H}$ and hence must belong to $\mathbb{R}$. Since $N(r^2) = 1$, the only possibility is $r^2 = -1$. Thus q and r play the same rôles as i and j in our original definition of $\mathbb{H}$, emphasising the fact that there is no 'canonical' embedding of $\mathbb{C}$ in $\mathbb{H}$.

Essentially the same idea can be used to show that, up to isomorphism, $\mathbb{H}$ is the only non-commutative division algebra over $\mathbb{R}$ with the property that each of its elements satisfies a polynomial equation with real coefficients. (For another application, see Taylor [**215**].) Variations on this theme will be used in the next two propositions and the next few sections.

For $q \in \mathbb{H}$, multiplication by q on the left defines the $\mathbb{R}$-linear transformation

$$L(q) : \mathbb{H} \to \mathbb{H} : h \mapsto qh$$

and multiplication on the right by $\overline{q}$ defines the $\mathbb{R}$-linear transformation

$$R(q) : \mathbb{H} \to \mathbb{H} : h \mapsto h\overline{q}.$$

If we define $\rho : \mathbb{H} \to \mathbb{H}$ by $\rho(h) = -\overline{h}$, then ρ is a reflection, $\rho^2 = 1$ and $\rho L(q)\rho = R(q)$ for all $q \in \mathbb{H}$.

Proposition 5.2. *For any element* $q \in \mathbb{H}$, $\det L(q) = \det R(q) = N(q)^2$.

Proof. If $q \in \mathbb{R}$, then $N(q) = q^2$ and $L(q) = R(q)$ is a scalar transformation with determinant $q^4 = N(q)^2$. Thus we may suppose that $q \notin \mathbb{R}$ and regard $\mathbb{H}$ as a vector space over $\mathbb{R}[q]$ with basis $1, r$. Then $1, q, r, qr$ is a basis for $\mathbb{H}$ as a real space and, by Proposition 5.1 (*vi*), the matrix of $L(q)$ with respect to this basis is

$$\begin{bmatrix} 0 & -N(q) & 0 & 0 \\ 1 & \mathrm{Tr}(q) & 0 & 0 \\ 0 & 0 & 0 & -N(q) \\ 0 & 0 & 1 & \mathrm{Tr}(q) \end{bmatrix},$$

hence $\det L(q) = \mathrm{N}(q)^2$. Finally, we have $\det R(q) = \det \rho L(q)\rho = \det L(q)$, as required. $\qquad\square$

The set $S^3 := \{\, q \in \mathbb{H} \mid \mathrm{N}(q) = 1\,\}$ is both the unit 3-sphere in $\mathbb{R}^4$ and a subgroup of the group $\mathbb{H}^*$ of non-zero quaternions; it is called the group of *unit quaternions*. The restriction of L (and R) to S^3 is a monomorphism from S^3 into $SL_4(\mathbb{R})$, the group of linear transformations of $\mathbb{H}$ of determinant 1. For $q, r \in \mathbb{H}$ the linear transformations $L(q)$ and $R(r)$ commute, hence

$$S^3 \times S^3 \to SL_4(\mathbb{R}) : (q, r) \mapsto L(q)R(r)$$

is a homomorphism. Its kernel is $\{(1, 1), (-1, -1)\}$ and its image is a central product of $L(S^3)$ and $R(S^3)$ which we denote by $L(S^3) \circ R(S^3)$. In the next section we shall see that $O_4(\mathbb{R}) = (L(S^3) \circ R(S^3))\langle \rho \rangle$.

Proposition 5.3. *For non-zero elements $q, r \in \mathbb{H}$, the following are equivalent:*

(*i*)   *q and r have the same norm and trace,*

(*ii*)   *$q = hrh^{-1}$ for some $h \in S^3$,*

(*iii*)   *$\mathrm{N}(q)$ is an eigenvalue of $L(q)R(r)$.*

Proof. If $q = hrh^{-1}$, it follows directly from (*iii*) and (*v*) of Proposition 5.1 that $\mathrm{N}(q) = \mathrm{N}(r)$ and $\mathrm{Tr}(q) = \mathrm{Tr}(r)$. Thus (*ii*) implies (*i*).

Now suppose that q and r have the same norm and trace. If $q \in \mathbb{R}$, then $\mathrm{N}(r-q) = 0$ and hence $r = q$. In this case we may take $h = 1$ so that both (*ii*) and (*iii*) are satisfied. If $q \notin \mathbb{R}$, we regard $\mathbb{H}$ as a 2-dimensional vector space over $K := \mathbb{R}[q]$. Then $R(r)$ is a K-linear transformation whose eigenvalues are the roots of $x^2 - \mathrm{Tr}(r)x + \mathrm{N}(r) = 0$ in K; namely q and $\bar{q}$. If h is an eigenvector of norm 1 corresponding to $\bar{q}$, then $h\bar{r} = \bar{q}h$ and so $q = hrh^{-1}$, hence (*ii*) holds. On the other hand, the equation $h\bar{r} = \bar{q}h$ is equivalent to $qh\bar{r} = \mathrm{N}(q)h$ and this completes the proof. $\qquad\square$

2. The groups $O_3(\mathbb{R})$ and $O_4(\mathbb{R})$

This section describes the structure of the groups $O_3(\mathbb{R})$ and $O_4(\mathbb{R})$ and establishes relationships between the special orthogonal groups in three and four dimensions and the group of quaternions of norm 1. (See Coxeter [**60**] for a somewhat different treatment.)

The norm function $\mathrm{N}(q)$ is the square of the usual Euclidean distance on $\mathbb{H}$, regarded as $\mathbb{R}^4$, and the associated Euclidean inner product is

$$q \cdot r = \tfrac{1}{2}\,\mathrm{Tr}(q\bar{r}).$$

The basis 1, i, j, k is orthonormal with respect to this inner product and the subspace V spanned by i, j, k is the orthogonal complement of 1. It is also the set of quaternions that have zero trace and its elements are called *pure quaternions*. From Proposition 5.1 (*vi*), $q \in V$ if and only if $q^2 \in \mathbb{R}$ and $q^2 \leq 0$.

Any non-zero $\mathbb{R}$-linear transformation of a Euclidean space is said to be *orthogonal* if it preserves the inner product. If M is the matrix of a linear transformation T with respect to an orthonormal basis, then T is orthogonal if and only if M is an *orthogonal matrix*; i.e. $MM^t = I$, where M^t denotes the transpose of M. Hence if M is orthogonal, $\det(M)^2 = 1$ and thus the determinant of an orthogonal transformation takes only the values 1 or -1.

The set of orthogonal transformations of $\mathbb{H}$ is the *orthogonal group* $O_4(\mathbb{R})$ and the subgroup of transformations of determinant 1 is the *special orthogonal group* $SO_4(\mathbb{R})$; its elements are called *rotations*. Similarly, the orthogonal transformations of V constitutes the orthogonal group $O_3(\mathbb{R})$ and those of determinant 1 the special orthogonal group $SO_3(\mathbb{R})$.

By Proposition 5.1 (*iii*) and Proposition 5.2, for $q \in S^3$, the linear transformations $L(q)$ and $R(q)$ belong to $SO_4(\mathbb{R})$. For $q \in S^3$, define the linear transformation $B(q)$ of V by $B(q)v := qvq^{-1} = L(q)R(q)v$. Then $B(q) \in SO_3(\mathbb{R})$.

Proposition 5.4. *The map $B : S^3 \rightarrow SO_3(\mathbb{R})$ is a homomorphism onto $SO_3(\mathbb{R})$ with kernel $\{\pm 1\}$.*

Proof. It is clear that for $q \in S^3$, $B(q) = B(-q)$ and conversely, if $qvq^{-1} = v$ for all $v \in V$, then $q \in \mathbb{R}$. Thus Ker $B = S^3 \cap \mathbb{R} = \{\pm 1\}$.

Now suppose that $T \in SO_3(\mathbb{R})$. By Proposition 5.3 there exists $h \in S^3$ such that $B(h)T(i) = i$. Replacing T by $B(h)T$ we may suppose that $T(i) = i$. Then $T(j)$ is orthogonal to i and so $T(j) = \alpha j$ for some $\alpha \in \mathbb{C}$. Choose $\beta \in \mathbb{C}$ such that $\alpha = \beta^2$ and observe that $B(\beta)j = \beta j \bar{\beta} = \beta^2 j = T(j)$. Thus $B(\beta)^{-1}T$ fixes i and j and since it is orthogonal and of determinant 1 it must be the identity. Thus $T = B(\beta)$. $\qquad \square$

The rotation $B(q)$ corresponding to $q \in S^3$ takes a particularly simple form. Because $N(q) = 1$ we can write $q = a + bu$ for some $u \in V$ such that $N(u) = 1$ and some $a, b \in \mathbb{R}$ such that $a^2 + b^2 = 1$. Now write $a = \cos\frac{1}{2}\theta$ and $b = \sin\frac{1}{2}\theta$ so that

$$q = \cos\tfrac{1}{2}\theta + u\sin\tfrac{1}{2}\theta.$$

Then $B(q)$ is the rotation of angle θ about the axis u. To see this, note that if v is a unit vector orthogonal to u, the matrix of $B(q)$ with respect to the orthonormal basis $u, v, \frac{1}{2}(uv - vu)$ is

$$\begin{bmatrix} 1 & 0 & 0 \\ 0 & \cos\theta & -\sin\theta \\ 0 & \sin\theta & \cos\theta \end{bmatrix}.$$

The centre of $O_3(\mathbb{R})$ is $\langle -I \rangle$, where I denotes the identity transformation, and since $\det(-I) = -1$ we have

Corollary 5.5. $O_3(\mathbb{R}) = \langle -I \rangle \times SO_3(\mathbb{R}) = \langle -I \rangle \times S^3/\{\pm 1\}$.

The set $S^2 := V \cap S^3 = \{ u \in V \mid u^2 = -1 \}$ is the 2-sphere in V and for $u \in S^2$ we see that $-B(u)v = uvu = v - 2(u \cdot v)u$ is the reflection in the plane orthogonal to u.

Proposition 5.6. *The element $r \in O_3(\mathbb{R})$ is a reflection if and only if $r = -B(u)$ for some $u \in S^2$.*

Proof. Any reflection in $O_3(\mathbb{R})$ must have order two and if r is a reflection, then $\det(r) = -1$. Thus $-r \in SO_3(\mathbb{R})$ and we can write $r = -B(u)$ for some $u \in S^3$. An element of this form has order two if and only if $u^2 = \pm 1$. If $u^2 = 1$, then $u = \pm 1$ and so the element is a reflection if and only if $u^2 = -1$. $\quad\square$

Proposition 5.7. *The map $S^3 \times S^3 \to SO_4(\mathbb{R}) : (q,r) \mapsto L(q)R(r)$ is a homomorphism onto $SO_4(\mathbb{R})$ with kernel $\{(1,1), (-1,-1)\}$.*

Proof. We have already seen that this is a homomorphism and that its kernel is $\{\pm(1,1)\}$. So suppose that $T \in SO_4(\mathbb{R})$ and set $q := T(1)$. Then $\mathrm{N}(q) = 1$ and so $R(q)T$ fixes 1 and hence acts on V as an element of $SO_3(\mathbb{R})$. By Proposition 5.4 we can find $h \in S^3$ such that $R(q)T = B(h) = L(h)R(h)$. Thus $T = L(h)R(q^{-1}h)$ has the required form. $\quad\square$

Corollary 5.8. $O_4(\mathbb{R}) = (L(S^3) \circ R(S^3))\langle \rho \rangle = \langle L(S^3), \rho \rangle$.

Proof. The map ρ taking q to $-\overline{q}$ is an element of $O_4(\mathbb{R})$ of determinant -1 and so $O_4(\mathbb{R})$ is the central product of two copies $L(S^3)$ and $R(S^3)$ of S^3 extended by the element ρ that interchanges them. $\quad\square$

3. The groups $SU_2(\mathbb{C})$ and $U_2(\mathbb{C})$

Throughout this section we regard $\mathbb{H}$ as a 2-dimensional vector space over $\mathbb{C} = \mathbb{R}[i]$. For $q, r \in \mathbb{H}$, write

$$q\overline{r} = (q,r) + [q,r]j$$

where (q,r) and $[q,r]$ belong to $\mathbb{C}$. Then

$$(q,r) = \tfrac{1}{2}(q\overline{r} - iq\overline{r}i)$$

and a further calculation shows that (q,r) is the unique hermitian inner product on $\mathbb{H}$ such that $(q,q) = \mathrm{N}(q)$ for all q. The group of all non-zero $\mathbb{C}$-linear transformations of $\mathbb{H}$ that preserve this form is the *unitary group* $U_2(\mathbb{C})$.

Proposition 5.9. *Given $q = \alpha + \beta j$, where $\alpha, \beta \in \mathbb{C}$, the matrix of $R(q)$, considered as a $\mathbb{C}$-linear transformation of the space $\mathbb{H}$ with respect to the basis $1, j$ is*
$$\begin{bmatrix} \overline{\alpha} & \overline{\beta} \\ -\beta & \alpha \end{bmatrix},$$ *with determinant $\mathrm{N}(q)$ and trace $\mathrm{Tr}(q)$.*

Proof. This follows directly from the definition of $R(q)$.   □

From this proposition we have $R(q) \in SU_2(\mathbb{C})$ for all $q \in S^3$ since it is clear that in this case $R(q)$ preserves the hermitian form.

Proposition 5.10. *The map $R : S^3 \to SU_2(\mathbb{C})$ is an isomorphism.*

Proof. Given $T \in SU_2(\mathbb{C})$, put $q := T(1)$ and observe that $R(q)T$ is an element of $SU_2(\mathbb{C})$ that fixes 1. The only map with this property is the identity and hence $T = R(q^{-1})$, thus showing that R is surjective. It is clear that R is a homomorphism and one-to-one.   □

The unit circle in $\mathbb{C}$ is $S^1 := \{\, \alpha \in \mathbb{C} \mid N(\alpha) = 1 \,\}$ and if $\alpha \in S^1$, the map $L(\alpha)$ is $\mathbb{C}$-linear and preserves the hermitian form.

Proposition 5.11. *The map $S^1 \times S^3 \to U_2(\mathbb{C}) : (\alpha, q) \mapsto L(\alpha)R(q)$ is a homomorphism onto $U_2(\mathbb{C})$ with kernel $\{(1, 1), (-1, -1)\}$.*

Proof. Given $T \in U_2(\mathbb{C})$, put $q := T(1)$. Then $R(q)T$ fixes 1 and hence leaves its orthogonal complement $\mathbb{C}j$ invariant. Consequently $R(q)T(j) = \alpha j$ for some $\alpha \in \mathbb{C}$ and thus $T = L(\alpha)R(q^{-1})$. In fact $\alpha \in S^1$ because $R(q)T$ preserves the hermitian form.   □

4. The finite subgroups of the quaternions

In this section we adapt an idea of Jordan to determine all finite subgroups of $\mathbb{H}^*$. The method was utilised by Dornhoff [86, §26] to describe the finite subgroups of $GL(\mathbb{C})$ and an outline of the process and further references can be found in Blichfeldt [26, §59]. The finite subgroups of $\mathbb{H}^*$ were first described by Stringham [**214**] in 1881.

If $g \in \mathbb{H}^*$ has finite order, it follows from Proposition 5.1 (*iii*) that $N(g) = 1$. Thus every finite subgroup of $\mathbb{H}^*$ is in fact a subgroup of S^3. Moreover, -1 is the only element of order two in S^3 and so one might expect the possible types of finite subgroups to be quite limited. As we shall see, this is indeed the case. Under the homomorphism $S^3 \to SO_3(\mathbb{R})$ of Proposition 5.4 the groups in question are the inverse images of cyclic groups, dihedral groups, and the rotation groups of the Platonic solids.

4.1. The binary dihedral groups. For each positive integer m, put $\zeta_m := \exp\dfrac{2\pi i}{m} = \cos\dfrac{2\pi}{m} + i\sin\dfrac{2\pi}{m}$. Let $\mathcal{C}_m$ denote the cyclic subgroup of order m generated by ζ_m and let $\mathcal{D}_m$ denote the subgroup of $\mathbb{H}^*$ generated by ζ_m and j. For $\alpha \in \mathcal{C}_m$ we have $j\alpha j^{-1} = \overline{\alpha} = \alpha^{-1}$ and so $\mathcal{C}_m$ is a normal subgroup of $\mathcal{D}_m$. If m is odd, $\mathcal{D}_m = \mathcal{D}_{2m}$ and if m is even, $\mathcal{C}_m$ is a subgroup of index 2 in $\mathcal{D}_m$. Therefore the order of $\mathcal{D}_{2m}$ is $4m$ and $x^2 = -1$ for all $x \in \mathcal{D}_{2m} \setminus \mathcal{C}_{2m}$. The quotient $\mathcal{D}_{2m}/\langle -1 \rangle$ is the dihedral group of order $2m$ and any group isomorphic to $\mathcal{D}_{2m}$

is called a *binary dihedral group*. (Coxeter [**62**] and others refer to these groups as *dicyclic* groups whereas, when m is a power of 2 they are also known as *generalised quaternion* groups.) The group $\mathcal{Q} := \mathcal{D}_4 = \{\pm 1, \pm i, \pm j, \pm k\}$ is usually known as the *quaternion* group.

4.2. The binary polyhedral groups $\mathcal{T}$, $\mathcal{O}$ and $\mathcal{I}$. From Proposition 5.3 two elements of S^3 are conjugate if and only if they have the same trace. In fact, knowledge of the trace of elements of small order will be useful in the proof of the main theorem and we record this information in the following table:

order	3	4	5	6	8	10
trace	-1	0	$-\tau, \tau^{-1}$	1	$\pm\sqrt{2}$	$\tau, -\tau^{-1}$

where $\tau = \frac{1}{2}(1 + \sqrt{5})$.

From this table we see that $\varpi := \frac{1}{2}(-1 + i + j + k)$ is an element of order 3. Direct calculation shows that ϖ normalises $\mathcal{Q}$. Thus $\mathcal{T} := \mathcal{Q}\langle\varpi\rangle$ is a group of order 24 and it is easily seen that $\mathcal{T} \setminus \mathcal{Q}$ consists of the 16 elements of the form $\frac{1}{2}(\pm 1 \pm i \pm j \pm k)$. The group $\mathcal{T}$ is the *binary tetrahedral group*; it can be generated by i and ϖ.

The element $\gamma := \frac{1}{\sqrt{2}}(1 + i)$ has order 8. From the description of $\mathcal{T}$ in the previous paragraph we see that γ normalises both $\mathcal{Q}$ and $\mathcal{T}$. Since $\gamma^2 = i$, $\mathcal{O} := \mathcal{T}\langle\gamma\rangle$ has order 48. The group $\mathcal{O}$ is the *binary octahedral group*; it can be generated by ϖ and γ. The set $\mathcal{O} \setminus \mathcal{T}$ consists of the 24 elements of the form $\frac{1}{\sqrt{2}}(\pm u \pm v)$, where u and v are distinct elements of $\{1, i, j, k\}$. We see from the table that 12 of these elements have order 4 and 12 have order 8.

The element $\sigma := \frac{1}{2}(\tau^{-1} + i + \tau j)$ has order 5 and acts on the space V of pure quaternions via the homomorphism $B : S^3 \to SO_3(\mathbb{R})$ of Proposition 5.4. The 12 vectors $\pm\tau i \pm j$, $\pm\tau j \pm k$ and $\pm i \pm \tau k$ form the vertices of a regular icosahedron in V and they are permuted among themselves by $B(q)$, for $q \in \mathcal{T}$, and by $B(\sigma)$. It is clear that only ± 1 can fix all six lines spanned by these vectors and thus the group $\mathcal{I}$ generated by $\mathcal{T}$ and σ is finite. The group $\mathcal{I}$ is the *binary icosahedral group*; it can be generated by σ and i.

In the course of the proof of the next theorem we shall see that the order of $\mathcal{I}$ is 120. Alternatively, this can be established directly as follows. The group $B(\mathcal{I})$ has index at most 6 in the group $\mathrm{Alt}(6)$ of even permutations of the six lines given above. If it were the case that $B(\mathcal{I}) = \mathrm{Alt}(6)$, then $\mathcal{I}$ would contain a non-cyclic subgroup of order 9. But if q and r are distinct elements of order 3 that commute and if qr has order 3, then $\mathrm{Tr}(q) = \mathrm{Tr}(r) = \mathrm{Tr}(qr) = -1$ and we can write $q = -\frac{1}{2} + u$ and $r = -\frac{1}{2} + v$ where u and v are commuting elements of V of norm $\frac{3}{4}$. On taking the trace of qr we see that $\mathrm{Tr}(uv) = -\frac{3}{2}$ and hence $uv = -\frac{3}{4}$; that is, $u = v$, which is a contradiction. The only possibility is that $B(\mathcal{I})$ has index 6 in $\mathrm{Alt}(6)$ and thus the order of $\mathcal{I}$ is 120.

The *binary polyhedral* groups $\mathcal{T}$, $\mathcal{O}$ and $\mathcal{I}$ take their names from the fact that their images in $SO_3(\mathbb{R})$ are the groups of rotations of a tetrahedron, octahedron and icosahedron, respectively. In [162] McKay showed that the finite subgroups of S^3 correspond to the 'simply laced affine Dynkin diagrams', $\tilde{A}_m$, $\tilde{D}_m$, $\tilde{E}_6$, $\tilde{E}_7$ and $\tilde{E}_8$, respectively. Further information about this intriguing correspondence can be found in [101, 137, 203, 204, 211, 212].

4.3. The finite subgroups of S^3. We are now able to prove that the groups described above account for all the finite subgroups of the quaternions.

Theorem 5.12. *Every finite subgroup of $\mathbb{H}^*$ is conjugate in S^3 to one of the following groups:*

(i)   *the* cyclic group $\mathcal{C}_m$,
(ii)   *the* binary dihedral group $\mathcal{D}_{2m}$,
(iii)   *the* binary tetrahedral group $\mathcal{T}$ *of order* 24,
(iv)   *the* binary octahedral group $\mathcal{O}$ *of order* 48,
(v)   *the* binary icosahedral group $\mathcal{I}$ *of order* 120.

Proof. Given a finite subgroup G of $\mathbb{H}^*$ we have $G \subseteq S^3$ and we may suppose that G has an abelian subgroup A that contains an element $q \ne \pm 1$. We shall determine the possibilities for its centraliser

$$C_G(A) := \{\, g \in G \mid ga = ag \text{ for all } a \in A \,\}$$

and its normaliser $N_G(A) := \{\, g \in G \mid gAg^{-1} = A \,\}$.

Considering $\mathbb{H}$ as a vector space over the field $\mathbb{R}[q]$, the linear transformation $R(\overline{q})$ has eigenvalues q and $\overline{q}$. The elements of $\mathbb{H}$ that commute with q form the eigenspace for q, namely $\mathbb{R}[q]$. Therefore $C_G(A)$ is cyclic since it is a subgroup of the multiplicative group of the field $\mathbb{R}[q] \simeq \mathbb{C}$ (Lang [142, p. 177]). It is also the unique maximal abelian subgroup of G that contains A and this shows that if B and C are distinct maximal abelian subgroups of G, then $B \cap C \subseteq \{\pm 1\}$.

If $r \in S^3$ and $rqr^{-1} \in A$, then rqr^{-1} is either q or $\overline{q}$, because these are the only roots of $x^2 - \mathrm{Tr}(q)x + 1 = 0$ in $\mathbb{R}[q]$. The elements $r \in \mathbb{H}$ such that $rqr^{-1} = \overline{q}$ form the eigenspace of $R(\overline{q})$ corresponding to $\overline{q}$ and if r is such an eigenvector of norm 1, then r^2 commutes with both r and q and so $r^2 = -1$. It follows that either $N_G(A) = C_G(A)$ or $N_G(A) = C_G(A)\langle r \rangle$, for some r such that $r^2 = -1$.

Given an element $q \in C_G(A)$ of order m we can find an element $\alpha \in \mathcal{C}_m$ with the same norm and trace and it follows from Proposition 5.3 that q is conjugate to α in S^3. Since $C_G(A)$ has even order we may suppose that $C_G(A) = \mathcal{C}_{2m}$ and our assumption that $A \ne \{\pm 1\}$ means that $m > 1$. If $r \in N_G(A) \setminus C_G(A)$, then $r^2 = -1$ and $r\alpha r^{-1} = \overline{\alpha}$ for all $\alpha \in A$. But then $r^{-1}j$ commutes with all $\alpha \in A$ and so $r^{-1}j \in C_G(A) \subseteq \mathbb{C}$. Thus we may write $j = \beta^2 r$ for some $\beta \in \mathbb{C}$ and then conjugation by β leaves all $\alpha \in A$ fixed and takes r to j. In this case we see that, up to conjugacy, $N_G(A)$ is $\mathcal{D}_{2m}$.

Suppose that A is a maximal abelian subgroup of G such that $N_G(A) = A$. Distinct conjugates of A have only the identity or ± 1 in common according to whether $|G|$ is odd or even. Thus the $|G|/|A|$ conjugates of A account for $(|G|/|A|)(|A| - 1)$ or $(|G|/|A|)(|A| - 2)$ elements of $G \setminus \{\pm 1\}$. In both cases this is more than half the elements of G and consequently there is no room for another maximal abelian subgroup B with $N_G(B) = B$.

If $|G|$ or $|G|/2$ is odd and if A is a maximal abelian subgroup of G, then $N_G(A) = A$. It follows from the previous paragraph that $G = A$. This proves that the only subgroups of S^3 of odd order or twice odd order are the cyclic subgroups.

The results established so far show that if G contains an abelian subgroup of index 2, then G is conjugate in S^3 to $\mathcal{C}_m$ or to $\mathcal{D}_{2m}$. From now on we shall assume that G is not a cyclic or binary dihedral group. In particular, G contains elements of order 4 and hence $-1 \in G$.

The elements of order 4 in $\mathbb{H}$ constitute the unit sphere S^2 in the space of pure quaternions and the group G acts on S^2 by conjugation. Let

$$\Omega := \{\, u \in S^2 \mid qu = uq \text{ for some } q \in G \setminus \{\pm 1\} \,\},$$

and let $\Omega_1, \Omega_2, \ldots, \Omega_t$ be the orbits of G acting on Ω. For each i choose a representative $u_i \in \Omega_i$. We have seen that for $q \in G \setminus \{\pm 1\}$, the centraliser of q in $\mathbb{H}$ is the field $\mathbb{R}[q]$ and therefore q commutes with exactly two elements of S^2.

By counting the set $\{\,(u, q) \in \Omega \times G \mid uq = qu\,\}$ in two ways we deduce the equation

$$t|G| = 2|\Omega| + 2(|G| - 2).$$

(This is just Burnside's formula for the number of orbits of a permutation group.)

Now write $k_i := |C_G(u_i)|$ so that $|\Omega_i| = |G|/k_i$. Then the previous formula can be written

$$\sum_{i=1}^{t} \frac{1}{k_i} = \frac{t}{2} - 1 + \frac{2}{|G|}.$$

Since $-1 \in G$ it follows that k_i is even and $k_i \geq 4$; therefore $t \leq 3$. If $t = 1$, then $|G| = 4k_1/(2 + k_1) < 4$, which is a contradiction. If $t = 2$, then $|G| = 2k_1 k_2/(k_1 + k_2) \leq \max\{k_1, k_2\}$, whence G is cyclic – a case we have excluded. Thus $t = 3$ and the equation becomes

$$\frac{1}{k_1} + \frac{1}{k_2} + \frac{1}{k_3} = \frac{1}{2} + \frac{2}{|G|}.$$

We may choose the notation so that $4 \le k_1 \le k_2 \le k_3$ and thus $\dfrac{3}{k_1} > \dfrac{1}{2}$ whence $k_1 = 4$ and the equation reduces to

$$\frac{1}{k_2} + \frac{1}{k_3} = \frac{1}{4} + \frac{2}{|G|}.$$

If $k_2 = 4$, then $|G| = 2k_3$ and $C_G(u_3)$ is an abelian subgroup of index 2 in G, whence G is conjugate to $\mathcal{D}_{k_3}$.

Thus from now on we may suppose that $k_2 = 6$. It follows immediately that the only possibilities for k_3 are 6, 8 and 10 and the corresponding orders for G are 24, 48 and 120, respectively. We shall show that up to conjugacy there is a unique solution for G in each case. First note that if A is an abelian subgroup of G and $A \ne \{\pm 1\}$, then A is cyclic and fixes an element of Ω. In particular, the order of A divides one of k_1, k_2 or k_3. We use the notation $A_i = C_G(u_i)$.

The case $k_1 = 4$, $k_2 = 6$, $k_3 = 6$. By the observation just made, A_1 is a maximal abelian (hence cyclic) subgroup of order 4. Consequently $\mathcal{Q} = N_G(A_1)$ is a non-abelian group of order 8. Thus, up to conjugacy, $\mathcal{Q}$ is the binary dihedral group $\mathcal{D}_4$ and we may replace G by a conjugate so that $\mathcal{Q} = \{\pm 1, \pm i, \pm j, \pm k\}$. The elements of order 4 in G belong to Ω_1 and so G has 6 elements of order 4 and they form a single conjugacy class. Furthermore they generate $\mathcal{Q}$ and thus $\mathcal{Q}$ is a normal subgroup of G. The elements of order 6 in S^3 have trace 1 and the elements of order 3 have trace -1. Thus the only elements of order 6 in S^3 that normalise $\mathcal{Q}$ are $1/2(1 \pm i \pm j \pm k)$ and their squares are the elements $1/2(-1 \pm i \pm j \pm k)$ of order 3. These 16 elements constitute $G \setminus \mathcal{Q}$ and this shows that G is uniquely identified as the binary tetrahedral group $\mathcal{T}$ of (iii).

The case $k_1 = 4$, $k_2 = 6$, $k_3 = 8$. In this case A_3 is a maximal abelian subgroup of order 8 and, since $|G| = 48$, its normaliser is a Sylow 2-subgroup. It follows that $N_G(A_3)$ is a binary dihedral group of order 16 and A_3 is its unique cyclic subgroup of order 8. This means that $N_G(A_3)$ cannot be normal in G and hence the kernel of the action of G on the 3 cosets of $N_G(A_3)$ is a normal subgroup of order 8. This latter group cannot be abelian and therefore it is conjugate to $\mathcal{Q} = \langle i, j \rangle$. Thus we may replace G by a conjugate so that $\mathcal{Q}$ is a normal subgroup of G. From the previous case we see that $\mathcal{Q}A_2 = \mathcal{T}$, the binary tetrahedral group.

The conjugates of A_3 contain 12 elements of order 8 and their squares are the 6 elements of order 4 in $\mathcal{T}$. An element of S^3 has order 8 if and only if its trace is $\pm\sqrt{2}$ and therefore the only elements of S^3 that square to i are $\pm\frac{1}{\sqrt{2}}(1 + i)$. It follows that G is the binary octahedral group $\mathcal{O}$ of (iv).

The case $k_1 = 4$, $k_2 = 6$, $k_3 = 10$. In this case A_1 is a maximal abelian subgroup and so $N_G(A_1)$ is a binary dihedral group of order 8. Thus we may replace

G by a conjugate so that $A_1 = \langle i \rangle$ and $N_G(A_1) = Q = \langle i, j \rangle$. Since G has only one conjugacy class of elements of order 4, there exists $h \in G$ such that $hih^{-1} = j$. But Q is the normaliser of both $\langle i \rangle$ and $\langle j \rangle$, hence $h \in N_G(Q)$. Since $N_G(Q)$ must be one of the finite subgroups of S^3 already determined and since 16 does not divide its order, we see that the only possibility for $N_G(Q)$ is the binary tetrahedral group $\mathcal{T}$.

The group $N_G(A_3)$ is isomorphic to $\mathcal{D}_{10}$ and hence it contains an element q of order 5 and an element r of order 4 such that $rqr^{-1} = q^{-1}$. Since all elements of order 4 in G are conjugate, we may replace A_3 by a conjugate so that $r = k$. The elements of order 5 in S^3 are those of trace $-\tau$ or τ^{-1}. It follows that we may write $q = \frac{1}{2}(a+bi+cj)$, where a is $-\tau$ or τ^{-1} and $a^2+b^2+c^2 = 4$. The products iq and jq have orders 3, 4, 5, 6, or 10 and on taking traces we see that b^2, $c^2 \in \{0, 1, \tau^2, \tau^{-2}\}$. Consequently, the only possibilities for q are $\frac{1}{2}(-\tau \pm i \pm \tau^{-1}j)$, $\frac{1}{2}(-\tau \pm \tau^{-1}i \pm j)$, $\frac{1}{2}(\tau^{-1} \pm i \pm \tau^{-1}j)$ and $\frac{1}{2}(\tau^{-1} \pm \tau^{-1}i \pm j)$.

The 96 elements of

$$\mathcal{T}q \cup \mathcal{T}q^2 \cup \mathcal{T}q^3 \cup \mathcal{T}q^4$$

are obtained from q by all possible sign changes and all *even* permutations of the coefficients of q. Thus G must be one of the groups

$$\langle \mathcal{T}, \tfrac{1}{2}(\tau^{-1} + i + \tau j) \rangle \quad \text{or} \quad \langle \mathcal{T}, \tfrac{1}{2}(\tau^{-1} + \tau i + j) \rangle$$

and as these two groups are interchanged by conjugation by $\frac{1}{\sqrt{2}}(i + j)$ we have proved that G is conjugate to the binary icosahedral group $\mathcal{I}$ of (v). $\qquad\square$

Further information about the groups $\mathcal{T}$, $\mathcal{O}$ and $\mathcal{I}$ can be gleaned from the preceding proof. For example, in $\mathcal{T}$ and $\mathcal{O}$ the normaliser of a subgroup of order 3 is a subgroup of index 4 and the permutation representation on its cosets shows that $\mathcal{T}/\langle -1 \rangle \simeq \mathrm{Alt}(4)$ and $\mathcal{O}/\langle -1 \rangle \simeq \mathrm{Sym}(4)$. Similarly, the permutation representation of $\mathcal{I}$ on the 5 cosets of its subgroup $\mathcal{T}$ shows that $\mathcal{I}/\langle -1 \rangle \simeq \mathrm{Alt}(5)$.

5. The finite subgroups of $SO_3(\mathbb{R})$ and $SU_2(\mathbb{C})$

The homomorphism of Proposition 5.4 allows us to recast the previous theorem in terms of subgroups of $SO_3(\mathbb{R})$.

Theorem 5.13. *Every finite subgroup of $SO_3(\mathbb{R})$ is conjugate in $SO_3(\mathbb{R})$ to one of the following groups:*

(i)   *the cyclic group generated by* $\begin{bmatrix} 1 & 0 & 0 \\ 0 & \cos(4\pi/m) & -\sin(4\pi/m) \\ 0 & \sin(4\pi/m) & \cos(4\pi/m) \end{bmatrix}$,

(ii)   *the dihedral group generated by*

$$\begin{bmatrix} 1 & 0 & 0 \\ 0 & \cos(4\pi/m) & -\sin(4\pi/m) \\ 0 & \sin(4\pi/m) & \cos(4\pi/m) \end{bmatrix} \quad \text{and} \quad \begin{bmatrix} -1 & 0 & 0 \\ 0 & 1 & 0 \\ 0 & 0 & -1 \end{bmatrix},$$

(iii)   *the* tetrahedral group $\mathcal{T}/\langle -1 \rangle \simeq \mathrm{Alt}(4)$ *generated by*

$$\begin{bmatrix} 0 & 1 & 0 \\ 0 & 0 & 1 \\ 1 & 0 & 0 \end{bmatrix} \quad and \quad \begin{bmatrix} 1 & 0 & 0 \\ 0 & -1 & 0 \\ 0 & 0 & -1 \end{bmatrix},$$

(iv)   *the* octahedral group $\mathcal{O}/\langle -1 \rangle \simeq \mathrm{Sym}(4)$ *generated by*

$$\begin{bmatrix} 0 & 1 & 0 \\ 0 & 0 & 1 \\ 1 & 0 & 0 \end{bmatrix} \quad and \quad \begin{bmatrix} 1 & 0 & 0 \\ 0 & 0 & -1 \\ 0 & 1 & 0 \end{bmatrix},$$

(v)   *the* icosahedral group $\mathcal{I}/\langle -1 \rangle \simeq \mathrm{Alt}(5)$ *generated by*

$$\frac{1}{2}\begin{bmatrix} 1 & \tau & \tau^{-1} \\ \tau & -\tau^{-1} & -1 \\ -\tau^{-1} & 1 & -\tau \end{bmatrix} \quad and \quad \begin{bmatrix} 1 & 0 & 0 \\ 0 & -1 & 0 \\ 0 & 0 & -1 \end{bmatrix}, \quad where\ \tau = \tfrac{1}{2}(1 + \sqrt{5}).$$

Similarly, the isomorphism of Proposition 5.10 provides an $SU_2(\mathbb{C})$ version of the theorem.

Theorem 5.14. *Every finite subgroup of $SU_2(\mathbb{C})$ is conjugate in $SU_2(\mathbb{C})$ to one of the following groups:*

(i)   *the* cyclic group $\mathcal{C}_m$ *of order m generated by*

$$\begin{bmatrix} \exp(2\pi i/m) & 0 \\ 0 & \exp(-2\pi i/m) \end{bmatrix},$$

(ii)   *the* binary dihedral group $\mathcal{D}_{2m}$ *of order $4m$ generated by*

$$\begin{bmatrix} \exp(\pi i/m) & 0 \\ 0 & \exp(-\pi i/m) \end{bmatrix} \quad and \quad \begin{bmatrix} 0 & -1 \\ 1 & 0 \end{bmatrix},$$

(iii)   *the* binary tetrahedral group $\mathcal{T}$ *of order 24 generated by*

$$\frac{1}{2}\begin{bmatrix} -1 - i & 1 - i \\ -1 - i & -1 + i \end{bmatrix} \quad and \quad \begin{bmatrix} -i & 0 \\ 0 & i \end{bmatrix},$$

(iv)   *the* binary octahedral group $\mathcal{O}$ *of order 48 generated by*

$$\frac{1}{2}\begin{bmatrix} -1 - i & 1 - i \\ -1 - i & -1 + i \end{bmatrix} \quad and \quad \frac{1}{\sqrt{2}}\begin{bmatrix} 1 - i & 0 \\ 0 & 1 + i \end{bmatrix},$$

(v)   *the* binary icosahedral group $\mathcal{I}$ *of order 120 generated by*

$$\frac{1}{2}\begin{bmatrix} \tau^{-1} - \tau i & 1 \\ -1 & \tau^{-1} + \tau i \end{bmatrix} \quad and \quad \begin{bmatrix} -i & 0 \\ 0 & i \end{bmatrix}.$$

5.1. Embeddings in finite matrix groups. The matrix representation given above for $\mathcal{I}$ leads directly to an isomorphism between $\mathcal{I}$ and the group $SL_2(\mathbb{F}_5)$ of 2×2 matrices of determinant 1 over the finite field $\mathbb{F}_5$. To see this, first observe that in $\mathbb{F}_5$ we have $x^2 + 1 = (x - 2)(x - 3)$ and $x^2 - x - 1 = (x - 3)^2$. Thus there is a homomorphism $\varphi : \mathbb{Z}[i, \tau] \rightarrow \mathbb{F}_5$ such that $\varphi(i) = \varphi(\tau) = 3$. The kernel of φ is a maximal ideal $\mathfrak{p}$, which does not contain 2. Therefore the generators of $\mathcal{I}$ are matrices over the localisation of $\mathbb{Z}[i, \tau]$ at $\mathfrak{p}$. Thus φ induces a homomorphism from $\mathcal{I}$ to $SL_2(\mathbb{F}_5)$ that takes the given generators to

$$\begin{bmatrix} 4 & 3 \\ 2 & 3 \end{bmatrix} \quad \text{and} \quad \begin{bmatrix} 2 & 0 \\ 0 & 3 \end{bmatrix}.$$

On comparing orders and noting that $\begin{bmatrix} -1 & 0 \\ 0 & -1 \end{bmatrix}$ is not in the kernel of the homomorphism we see that $\mathcal{I} \simeq SL_2(\mathbb{F}_5)$.

When we carry out this process modulo 3 we obtain an embedding of $\mathcal{I}$ in the group $SL_2(\mathbb{F}_9)$. This is because $x^2 + 1$ is irreducible over $\mathbb{F}_3$ and so the field of 9 elements can be written as $\mathbb{F}_9 = \mathbb{F}_3[\theta]$, where $\theta^2 + 1 = 0$. Thus there is a homomorphism $\psi : \mathbb{Z}[i, \tau] \rightarrow \mathbb{F}_9$ such that $\psi(i) = \theta$ and $\psi(\tau) = \theta + 2$; and ψ induces an embedding of $\mathcal{I}$ in $SL_2(\mathbb{F}_9)$ as a subgroup of index 6.

Under this embedding the generators of $\mathcal{T}$ are mapped to

$$\begin{bmatrix} \theta + 1 & \theta - 1 \\ \theta + 1 & -\theta + 1 \end{bmatrix} \quad \text{and} \quad \begin{bmatrix} -\theta & 0 \\ 0 & \theta \end{bmatrix}.$$

Conjugation by $\begin{bmatrix} -\theta + 1 & \theta + 1 \\ -1 & \theta \end{bmatrix}$ transforms these matrices to the elements

$$\begin{bmatrix} -1 & -1 \\ 1 & 0 \end{bmatrix} \quad \text{and} \quad \begin{bmatrix} 0 & 1 \\ -1 & 0 \end{bmatrix}$$

of $SL_2(\mathbb{F}_3)$. On comparing orders we see that $\mathcal{T} \simeq SL_2(\mathbb{F}_3)$. Thus $\mathcal{T}$ occurs as a subgroup of index 2 in $GL_2(\mathbb{F}_3)$ and in $\mathcal{O}$. But these extensions are not isomorphic because $\mathcal{O}$ has only one element of order two whereas this is not the case for $GL_2(\mathbb{F}_3)$.

We leave it as an exercise to show that $SL_2(\mathbb{F}_7)$ is the smallest group of the form $SL_2(\mathbb{F}_q)$ that contains $\mathcal{O}$.

6. Quaternions, reflections and root systems

It was observed by Witt [**226**] (see also Humphreys [**119**, §2.13]) that every finite subgroup of S^3 of even order is a root system. In particular it is possible to obtain the Euclidean reflection groups of types F_4 and H_4 in this way. In this section we describe this construction and extend it to obtain several other reflection subgroups of $O_4(\mathbb{R})$ and $U_2(\mathbb{C})$.

If $T \in O_4(\mathbb{R})$, then $\det T = \pm 1$ and therefore every reflection in $O_4(\mathbb{R})$ has order two. For $a \in S^3$ the linear transformation $r_a : \mathbb{H} \to \mathbb{H}$ defined by $r_a(h) = -a\bar{h}a$ is a reflection of order two with root a.

The following proposition shows that every reflection in $O_4(\mathbb{R})$ is conjugate to $\rho := r_1$.

Proposition 5.15. *An element of $O_4(\mathbb{R})$ is a reflection if and only if it has the form $L(a)\rho L(a)^{-1} = L(a)R(\bar{a})\rho : h \mapsto -a\bar{h}a$, for some $a \in S^3$.*

Proof. The elements of $O_4(\mathbb{R}) \setminus SO_4(\mathbb{R})$ have the form $L(q)R(r)\rho = L(q)\rho L(r)$ for some $q, r \in S^3$. Such an element has order two if and only if $qr = \pm 1$. It is a reflection only when $qr = 1$ because only in this case does the eigenvalue -1 have multiplicity 1 (with eigenvector q). $\qquad\square$

Proposition 5.16. *For all $a, b \in S^3$ we have*

(i) $L(aba^{-1}b^{-1}) = r_a r_1 r_{\bar{b}\bar{a}} r_b,$
(ii) $L(a)r_b L(a)^{-1} = r_{ab},$
(iii) $R(a)^{-1}r_b R(a) = r_{ba},$
(iv) $r_a L(b)r_a = R(\bar{a}ba),$ *and*
(v) $r_a R(b)r_a = L(ab\bar{a}).$
(vi) $r_a r_b r_a = r_{a\bar{b}a}.$

Proof. All parts of this proposition follow at once by direct computation using the definitions of $L(a)$, $R(a)$ and r_a. As an example, we prove (i), first noting that $\bar{a} = a^{-1}$ for all $a \in S^3$.

For $h \in \mathbb{H}$ we have

$$r_a r_1 r_{\bar{b}\bar{a}} r_b(h) = r_a r_1 r_{\bar{b}\bar{a}}(-b\bar{h}b) = r_a r_1(\bar{b}\bar{a}b h \bar{a})$$
$$= r_a(-a\bar{h}ba b) = ab\bar{a}\bar{b}h = L(ab\bar{a}\bar{b})h$$

and therefore $L(aba^{-1}b^{-1}) = r_a r_1 r_{\bar{b}\bar{a}} r_b.$ $\qquad\square$

Theorem 5.17. *Suppose that Σ is a finite subgroup of S^3 of even order and suppose that $\Delta \subseteq \Sigma$ is a union of conjugacy classes of elements of order 4. Let A be the ring of integers of the subfield of $\mathbb{R}$ generated by the elements $a + a^{-1}$, for $a \in \Sigma$. Then $\mu(A) = \{\pm 1\}$ and*

(i) $W(\Sigma) := \langle\, r_a \mid a \in \Sigma \,\rangle$ *is a finite reflection subgroup of $O_4(\mathbb{R})$ and Σ is an A-root system for $W(\Sigma)$. Furthermore,*

$$L(\Sigma')R(\Sigma') \trianglelefteq W(\Sigma) \trianglelefteq L(\Sigma)R(\Sigma)\langle\rho\rangle,$$

where $|L(\Sigma)R(\Sigma)\langle\rho\rangle| = |\Sigma|^2$ and Σ' is the derived group of Σ.
(ii)   *if $\Delta = -\Delta$, then $W(\Delta) := \langle\, r_a \mid a \in \Delta \,\rangle$ is a finite reflection subgroup of $O_3(\mathbb{R})$ and Δ is an A-root system for $W(\Delta)$.*

Proof. Since $|\Sigma|$ is even, it follows that $-1 \in \Sigma$ and therefore $a \in \Sigma$ if and only if $-a \in \Sigma$. Since Σ is finite, $a \in \Sigma$ is a root of unity and so $a + a^{-1} \in \mathbb{R}$ is an algebraic integer and hence belongs to A. The Cartan coefficient of $a, b \in \Sigma$ is $\langle a \,|\, b \rangle = 2\, a \cdot b = a\bar{b} + b\bar{a}$ and so $\langle a \,|\, b \rangle \in A$.

(*i*) For $a, b \in \Sigma$ we have $r_a(b) = -a\bar{b}a \in \Sigma$ and therefore, according to Definition 1.43, Σ is an A-root system in the $\mathbb{R}$-linear span of Σ. It follows from Proposition 5.16 (*i*), (*iv*) and (*v*) that $L(\Sigma')R(\Sigma')$ is a normal subgroup of $W(\Sigma)$ and from parts (*ii*) and (*iii*) that $W(\Sigma)$ is a normal subgroup of $L(\Sigma)R(\Sigma)\langle \rho \rangle$.

(*ii*) If $a, b \in \Delta$, then $r_a(b) = ab^{-1}a^{-1} \in \Delta$ and therefore Δ is also an A-root system in the sense of Definition 1.43. For $a \in \Delta$ the reflection r_a fixes 1 and so $W(\Delta)$ is a finite reflection subgroup of $O_3(\mathbb{R})$. $\qquad\square$

The only finite subgroups of S^3 of odd order are cyclic and since $r_a = r_{-a}$ there is no loss of generality in considering the reflection groups $W(\Sigma)$ where Σ is a group of even order. Then $W(\Sigma)$ is a Euclidean reflection group, as defined in Humphreys [**119**] or Bourbaki [**33**]. In the following paragraphs we shall identify these groups using the standard notation. They will be placed in a broader context in Chapters 7 and 8.

The groups $W(\mathcal{C}_{2m})$ and $W(\mathcal{D}_{2m})$. If $a = \cos\theta + i\sin\theta \in S^3$, then the matrices of r_a and r_{aj} with respect to the basis $1, i, j, k$ of $\mathbb{H}$ are

$$
\begin{bmatrix}
-\cos 2\theta & -\sin 2\theta & 0 & 0 \\
-\sin 2\theta & \cos 2\theta & 0 & 0 \\
0 & 0 & 1 & 0 \\
0 & 0 & 0 & 1
\end{bmatrix}
\quad \text{and} \quad
\begin{bmatrix}
1 & 0 & 0 & 0 \\
0 & 1 & 0 & 0 \\
0 & 0 & -\cos 2\theta & -\sin 2\theta \\
0 & 0 & -\sin 2\theta & \cos 2\theta
\end{bmatrix}.
$$

It follows that $W(\mathcal{C}_{2m})$ is isomorphic to the dihedral group $G(m, m, 2)$ and that $W(\mathcal{D}_{2m})$ is isomorphic to $G(m, m, 2) \times G(m, m, 2)$. In particular, $W(\mathcal{Q})$ is an elementary abelian group of order 16.

The group $W(\mathcal{T})$. The derived group of $\mathcal{T}$ is $\mathcal{Q}$ and therefore $L(\mathcal{Q})R(\mathcal{Q})$ is a normal subgroup of $W(\mathcal{T})$. Direct calculation, using Proposition 5.16 (*iv*) and (*v*) shows that

$$
\begin{aligned}
A_i &:= \langle L(i)R(i), L(j)R(k), L(k)R(j) \rangle, \\
A_j &:= \langle L(j)R(j), L(k)R(i), L(i)R(k) \rangle \quad \text{and} \\
A_k &:= \langle L(k)R(k), L(i)R(j), L(j)R(i) \rangle
\end{aligned}
$$

are normal elementary abelian subgroups of $W(\mathcal{T})$ of order 8. Therefore, by Theorem 2.3, $W(\mathcal{T})$ is imprimitive. From the classification of the imprimitive groups in

Chapter 2 the only possibility for $W(\mathcal{T})$ is the group $G(2,2,4)$ of order 192 – this group is also known as the Weyl group of type D_4.

The group $L(\mathcal{T})R(\mathcal{T})\langle\rho\rangle$ acts on the set $\{A_i, A_j, A_k\}$ as the symmetric group $\mathrm{Sym}(3)$ and $W(\mathcal{T})$ is the normal subgroup of index 3 fixing A_i, A_j and A_k. The elements of order 3 permuting A_i, A_j, A_k are known as *triality* automorphisms of $W(\mathcal{T})$.

The group $W(\mathcal{O})$. The derived group of $\mathcal{O}$ is $\mathcal{T}$ and therefore $L(\mathcal{T})R(\mathcal{T})$ is a normal subgroup of $W(\mathcal{O})$. The index of $\mathcal{T}$ in $\mathcal{O}$ is 2 and hence, from Proposition 5.16 (vi), $W(\mathcal{O})$ has two orbits on $\mathcal{O}$, namely $\mathcal{T}$ and $\mathcal{O}\setminus\mathcal{T}$. In particular, $W(\mathcal{T})$ is a normal subgroup of $W(\mathcal{O})$. On the other hand, it is clear that $L(\mathcal{O})R(\mathcal{O})\langle\rho\rangle$ acts transitively on $\mathcal{O}$ and if $a \in \mathcal{O}\setminus\mathcal{T}$, then $r_a \notin L(\mathcal{T})R(\mathcal{T})\langle\rho\rangle$. Thus we have proper containments

$$L(\mathcal{Q})R(\mathcal{Q}) \subset W(\mathcal{T}) \subset L(\mathcal{T})R(\mathcal{T})\langle\rho\rangle \subset W(\mathcal{O}) \subset L(\mathcal{O})R(\mathcal{O})\langle\rho\rangle$$

and on comparing orders we find that $W(\mathcal{O})$ has index 2 in $L(\mathcal{O})R(\mathcal{O})\langle\rho\rangle$. Consequently $W(\mathcal{O})$ is a primitive reflection group of order 1152, also known as the Weyl group of type F_4.

The set $\mathcal{O}$ is a $\mathbb{Z}[\sqrt{2}]$-root system for $W(\mathcal{O})$ in which all roots are short. However, there is a $\mathbb{Z}$-root system for $W(\mathcal{O})$ which has 24 short and 24 long roots: namely, $\mathcal{T} \cup \{\sqrt{2}\,a \mid a \in \mathcal{O}\setminus\mathcal{T}\}$.

If Δ is the set of all elements of order 4 in $\mathcal{O}$, then $|\Delta| = 18$ and $W(\Delta)$ is the group $G(2,1,3)$, also known as the Weyl group of type B_3.

The group $W(\mathcal{I})$. The group $\mathcal{I}$ is its own derived group and therefore $W(\mathcal{I}) = L(\mathcal{I})R(\mathcal{I})\langle\rho\rangle$ is a primitive unitary reflection group of order 14400, also known as the Weyl group of type H_4. The set $\mathcal{I}$ is a $\mathbb{Z}[\tau]$-root system for $W(\mathcal{I})$.

The 15 elements of order 4 in $\mathcal{I}$ form a single conjugacy class Δ. The group $W(\Delta) \simeq \mathcal{C}_2 \times \mathrm{Alt}(5)$ is a primitive reflection group of rank 3, known as the Weyl group of type H_3; it is the subgroup of $W(\mathcal{I})$ that fixes 1.

If $u \in \mathbb{H}\setminus\mathbb{R}$, then $\mathbb{R}[u]$ is a copy of the complex numbers and $\mathbb{H}$ is a 2-dimensional vector space over $\mathbb{R}[u]$. In section 3 we used the special case $u = i$ to establish a connection between $\mathbb{H}$ and $U_2(\mathbb{C})$. In this context, a variation of the construction given above produces finite unitary reflection subgroups of $U_2(\mathbb{C})$ from finite subgroups of the quaternions. The only groups considered here will be generated by reflections of order two. However, in the next chapter we shall give a complete account of *all* finite reflection groups of rank 2.

Consider $\mathbb{H}$ as a vector space over $\mathbb{C} = \mathbb{R}[i]$. Proposition 5.11 shows that $U_2(\mathbb{C}) = L(S^1)R(S^3)$ and consequently it is an easy matter to determine which elements of $U_2(\mathbb{C})$ are reflections.

Proposition 5.18. *For* $(\alpha, q) \in S^1 \times S^3 \setminus \{\pm(1,1)\}$, *the linear transformation* $L(\alpha)R(q)$ *is a reflection if and only if* $\mathrm{Tr}(\alpha) = \mathrm{Tr}(q)$ *if and only if* $q = \overline{a}\alpha a$ *for some* $a \in S^3$.

Proof. Since $\mathrm{N}(\alpha) = \mathrm{N}(q) = 1$ this result is an immediate consequence of Proposition 5.3.   □

Suppose that $\alpha \in S^1$ is an mth root of unity and that $a \in S^3$. If m is odd, then $L(\alpha)R(\overline{a}\,\overline{\alpha}a)$ is a reflection of order m with a root a; if m is even, it is a reflection of order $m/2$. Thus the reflections of order two in $U_2(\mathbb{C})$ are the $\mathbb{C}$-linear transformations of $\mathbb{H}$ of the form $\hat{r}_a(h) = ih\overline{a}ia$.

If Σ is a finite subgroup of $\mathbb{H}$, and $i \in \Sigma$, then Σ is a root system with Weyl group $W_{\mathbb{C}}(\Sigma) := \langle \hat{r}_a \mid a \in \Sigma \rangle$. The quaternion group $\mathcal{Q}$ is a normal subgroup of $\mathcal{T}$ and $\mathcal{O}$ and we have $\hat{r}_1\hat{r}_a = R(i^{-1}a^{-1}ia)$. Therefore $W_{\mathbb{C}}(\mathcal{T}) = W_{\mathbb{C}}(\mathcal{O}) \simeq \mathcal{C}_4 \circ \mathcal{Q} \simeq G(4,2,2)$. Similarly $W_{\mathbb{C}}(\mathcal{I}) \simeq \mathcal{C}_4 \circ \mathcal{I}$.

Exercises

1. Use Proposition 5.1 (*vi*) to verify the entries in the table of traces of Section 4.
2. Given that the algebra $\mathbb{H}$ is defined as the two-dimensional vector space over $\mathbb{C}$ with basis 1 and j, and a multiplication such that 1 is the identity and where j satisfies $j^2 = 1$ and $ij = -ji$, show that $\mathbb{H}$ is associative.
3. Prove that $\{(1,1), (-1,-1)\}$ is the kernel of the homomorphism $S^3 \times S^3 \to SO_4(\mathbb{R})$ that sends (q,r) to $L(q)R(r)$.
4. In the notation of Section 4, show that the subgroup of the binary icosahedral group that fixes the vector $\tau i + j$ is the cyclic group of order 10 generated by $-\sigma$. Deduce that the order of the binary icosahedral group is 120.
5. Show that $SL_2(\mathbb{F}_7)$ is the smallest group of the form $SL_2(\mathbb{F}_q)$ that contains $\mathcal{O}$.
6. Show that the group $\langle r_a \mid a \in \mathcal{Q} \cup (\mathcal{O} \setminus \mathcal{T}) \rangle$ is isomorphic to the imprimitive reflection group $G(2,1,4)$, also known as the Weyl group of type B_4.
7. Prove that $W(\mathcal{T}) \cap W(\mathcal{O} \setminus \mathcal{T}) = L(\mathcal{Q})R(\mathcal{Q})$.

CHAPTER 6

Finite unitary reflection groups of rank two

In this chapter we determine all finite unitary reflection subgroups of $U_2(\mathbb{C})$ and describe their structure and their invariants. The imprimitive reflection subgroups of $U_2(\mathbb{C})$ are the groups $G(m, p, 2)$ of Chapter 2 and thanks to the results of the previous chapter we are now in a position to determine all *primitive* reflection subgroups of $U_2(\mathbb{C})$.

1. The primitive reflection subgroups of $U_2(\mathbb{C})$

Let $\mathcal{C}_m$ denote the cyclic group of order m of scalar matrices in $U_2(\mathbb{C})$. We know from Proposition 5.11 that $U_2(\mathbb{C})$ is the central product $L(S^1) \circ R(S^3)$ and from Proposition 5.10 $R(S^3) \simeq SU_2(\mathbb{C})$. We make this decomposition explicit as follows.

If G is a finite subgroup of $U_2(\mathbb{C})$ and if $g \in G$, we choose $\lambda_g \in \mathbb{C}$ such that $\lambda_g^2 = \det(g)$. Then

$$\widehat{G} = \{ \pm \lambda_g^{-1} g \mid g \in G \}$$

is a finite subgroup of $SU_2(\mathbb{C})$ and $G \subseteq \mathcal{C}_m \circ \widehat{G}$, where m is the exponent of G. By construction we have $G/Z(G) \simeq \widehat{G}/Z(\widehat{G})$.

If G is primitive, then (up to conjugacy in $U_2(\mathbb{C})$) $\widehat{G}$ is a binary tetrahedral, binary octahedral or binary icosahedral group, namely one of $\mathcal{T}$, $\mathcal{O}$ or $\mathcal{I}$ in its two-dimensional representation described in Theorem 5.14.

Define the *type* of G to be $\mathcal{T}$, $\mathcal{O}$ or $\mathcal{I}$ according to whether $\widehat{G}$ is isomorphic to $\mathcal{T}$, $\mathcal{O}$ or $\mathcal{I}$, respectively. We shall see that there are 19 primitive reflection subgroups of $U_2(\mathbb{C})$: 4 of type $\mathcal{T}$, 8 of type $\mathcal{O}$ and 7 of type $\mathcal{I}$.

Theorem 6.1. *If G is a finite primitive reflection subgroup of $U_2(\mathbb{C})$ then (up to conjugacy) G is a normal subgroup of $\mathcal{C}_{12} \circ \mathcal{T}$, $\mathcal{C}_{24} \circ \mathcal{O}$ or $\mathcal{C}_{60} \circ \mathcal{I}$ according to whether G is of type $\mathcal{T}$, $\mathcal{O}$ or $\mathcal{I}$, respectively.*

Proof. If $g \in \widehat{G}$ has eigenvalues θ and θ^{-1}, then θg and $\theta^{-1} g$ are reflections. Moreover, every reflection in G can be obtained in this way. Thus G is a normal subgroup of $\mathcal{C}_m \circ \widehat{G}$, where m is the exponent of $\widehat{G}$. To complete the proof note that the exponents of $\mathcal{T}$, $\mathcal{O}$ and $\mathcal{I}$ are 12, 24 and 60, respectively. $\qquad \square$

Since $\mathbf{T} := \mathcal{C}_{12} \circ \mathcal{T}$, $\mathbf{O} := \mathcal{C}_{24} \circ \mathcal{O}$ and $\mathbf{I} := \mathcal{C}_{60} \circ \mathcal{I}$ are reflection groups and as $\mathbf{T} = \mathbf{O} \cap \mathbf{I}$, every finite reflection subgroup G of $U_2(\mathbb{C})$ is a subgroup of $\mathbf{O}$ or $\mathbf{I}$.

It follows from the previous theorem that $GZ(H) = H$, where H is one of $\mathbf{T}$, $\mathbf{O}$ or $\mathbf{I}$. Thus elements g_1 and $g_2 \in G$ are conjugate in G if and only if they are conjugate in H. Therefore, to determine all possibilities for G, it is enough to find representative generators for the conjugacy classes of cyclic subgroups generated by reflections in H and then compute the normal closures of all subsets of these representatives. (We denote the conjugacy class of $r \in H$ by r^H and therefore the normal closure of $\langle r \rangle$ is $\langle r^H \rangle$.)

As we saw in the proof of the theorem, each element $g \in H$ determines a pair of reflections $r_1 := \theta_1^{-1} g$ and $r_2 := \theta_2^{-1} g$, where θ_1 and θ_2 are the eigenvalues of g. Furthermore, if $z \in Z(H)$, then g and gz determine the same pair of reflections. Thus to find the conjugacy classes of reflections in H we obtain a pair of reflections r_1, r_2 from each non-trivial conjugacy class in $H/Z(H)$ and then check whether or not r_1 and r_2 are conjugate in H.

2. The reflection groups of type $\mathcal{T}$

The generators of $\mathcal{T}$ given in Theorem 5.14 are

$$a := \frac{1}{2} \begin{bmatrix} -1-i & 1-i \\ -1-i & -1+i \end{bmatrix} \quad \text{and} \quad b := \begin{bmatrix} -i & 0 \\ 0 & i \end{bmatrix}$$

of orders 3 and 4, respectively, and the non-trivial conjugacy classes of the group $\overline{\mathcal{T}} := \mathcal{T}/Z(\mathcal{T}) \simeq \mathrm{Alt}(4)$ are represented by $\bar{a}$, $\bar{a}^2$ and $\bar{b}$. Since b is conjugate to $-b$ in $\mathcal{T}$, the reflections ib and $-ib$ are conjugate in $\mathbf{T}$. On the other hand, the reflections ωa and $\omega^2 a$, where $\omega = \exp(2\pi i/3)$, are not conjugate in $\mathbf{T}$. Thus $\mathbf{T}$ has five conjugacy classes of reflections with representatives $r := ib$, $r_1 := \omega a$, $r_2 := \omega a^2$, $r_1^2 := \omega^2 a^2$, and $r_2^2 = \omega^2 a$. As matrices these reflections are

$$r = \begin{bmatrix} 1 & 0 \\ 0 & -1 \end{bmatrix}, \quad r_1 = \frac{\omega}{2} \begin{bmatrix} -1-i & 1-i \\ -1-i & -1+i \end{bmatrix} \quad \text{and} \quad r_2 = \frac{\omega}{2} \begin{bmatrix} -1+i & -1+i \\ 1+i & -1-i \end{bmatrix}$$

and their orders are 2, 3 and 3, respectively.

The normal closure of b in $\mathcal{T}$ is the quaternion group $\mathcal{Q}$ and therefore the normal closure of r in $\mathbf{T}$ is the imprimitive group $G(4,2,2) = \mathcal{C}_4 \circ G(2,1,2) = \mathcal{C}_4 \circ G(4,4,2) = \mathcal{C}_4 \circ \mathcal{Q}$. It is the Sylow 2-subgroup of $\mathbf{T}$ and it is a normal subgroup of $\mathbf{O}$. Thus $\mathbf{O}$ is the largest reflection subgroup of $U_2(\mathbb{C})$ that contains $G(4,2,2)$ as a normal subgroup.

If G contains the conjugacy class of r_1 or r_2, then $G/Z(G) \simeq \mathcal{T}/Z(\mathcal{T})$ and so the reflection subgroups of $\mathbf{T}$ are $\langle r_1^{\mathbf{T}} \rangle$, $\langle r_2^{\mathbf{T}} \rangle$, $\langle r_1^{\mathbf{T}}, r_2^{\mathbf{T}} \rangle$, $\langle r^{\mathbf{T}}, r_1^{\mathbf{T}} \rangle$, $\langle r^{\mathbf{T}}, r_2^{\mathbf{T}} \rangle$ and $\langle r^{\mathbf{T}}, r_1^{\mathbf{T}}, r_2^{\mathbf{T}} \rangle$. However, $r_1^{\mathbf{T}}$ and $r_2^{\mathbf{T}}$ are interchanged by the action of any element of $\mathbf{O} \setminus \mathbf{T}$. Thus (up to conjugacy) the only possibilities for G are the normal closures $G_4 := \langle r_1^{\mathbf{T}} \rangle$, $G_5 := \langle r_1^{\mathbf{T}}, r_2^{\mathbf{T}} \rangle$, $G_6 := \langle r^{\mathbf{T}}, r_1^{\mathbf{T}} \rangle$ and $G_7 := \langle r^{\mathbf{T}}, r_1^{\mathbf{T}}, r_2^{\mathbf{T}} \rangle = \mathbf{T}$.

TABLE 6.1. The rank 2 reflection groups of type $\mathcal{T}$

G	Structure	$|G|$	$|Z(G)|$		Degrees
G_4	$\langle r_1, r_1' \rangle \simeq SL_2(\mathbb{F}_3)$	24	2	3[3]3	4, 6
G_5	$\langle r_1, r_2' \rangle = \mathcal{C}_3 \times \mathcal{T}$	72	6	3[4]3	6, 12
G_6	$\langle r, r_1 \rangle = \mathcal{C}_4 \circ G_4$	48	4	2[6]3	4, 12
G_7	$\langle r, r_1, r_2 \rangle = \mathcal{C}_3 \times (\mathcal{C}_4 \circ \mathcal{T})$	144	12	—	12, 12

(We use the notation G_k to denote the kth group from the tables of Shephard and Todd [**193**].)

To determine the orders of the groups we use the fact that $|r^{\mathbf{T}}| = 6$, $|r_1^{\mathbf{T}}| = |r_2^{\mathbf{T}}| = 4$ and hence G_4, G_5, G_6 and G_7 contain 8, 16, 14 and 22 reflections, respectively. Then from Theorem 4.14, the order of G_i is $d_1 d_2$, where $d_1 + d_2 - 2$ is the number of reflections and $d_1 \leq d_2$ are the degrees of G_i. Since 24 divides $|G_i|$ and since all four groups are subgroups of $\mathbf{T}$, of order 144, this is sufficient information to determine their orders. The results of these calculations are recorded in Table 6.1. In particular no two groups have the same order and hence no two of them are conjugate. Furthermore, $\mathcal{T}$ is not a subgroup of G_4 nor of G_6.

The containment relation between these reflection groups is depicted in the following diagram, in which the groups that require 3 generators are shown boxed.

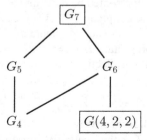

The generators r_1' and r_2' are the reflections of order 3, where

$$r_1' := r r_1 r = \frac{\omega}{2} \begin{bmatrix} -1-i & -1+i \\ 1+i & -1+i \end{bmatrix} \quad \text{and} \quad r_2' := r r_2 r = \frac{\omega}{2} \begin{bmatrix} -1+i & 1-i \\ -1-i & -1-i \end{bmatrix}.$$

2.1. Shephard groups. The symbols $k[m]\ell$ which appear in the tables indicate that the given generators r and s have orders k and ℓ respectively and satisfy the 'braid relation'

$$\underbrace{rsr\cdots}_{m} = \underbrace{srs\cdots}_{m}.$$

In 1952 Shephard [**191**] initiated the study of regular complex polytopes and ten years later Coxeter [**61**] showed that every finite group $k[m]\ell$ is the symmetry group of a regular complex polygon. Following Orlik and Solomon [**181**], we define a *Shephard group* to be the symmetry group of a regular complex polytope.

In [**61**] (see also [**63**, p. 79]) it is shown that the abstract group $k[m]\ell$ is finite if and only if $(k + \ell)m > k\ell(m - 2)$, in which case its order is

$$\frac{8}{m}\left(\frac{1}{k} + \frac{2}{m} + \frac{1}{\ell} - 1\right)^{-2}.$$

The group $2[m]2$ is the dihedral group $G(m, m, 2)$ of order $2m$ and the group $2[4]m$ is $G(m, 1, 2)$. We shall see that the Shephard groups $k[m]\ell$ are precisely the reflection subgroups of $U_2(\mathbb{C})$ that can be generated by two reflections.

3. The reflection groups of type $\mathcal{O}$

The generators of $\mathcal{O}$ given in Theorem 5.14 are

$$a := \frac{1}{2}\begin{bmatrix} -1 - i & 1 - i \\ -1 - i & -1 + i \end{bmatrix} \quad \text{and} \quad c := \frac{1}{\sqrt{2}}\begin{bmatrix} 1 - i & 0 \\ 0 & 1 + i \end{bmatrix},$$

of orders 3 and 8, respectively. Thus the non-trivial conjugacy classes of $\overline{\mathcal{O}} := \mathcal{O}/Z(\mathcal{O}) \simeq \mathrm{Sym}(4)$ are represented by $\bar{a}$, $\bar{b}$, $\bar{c}$ and $\bar{d}$, where $b := c^2$ and $d := c^3a^2c^2$. (We have chosen d so that $d^{-1}ad = a^{-1}$.) The elements $\bar{a}$, $\bar{b}$, $\bar{c}$ and $\bar{d}$ correspond to the permutations $(2, 3, 4)$, $(1, 3)(2, 4)$, $(1, 2, 3, 4)$ and $(2, 3)$ in $\mathrm{Sym}(4)$.

The reflections ib and $-ib$ are conjugate in **T**. Similarly, the reflections id and $-id$ are conjugate in **O** since d is conjugate to $-d$ in $\mathcal{O}$. Thus **O** has six conjugacy classes of reflections, with representatives ib, ωa, $\omega^2 a$, id, $(1 - i)c\sqrt{2}$, and $(1 + i)c/\sqrt{2}$.

The conjugacy classes of cyclic subgroups of **O** generated by reflections are represented by $r := ib$, $r_1 := \omega a$, $r_3 := id$ and $r_4 := (1 + i)c/\sqrt{2}$ of orders 2, 3, 2 and 4, respectively. The matrices for r and r_1 are given above and we have

$$r_3 = \frac{1}{\sqrt{2}}\begin{bmatrix} 1 & -1 \\ -1 & -1 \end{bmatrix} \quad \text{and} \quad r_4 = \begin{bmatrix} 1 & 0 \\ 0 & i \end{bmatrix}.$$

In order to have $G/Z(G) \simeq \mathcal{O}/Z(\mathcal{O})$, the conjugacy class of either r_3 or r_4 must be a subset of G. Since $r_4^2 = r$ we need to consider the normal closures of just eight subsets of $\{r, r_1, r_3, r_4\}$. Direct calculation leads to the groups listed in Table 6.2, where $r_3' = rr_3r$, $r_3'' = (r_1^2r_4)^{-1}r_3(r_1^2r_4)$ and $r_4' = r_3^{-1}r_4r_3$. That is, we put

$$r_3' := \frac{1}{\sqrt{2}}\begin{bmatrix} 1 & 1 \\ 1 & -1 \end{bmatrix}, \quad r_3'' := \frac{1}{\sqrt{2}}\begin{bmatrix} 0 & 1 + i \\ 1 - i & 0 \end{bmatrix}$$

and

$$r_4' := \frac{1}{2}\begin{bmatrix} 1 + i & -1 + i \\ -1 + i & 1 + i \end{bmatrix}.$$

TABLE 6.2. The rank 2 reflection groups of type $\mathcal{O}$

G	Structure	$|G|$	$|Z(G)|$		Degrees
G_8	$\langle r_4, r_4' \rangle = \mathcal{T}\,\mathcal{C}_4$	96	4	4[3]4	8, 12
G_9	$\langle r_3, r_4 \rangle = \mathcal{C}_8 \circ \mathcal{O}$	192	8	2[6]4	8, 24
G_{10}	$\langle r_1, r_4' \rangle = \mathcal{C}_3 \times \mathcal{T}\,\mathcal{C}_4$	288	12	3[4]4	12, 24
G_{11}	$\langle r_1, r_3, r_4 \rangle = \mathcal{C}_3 \times (\mathcal{C}_8 \circ \mathcal{O})$	576	24	—	24, 24
G_{12}	$\langle r_3, r_3', r_3'' \rangle \simeq GL_2(\mathbb{F}_3)$	48	2	—	6, 8
G_{13}	$\langle r, r_3, r_3'' \rangle = \mathcal{C}_4 \circ \mathcal{O}$	96	4	—	8, 12
G_{14}	$\langle r_1, r_3' \rangle = \mathcal{C}_3 \times G_{12}$	144	6	3[8]2	6, 24
G_{15}	$\langle r, r_1, r_3 \rangle = \mathcal{C}_3 \times (\mathcal{C}_4 \circ \mathcal{O})$	288	12	—	12, 24

The groups G in the table are all subgroups of the group $\mathbf{O}$ of order 576 and as before we have $|G| = d_1 d_2$, where $d_1 + d_2 - 2$ is the number of reflections and $d_1 \le d_2$ are the degrees of G. We also know from Lemma 3.24 that $|Z(G)| = \gcd(d_1, d_2)$. Therefore, as $G/Z(G) \simeq \mathcal{O}/Z(\mathcal{O})$, we see that $\mathrm{lcm}(d_1, d_2) = 24$. These facts and the values $|r^{\mathbf{O}}| = 6$, $|r_1^{\mathbf{O}}| = 8$, $|r_3^{\mathbf{O}}| = 12$ and $|r_4^{\mathbf{O}}| = 6$ suffice to determine the orders and the degrees given in the table.

The containment relation between the reflection subgroups of G_{11} of type $\mathcal{O}$ is as follows. As before, the 3-generator groups are shown boxed.

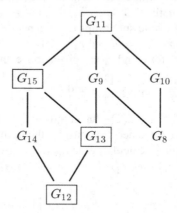

In order to combine this with the previous diagram observe that G_5 is a subgroup of G_{14}, G_7 is a subgroup of G_{10} and G_{15} and that $G(4, 2, 2)$ is a subgroup of G_8 and G_{13}.

The notation $\mathcal{T}\,\mathcal{C}_4$ means that G_8 is the semidirect product of $\mathcal{T}$ by the cyclic group $\langle r_4 \rangle$ of order 4. The derived group of $\mathcal{O}$ is $\mathcal{T}$ and therefore $\mathcal{T}$ is the derived group of every group in the table. The group $\mathcal{O}$ is contained in G_9, G_{11}, G_{13} and G_{15} but not in G_8, G_{10}, G_{12} nor G_{14}. We have $\mathcal{C}_4 \circ \mathcal{O} = \mathcal{C}_4 \circ G_{12}$ and from this we see that $G_{12} \simeq GL_2(\mathbb{F}_3)$ is a subgroup of all the groups except G_8 and G_{10}. In particular, G_8 is not isomorphic to G_{13} and G_{10} is not isomorphic to G_{15}.

4. The reflection groups of type $\mathcal{I}$

The generators of $\mathcal{I}$ given in Theorem 5.14 are

$$e := \frac{1}{2} \begin{bmatrix} \tau^{-1} - \tau i & 1 \\ -1 & \tau^{-1} + \tau i \end{bmatrix} \quad \text{and} \quad b := \begin{bmatrix} -i & 0 \\ 0 & i \end{bmatrix}$$

of orders 5 and 4 respectively. Thus the non-trivial conjugacy classes of the group $\bar{\mathcal{I}} := \mathcal{I}/Z(\mathcal{I}) \simeq \text{Alt}(5)$ are represented by $\bar{a}$, $\bar{b}$, $\bar{e}$ and $\bar{e}^2$, where $a := (be)^3 e^2 b^3$. The elements a and b are the previous generators of $\mathcal{T}$ and the elements $\bar{a}$, $\bar{b}$ and $\bar{e}$ correspond to the permutations $(2,4,3)$, $(2,3)(4,5)$ and $(1,2,4,3,5)$ in $\text{Alt}(5)$.

Let ζ be an eigenvalue of e. Then ζ is a fifth root of unity and we may take $\zeta := \exp(2\pi i/5)$ so that $\tau = \zeta + \zeta^{-1} + 1$. The group $\mathbf{I}$ has seven conjugacy classes of reflections with representatives ib, ωa, $\omega^2 a$, ζe, $\zeta^{-1} e$, $\zeta^2 e^2$ and $\zeta^{-2} e^2$.

As representatives for the conjugacy classes of cyclic subgroups of $\mathbf{I}$ we take $r := ib$, $r_1 := \omega a$ and $r_5 := \zeta^2 e^3$. Their orders are 2, 3 and 5 respectively and their matrices are

$$r = \begin{bmatrix} 1 & 0 \\ 0 & -1 \end{bmatrix}, \quad r_1 = \frac{\omega}{2} \begin{bmatrix} -1 - i & 1 - i \\ -1 - i & -1 + i \end{bmatrix}$$

$$\text{and} \quad r_5 = \frac{\zeta^2}{2} \begin{bmatrix} -\tau + i & -\tau + 1 \\ \tau - 1 & -\tau - i \end{bmatrix}.$$

It turns out that the normal closures of the seven non-empty subsets of $\{r, r_1, r_5\}$ are distinct. Their structure is given in Table 6.3, where $r' := r_1^{-2} r r_1^2$, $r'' := r_5^{-1} r r_5$, $r_1'' := r_5 r_1 r_5^{-1}$ and $r_5' := r^{-1} r_5 r$. That is,

$$r' := \begin{bmatrix} 0 & 1 \\ 1 & 0 \end{bmatrix}, \quad r'' := \frac{1}{2} \begin{bmatrix} \tau & (\tau - 1)i + 1 \\ (-\tau + 1)i + 1 & -\tau \end{bmatrix},$$

$$r_1'' := \frac{\omega}{2} \begin{bmatrix} -\tau i - 1 & (-\tau + 1)i \\ (-\tau + 1)i & \tau i - 1 \end{bmatrix} \quad \text{and} \quad r_5' := -\frac{\zeta^2}{2} \begin{bmatrix} \tau - i & -\tau + 1 \\ \tau - 1 & \tau + i \end{bmatrix}.$$

The groups G in the table are all subgroups of the group $\mathbf{I}$ of order 3600 and as for the reflection subgroups of $\mathbf{T}$ and $\mathbf{O}$ we have $|G| = d_1 d_2$, where $d_1 + d_2 - 2$ is the number of reflections, and $d_1 \leq d_2$ are the degrees of G. In the present case we have $\text{lcm}(d_1, d_2) = 60$, $|r^{\mathbf{I}}| = 30$, $|r_1^{\mathbf{I}}| = 20$ and $|r_5^{\mathbf{I}}| = 12$; this suffices to determine the orders and the degrees given in the table.

TABLE 6.3. The rank 2 reflection groups of type $\mathcal{I}$

G	Structure		$|G|$	$|Z(G)|$		Degrees
G_{16}	$\langle r_5, r_5' \rangle = \mathcal{C}_5 \times \mathcal{I}$		600	10	5[3]5	20, 30
G_{17}	$\langle r, r_5 \rangle = \mathcal{C}_5 \times (\mathcal{C}_4 \circ \mathcal{I})$		1200	20	2[6]5	20, 60
G_{18}	$\langle r_1^2, r_5 \rangle = \mathcal{C}_{15} \times \mathcal{I}$		1800	30	3[4]5	30, 60
G_{19}	$\langle r, r_1, r_5 \rangle = \mathcal{C}_{15} \times (\mathcal{C}_4 \circ \mathcal{I})$		3600	60	—	60, 60
G_{20}	$\langle r_1, r_1'' \rangle = \mathcal{C}_3 \times \mathcal{I}$		360	6	3[5]3	12, 30
G_{21}	$\langle r, r_1'' \rangle = \mathcal{C}_3 \times (\mathcal{C}_4 \circ \mathcal{I})$		720	12	2[10]3	12, 60
G_{22}	$\langle r, r', r'' \rangle = \mathcal{C}_4 \circ \mathcal{I}$		240	4	—	12, 20

The containment relation between these seven groups is isomorphic to the partially ordered set of non-empty subsets of $\{r, r_1, r_5\}$.

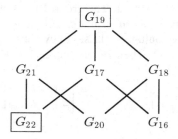

The group $G_{22} = \mathcal{C}_4 \circ \mathcal{I}$ is isomorphic to the group $SL^{\pm}(2, 5)$ of 2×2 matrices of determinant ± 1 over the field of 5 elements (Exercise 4 at the end of the chapter).

5. Cartan matrices and the ring of definition

Suppose that G is a reflection group of rank two generated by two non-commuting reflections r and s. If the order of r is p, then r has an eigenvector a with eigenvalue α, which is a primitive p^{th} root of unity. Similarly, if the order of s is q, then s has an eigenvector b with eigenvalue β, which is a qth root of unity. We may scale b so that $sa = a - b$ and then $rb = b - \theta a$ for some θ. The matrices of r, s and rs with respect to the basis a, b are of the form

$$\begin{bmatrix} \alpha & -\theta \\ 0 & 1 \end{bmatrix}, \quad \begin{bmatrix} 1 & 0 \\ -1 & \beta \end{bmatrix} \quad \text{and} \quad \begin{bmatrix} \alpha + \theta & -\theta\beta \\ -1 & \beta \end{bmatrix}$$

and the Cartan matrix of r and s with respect to a and b is

$$\begin{bmatrix} 1-\alpha & 1 \\ \theta & 1-\beta \end{bmatrix}.$$

It follows that $\alpha, \beta, \theta \in \mathbb{Q}(G)$ and since α, β and θ are algebraic integers we have the stronger result that $\alpha, \beta, \theta \in \mathbb{Z}(G)$. Therefore G can be represented by matrices over $\mathbb{Z}(G)$.

5.1. Cartan matrices for the Shephard groups. Of the 19 finite primitive reflection groups of rank 2, the 12 Shephard groups can be generated by two reflections. For the Shephard groups of type $\mathcal{T}$, the Cartan matrices are

$$C_4 = \begin{bmatrix} 1-\omega & 1 \\ -\omega & 1-\omega \end{bmatrix}, \quad C_5 = \begin{bmatrix} 1-\omega & 1 \\ -2\omega & 1-\omega \end{bmatrix}$$

$$\text{and} \quad C_6 = \begin{bmatrix} 2 & 1 \\ 1-\omega+i\omega^2 & 1-\omega \end{bmatrix},$$

where C_i is a Cartan matrix for G_i.

For the Shephard groups of type $\mathcal{O}$ we have

$$C_8 = \begin{bmatrix} 1-i & 1 \\ -i & 1-i \end{bmatrix}, \quad C_9 = \begin{bmatrix} 2 & 1 \\ (1+\sqrt{2})\zeta_8 & 1+i \end{bmatrix},$$

$$C_{10} = \begin{bmatrix} 1-\omega & 1 \\ -i-\omega & 1-i \end{bmatrix} \quad \text{and} \quad C_{14} = \begin{bmatrix} 1-\omega & 1 \\ 1-\omega+i\omega^2\sqrt{2} & 2 \end{bmatrix},$$

where $\zeta_8 := (1+i)/\sqrt{2}$ is a primitive 8th root of unity.

Let ζ_5 be a primitive 5th root of unity and put $\tau = \zeta_5 + \bar{\zeta}_5 + 1$. Then the Cartan matrices for the Shephard groups of type $\mathcal{I}$ are

$$C_{16} = \begin{bmatrix} 1-\zeta_5 & 1 \\ -\zeta_5 & 1-\zeta_5 \end{bmatrix}, \quad C_{17} = \begin{bmatrix} 2 & 1 \\ 1-\zeta_5-i\zeta_5^3 & 1-\zeta_5 \end{bmatrix},$$

$$C_{18} = \begin{bmatrix} 1-\omega & 1 \\ -\omega-\zeta_5 & 1-\zeta_5 \end{bmatrix}, \quad C_{20} = \begin{bmatrix} 1-\omega & 1 \\ \omega(\tau-2) & 1-\omega \end{bmatrix},$$

$$\text{and} \quad C_{21} = \begin{bmatrix} 2 & 1 \\ 1-\omega-i\omega^2\tau & 1-\omega \end{bmatrix}.$$

In every case the ring of definition of the reflection group is generated by the entries of the Cartan matrix.

5.2. Cartan matrices for the 3-generator groups. The remaining 7 primitive reflection groups of rank 2 cannot be generated by two reflections. However, in each case the group is generated by 3 reflections r, s and t with roots a, b and c such that the entries of the Cartan matrix C belong to $\mathbb{Z}(G)$. Moreover, there is a vector

$v := (\mu, \nu, 1)$ with coordinates in $\mathbb{Z}(G)$ such that $vC = 0$. Therefore, using the construction of reflections from the Cartan matrix described in Chapter 1 §6, it follows that G can be represented by matrices with entries in $\mathbb{Z}(G)$.

For the groups of type $\mathcal{T}$ only G_7 requires 3 generators and for its Cartan matrix we take

$$C_7 = \begin{bmatrix} 1 - \omega & 1 & 0 \\ 1 - \omega + i\omega^2 & 2 & 1 \\ 0 & 1 - \omega - i\omega^2 & 1 - \omega \end{bmatrix}.$$

The vector $v := (1 - \omega - \omega^2 i, -1 + \omega, 1)$ satisfies $vC_7 = 0$.

The 3-generator groups of type $\mathcal{O}$ are G_{11}, G_{12}, G_{13} and G_{15}. The Cartan matrices are:

$$C_{11} := \begin{bmatrix} 1 + i & 1 & 1 \\ 1 + i + \zeta_8 & 2 & \bar{\zeta}_8 - \omega\zeta_8 + 1 \\ i - \omega & -\omega^2\bar{\zeta}_8 - \omega - i\omega^2 & 1 - \omega \end{bmatrix},$$

where $\zeta_8 := (1 + i)/\sqrt{2}$ and $v := (\omega + \omega^2(i - \bar{\zeta}_8), \omega^2\bar{\zeta}_8, 1)$ is such that $vC_{11} = 0$;

$$C_{12} := \begin{bmatrix} 2 & -1 - i\sqrt{2} & 1 \\ -1 + i\sqrt{2} & 2 & i\sqrt{2} \\ 1 & -i\sqrt{2} & 2 \end{bmatrix}$$

where $v := (i\sqrt{2}, -1 + i\sqrt{2}, 1)$ satisfies $vC_{12} = 0$;

$$C_{13} := \begin{bmatrix} 2 & 1 & i \\ 2 + \sqrt{2} & 2 & \zeta_8^3(1 - i\sqrt{2}) \\ -(2 + \sqrt{2})i & -\zeta_8(1 + i\sqrt{2}) & 2 \end{bmatrix}$$

where $v := (\zeta_8(1 + \sqrt{2}), -1, 1)$ satisfies $vC_{13} = 0$.

The group $G_{11} := W(C_{11})$ is generated by reflections r, s and t of orders 4, 2 and 3, respectively. The group G_{15} is the subgroup $\langle r^2, s, t \rangle$ of G_{11}. Therefore, we obtain a Cartan matrix for G_{15} by multiplying the first column of C_{11} by $1 - i$. Thus the (row) null space of C_{15} coincides with the (row) null space of C_{11}:

$$C_{15} = \begin{bmatrix} 2 & 1 & 1 \\ 2 + (1 - i)\zeta_8 & 2 & \bar{\zeta}_8 - \omega\zeta_8 + 1 \\ (1 - i)(i - \omega) & -\omega^2\bar{\zeta}_8 - \omega - i\omega^2 & 1 - \omega \end{bmatrix}.$$

For the 3-generator groups of type $\mathcal{I}$ we have:

$$C_{19} := \begin{bmatrix} 1 - \zeta_5 & 1 & 1 \\ -\omega - \zeta_5 & 1 - \omega & \omega^2(i\zeta_5 - \zeta_5^4) + 1 \\ 1 - \zeta_5 - i\zeta_5^3 & i\omega^2\zeta_5^4 - i\zeta_5^3 + 1 & 2 \end{bmatrix}$$

where the vector $v := (-i\omega\zeta_5^4 + i\zeta_5^3 - 1, i\omega\zeta_5^4, 1)$ satisfies $vC_{19} = 0$;

$$C_{22} = \begin{bmatrix} 2 & \tau + i - 1 & -i - 1 \\ \tau - i - 1 & 2 & i \\ i - 1 & -i & 2 \end{bmatrix}$$

where the vector $v := (2 - \tau - i, \tau^{-1}(i - 1), 1)$ satisfies $vC_{22} = 0$.

Theorem 6.2. *If G is a finite primitive unitary reflection group of rank 2, then G can be represented by matrices with entries in its ring of definition $\mathbb{Z}(G)$. In each case G is the Weyl group of a $\mathbb{Z}(G)$-root system.*

Proof. Suppose first that G is a Shephard group. We may suppose that $G = W(C)$, where C is one of the Cartan matrices listed in section 5.1. Then G is generated by reflections r and s with roots a and b and every root of a reflection in G is in the orbit of a scalar multiple of a or b. Therefore, for the root system of G we take the union of the orbits of a and b. In each case the entries of the Cartan matrix belong to the ring $\mathbb{Z}(G)$.

Now suppose that G is defined by a 3×3 Cartan matrix of rank 2 and that the reflections r, s and t with roots a, b and c are obtained by the method of Construction 1.35. From §5.2, the Cartan matrices and the vector $v := (\mu, \nu, 1)$ such that $vC = 0$ have the property that their entries belong to $\mathbb{Z}(G)$. We have $c = -\mu a - \nu b$ and therefore the entries in r, s and t, with respect to the basis a and b, belong to $\mathbb{Z}(G)$. The $\mathbb{Z}(G)$-root system is the union of the G-orbits of a, b and c. $\square$

6. Invariants

In this section we determine the basic invariants of the reflection subgroups of $U_2(\mathbb{C})$ and use them to illustrate several of the theorems from the previous chapter.

Definition 6.3. Given a group G acting on a vector space V, a *semi-invariant* for G is a polynomial function $P \in S(V^*)$ such that for all $g \in G$ there exists $\chi(g) \in \mathbb{C}^\times$ such that $gP = \chi(g)P$.

It follows immediately that $\chi : G \to \mathbb{C}^\times$ is a homomorphism, or in other words, a linear character of G. When χ is the trivial character, i.e., when $\chi(g) = 1$ for all $g \in G$, the semi-invariant P is what up until now we have called simply an invariant of G.

Our strategy is to find semi-invariants for the binary tetrahedral, octahedral and icosahedral groups and then find suitable powers of these invariants that are invariants of the various reflection subgroups of $U_2(\mathbb{C})$. We note that if G_1 and G_2 are subgroups of $U_n(\mathbb{C})$ such that $Z \circ G_1 = Z \circ G_2$ for some group Z of scalar matrices, then G_1 and G_2 have the same (homogeneous) semi-invariants.

Lemma 6.4. *Suppose that Δ is an orbit of G on the 1-dimensional subspaces of $S(V^*)$ and that for each $L \in \Delta$ we choose a non-zero element $\varphi_L \in L$. Then $\prod_{L \in \Delta} \varphi_L$ is a semi-invariant of G.*

Proof. For $L \in \Delta$ and $g \in G$ we have $g\varphi_L = \lambda\varphi_M$, where $M := gL \in \Delta$. As L varies over Δ the same is true of gL and putting $P := \prod_{L \in \Delta} \varphi_L$ we see that $gP = \chi(g)P$ for some $\chi(g) \in \mathbb{C}$. That is, P is a semi-invariant of G. $\qquad\square$

In order to obtain semi-invariants of small degree in the applications of this lemma we generally take Δ to be an orbit of subspaces spanned by linear forms.

Once we have at least one semi-invariant there are some standard ways to create others.

Definition 6.5.

(i)   Given n functions $f_1(x_1, x_2, \ldots, x_n), \ldots, f_n(x_1, x_2, \ldots, x_n)$ of the n variables $x_1, x_2, \ldots, x_n$, their *Jacobian* is the determinant

$$\frac{\partial(f_1, f_2, \ldots, f_n)}{\partial(x_1, x_2, \ldots, x_n)} := \det\left(\frac{\partial f_i}{\partial x_j}\right).$$

(ii)   The *Hessian* of a function $f(x_1, x_2, \ldots, x_n)$ is the Jacobian of the first derivatives of f, namely

$$\det\left(\frac{\partial^2 f}{\partial x_i \partial x_j}\right).$$

Lemma 6.6. *If P is a polynomial function on V and if the matrix of $g \in G$ with respect to the basis $v_1, v_2, \ldots, v_n$ is (a_{ij}), then*

$$g\frac{\partial P}{\partial X_j} = \sum_{i=0}^{n} a_{ij}\frac{\partial(gP)}{\partial X_i}.$$

Proof. For $v \in V$ we have

$$\left(g\frac{\partial P}{\partial X_j}\right)(v) = \frac{\partial P}{\partial X_j}(g^{-1}v) = \lim_{h \to 0}\frac{P(g^{-1}v + hv_j) - P(g^{-1}v)}{h}$$

$$= \lim_{h \to 0}\frac{(gP)(v + hgv_j) - (gP)(v)}{h}$$

$$= \sum_{i=0}^{n} a_{ij}\frac{\partial(gP)}{\partial X_i}(v). \qquad\square$$

Corollary 6.7. *If $f_1(X_1, \ldots, X_n), \ldots, f_n(X_1, \ldots, X_n)$ are semi-invariants of a group G, then their Jacobian is also a semi-invariant. Similarly, the Hessian of a semi-invariant $f(X_1, X_2, \ldots, X_n)$ is again a semi-invariant of G.*

Proof. By assumption there are linear characters $\theta_1, \theta_2, \ldots, \theta_n$ of G such that $g f_i = \theta_i(g) f_i$ for all $g \in G$.

If the matrix of $g \in G$ with respect to the basis $v_1, v_2, \ldots, v_n$ is (a_{ij}) and if
$$\Delta := \frac{\partial(f_1, f_2, \ldots, f_n)}{\partial(X_1, X_2, \ldots, X_n)}, \text{ then}$$

$$g\Delta = \det\left(\sum_k a_{kj} \frac{\partial(gf_i)}{\partial X_k}\right)$$
$$= \det(g) \det\left(\theta_i(g) \frac{\partial f_i}{\partial X_k}\right)$$
$$= \det(g) \prod_{i=1}^{n} \theta_i(g) \, \Delta.$$

Now suppose that $gf = \theta(g)f$. Carrying out the previous calculation with $f_i = \frac{\partial f}{\partial X_i}$ we find that $g\Theta = (\det g)^2 \theta(g)^n \Theta$, where Θ is the Hessian of f. $\square$

For each of the groups $\mathcal{T}$, $\mathcal{O}$ and $\mathcal{I}$ we shall construct a semi-invariant f of minimal degree and then show that the Hessian h of f and the Jacobian t of f and h suffice to determine basic invariants for the associated reflection groups. It turns out that the semi-invariants f, h and t correspond to products of the linear forms associated with the orbits of lengths $|G|/k_3$, $|G|/k_2$ and $|G|/k_1$ whose existence was established in the course of proving Theorem 5.12.

In what follows, let V be the 2-dimensional space on which the group acts and let X_1 and X_2 be a basis for the dual space. Then $S := S(V^*)$ is the algebra of polynomials in X_1 and X_2.

The semi-invariants of $\mathcal{T}$. According to Theorem 5.14 the binary tetrahedral group $\mathcal{T}$ is generated by the matrices
$$a := \frac{1}{2}\begin{bmatrix} -1-i & 1-i \\ -1-i & -1+i \end{bmatrix} \quad \text{and} \quad b := \begin{bmatrix} -i & 0 \\ 0 & i \end{bmatrix}$$
of orders 3 and 4 respectively. The polynomial $X_1 + (-i+(1-i)\omega)X_2$ is an eigenvector for the action of a on the dual space and the 1-dimensional subspace spanned by this polynomial has four images under the action of $\mathcal{T}$. By Lemma 6.4 the product of representatives for these four subspaces is a semi-invariant of $\mathcal{T}$. A direct calculation shows that, up to a scalar factor, this semi-invariant is
$$f_T := X_1^4 + 2i\sqrt{3}\, X_1^2 X_2^2 + X_2^4.$$

Up to a scalar factor, the Hessian of f_T is
$$h_T := X_1^4 - 2i\sqrt{3}\, X_1^2 X_2^2 + X_2^4$$

and the Jacobian (up to a scalar factor) of f_T and h_T is

$$t_T := X_1^5 X_2 - X_1 X_2^5.$$

Furthermore, these three semi-invariants satisfy the relation

$$f^3 - h^3 = 12i\sqrt{3}\,t^2.$$

where, for convenience, we have dropped the subscript T.

It is clear that f, h and t are fixed by b and a simple calculation shows that $af = \omega f$ and $ah = \omega^2 f$, whence a fixes t. From this we can determine the action of the reflections r, r_1, and r_2 introduced in §1 and thus arrive at the following table of basic invariants for the groups of type T.

	Order	Invariants	Degrees
G_4	24	f, t	4, 6
G_5	72	f^3, t	12, 6
G_6	48	f, t^2	4, 12
G_7	144	f^3, t^2	12, 12

The semi-invariants of $\mathcal{O}$. The binary octahedral group $\mathcal{O}$ is generated by the matrices

$$a := \frac{1}{2}\begin{bmatrix} -1-i & 1-i \\ -1-i & -1+i \end{bmatrix} \quad \text{and} \quad c := \frac{1}{\sqrt{2}}\begin{bmatrix} 1-i & 0 \\ 0 & 1+i \end{bmatrix}$$

of orders 3 and 8 respectively. The polynomial X_1 is an eigenvector for c and the line it spans has six images under the action of $\mathcal{O}$. Representatives for these lines are X_1, X_2, $X_1 \pm X_2$ and $X_1 \pm iX_2$. Their product is the semi-invariant

$$f_O := X_1^5 X_2 - X_1 X_2^5,$$

which we have already seen in connection with the binary tetrahedral group.

Up to a scalar factor the Hessian of f_O is

$$h_O := X_1^8 + 14X_1^4 X_2^4 + X_2^8 = f_T h_T$$

and the Jacobian (up to a scalar factor) of f and h is

$$t_O := X_1^{12} - 33X_1^8 X_2^4 - 33X_1^4 X_2^8 + X_2^{12}.$$

Furthermore, these three semi-invariants satisfy the relation

$$h^3 - t^2 = 108f^4.$$

where, as before, we have omitted the subscripts.

By calculating the action of the reflections r, r_1, r_3 and r_4 on these semi-invariants we find the following table of basic invariants for the reflection groups of type $\mathcal{O}$.

	Order	Invariants	Degrees
G_8	96	h, t	8, 12
G_9	192	h, t^2	8, 24
G_{10}	288	h^3, t	24, 12
G_{11}	576	h^3, t^2	24, 24
G_{12}	48	f, h	6, 8
G_{13}	96	f^2, h	12, 8
G_{14}	144	f, t^2	6, 24
G_{15}	288	f^2, t^2	12, 24

The semi-invariants of $\mathcal{I}$. The binary icosahedral group $\mathcal{I}$ is generated by the matrices

$$b := \begin{bmatrix} -i & 0 \\ 0 & i \end{bmatrix} \quad \text{and} \quad e := \frac{1}{2} \begin{bmatrix} \tau^{-1} - \tau i & 1 \\ -1 & \tau^{-1} + \tau i \end{bmatrix}$$

of orders 4 and 5 respectively. If ζ is a suitable fifth root of unity, the polynomial $X_1 + (2\zeta - \tau^{-1} + \tau i)X_2$ is an eigenvector for e and the line it spans has 12 images under the action of $\mathcal{I}$. The product of representatives for these 12 subspaces is a semi-invariant of $\mathcal{I}$. Up to a scalar factor, this semi-invariant is

$$f_I := X_1^{12} + \frac{22}{\sqrt{5}} X_1^{10} X_2^2 - 33 X_1^8 X_2^4 - \frac{44}{\sqrt{5}} X_1^6 X_2^6 - 33 X_1^4 X_2^8 + \frac{22}{\sqrt{5}} X_1^2 X_2^{10} + X_2^{12}.$$

Up to a scalar factor, the Hessian of f_I is

$$\begin{aligned}
h_I := & X_1^{20} - \frac{38}{3}\sqrt{5} X_1^{18} X_2^2 - 19 X_1^{16} X_2^4 - 152\sqrt{5} X_1^{14} X_2^6 \\
& - 494 X_1^{12} X_2^8 + \frac{988}{3}\sqrt{5} X_1^{10} X_2^{10} - 494 X_1^8 X_2^{12} \\
& - 152\sqrt{5} X_1^6 X_2^{14} - 19 X_1^4 X_2^{16} - \frac{38}{3}\sqrt{5} X_1^2 X_2^{18} + X_2^{20}
\end{aligned}$$

and the Jacobian (up to a scalar factor) of f_I and h_I is

$$\begin{aligned}
t_I := & X_1^{29} X_2 - \frac{116}{9\sqrt{5}} X_1^{27} X_2^3 + \frac{1769}{25} X_1^{25} X_2^5 + \frac{464}{\sqrt{5}} X_1^{23} X_2^7 + \frac{2001}{5} X_1^{21} X_2^9 \\
& - \frac{2668}{3\sqrt{5}} X_1^{19} X_2^{11} + \frac{12673}{5} X_1^{17} X_2^{13} - \frac{12673}{5} X_1^{13} X_2^{17} + \frac{2668}{3\sqrt{5}} X_1^{11} X_2^{19} \\
& - \frac{2001}{5} X_1^9 X_2^{21} - \frac{464}{\sqrt{5}} X_1^7 X_2^{23} - \frac{1769}{25} X_1^5 X_2^{25} + \frac{116}{9\sqrt{5}} X_1^3 X_2^{27} - X_1 X_2^{29}.
\end{aligned}$$

Furthermore, dropping subscripts, these three semi-invariants satisfy the relation

$$f^5 - h^3 = 60\sqrt{5}\,t^2.$$

These polynomials are all absolute invariants because $\mathcal{I}$ is its own derived group. Thus it is quite easy to determine the actions of the reflections r, r_1 and r_5 of §1 and arrive at the following table of basic invariants for the reflection groups of type $\mathcal{I}$.

	Order	Invariants	Degrees
G_{16}	600	h, t	20, 30
G_{17}	1200	h, t^2	20, 60
G_{18}	1800	h^3, t	60, 30
G_{19}	3600	h^3, t^2	60, 60
G_{20}	360	f, t	12, 30
G_{21}	720	f, t^2	12, 60
G_{22}	240	f, h	12, 20

Exercises

1. Show that there is a homomorphism $\varphi : \mathbf{T} \to \mathrm{Alt}(4)$ such that $\varphi(r) = (1,2)(3,4)$, $\varphi(r_1) = (2,3,4)$ and $\varphi(r_2) = (2,4,3)$. Deduce that the groups G_4, G_5, G_6 and G_7 are generated by the reflections given in the first table of §1.

2. Verify directly that the orders of the reflection subgroups given in the tables of §1 are correct.

3. Verify directly that the given generators for the Shephard groups satisfy the relevant braid relation.

4. Prove that the group $\mathcal{C}_4 \circ \mathcal{I}$ is isomorphic to the group $SL^{\pm}(2,5)$ of 2×2 matrices of determinant ± 1 over the field of 5 elements.

5. Use the table of invariants for groups of type $\mathcal{T}$ to deduce that $\mathcal{T}$ is not contained in G_4 or G_6.

6. For the group $\mathcal{O}$, show that $cf_O = -f_O$ and $ct_O = -t_O$. Deduce that $\mathcal{O}$ is not contained in G_8, G_{10}, G_{12} or G_{14}.

7. Using a computer algebra system, or otherwise, check directly that the polynomials f, h and t (defined for each of the groups $\mathcal{T}$, $\mathcal{O}$ and $\mathcal{I}$) are semi-invariants.

8. Express f_I as a polynomial in t_T and t_O.

9. Show that, after a change of basis, $f_I = XY(X^{10} + 11X^5Y^5 + Y^{10})$.

Line systems

Definition 7.1. A *line system* is a collection of lines through the origin of $\mathbb{C}^n$.

It will be convenient, as in Coxeter [62], to define the angle between two lines $\langle u \rangle$ and $\langle v \rangle$ as the (acute) angle θ such that

$$\cos\theta = \frac{|(u,v)|}{\sqrt{(u,u)(v,v)}}.$$

We shall shortly see that if the set of angles between the lines in a system is finite, then the system is finite. Restrictions on the set of angles arise in our context from the fact that the reflections in hyperplanes orthogonal to the lines in a system are assumed to generate a finite group.

Given a unitary reflection group G acting on $\mathbb{C}^n$, the set $\mathfrak{L}(G)$ of lines spanned by the roots of the reflections of G (as defined in Chapter 1) is a line system. If G is a finite primitive reflection group and if the rank of G is at least five, then in Chapter 8 we shall prove that every reflection in G has order two and the angle between two distinct roots is 60° or 90°. In lower dimensions other angles occur but the possibilities are quite limited. This motivates the study of line systems with restrictions on the angles between lines.

The aim of this chapter is the classification of all 'star-closed' line systems in which the angles between lines are 45°, 60° or 90°. These correspond precisely to the unitary reflection groups in which all reflections have order two and in which the order of the product of any two reflections is at most four. This classification is achieved in Theorems 7.31 and 7.42. It is an essential step in the classification of all finite unitary reflection groups, which is carried out in the next chapter. In effect, this classification is an amalgamation and simplification of a theorem of Mitchell [166] dating from 1914 and results of Cohen [54] published in 1976.

1. Bounds on line systems

In this section we derive a bound on the size of a line system in $\mathbb{C}^n$ having a prescribed number of angles. The result is due to Delsarte, Goethals and Seidel [75] but the short proof presented here is due to Koornwinder [136].

99

Theorem 7.2. *Let* Σ *be a set of unit vectors that span distinct one-dimensional subspaces in* $\mathbb{C}^n$. *If* $A := \{ |(u, v)|^2 \mid u, v \in \Sigma, u \neq v \}$ *and* $s := |A|$, *then*

$$
|\Sigma| \leq \begin{cases} \dbinom{n + s - 1}{s} \dbinom{n + s - 2}{s - 1} & \text{if } 0 \in A \\[2ex] \dbinom{n + s - 1}{s}^2 & \text{if } 0 \notin A. \end{cases}
$$

Proof. For each $u \in \Sigma$, define a function $F_u : V \to \mathbb{C}$ by

$$
F_u(v) := (v, u)^\varepsilon \prod_{\alpha \in A \setminus \{0\}} \frac{|(v, u)|^2 - \alpha(v, v)}{1 - \alpha},
$$

where ε is 1 if $0 \in A$ and 0 otherwise.

If $e_1, e_2, \ldots, e_n$ is a basis for V and if $v := x_1 e_1 + \cdots + x_n e_n$, then F_u is a polynomial which is homogeneous of degree s in $x_1, \ldots, x_n$, and homogeneous of degree $s - \varepsilon$ in $\bar{x}_1, \ldots, \bar{x}_n$. Furthermore, for $u, v \in \Sigma$ we have $F_u(v) = \delta_{uv}$ and therefore the functions F_u ($u \in \Sigma$) are linearly independent. The complex dimension of the space of all polynomials that are homogeneous of degree s in the x_j and homogeneous of degree $s - \varepsilon$ in the $\bar{x}_j$ is

$$
\binom{n + s - 1}{s} \binom{n + s - \varepsilon - 1}{s - \varepsilon}.
$$

This is the required bound. $\square$

The reason for distinguishing the cases $0 \in A$ and $0 \notin A$ is simply to provide a sharper bound when $0 \in A$. In low dimensions there are line systems that meet these bounds. For example, the line system $\mathcal{D}_3^{(3)}$ described later in the chapter consists of 9 lines, each pair of which are at $60°$ thus meeting the bound for $n = 3$, $s = 1$. In §6 we shall meet the line system $\mathcal{K}_6$, which has 126 lines at $90°$ and $60°$ and which meets the bound for $n = 6$ and $s = 2$.

The same method of proof provides similar but tighter bounds on the number of lines through the origin of $\mathbb{R}^n$.

Remark 7.3. An important consequence of the above theorem is that if the set of angles between the lines in a system is finite, then the system itself is finite.

2. Star-closed Euclidean line systems

In a fundamental paper [45], Cameron, Goethals, Seidel and Shult consider line systems in $\mathbb{R}^n$ in which the angle between every pair of lines is $60°$ or $90°$. In this context a *star* is a set of three coplanar lines such that the angle between each pair is $60°$. Such a line system is *star-closed* if for each pair of lines at $60°$ the set contains the third line of the star; it is *indecomposable* if it cannot be written as the disjoint union of non-empty mutually orthogonal subsets. The indecomposable star-closed sets of lines in $\mathbb{R}^n$ are the lines of the root systems of types A_n, D_n, E_6, E_7, and E_8 (see Humphreys [**119**]). The classification of these Euclidean star-closed line systems was a major tool in the determination of the graphs whose adjacency matrix has least eigenvalue -2. Cvetković, Rowlinson and Simić give a complete account of this result in their book [**68**] and they determine the embeddings between Euclidean star-closed line systems. (See also [**43**, §3.10] and [**46**, Chap. 3].)

In what follows we extend the work on the classification and embedding of Euclidean line systems to line systems in $\mathbb{C}^n$. Some of this work has appeared in [**140**].

3. Reflections and star-closed line systems

Recall (from Definition 1.8 in Chapter 1) that $a \in \mathbb{C}^n$ is a long root of the line $\ell := \langle a \rangle$ if $(a, a) = 2$.

Lines ℓ and m with (long) roots a and b are said to be at $90°$ if $(a, b) = 0$, at $60°$ if $|(a, b)| = 1$, and at $45°$ if $|(a, b)| = \sqrt{2}$. This coincides with the usual notion in $\mathbb{R}^n$.

It is clear from Chapter 1 that, given a line ℓ, there is a unique reflection r_ℓ of order two which fixes $\ell^\perp$ pointwise. Its action on $v \in \mathbb{C}^n$ is given by

$$r_\ell(v) := v - (v, a)a,$$

where a is any long root of ℓ. If $\ell = \langle u \rangle$ for $u \in \mathbb{C}^n$, we also write $r_u := r_\ell$.

Definition 7.4. A line system $\mathfrak{L}$ is *star-closed* if $r_\ell(m) \in \mathfrak{L}$ for all lines ℓ, m of $\mathfrak{L}$. A *star* in $\mathbb{C}^n$ is a finite, coplanar star-closed line system. A star of k lines is called a *k-star*.

The next lemma shows that this definition of a star generalises the notion of the previous section.

Lemma 7.5. *Given lines ℓ and m in $\mathbb{C}^n$ at $60°$, there is a unique line n in the plane of ℓ and m at $60°$ to both ℓ and m. If a is a root of ℓ and b is a root of m, then $r_b(a) = -(a, b)r_a(b)$ is a root of n and hence $\{\ell, m, n\}$ is a 3-star.*

Proof. We have $r_a(b) = b - (b, a)a$ and thus $-(a, b)r_a(b) = -(a, b)b + a = r_b(a)$. Furthermore $(r_a(b), b) = 1$ and $(r_a(b), a) = -(b, a)$, whence $r_a(b)$ is at $60°$ to a and b.

Conversely, suppose that $c := \alpha a + \beta b$ is a root and that c is at $60°$ to both a and b. Scale b so that $(a, b) = -1$ and scale c so that $(a, c) = -1$. From the equations $(c, c) = 2$, $(a, c) = -1$ and $|(b, c)| = 1$, we find that $\alpha = \beta = -1$ and hence $c = -a - b = -r_a(b)$. $\square$

Definition 7.6. A line system $\mathfrak{L}$ is a *k-system* if

(i) $\mathfrak{L}$ is star-closed,

(ii)   for all $\ell, m \in \mathfrak{L}$ the order of $r_\ell r_m$ is at most k, and

(iii)   there exist $\ell, m \in \mathfrak{L}$ such that the order of $r_\ell r_m$ equals k.

We shall show that the classification of k-systems is equivalent to the classification of finite unitary reflection groups which are generated by reflections of order two. It will transpire that for a primitive group of rank at least three, this requires only k-systems for $k \leq 5$.

Lemma 7.7. *If a and b are long roots such that the order of $r_a r_b$ is a given integer m, then there are only finitely many values possible for $|(a, b)|$. In particular,*

(i) $|r_a r_b| = 2$ *if and only if* $(a, b) = 0$,

(ii) $|r_a r_b| = 3$ *if and only if* $|(a, b)| = 1$,

(iii) $|r_a r_b| = 4$ *if and only if* $|(a, b)| = \sqrt{2}$, *and*

(iv) $|r_a r_b| = 5$ *if and only if* $|(a, b)| = \tau$ *or* τ^{-1}, *where* $\tau = \frac{1}{2}(1 + \sqrt{5})$.

Proof. We may suppose that a and b are linearly independent. The reflections r_a and r_b fix the subspace $\langle a, b \rangle = \mathbb{C}a \oplus \mathbb{C}b$ and act as the identity on $\langle a, b \rangle^{\perp}$. The matrices of the restrictions of r_a and r_b to $\langle a, b \rangle$, with respect to the basis a, b, are

$$A := \begin{bmatrix} -1 & -(b, a) \\ 0 & 1 \end{bmatrix} \quad \text{and} \quad B := \begin{bmatrix} 1 & 0 \\ -(a, b) & -1 \end{bmatrix}.$$

Consequently

$$AB = \begin{bmatrix} |(a, b)|^2 - 1 & (b, a) \\ -(a, b) & -1 \end{bmatrix}$$

and the order m of $r_a r_b$ is also the order of AB.

The eigenvalues of AB are $\exp(2\pi i h/m)$ and $\exp(-2\pi i h/m)$ for some h coprime to m, where $1 \leq h < m$. On taking the trace we see that $|(a, b)|^2 - 2 = 2\cos(2\pi h/m)$ and so $|(a, b)| = 2|\cos(\pi h/m)|$. The result follows. $\square$

It is a simple consequence of the previous lemma and Theorem 7.2 that if $\mathfrak{L}$ is a k-system, then $\mathfrak{L}$ is finite.

Definition 7.8. A line system $\mathfrak{L}$ is *decomposable* if there is a partition of $\mathfrak{L}$ into a pair of non-empty subsets A and B such that every line of A is orthogonal to every line of B. In this case we write $\mathfrak{L} = A \perp B$. The line system is *indecomposable* if it is not decomposable. We write $\mathfrak{L} = k\,\mathfrak{M}$ to indicate that $\mathfrak{L}$ is the orthogonal sum of k copies of $\mathfrak{M}$.

It follows from the Lemma 7.7 that the lines of a 3-system are either orthogonal or at 60°. In the Euclidean case the indecomposable 3-systems correspond to so-called *simply laced* root systems.

Let $W(\mathfrak{L})$ denote the group generated by the reflections r_{ℓ}, where $\ell \in \mathfrak{L}$. Every line system $\mathfrak{L}$ is an orthogonal sum $\mathfrak{L} = \mathfrak{L}_1 \perp \mathfrak{L}_2 \perp \cdots \perp \mathfrak{L}_k$ of indecomposable line systems $\mathfrak{L}_j$. The $\mathfrak{L}_j$ are characterised as the maximal indecomposable subsystems of $\mathfrak{L}$, and $\mathfrak{L}$ is star-closed if and only if every $\mathfrak{L}_j$ is star-closed. These statements are the line system analogues of Theorem 1.27, and so $W(\mathfrak{L})$ is the direct product of the groups $W(\mathfrak{L}_j)$.

Theorem 7.9. *If $\mathfrak{L}$ is a k-system, then $W(\mathfrak{L})$ is a finite unitary reflection group. Conversely, if G is a finite unitary reflection group generated by reflections of order two and if k is the maximum of the orders of products of pairs of reflections of G, then $G = W(\mathfrak{L})$ for some k-system $\mathfrak{L}$.*

Proof. If $\mathfrak{L}$ is a k-system in $\mathbb{C}^n$, then by the remark above, $\mathfrak{L}$ is finite. The group $W(\mathfrak{L})$ acts on $\mathfrak{L}$ as a group of permutations and so to prove that $W(\mathfrak{L})$ is finite it suffices to show that the subgroup fixing every line of $\mathfrak{L}$ is finite. Furthermore, we may assume that $\mathfrak{L}$ is indecomposable. Then an element $g \in W(\mathfrak{L})$ fixes every line in $\mathfrak{L}$ if and only if g is a scalar matrix. If $r \in W(\mathfrak{L})$ is a reflection, then $\det(r) = -1$ and therefore $\det(g) = \pm 1$. It follows that $|g|$ divides $2n$ and hence $W(\mathfrak{L})$ is finite.

Conversely, if G is a finite reflection group generated by reflections of order two, the set $\mathfrak{L}$ of lines spanned by the roots of the reflections of G is a k-system, where k is the maximum of the orders of products of pairs of reflections. By construction $G = W(\mathfrak{L})$. $\qquad\square$

Theorem 7.10. *A 3-system $\mathfrak{L}$ is indecomposable if and only if $W(\mathfrak{L})$ is transitive on $\mathfrak{L}$.*

Proof. Suppose that ℓ and m are lines of $\mathfrak{L}$ with roots a and b, respectively. If $(a, b) \neq 0$, then the reflection whose root spans the third line of the 3-star containing a and b takes ℓ to m. If $\mathfrak{L}$ is indecomposable, then there are roots $a = a_0, a_1, \ldots, a_k = b$ such that $(a_j, a_{j+1}) \neq 0$. It follows that there is an element of $W(\mathfrak{L})$ that takes ℓ to m and hence $W(\mathfrak{L})$ is transitive on $\mathfrak{L}$.

Conversely, suppose that $\mathfrak{L} = A \perp B$, where A and B are star-closed and non-empty. If a is the root of an element of A, then r_a fixes A and B. It follows that $W(\mathfrak{L})$ is not transitive. $\qquad\square$

4. Extensions of line systems

Definition 7.11. Given a star-closed line system $\mathfrak{L}$, the *star-closure* X^* of a subset X of $\mathfrak{L}$ is the intersection of all star-closed line systems in $\mathfrak{L}$ that contain X. In the case of a 3-system it can be obtained from X by successively adjoining the third line of the 3-star of each pair of lines at $60°$ and so $W(X^*)$ is generated by the reflections r_a where a runs through the roots of X. The star-closure of a set of roots is defined to be the star-closure of the lines that they span.

Every line system in $\mathbb{R}^n$ in which the lines are at angles $60°$ or $90°$ is a subset of a 3-system ([**45**, Lemma 2.3]). However, in $\mathbb{C}^n$ this is no longer true. For example, consider the star with roots $a = (1, -1, 0)$, $b = (0, 1, -1)$ and $c = (-1, 0, 1)$. If $v = (1, 0, \omega)$, where ω is a primitive cube root of unity, then the lines $\langle a \rangle$, $\langle b \rangle$ and $\langle v \rangle$ are pairwise at $60°$ but they are not part of a 3-system because $\langle c \rangle$ and $\langle v \rangle$ are neither at $60°$ nor at $90°$.

Definition 7.12. Given star-closed line systems $\mathfrak{L} \subseteq \mathfrak{M}$ in $\mathbb{C}^n$, we say that $\mathfrak{M}$ is a *simple extension* of $\mathfrak{L}$ if $\mathfrak{M} \neq \mathfrak{L}$ and if $\mathfrak{M}$ is the star-closure of $\mathfrak{L} \cup \{\ell\}$ for some line $\ell \in \mathfrak{M}$; in [68, p. 72] this is called a *one-line* extension of $\mathfrak{L}$. We say that $\mathfrak{M}$ is a *minimal extension* of $\mathfrak{L}$ if for all star-closed line systems $\mathfrak{U}$ such that $\mathfrak{L} \subseteq \mathfrak{U} \subsetneq \mathfrak{M}$ we have $\mathfrak{L} = \mathfrak{U}$. (See §8 and the exercises for examples of simple extensions that are not minimal.)

The *dimension* dim $\mathfrak{L}$ of the line system $\mathfrak{L}$ is the dimension of the linear span $\langle \mathfrak{L} \rangle$ of $\mathfrak{L}$.

Two line systems $\mathfrak{L}_1$ and $\mathfrak{L}_2$ are said to be *equivalent* if there is a unitary transformation $\psi : \langle \mathfrak{L}_1 \rangle \to \langle \mathfrak{L}_2 \rangle$ such that $\psi(\mathfrak{L}_1) = \mathfrak{L}_2$. In this case we write $\mathfrak{L}_1 \simeq \mathfrak{L}_2$.

Theorem 7.13. *If $\mathfrak{L}$ and $\mathfrak{M}$ are indecomposable k-systems and $\mathfrak{L} \subset \mathfrak{M}$, then there is a sequence $\mathfrak{L} = \mathfrak{L}_0 \subset \mathfrak{L}_1 \subset \cdots \subset \mathfrak{L}_h = \mathfrak{M}$ of indecomposable star-closed line systems such that for $1 \leq j \leq h$, $\mathfrak{L}_j$ is a simple indecomposable extension of $\mathfrak{L}_{j-1}$.*

Proof. Since $\mathfrak{M}$ is indecomposable, there is a line $\ell \in \mathfrak{M}$ that is neither in $\mathfrak{L}$ nor orthogonal to $\mathfrak{L}$. Then the star-closure $\mathfrak{L}_1$ of $\mathfrak{L} \cup \{\ell\}$ is indecomposable. From Theorem 7.2 we know that $|\mathfrak{M}|$ is finite. Thus $|\mathfrak{M} \setminus \mathfrak{L}_1| < |\mathfrak{M} \setminus \mathfrak{L}|$ and the result follows by induction. $\qquad\qquad\square$

5. Line systems for imprimitive reflection groups

We consider the imprimitive reflection groups $G(m, p, n)$, where $n > 1$, introduced in Chapter 2. It is immediate that the reflections in $G(m, p, n)$ have order two if and only if $m = p$ or $m = 2p$. Furthermore, the reflections have order two and the order of the product of any two of them is at most three if and only if $m = p$ and $m \leq 3$.

Let $\boldsymbol{\mu}_m$ denote the group of all m^{th} roots of unity in $\mathbb{C}$. For $n \geq 2$, let $\mathcal{D}_n^{(m)}$ be the line system for the group $G(m, m, n)$. If $m \geq 3$, then $\mathcal{D}_n^{(m)}$ is an m-system. If e_1, $e_2, \ldots, e_n$ is an orthonormal basis for $\mathbb{C}^n$, we may take the roots of $\mathcal{D}_n^{(m)}$ to be the $m\binom{n}{2}$ vectors $e_h - \zeta e_j$, for $1 \leq h < j \leq n$, where $\zeta \in \boldsymbol{\mu}_m$. For example, the roots of $\mathcal{D}_n^{(2)}$ are the vectors $e_h \pm e_j$ for $1 \leq h < j \leq n$, and the roots of $\mathcal{D}_n^{(3)}$ are the vectors $e_h - e_j, e_h - \omega e_j$ and $e_h - \omega^2 e_j$ for $1 \leq h < j \leq n$, where ω is a primitive cube root of unity.

When $m = 1$ the group $G(1, 1, n)$ is the symmetric group $\text{Sym}(n)$ and its line system is denoted by $\mathcal{A}_{n-1}$. Its roots are the $\binom{n}{2}$ vectors $e_h - e_j$, where $h < j$. The group $G(1, 1, n)$ acts irreducibly on the hyperplane orthogonal to $e_1 + e_2 + \cdots + e_n$, and this action is primitive if and only if $n \geq 4$.

The line systems $\mathcal{D}_n^{(2)}$ ($n \geq 3$), $\mathcal{D}_n^{(3)}$ ($n \geq 2$) and $\mathcal{A}_n$ ($n \geq 1$) are indecomposable 3-systems; moreover $\mathcal{A}_{n-1}$ is a subset of $\mathcal{D}_n^{(2)}$ and $\mathcal{D}_n^{(3)}$. In low dimensions some of these line systems are equivalent: $\mathcal{A}_2 \simeq \mathcal{D}_2^{(3)}$, $\mathcal{A}_3 \simeq \mathcal{D}_3^{(2)}$ and $\mathcal{D}_2^{(2)} \simeq 2\mathcal{A}_1$.

The group $W(\mathcal{A}_{n-1}) = \mathrm{Sym}(n)$ acts transitively on the lines of $\mathcal{D}_n^{(2)}$ not in $\mathcal{A}_{n-1}$ and transitively on the lines of $\mathcal{D}_n^{(3)}$ not in $\mathcal{A}_{n-1}$. Thus $\mathcal{D}_n^{(2)}$ and $\mathcal{D}_n^{(3)}$ are minimal extensions of $\mathcal{A}_{n-1}$.

If the reflections in $G(m, p, n)$ have order two and if the product of any two reflections has order at most four, then $m = p$ or $m = 2p$ and $m \leq 4$. For given n only the groups $G(2, 1, n)$, $G(4, 4, n)$ and $G(4, 2, n)$ satisfy these requirements and contain a pair of reflections whose product has order 4. We extend the previous notation and let $\mathcal{B}_n^{(2p)}$ denote the line system of $G(2p, p, n)$; this line system is the union of $\mathcal{D}_n^{(2p)}$ with the set of coordinate axes. Convenient choices for the additional roots are the vectors $\sqrt{2}\, e_j$ $(1 \leq j \leq n)$.

The group $G(2, 1, n) = W(\mathcal{B}_n^{(2)}) = C_2 \wr \mathrm{Sym}(n)$ is the Coxeter group of type B_n. It acts on $\mathbb{C}^n$ by permuting the coordinates and changing their signs.

The group $G(2, 2, n) = W(\mathcal{D}_n^{(2)})$, the Coxeter group of type D_n, is a subgroup of index two in $G(2, 1, n)$. It acts on $\mathbb{C}^n$ by permuting the coordinates and effecting an even number of sign changes. The groups $G(2, 1, 2)$ and $G(4, 4, 2)$ are conjugate in $U_2(\mathbb{C})$ and consequently $\mathcal{B}_2^{(2)} \simeq \mathcal{D}_2^{(4)}$.

6. Line systems for primitive reflection groups

In addition to $\mathcal{A}_n$ and the line systems of imprimitive reflection groups defined in the previous section there are twelve other indecomposable 3-, 4- or 5-systems that we shall encounter in this chapter. Our notation for the line systems is based on that of Cohen [54] for his 'vector graphs'.

The 3-systems arise from the Coxeter groups of types E_6, E_7 and E_8 in Euclidean space, and from the reflection groups of (Cohen's) types K_5 and K_6 in complex space. The latter were discovered by Mitchell [166]. The 4-systems arise from the Coxeter group of type F_4 in Euclidean space and the reflection groups of (Cohen's) types $J_3(4)$, N_4 and EN_4 in complex space. The 5-systems arise from the Coxeter groups of types H_3 and H_4 in Euclidean space and the reflection group of type $J_3(5)$ in complex space.

To discuss the examples, let $e_1, e_2, \ldots, e_n$ be an orthonormal basis of $\mathbb{C}^n$. The vector $\lambda_1 e_1 + \lambda_2 e_2 + \cdots + \lambda_n e_n$ may also be referred to in terms of its coordinate vector $(\lambda_1, \lambda_2, \ldots, \lambda_n)$. With our established convention, in which matrices act on the left, the matrix of a linear transformation of $\mathbb{C}^n$ will then act by multiplication on the transpose of the coordinate vector.

6.1. The indecomposable 3-systems. If $\mathfrak{L}$ is a star-closed line system in $\mathbb{C}^n$ and ℓ is any line in $\mathbb{C}^n$, then the set of lines of $\mathfrak{L}$ orthogonal to ℓ is also a star-closed line system. The line systems described in this section will be obtained by applying this construction to the Euclidean line system $\mathcal{E}_8$ and the complex line system $\mathcal{K}_6$, described below.

Checking that $\mathcal{E}_8$ and $\mathcal{K}_6$ are star-closed is facilitated by the observation that in both cases we obtain them from a line system $\mathfrak{L}$ of an imprimitive reflection group by adjoining the images of a line $\ell \notin \mathfrak{L}$ under the action of $W(\mathfrak{L})$.

The Euclidean 3-systems are described in the book by Cvetković, Rowlinson and Simić [**68**] and descriptions of the complex 3-systems can be found in the *Atlas of Finite Groups* [**56**, pp. 26, 52]. There is a wealth of material on both Euclidean and complex line systems and the lattices they generate in the book by Conway and Sloane [**58**].

The line system $\mathcal{E}_8$. Let X be the set of vectors $\frac{1}{2}(e_1 \pm e_2 \pm \cdots \pm e_8)$, where the number of positive coefficients is even. The 56 vectors $e_h \pm e_j$ $(1 \leq h < j \leq 8)$ are the roots of $\mathcal{D}_8^{(2)}$. The line system $\mathcal{E}_8$ is the union of $\mathcal{D}_8^{(2)}$ and the set of 64 lines spanned by the elements of X; these sets are the orbits of $W(\mathcal{D}_8^{(2)})$ on $\mathcal{E}_8$.

If $a := e_7 - e_8$ and $b := e_7 + e_8$, the stabiliser of the line $\ell := \langle a \rangle$ is $H := \langle r_a, r_b \rangle \times W(\mathcal{D}_6^{(2)})$. The group H has two orbits on the lines spanned by the elements of X: the 32 lines orthogonal to ℓ and the 32 lines at $60°$ to ℓ. It is now an easy calculation to check that $\mathcal{E}_8$ is star-closed.

Since $\mathcal{D}_8^{(2)} \subset \mathcal{E}_8$, the line system $\mathcal{E}_8$ is indecomposable and therefore, from Theorem 7.10, $W(\mathcal{E}_8)$ is transitive on lines.

The matrix $\mathrm{diag}(1,1,1,1,1,1,1,-1)$ is orthogonal and fixes $\mathcal{D}_8^{(2)}$. It transforms $\mathcal{E}_8$ into the line system consisting of the lines of $\mathcal{D}_8^{(2)}$ together with the lines spanned by the 64 vectors $\frac{1}{2}(e_1 \pm e_2 \pm \cdots \pm e_8)$, where the number of positive coefficients is odd.

Another useful variant of $\mathcal{E}_8$ is obtained as follows. Let $\mathbf{j}_n$ denote the row vector in which every entry is 1 and let J_n denote the $n \times n$ matrix in which every entry is 1. The matrix

$$U = I - \tfrac{1}{6}J_9 + \tfrac{1}{2}\begin{bmatrix} 0_8 & \mathbf{j}_8{}^t \\ \mathbf{j}_8 & -1 \end{bmatrix}$$

is orthogonal and transforms the second variant of $\mathcal{E}_8$, considered as a line system in $\mathbb{C}^9$, into the line system $\overline{\mathcal{E}}_8$ consisting of the 36 lines of $\mathcal{A}_8$ together with the lines spanned by the 84 images of $\frac{1}{3}(2,2,2,-1,-1,-1,-1,-1,-1)$ under the action of $W(\mathcal{A}_8)$. This variant of $\mathcal{E}_8$ lies in the hyperplane of $\mathbb{C}^9$ orthogonal to $\mathbf{j}_9$.

The line system $\mathcal{E}_7$. Let $z = \frac{1}{2}(e_1 + e_2 + \cdots + e_8) = \frac{1}{2}\mathbf{j}_8$. The line system $\mathcal{E}_7$ is the subsystem of $\mathcal{E}_8$ orthogonal to z.

The group $W(\mathcal{E}_8)$ is transitive on lines and therefore for each line ℓ of $\mathcal{E}_8$ the 63 lines of $\mathcal{E}_8$ orthogonal to ℓ span the lines of a 3-system equivalent to $\mathcal{E}_7$. In particular, $\mathcal{A}_7 \subset \mathcal{E}_7$ and therefore $\mathcal{E}_7$ is indecomposable.

The line system $\mathcal{E}_6$. The stabiliser of z in $W(\mathcal{D}_8^{(2)})$ is $\mathrm{Sym}(8)$ and this group has two orbits on the lines of $\mathcal{E}_8$ at $60°$ to z: the 28 lines of $\mathcal{D}_8^{(2)}$ with roots $e_h + e_j$

$(1 \leq h < j \leq 8)$ and the 28 lines of X spanned by the roots $\frac{1}{2}(\pm e_1 \pm e_2 \cdots \pm e_8)$ with exactly two negative coordinates. These orbits are interchanged by the action of r_z. It follows that $W(\mathcal{E}_8)$ acts transitively on the 3-stars of $\mathcal{E}_8$.

The set of 36 lines of $\mathcal{E}_8$ orthogonal to the star with roots $e_1 - e_2$, $e_2 - e_3$ and $e_1 - e_3$ is the line system $\mathcal{E}_6$. We have $z \in \mathcal{E}_6$ and the 15 lines of $\mathcal{E}_6$ orthogonal to z form a subsystem equivalent to $\mathcal{A}_5$. Thus $\mathcal{A}_5 \perp \mathcal{A}_1 \subset \mathcal{E}_6$ and $\mathcal{E}_6$ is indecomposable.

The line system $\mathcal{K}_6$. Let ω denote the cube root of unity $\frac{1}{2}(-1 + i\sqrt{3})$ and set $\theta := \omega - \omega^2 = i\sqrt{3}$. Let X be the set of vectors $\theta^{-1}(1, \alpha_2, \ldots, \alpha_6)$, where α_h is a power of ω and $\prod_{h=2}^{6} \alpha_h = 1$. The line system $\mathcal{K}_6$ is the union of $\mathcal{D}_6^{(3)}$ with the set of 81 lines spanned by the elements of X. These 126 lines form a 3-system in $\mathbb{C}^6$ and since $\mathcal{D}_6^{(3)} \subset \mathcal{K}_6$, it is indecomposable. As an aid to the calculation checking that $\mathcal{K}_6$ is star-closed note that the stabiliser H of the line ℓ spanned by $a := e_5 - e_6$ is $\langle r_a \rangle \times \langle \omega I \rangle \times W(\mathcal{D}_4^{(3)})$ and that H has two orbits on the lines spanned by the elements of X: the 27 lines orthogonal to ℓ and the 54 lines at $60°$ to ℓ.

Further information about the line system can be found in an article [57] and book [58, section 4.9] by Conway and Sloane. Earlier investigations of the group $W(\mathcal{K}_6)$ can be found in the papers of Hamill [108], Hartley [109], Shephard [192], and Coxeter and Todd [218, 64].

Let $\overline{\mathcal{K}}_6$ denote the line system which is the union of the 30 lines of $\mathcal{D}_6^{(2)}$ with the 96 lines spanned by the images of $\frac{1}{2}(1, 1, 1, 1, 1, \theta)$ under the action of $W(\mathcal{D}_6^{(2)})$. The unitary matrix

$$U := \mathrm{diag}(1, 1, 1, 1, 1, \omega)(\frac{1}{6}\omega\theta J_6 + I)$$

defines an equivalence between $\overline{\mathcal{K}}_6$ and $\mathcal{K}_6$.

The line system $\mathcal{K}_5$. The subsystem of $\mathcal{K}_6$ orthogonal to $\theta^{-1}(1, 1, \ldots, 1)$ is the line system $\mathcal{K}_5$; it is the union of $\mathcal{A}_5$ and the 30 lines spanned by the images of $\theta^{-1}(1, \omega, \omega^2, 1, \omega, \omega^2)$ under the action of $W(\mathcal{A}_5)$. It follows that $\mathcal{K}_5$ is indecomposable.

The group $W(\mathcal{K}_6)$ is transitive on lines and thus for all $\ell \in \mathcal{K}_6$, the 45 lines of $\mathcal{K}_6$ orthogonal to ℓ form a line system equivalent to $\mathcal{K}_5$. In particular, since the lines of $\mathcal{D}_4^{(3)}$ are orthogonal to $e_1 - e_2$, we see that $\mathcal{K}_5$ contains line systems of type $\mathcal{D}_4^{(3)}$.

It will be useful later to know that $\mathcal{K}_5$ is equivalent to the line system $\overline{\mathcal{K}}_5$ in $\mathbb{C}^5$ that is the union of the 12 lines of $\mathcal{D}_4^{(2)}$ (with fifth coordinate zero), the 32 lines spanned by the images of $\frac{1}{2}(1, 1, 1, \theta, \sqrt{2})$ under the action of $W(\mathcal{D}_4^{(2)})$, and the line spanned by $(0, 0, 0, 0, \sqrt{2})$. The linear transformation $\psi : \mathbb{C}^5 \to \mathbb{C}^6$ such that $\psi(e_h) = e_h$ $(1 \leq h \leq 4)$ and $\psi(e_5) = \frac{1}{\sqrt{2}}(e_5 + e_6)$ preserves the hermitian inner product and defines an equivalence between $\overline{\mathcal{K}}_5$ and the $\mathcal{K}_5$-subsystem of $\overline{\mathcal{K}}_6$ orthogonal to $e_5 - e_6$.

Subsystems of $\mathcal{K}_6$. The vectors $e_h - e_j$, where $1 \le h < j \le 5$, are the roots of the line system $\mathcal{A}_4$ in both $\overline{\mathcal{K}}_6$ and $\mathcal{K}_6$. We see that the matrix U defined above sends the subsystem $\mathcal{A}_5$ of $\overline{\mathcal{K}}_6$ to the star-closure in $\mathcal{K}_6$ of $\mathcal{A}_4$ with the line $\langle e_5 - \omega e_6 \rangle$.

In $\mathcal{K}_6$, the line system $\mathcal{A}_4$ is contained in three line systems of type $\mathcal{A}_5$. They represent the orbits of $W(\mathcal{K}_6)$ on subsystems of type $\mathcal{A}_5$ and are obtained by taking the star-closures of $\mathcal{A}_4$ with $\langle e_5 - e_6 \rangle$, $\langle e_5 - \omega e_6 \rangle$ and $\langle e_5 - \omega^2 e_6 \rangle$, respectively.

In $\overline{\mathcal{K}}_6$, the 20 lines orthogonal to e_6 is a line system equivalent to $\mathcal{D}_5^{(2)}$. These lines together with the 16 lines spanned by the images of $\frac{1}{2}(1, 1, 1, 1, 1, \theta)$ under the action of $W(\mathcal{D}_5^{(2)})$ form a line system equivalent to $\mathcal{E}_6$. The equivalence is given by the transformation $\varphi : \mathbb{C}^6 \to \mathbb{C}^8$ where $\varphi(e_h) = e_h$ $(1 \le h \le 5)$ and $\varphi(e_6) = \theta^{-1}(e_6 + e_7 + e_8)$.

In $\mathcal{K}_6$ the $\mathcal{A}_5$ subsystem which is the star-closure of $\mathcal{A}_4$ and $\langle e_5 - e_6 \rangle$ is orthogonal to $\theta^{-1}(1, 1, \ldots, 1)$ and thus lies in $\mathcal{K}_5$. However, the other $\mathcal{A}_5$ subsystems extending $\mathcal{A}_4$ are interchanged by complex conjugation and furthermore there is no line of $\mathcal{K}_6$ orthogonal to either of them.

Thus the first $\mathcal{A}_5$ subsystem has simple extensions to $\mathcal{A}_5 \perp \mathcal{A}_1$, $\mathcal{E}_6$, $\mathcal{K}_5$, and $\mathcal{D}_6^{(3)}$ whereas the second and third subsystems have simple extensions to $\mathcal{A}_6$, $\mathcal{D}_6^{(2)}$, $\mathcal{D}_6^{(3)}$ and to $\mathcal{K}_6$ itself. The $\mathcal{A}_6$ extension can be seen directly as the star-closure of the lines $\langle e_1 - e_2 \rangle$, $\langle e_2 - e_3 \rangle$, $\langle e_3 - e_4 \rangle$, $\langle e_4 - e_5 \rangle$, $\langle e_5 - \omega e_6 \rangle$ and $\langle \theta^{-1}(e_1 + e_2 + \cdots + e_6) \rangle$.

6.2. The indecomposable 4-systems. Constructions for the indecomposable 4-systems follow a common pattern. In each case we begin with the line system $\mathfrak{L}$ of an imprimitive reflection group and a line ℓ not in $\mathfrak{L}$, then adjoin the images of ℓ under the action of $W(\mathfrak{L})$.

The line system $\mathcal{J}_3^{(4)}$. Let $\lambda := -\frac{1}{2}(1 + i\sqrt{7})$. Then $\lambda^2 + \lambda + 2 = 0$ and $|\lambda| = \sqrt{2}$. As roots of the line system $\mathfrak{L} = \mathcal{B}_3^{(2)}$ we may take $\frac{1}{2}(\lambda^2, \pm\lambda^2, 0)$, $\frac{1}{2}(\lambda^2, 0, \pm\lambda^2)$, $\frac{1}{2}(0, \lambda^2, \pm\lambda^2)$, $(\lambda, 0, 0)$, $(0, \lambda, 0)$ and $(0, 0, \lambda)$. The line system $\mathcal{J}_3^{(4)}$ is the union of the 9 lines of $\mathcal{B}_3^{(2)}$ with the set X of 12 lines spanned by the vectors $\frac{1}{2}(\pm\lambda, \pm\lambda, 2)$, $\frac{1}{2}(\pm\lambda, 2, \pm\lambda)$ and $\frac{1}{2}(2, \pm\lambda, \pm\lambda)$. The set X is an orbit of $G(2, 1, 3)$ and we have $|\mathcal{J}_3^{(4)}| = 21$. This choice of roots ensures that the inner products of pairs of roots belong to the ring $\mathbb{Z}[\lambda]$.

In §12 we prove that there is just one other extension of $\mathcal{B}_3^{(2)}$ in $\mathbb{C}^3$ equivalent to $\mathcal{J}_3^{(4)}$; namely its conjugate $\overline{\mathcal{J}}_3^{(4)}$, the union of $\mathcal{B}_3^{(2)}$ with the 12 lines spanned by the vectors $\frac{1}{2}(\pm\bar{\lambda}, \pm\bar{\lambda}, 2)$, $\frac{1}{2}(\pm\bar{\lambda}, 2, \pm\bar{\lambda})$ and $\frac{1}{2}(2, \pm\bar{\lambda}, \pm\bar{\lambda})$. An equivalence between

$\overline{\mathcal{J}}_3^{(4)}$ and $\mathcal{J}_3^{(4)}$ is given by the unitary transformation with matrix

$$\frac{1}{2} \begin{bmatrix} \lambda & 0 & \lambda \\ 0 & 2 & 0 \\ \lambda & 0 & -\lambda \end{bmatrix}.$$

The line system $\mathcal{F}_4$. In this case $\mathfrak{L} = \mathcal{B}_4^{(2)}$. As roots of $\mathfrak{L}$ we may take the 12 vectors $e_h \pm e_j$ $(1 \le h < j \le 4)$ and the 4 vectors $\sqrt{2}\, e_h$ $(1 \le h \le 4)$. The 8 lines $\langle \frac{1}{\sqrt{2}}(1, \pm 1, \pm 1, \pm 1) \rangle$ form a single orbit under the action of $G(2, 1, 4)$ and $\mathcal{F}_4$ is the union of these 8 lines with the 16 lines of $\mathcal{B}_4^{(2)}$.

The line system $\mathcal{N}_4$. In this case $\mathfrak{L} = \mathcal{D}_4^{(4)}$. The 24 lines of $\mathfrak{L}$ are spanned by the roots $e_j \pm e_k$ and $e_j \pm i e_k$ $(1 \le j < k \le 4)$. The line system $\mathcal{N}_4$ contains 40 lines: it is the union of $\mathcal{D}_4^{(4)}$ and the 16 images of $\langle \frac{1+i}{2}(1, 1, 1, 1) \rangle$ under the action of $G(4, 4, 4) = W(\mathcal{D}_4^{(4)})$.

The line system $\mathcal{N}_4$ is equivalent to the line system $\overline{\mathcal{N}}_4$, which is the union of the 16 lines of $\mathcal{B}_4^{(2)}$ and the 24 images of the line $\langle \frac{1+i}{2}(1, 1, i, i) \rangle$ under the action of $G(2, 1, 4) = W(\mathcal{B}_4^{(2)})$.

The set of 6 lines of $\mathcal{N}_4$ spanned by the roots

$$(1, i, 0, 0), \quad (1, 0, i, 0), \quad (1, 0, 0, i),$$
$$(0, 1, -1, 0), \quad (0, 1, 0, -1) \quad \text{and} \quad (0, 0, 1, -1)$$

is a line system of type $\mathcal{A}_3$. It has a unique extension to a line system $\mathfrak{M}$ of type $\mathcal{A}_4$; the additional lines are spanned by the roots

$$\tfrac{1+i}{2}(1, 1, 1, 1), \quad \tfrac{1+i}{2}(1, -1, -i, -i),$$
$$\tfrac{1+i}{2}(1, -i - 1, -i) \quad \text{and} \quad \tfrac{1+i}{2}(1, -i, -i, -1).$$

Complex conjugation fixes $\mathcal{N}_4$ but takes $\mathfrak{M}$ to $\overline{\mathfrak{M}}$. The line system $\overline{\mathfrak{M}}$ is not in the $W(\mathcal{N}_4)$ orbit of $\mathfrak{M}$.

The line system $\mathcal{O}_4$. In this case $\mathfrak{L} = \mathcal{B}_4^{(4)}$. The line system $\mathcal{B}_4^{(4)}$ contains 28 lines: it is the union of $\mathcal{D}_4^{(4)}$ with the coordinate axes (with roots $(1 + i)e_j$, $1 \le j \le 4$). The orbit X of $\langle \frac{1+i}{2}(1, 1, 1, 1) \rangle$ under the action of $G(4, 2, 4) = W(\mathcal{B}_4^{(4)})$ contains 32 lines. The line system $\mathcal{O}_4$ is the union of $\mathcal{B}_4^{(4)}$ and X; it contains 60 lines. The group $W(\mathcal{O}_4)$ is given the symbol $EW(N_4)$ in Cohen [**54**].

The matrix $\mathrm{diag}(1, 1, 1, i) \in G(4, 1, 4)$ fixes $\mathcal{B}_4^{(4)}$ and transforms $\mathcal{O}_4$ into the equivalent line system $\overline{\mathcal{O}}_4$ obtained as the union of $\mathcal{B}_4^{(4)}$ with the 32 images of $\langle \frac{1+i}{2}(1, 1, 1, i) \rangle$ under the action of $G(4, 2, 4)$.

The coordinates of the roots of $\mathcal{O}_4$ and $\mathcal{N}_4$ belong to the field $\mathbb{Q}[i]$. Since $|1+i| = \sqrt{2}$ we may scale the roots of $\mathcal{F}_4$ so that they too are defined over $\mathbb{Q}[i]$ and then it is evident that $\mathcal{F}_4$ and $\mathcal{N}_4$ are subsystems of $\mathcal{O}_4$.

Let Σ be the images of $(1 + i, 0, 0, 0)$, $(1, 1, 0, 0)$ and $\frac{1+i}{2}(1, 1, 1, 1)$ under the action of $G(4, 2, 4)$. Then $|\Sigma| = 240$ and the elements of Σ span the 60 lines of $\mathcal{O}_4$: the line spanned by a is also spanned by $\pm a$ and $\pm ia$. If we regard $\mathbb{C}^4$ as the real space $\mathbb{R}^8$, in the obvious way, then Σ is the root system of $\mathcal{E}_8$. For example, $(1 + i, 0, 0, 0) \in \mathbb{C}^4$ corresponds to $(1, 1, 0, 0, 0, 0, 0, 0) \in \mathbb{R}^8$.

We see immediately that each line $\ell \in \mathcal{O}_4$ corresponds to a pair ℓ_1, ℓ_2 of orthogonal lines of $\mathcal{E}_8$. Furthermore, the linear transformation of $\mathbb{R}^8$ induced by the reflection $r_\ell \in W(\mathcal{O}_4)$ is the product $r_{\ell_1} r_{\ell_2} \in W(\mathcal{E}_8)$. Thus we have an embedding $W(\mathcal{O}_4) \subset W(\mathcal{E}_8)$.

6.3. The indecomposable 5-systems. There are just three indecomposable 5-systems. They are associated with primitive unitary reflection groups acting on $\mathbb{C}^3$ and $\mathbb{C}^4$. These line systems are listed here for completeness but they will not appear again until the next chapter. Furthermore, contrary to the assumptions that hold throughout this chapter, their roots are *short*: that is, $(a, a) = 1$ for all roots a.

The line system $\mathcal{H}_3$. Let $\tau := \frac{1}{2}(1+\sqrt{5})$; then $\tau^2 = \tau+1$. The line system $\mathcal{H}_3$ is the set of 15 lines joining the mid-points of opposite edges of a regular dodecahedron. The vectors $(1, 0, 0)$, $\frac{1}{2}(1, \pm\tau, \pm\tau^{-1})$ and their cyclic shifts are roots of these lines.

The line system $\mathcal{H}_4$. The wreath product $A := C_2 \wr \mathrm{Alt}(4)$ acts on $\mathbb{C}^4$ by even permutations of the coordinates and arbitrary sign changes. The line system $\mathcal{H}_4$ is the collection of 60 lines spanned by the images of $(1, 0, 0, 0)$, $\frac{1}{2}(1, \tau, \tau^{-1}, 0)$ and $\frac{1}{2}(1, 1, 1, 1)$ under the action of A. The subsystem of lines orthogonal to any given line of $\mathcal{H}_4$ is equivalent to $\mathcal{H}_3$.

The line system $\mathcal{J}_3^{(5)}$. This is the union of $\mathcal{H}_3$ and the 30 lines spanned by the images of $\frac{1}{2}(\tau + \omega, \tau^{-1}\omega - 1, 0)$ under the action of $W(\mathcal{H}_3)$. One root for each line of $\mathcal{J}_3^{(5)}$ not in $\mathcal{H}_3$ can be obtained by applying cyclic shifts and sign changes to the vectors $\frac{1}{2}(\tau + \omega, \tau^{-1}\omega - 1, 0)$, $\frac{1}{2}(\tau\omega, 1, \tau^{-1}\omega^2)$ and $\frac{1}{2}(\omega^2, 1, \tau\omega + 1)$.

In addition to $\mathcal{H}_3$ there are several other line systems of dimension three contained in $\mathcal{J}_3^{(5)}$ that are evident from this description. The nine lines spanned by the roots that have at least one coordinate zero is a 3-system equivalent to $\mathcal{B}_3^{(2)}$. The star-closure of $(1, 0, 0)$, $(0, 1, 0)$ and $\frac{1}{2}(1, \omega, \tau^{-1} + \tau)$ is equivalent to $\mathcal{A}_3$ and the star-closure of $(0, 1, 0)$, $\frac{1}{2}(-\tau^{-1}, 1, \tau)$ and $\frac{1}{2}(\tau\omega, 1, \tau^{-1}\omega^2)$ has nine lines and is equivalent to $\mathcal{D}_3^{(3)}$. The centre of $W(\mathcal{D}_3^{(3)}) = G(3, 3, 3)$ is generated by the scalar matrix ωI and the centre of $W(\mathcal{B}_3^{(2)}) = G(2, 1, 3)$ is generated by $-I$. Thus the element $-\omega I$, of order 6, belongs to the centre of $W(\mathcal{J}_3^{(5)})$.

In the next chapter we shall prove that there is just one other simple extension of $\mathcal{H}_3$ in $\mathbb{C}^3$; namely $\overline{\mathcal{J}}_3^{(5)}$, the star-closure of $\frac{1}{2}(\tau + \omega^2, \tau^{-1}\omega^2 - 1, 0)$ and $\mathcal{H}_3$. Its roots are the complex conjugates of the roots of $\mathcal{J}_3^{(5)}$. The unitary transformation $\mathbb{C}^3 \to \mathbb{C}^3$ with matrix

$$\frac{1}{2}\begin{bmatrix} \tau^{-1}\omega + \tau & 0 & \tau^{-1} - \omega \\ 0 & 2 & 0 \\ \tau^{-1} - \omega & 0 & \tau\omega + 1 \end{bmatrix}$$

defines an equivalence between $\mathcal{J}_3^{(5)}$ and $\overline{\mathcal{J}}_3^{(5)}$.

7. The Goethals–Seidel decomposition for 3-systems

Throughout this section suppose that $\mathcal{L}$ is an indecomposable 3-system in $\mathbb{C}^n$ and that Σ is a set of long roots (cf. Definition 7.1), one for each line of $\mathcal{L}$. We refine the choice of Σ as we proceed.

Choose long roots a and b corresponding to a pair of lines at $60°$ and scale b so that $(a, b) = -1$. Then $c = -a - b$ is a root that spans the third line of the 3-star of a and b and we may suppose that $a, b, c \in \Sigma$. Extending the notation of [**45**], the Goethals–Seidel decomposition of Σ has components

$$\begin{aligned} \Gamma_a &= \{\, x \in \Sigma \mid (a, x) = 0 \text{ and } (b, x) \neq 0 \,\}, \\ \Gamma_b &= \{\, x \in \Sigma \mid (b, x) = 0 \text{ and } (c, x) \neq 0 \,\}, \\ \Gamma_c &= \{\, x \in \Sigma \mid (c, x) = 0 \text{ and } (a, x) \neq 0 \,\}, \\ \Delta &= \{\, x \in \Sigma \mid (a, x) = 0 \text{ and } (b, x) = 0 \,\}, \quad \text{and} \\ \Lambda &= \{\, x \in \Sigma \setminus \{a, b, c\} \mid (a, x)(b, x)(c, x) \neq 0 \,\}. \end{aligned}$$

(7.14)

It is easy to see that Σ is the disjoint union of the sets $\{a, b, c\}$, Γ_a, Γ_b, Γ_c, Δ and Λ. Some of these sets may be empty; indeed, we shall see that $\mathcal{L}$ is a Euclidean 3-system if and only if $\Lambda = \emptyset$. Also, in $\mathcal{D}_n^{(3)}$ every line is in a unique 3-star such that $\Gamma_a = \Gamma_b = \Gamma_c = \emptyset$.

Scale the roots $x \in \Gamma_a$ so that $(b, x) = 1$, and hence $(c, x) = -1$. Similarly, scale $x \in \Gamma_b$ so that $(c, x) = 1$ and scale $x \in \Gamma_c$ so that $(a, x) = 1$.

Lemma 7.15. *We have*

$$\Gamma_b = r_c(\Gamma_a) = \{\, x + c \mid x \in \Gamma_a \,\} \quad and \quad \Gamma_c = r_b(\Gamma_a) = \{\, x - b \mid x \in \Gamma_a \,\}.$$

Proof. If $x \in \Gamma_a$, then $(x, c) = -1$ and thus $r_c(x) = x + c$. Furthermore, $(b, x + c) = 0$ and $(c, x + c) = 1$, hence $x + c \in \Gamma_b$. Similarly $r_b(x) = x - b \in \Gamma_c$. Conversely, for $y \in \Gamma_b$ we have $r_c(y) = y - c \in \Gamma_a$ and for $z \in \Gamma_c$ we have $r_b(z) = z + b$. Thus $\Gamma_b = r_c(\Gamma_a)$ and $\Gamma_c = r_b(\Gamma_a)$. $\qquad\square$

The following lemma shows that cube roots of unity enter into our calculations in an essential way.

Lemma 7.16. *Suppose that σ is a 3-star and that $\ell \notin \sigma$ is a line at 60° to a line of σ with root a. Let v be the root of ℓ such that $(a, v) = 1$ and let V be the subspace spanned by σ and ℓ.*

(i)   *If ℓ is orthogonal to a line of σ with root b, where $(a, b) = -1$, then there is an orthonormal basis of V with respect to which a, b and v have coordinates $(1, -1, 0)$, $(0, 1, -1)$ and $(0, -1, -1)$. In this case the star-closure of σ and ℓ is $\mathcal{D}_3^{(2)}$.*

(ii)   *If ℓ is at 60° to all three lines $\langle a \rangle$, $\langle b \rangle$ and $\langle c \rangle$ of σ, where $a + b + c = 0$ and $(a, b) = -1$, then $(b, v) \in \{\omega, \omega^2\}$ and there is an orthonormal basis of V with respect to which a, b and v have coordinates $(1, -1, 0)$, $(0, 1, -1)$ and $(0, -1, (b, v))$. In this case the star-closure of σ and ℓ is $\mathcal{D}_3^{(3)}$.*

Proof. First choose a vector $e \in V$ orthogonal to a and b and such that $(e, e) = 3$. Since $(a, b) = -1$ the vectors $\frac{1}{3}(e + 2a + b)$, $\frac{1}{3}(e - a + b)$ and $\frac{1}{3}(e - a - 2b)$ form an orthonormal basis of V with respect to which a and b have coordinates $(1, -1, 0)$ and $(0, 1, -1)$. If v has coordinates (α, β, γ) with respect to this basis, then $\beta = \alpha - 1$.

If $(b, v) = 0$, then $\gamma = \beta = \alpha - 1$. On the other hand, if ℓ is at 60° to b and c, and if $\theta = (b, v)$, then $|\theta| = |1 + \theta| = 1$ and hence θ is ω or ω^2. In both cases it follows that v has coordinates $(\alpha, \alpha - 1, \alpha + \theta)$, where $\theta = -1$ if $(b, v) = 0$. Since $(v, v) = 2$ we have $3|\alpha|^2 + (\bar{\theta} - 1)\alpha + (\theta - 1)\bar{\alpha} = 0$. The matrix $I - \bar{\alpha}(1 - \bar{\theta})^{-1}J$ is unitary, fixes a and b and takes v to $(0, -1, \theta)$. Thus the star-closure of σ and ℓ exists: in case (i) it is $\mathcal{D}_3^{(2)}$ whereas in case (ii) it is $\mathcal{D}_3^{(3)}$.   □

Corollary 7.17 (Cameron et al. [45]). *If $\mathfrak{L}$ is a line system in $\mathbb{R}^k$ in which the lines are at 60° or 90°, then $\mathfrak{L}$ is a subset of a 3-system and every line of $\mathfrak{L}$ not in a given 3-star σ of $\mathfrak{L}$ is orthogonal to at least one line of σ.*

Proof. Case (ii) of the lemma cannot occur. Therefore, if m and n are lines of $\mathfrak{L}$ at 60° and if p is the third line of the 3-star of m and n in $\mathbb{R}^k$, then any line $\ell \neq p$ of $\mathfrak{L}$ is at 60° or 90° to p. Thus $\mathfrak{L} \cup \{p\}$ is again a line system and we obtain a 3-system by successively adjoining the third line of the 3-star of each pair of lines at 60°.   □

Scale the roots $v \in \Lambda$ so that $(a, v) = 1$. Then, by Lemma 7.16, $(b, v) = \omega$ and $(c, v) = \omega^2$ or else $(b, v) = \omega^2$ and $(c, v) = \omega$. In both cases the star-closure of the lines spanned by a, b and v is equivalent to $\mathcal{D}_3^{(3)}$.

The results obtained so far allow us to determine all 2- and 3-systems up to dimension three.

Theorem 7.18. *Let $\mathfrak{L}$ be a 2- or 3-system of dimension at most three.*

(i)   *If $\dim \mathfrak{L} = 1$, then $\mathfrak{L} \simeq \mathcal{A}_1$ and $\mathfrak{L}$ consists of a single line.*

(ii)   *If $\dim \mathfrak{L} = 2$, then $\mathfrak{L}$ is equivalent to $\mathcal{D}_2^{(2)} \simeq 2\mathcal{A}_1$ or to a 3-star $\mathcal{D}_3^{(3)} \simeq \mathcal{A}_2$.*

(iii)   *If $\dim \mathfrak{L} = 3$, then $\mathfrak{L}$ is equivalent to $3\mathcal{A}_1$, $\mathcal{A}_1 \perp \mathcal{A}_2$, $\mathcal{A}_3 \simeq \mathcal{D}_3^{(2)}$ or $\mathcal{D}_3^{(3)}$.*

Proof. (*i*) and (*ii*). The result is clear when dim $\mathcal{L} = 1$. In $\mathbb{C}^2$, there is a unique 3-star containing a pair of lines at $60°$ and a unique line orthogonal to a given line: thus $\mathcal{L}$ is either a pair of orthogonal lines or a 3-star.

(*iii*) If $\mathcal{L}$ is decomposable, then $\mathcal{L} = \mathcal{A}_1 \perp \mathfrak{M}$, where $\mathfrak{M}$ is a star-closed line system of dimension two. In this case $\mathcal{L}$ is equivalent to $3\mathcal{A}_1$ or $\mathcal{A}_1 \perp \mathcal{A}_2$.

If $\mathcal{L}$ is indecomposable and a simple extension of $\mathcal{A}_2$, then by Lemma 7.16 $\mathcal{L}$ is equivalent to $\mathcal{D}_3^{(2)}$ or $\mathcal{D}_3^{(3)}$. The line system $\mathcal{A}_3$ is a simple extension of $\mathcal{A}_2$ and therefore it is equivalent to $\mathcal{D}_3^{(2)}$.

It follows from Theorem 7.13 that to complete the proof it is enough to show that neither $\mathcal{D}_3^{(2)}$ nor $\mathcal{D}_3^{(3)}$ have simple extensions in $\mathbb{C}^3$. By way of contradiction, suppose that $\mathcal{L}$ contains a line system $\mathfrak{M}$ and a root $v := (\alpha, \beta, \gamma)$ such that $\langle v \rangle \notin \mathfrak{M}$, where $\mathfrak{M}$ is either $\mathcal{D}_3^{(2)}$ or $\mathcal{D}_3^{(3)}$. If v is orthogonal to $a := (1, -1, 0)$ and $b := (0, 1, -1)$ then $\alpha = \beta = \gamma$ and on taking the inner product of v with $(1, 1, 0)$ or $(1, -\omega, 0)$ we reach a contradiction. Thus we may suppose that $(a, v) = 1$ and hence $\beta = \alpha - 1$. If $(b, v) = 0$, then $\gamma = \alpha - 1$. If $\theta := (b, v) \neq 0$, then $|\theta| = |1 + \theta| = 1$ and hence θ is ω or ω^2. Thus $v = (\alpha, \alpha - 1, \alpha + \theta)$, where θ is -1, ω or ω^2. As in Lemma 7.16 we have $3|\alpha|^2 + (\bar{\theta} - 1)\alpha + (\theta - 1)\bar{\alpha} = 0$.

If $\mathfrak{M} = \mathcal{D}_3^{(2)}$, then $d := (1, 1, 0)$ is a root of $\mathfrak{M}$, which is not orthogonal to v. Therefore $|(d, v)| = 1$ and hence $2|\alpha|^2 - \alpha - \bar{\alpha} = 0$. Combined with the previous equation this implies that α is 0 or 1. But then $\mathcal{L}$ contains $\mathcal{D}_3^{(3)}$ and this is a contradiction since the lines $\langle (1, 1, 0) \rangle$ and $\langle (1, -\omega, 0) \rangle$ are neither at $60°$ nor $90°$.

If $\mathfrak{M} = \mathcal{D}_3^{(3)}$, then $(1, -\omega, 0)$ and $(1, -\omega^2, 0)$ are roots of $\mathfrak{M}$. If both these roots are at $60°$ to v, then

$$3|\alpha|^2 + (\omega - 1)\alpha + (\omega^2 - 1)\bar{\alpha} = 0 \quad \text{and}$$
$$3|\alpha|^2 + (\omega^2 - 1)\alpha + (\omega - 1)\bar{\alpha} = 0.$$

On adding these equations we find that $2|\alpha|^2 - \alpha - \bar{\alpha} = 0$. As before $\alpha = 0$ or $\alpha = 1$ and again we arrive at a contradiction. The remaining possibility is that v is orthogonal to exactly one of $(1, -\omega, 0)$ or $(1, -\omega^2, 0)$. But in neither case is there a solution for α. $\qquad\qquad\square$

Lemma 7.19. *In the notation of (7.14), suppose that $x, y \in \Gamma_a$, where $x \neq y$, and set $\mu = (x, y)$. Then $\mu \in \{0, 1, -\omega, -\omega^2\}$ and the vectors a, b, x and y are linearly independent.*

(*i*)   *If $\mu = 0$, then $z = b - c - x - y \in \Gamma_a$ and $(x, z) = (y, z) = 0$. The star-closure of a, b, x and y is equivalent to $\mathcal{D}_4^{(2)}$. Furthermore, if $w \in \Gamma_a$, then*

$$(w, x) + (w, y) + (w, z) = 2.$$

(*ii*)   *If $\mu = 1$, then a multiple of $x - y$ belongs to Δ. The star-closure of a, b, x and y is equivalent to $\mathcal{A}_4$.*

(iii)   *If $\mu = -\omega$ or $\mu = -\omega^2$, then $\mu x + \bar{\mu}y \in \Gamma_a$ and $a + b - \mu x - y \in \Lambda$. The star-closure of a, b, x and y is equivalent to $\mathcal{D}_4^{(3)}$.*

Proof. It follows from Lemma 7.16 applied to $-y$ and the 3-star of c and x that $\mu \in \{0, 1, -\omega, -\omega^2\}$. The Gram matrix of the vectors a, b, x and y is

$$\begin{bmatrix} 2 & -1 & 0 & 0 \\ -1 & 2 & 1 & 1 \\ 0 & 1 & 2 & \mu \\ 0 & 1 & \bar{\mu} & 2 \end{bmatrix}$$

and its determinant is $4 + 2(\mu + \bar{\mu}) - 3|\mu|^2$, which is never 0. Therefore a, b, x and y are linearly independent. In particular, the star-closure $\mathfrak{M}$ of a, b, x and y is a line system of dimension four.

The star-closure of the roots $r = (1, -1, 0, 0, 0)$, $s = (0, 1, -1, 0, 0)$ and $t = (0, 0, 1, -1, 0)$ is $\mathcal{A}_3$, represented as a line system in $\mathbb{C}^5$. The linear map φ such that $\varphi(r) = a$, $\varphi(s) = b$ and $\varphi(t) = -x$ preserves inner products and extends to an isometry from $\mathcal{A}_3$ to the star-closure of a, b and x. For each possibility for μ we define a vector $u \in \mathbb{C}^5$ such that φ extends to an isometry between the star-closure of r, s, t and u and $\mathfrak{M}$.

(i)   If $\mu = 0$, put $u = (1, 1, 0, 0, 0)$ and define $\varphi(u) = y$. Then the star-closure of r, s, t and u is $\mathcal{D}_4^{(2)}$ and isometric to $\mathfrak{M}$. From Lemma 7.15 we have $x + c$, $y - b \in \Sigma$ and since $(x + c, y - b) = -1$ it follows that $x + c + y - b$ is a root. Putting $z = b - c - x - y$ we find that $z \in \Gamma_a$ and $(x, z) = (y, z) = 0$. Thus, $x + y + z = b - c$ and so $(w, x) + (w, y) + (w, z) = 2$ for all $w \in \Gamma_a$.

(ii)   If $\mu = 1$, put $u = (0, 0, 1, 0, -1)$ and define $\varphi(u) = -y$. The star-closure of r, s, t and u is $\mathcal{A}_4$ and isometric to $\mathfrak{M}$. Furthermore $\varphi(0, 0, 0, 1, -1) = x - y$ and so $x - y$ is a root of $\mathfrak{L}$. Since $(a, x - y) = (b, x - y) = 0$, a multiple of $x - y$ belongs to Δ.

(iii) If μ is $-\omega$ or $-\omega^2$, put $u = (0, 0, 1, \mu, 0)$ and define $\varphi(u) = -y$. Then the star-closure of r, s, t and u is $\mathcal{D}_4^{(3)}$ and isometric to $\mathfrak{M}$. Then $x - \mu y$ is a root of $\mathfrak{L}$ spanning the third line of the 3-star of x and y. Its multiple $z = \mu x + \bar{\mu}y$ satisfies $(a, z) = 0$ and $(b, z) = 1$ and thus $z \in \Gamma_a$. Since $y + c \in \Gamma_b$ and $(x, y + c) = -\bar{\mu}$ we see that $x + \bar{\mu}(y + c)$ is a root of $\mathfrak{L}$. If $v = -\mu x - y - c$, then $(a, v) = 1$, $(b, v) = -\bar{\mu}$ and $(c, v) = -\mu$; hence $v \in \Lambda$.   □

Let $\theta = \omega - \omega^2$. In the line system $\mathcal{K}_6$ the 12 lines orthogonal to the roots $\theta^{-1}(1, 1, 1, 1, 1, 1)$ and $(1, -1, 0, 0, 0, 0)$ are spanned by the six vectors $e_h - e_j$ where $3 \le h < j \le 6$ and the six vectors $\theta^{-1}(1, 1, \omega^c, \omega^d, \omega^e, \omega^f)$, where c, d, e $f \in \{1, 2\}$ and $c + d + e + f \equiv 0 \pmod 3$. It follows from the previous lemma that this subsystem is equivalent to $\mathcal{D}_4^{(2)}$.

8. Extensions of $\mathcal{D}_n^{(2)}$ and $\mathcal{D}_n^{(3)}$

In preparation for the theorems in this and later sections we characterise certain simple extensions of $\mathcal{D}_n^{(k)}$, where $k \leq 5$ and $n \geq 2$. To this end, let $V = \mathbb{C}^{n+1}$ and let $e_1, e_2, \ldots, e_{n+1}$ be an orthonormal basis of V. For the roots of $\mathcal{D}_n^{(k)}$ we take the vectors $e_h - \zeta e_j$, where ζ is a kth root of unity and $1 \leq h < j \leq n$.

If $\mathcal{L}$ is a simple extension of $\mathcal{D}_n^{(k)}$, we may suppose that $\mathcal{L}$ is the star-closure in V of $\mathcal{D}_n^{(k)}$ and a line ℓ with long root x. We write $x = (\alpha_1, \alpha_2, \ldots, \alpha_{n+1})$ using coordinates with respect to the basis $e_1, e_2, \ldots, e_{n+1}$.

Lemma 7.20. *Suppose that $2 \leq k \leq m \leq 5$ and that $\mathcal{L}$ is an indecomposable m-system and the star-closure in V of $\mathcal{D}_n^{(k)}$ and a line $\ell \notin \mathcal{D}_n^{(k)}$ with root $x = (\alpha_1, \alpha_2, \ldots, \alpha_{n+1})$ as above. If $\alpha_s = 0$ for some $s \leq n$, then either $|\alpha_h| = \sqrt{2}$ for some $h \leq n$ and the extension is equivalent to*

$$\mathcal{D}_n^{(2)} \subset \mathcal{B}_n^{(2)} \quad or \quad \mathcal{D}_n^{(4)} \subset \mathcal{B}_n^{(4)}$$

or $|\alpha_h| = 1$ for two values of $h \leq n$ and the extension is equivalent to

$$\mathcal{D}_n^{(k)} \subset \mathcal{D}_{n+1}^{(k)} \quad or \quad \mathcal{D}_n^{(2)} \subset \mathcal{D}_n^{(4)}.$$

Proof. The group $G(k, 1, n)$ acts on $\mathbb{C}^{n+1}$ by permuting the first n coordinates and multiplying their values by k^{th} roots of unity thus fixing the line system $\mathcal{D}_n^{(k)}$. In particular, we may suppose that $\alpha_1 = 0$ and since $\mathcal{L}$ is indecomposable we have $\alpha_j \neq 0$ for some $j \leq n$.

Set $a := e_1 - e_j$ and $b := e_1 - \zeta e_j$, where ζ is a kth root of unity. Then $(a, x) = -\bar{\alpha}_j$ and $(x, x) = 2$, hence $|\alpha_j| \in \{1, \sqrt{2}, \tau^{-1}\}$. Furthermore, $(b, r_x(a)) = (b, a) + \alpha_j(b, x) = 1 + \zeta(1 - |\alpha_j|^2)$ and we have $|(b, r_x(a))| \in \{0, 1, \sqrt{2}, \tau, \tau^{-1}, 2\}$. It follows that $|\alpha_j| = 1$ or $|\alpha_j| = \sqrt{2}$. If $|\alpha_j| = \sqrt{2}$, scale x so that $x = \sqrt{2}\, e_j$. In this case $\mathcal{L}$ is a 4-system in $\mathbb{C}^n$ and either $k = 2$ and $\mathcal{L} = \mathcal{B}_n^{(2)}$ or $k = 4$ and $\mathcal{L} = \mathcal{B}_n^{(4)}$.

From now on we may suppose that $\alpha_h = 0$ or $|\alpha_h| = 1$ for $h \leq n$. Since $(x, x) = 2$ it follows that $\alpha_h \neq 0$ for exactly two values of $h \leq n + 1$. If these values are h and j with $h < j$, we scale x so that $\alpha_h = 1$.

If $j = n + 1$, the $(n + 1) \times (n + 1)$ diagonal matrix $\text{diag}(1, 1, \ldots, 1, -\alpha_{n+1}^{-1})$ is unitary. It fixes every root of $\mathcal{D}_n^{(k)}$ and sends x to $e_h - e_{n+1}$; thus $\mathcal{L} \simeq \mathcal{D}_{n+1}^{(k)}$. Thus from now on we may suppose that $j \leq n$ and $\mathcal{L}$ is contained in $\mathbb{C}^n$. We shall show that in this case $k = 2$, $m = 4$ and $\mathcal{L} = \mathcal{D}_n^{(4)}$.

If $\alpha_j = 1$, then $k \neq 2$ or 4 because, by assumption, x is not a root of $\mathcal{D}_n^{(k)}$. If $k = 3$ or 5 and $c := e_h - \zeta e_j$, then $(c, x) = 1 - \zeta$, which is a contradiction. Thus $\alpha_j \neq 1$ and if $d := e_h - e_j$, then $|(d, x)| = |1 - \alpha_j| \in \{1, \sqrt{2}, \tau, \tau^{-1}\}$.

If $|1 - \alpha_j| = 1$, then α_j is $-\omega$ or $-\omega^2$ and hence $\ell \in \mathcal{D}_n^{(3)}$, contrary to the choice of ℓ.

If $|1 - \alpha_j| = \sqrt{2}$, then $\alpha_j = \pm i$ and in this case we have $k = 2$, $m = 4$ and $\mathfrak{L} = \mathcal{D}_n^{(4)}$, as required.

If $|1 - \alpha_j| = \tau$ or τ^{-1}, then consideration of the inner products of x with $e_h + e_j$ and $e_h - \omega e_j$ shows that $k = m = 5$. If $n_i := |(e_h - \zeta^i e_j, x)|^2 = |1 - \zeta^i \bar{\alpha}_j|^2$, then $n_i \in \{0, 1, 2, \tau^2, \tau^{-2}\}$ and $\sum_{i=0}^4 n_i = 10$, which is a contradiction, since n_0 is τ^2 or τ^{-2}. $\square$

Theorem 7.21. *If $\mathfrak{L}$ is an indecomposable 3-system and a simple extension of $\mathcal{D}_n^{(2)}$ for $n \geq 3$, the extension is equivalent to one of*

$$\mathcal{D}_n^{(2)} \subset \mathcal{D}_{n+1}^{(2)}, \quad \mathcal{D}_3^{(2)} \subset \mathcal{A}_4, \quad \mathcal{D}_3^{(2)} \subset \mathcal{D}_4^{(3)}, \quad \mathcal{D}_4^{(2)} \subset \mathcal{K}_5, \quad \mathcal{D}_5^{(2)} \subset \mathcal{E}_6,$$

$$\mathcal{D}_5^{(2)} \subset \mathcal{K}_6, \quad \mathcal{D}_6^{(2)} \subset \mathcal{K}_6, \quad \mathcal{D}_6^{(2)} \subset \mathcal{E}_7, \quad \mathcal{D}_7^{(2)} \subset \mathcal{E}_8 \quad \text{or} \quad \mathcal{D}_8^{(2)} \subset \mathcal{E}_8.$$

These extensions are minimal except for

$$\mathcal{D}_4^{(2)} \subset \mathcal{D}_4^{(2)} \perp \mathcal{A}_1 \subset \mathcal{K}_5, \quad \mathcal{D}_5^{(2)} \subset \mathcal{D}_6^{(2)} \subset \mathcal{K}_6, \quad \mathcal{D}_5^{(2)} \subset \mathcal{E}_6 \subset \mathcal{K}_6,$$

$$\mathcal{D}_6^{(2)} \subset \mathcal{D}_6^{(2)} \perp \mathcal{A}_1 \subset \mathcal{E}_7 \quad \text{and} \quad \mathcal{D}_7^{(2)} \subset \mathcal{D}_8^{(2)} \subset \mathcal{E}_8.$$

Proof. Using the notation introduced above we may suppose that $\mathfrak{L}$ is a line system in $\mathbb{C}^{n+1}$ and that $\mathfrak{L}$ is the star-closure of $\mathcal{D}_n^{(2)}$ and a line ℓ with root $x = (\alpha_1, \alpha_2, \ldots, \alpha_{n+1})$. From Lemma 7.20 we may suppose $\alpha_h \neq 0$ for all $h \leq n$.

Suppose that for some $h, j, k \leq n$, the six quantities $\pm \alpha_h$, $\pm \alpha_j$ and $\pm \alpha_k$ are distinct. On taking inner products with the roots $e_h \pm e_j$ we find that $|\alpha_h - \alpha_j| = |\alpha_h + \alpha_j| = 1$. From this and the corresponding calculations with $e_h \pm e_k$ and $e_j \pm e_k$ it follows that

$$|\alpha_h|^2 \pm (\bar{\alpha}_h \alpha_j + \alpha_h \bar{\alpha}_j) + |\alpha_j|^2 = 1,$$

$$|\alpha_h|^2 \pm (\bar{\alpha}_h \alpha_k + \alpha_h \bar{\alpha}_k) + |\alpha_k|^2 = 1, \quad \text{and}$$

$$|\alpha_j|^2 \pm (\bar{\alpha}_j \alpha_k + \alpha_j \bar{\alpha}_k) + |\alpha_k|^2 = 1.$$

Thus $|\alpha_h|^2 = |\alpha_j|^2 = |\alpha_k|^2 = \frac{1}{2}$. Putting $\theta = \alpha_h \bar{\alpha}_j$ we find that $\theta + \bar{\theta} = 0$ and $\theta \bar{\theta} = \frac{1}{4}$, whence $\theta = \pm \frac{1}{2} i$. Thus $\alpha_h = \pm i \alpha_j$ and similarly $\alpha_h = \pm i \alpha_k$. But then $\alpha_j = \pm \alpha_k$, which is a contradiction.

We have shown that there exist α and β with $\alpha \neq \pm \beta$ such that for all $h \leq n$, $\alpha_h \in \{\alpha, -\alpha, \beta, -\beta\}$. On replacing x by an image under $G(2, 1, n)$ we may suppose that for all $h \leq n$, α_h is either α or β. If there are k values of h such that $\alpha_k = \alpha$, then without loss of generality $k \geq n - k$. Since $n \geq 3$ we have $k \geq 2$ and therefore $|\alpha| = \frac{1}{2}$.

Suppose first that $k \neq n$. Then $|\alpha + \beta| = |\alpha - \beta| = 1$ and consequently $|\alpha|^2 \pm (\bar{\alpha}\beta + \alpha\bar{\beta}) + |\beta|^2 = 1$. It follows that $|\alpha|^2 + |\beta|^2 = 1$, $\bar{\alpha}\beta + \alpha\bar{\beta} = 0$ and therefore $|\beta|^2 = \frac{3}{4}$. If $n - k \geq 2$, then $|\beta| = \frac{1}{2}$, which is a contradiction. Thus $n = k + 1$ and $\beta = \pm i\sqrt{3}\,\alpha$.

Scale x so that $\alpha = \frac{1}{2}$ and $\beta = \frac{1}{2}i\sqrt{3}$. Thus $x = \frac{1}{2}(1,1,\ldots,1,i\sqrt{3},2\alpha_{n+1})$ and it follows that $2 = (x,x) = \frac{1}{4}k + \frac{3}{4} + |\alpha_{n+1}|^2$. That is, $k + 4|\alpha_{n+1}|^2 = 5$ and hence $k \leq 5$. We consider each possibility for k in turn, noting that multiplying the last coordinate of x by a root of unity does not change the equivalence class of $\mathcal{L}$.

The case $k = 2$, $n = 3$. In this case we have $|\alpha_{n+1}| = \frac{1}{2}\sqrt{3}$ and we take $x = -\frac{1}{2}\omega^2(1,1,i\sqrt{3},i\sqrt{3})$. If $y = (1,1,0,0)$, then $(x,y) = -\omega^2$ and so from Lemma 7.19 with $a = (1,-1,0,0)$ and $b = (0,1,-1,0)$ it follows that $\mathcal{L}$ is equivalent to $\mathcal{D}_4^{(3)}$. The group $W(\mathcal{D}_3^{(2)})$ is transitive on the 12 roots of $\mathcal{D}_4^{(3)}$ not in $\mathcal{D}_3^{(2)}$ and therefore the extension is minimal.

The case $k = 3$, $n = 4$. We have $|\alpha_{n+1}| = \frac{1}{\sqrt{2}}$ and so without loss of generality $x = \frac{1}{2}(1,1,1,i\sqrt{3},\sqrt{2})$. It follows that $\mathcal{L}$ is the line system $\overline{\mathcal{K}}_5 \simeq \mathcal{K}_5$ described in §6. In particular, $(0,0,0,0,\sqrt{2})$ is a root of $\mathcal{L}$ and so $\mathcal{D}_4^{(2)} \perp \mathcal{A}_1 \subset \mathcal{L}$ whence the extension of $\mathcal{D}_4^{(2)}$ to $\mathcal{K}_5$ is simple but not minimal.

The case $k = 4$, $n = 5$. In this case $|\alpha_{n+1}| = \frac{1}{2}$ and therefore we take $x = \frac{1}{2}(1,1,1,1,i\sqrt{3},1)$. The vectors x and $\frac{1}{2}(-1,-1,-1,1,-i\sqrt{3},1)$ are roots of $\mathcal{L}$ and the third line of their 3-star is spanned by $(0,0,0,1,0,1)$. The star-closure of this vector with $\mathcal{D}_5^{(2)}$ is $\mathcal{D}_6^{(2)}$. It follows that $\mathcal{L}$ is the line system $\overline{\mathcal{K}}_6$ defined in §6 and thus $\mathcal{L}$ is equivalent to $\mathcal{K}_6$. We have $\mathcal{D}_5^{(2)} \subset \mathcal{D}_6^{(2)} \subset \overline{\mathcal{K}}_6$ and so the extension is simple but not minimal.

As shown in §6, $\mathcal{D}_5^{(2)} \subset \mathcal{E}_6 \subset \overline{\mathcal{K}}_6$. The group $W(\mathcal{D}_5^{(2)}) = G(2,2,5)$ has three orbits on the roots of $\overline{\mathcal{K}}_6$ not in $\mathcal{D}_5^{(2)}$. Their lengths are 10, 16 and 80 their star-closures with $\mathcal{D}_5^{(2)}$ are equivalent to the line systems $\mathcal{D}_6^{(2)}$, $\mathcal{E}_6$ and $\mathcal{K}_6$.

The case $k = 5$, $n = 6$. In this case $\alpha_{n+1} = 0$ and therefore we regard $\mathcal{L}$ as a line system in $\mathbb{C}^6$. That is, we take $x = \frac{1}{2}(1,1,1,1,1,i\sqrt{3})$. As in the previous case $\mathcal{L}$ is the line system $\overline{\mathcal{K}}_6$, which is equivalent to $\mathcal{K}_6$.

The case $k = n$. In this case $\frac{1}{4}n + |\alpha_{n+1}|^2 = 2$ and so $n \leq 8$. Therefore we may take $x = \frac{1}{2}(1,1,\ldots,1,\sqrt{8-n})$ and consequently the coordinate values are real. This case is covered by [**68**, Theorem 3.3.3] but for completeness we give the details here.

The orbit of x under $W(\mathcal{D}_n^{(2)})$ is the set of 2^{n-1} roots obtained by applying an even number of sign changes to the first n coordinates of x. Let X denote the set of lines spanned by these vectors.

If $n = 3$, it follows from Lemma 7.19 that $\mathcal{L}$ is equivalent to $\mathcal{A}_4$. If $n = 4$, the union of $\mathcal{D}_4^{(2)}$ and X is a line system equivalent to $\mathcal{D}_5^{(2)}$.

For $3 \leq n < 8$, there is a unitary transformation $\varphi : \mathbb{C}^{n+1} \to \mathbb{C}^8$ such that $\varphi(e_h) = e_h$ for $1 \leq h \leq n$ and $\varphi(e_{n+1}) = \frac{1}{\sqrt{8-n}}(e_{n+1} + \cdots + e_8)$. In particular, $\varphi(x)$ is the root $z = \frac{1}{2}(e_1 + e_2 + \cdots + e_8)$ of §6 and $\varphi(S)$ is a subsystem of $\mathcal{E}_8$.

If $n = 5$, the image of φ is orthogonal to $e_6 - e_7$ and $e_7 - e_8$. Thus $\mathfrak{L}$ is the union of $\mathcal{D}_5^{(2)}$ and the 16 lines of X and hence equivalent to $\mathcal{E}_6$.

If $n = 6$, the image of φ is orthogonal to $e_7 - e_8$ and so the union of $\mathcal{D}_6^{(2)}$ and X provides 62 of the 63 lines of $\mathcal{E}_7$. The remaining line is orthogonal to $\mathcal{D}_6^{(2)}$ and spanned by $(0, 0, 0, 0, 0, 0, \sqrt{2})$. Thus $\mathfrak{L}$ is equivalent to $\mathcal{E}_7$.

If $n = 7$, the map φ is the identity and the root $e_7 + e_8$ is in the star-closure of the orbit of x. Thus $\mathfrak{L}$ contains the 56 lines of $\mathcal{D}_8^{(2)}$ and the 64 lines of X, i.e. $\mathfrak{L} = \mathcal{E}_8$.

The case $n = 8$ reduces to the previous one and we have $\mathfrak{L} \simeq \mathcal{E}_8$.

In particular, we obtain the extensions

$$\mathcal{D}_3^{(2)} \subset \mathcal{A}_4, \qquad \mathcal{D}_4^{(2)} \subset \mathcal{D}_5^{(2)}, \qquad \mathcal{D}_5^{(2)} \subset \mathcal{E}_6,$$

$$\mathcal{D}_6^{(2)} \subset \mathcal{E}_7, \qquad \mathcal{D}_7^{(2)} \subset \mathcal{E}_8 \quad \text{and} \quad \mathcal{D}_8^{(2)} \subset \mathcal{E}_8.$$

These extensions are minimal except for $\mathcal{D}_6^{(2)} \subset \mathcal{E}_7$ and $\mathcal{D}_7^{(2)} \subset \mathcal{E}_8$, where we have $\mathcal{D}_6^{(2)} \subset \mathcal{D}_6^{(2)} \perp \mathcal{A}_1 \subset \mathcal{E}_7$ and $\mathcal{D}_7^{(2)} \subset \mathcal{D}_8^{(2)} \subset \mathcal{E}_8$.   □

Theorem 7.22. *If $\mathfrak{L}$ is an indecomposable 3-system and a simple extension of $\mathcal{D}_n^{(3)}$, where $n \geq 2$, the extension is equivalent to one of*

$$\mathcal{D}_n^{(3)} \subset \mathcal{D}_{n+1}^{(3)}, \quad \mathcal{D}_2^{(3)} \subset \mathcal{D}_3^{(2)}, \quad \mathcal{D}_4^{(3)} \subset \mathcal{K}_5, \quad \mathcal{D}_5^{(3)} \subset \mathcal{K}_6 \quad or \quad \mathcal{D}_6^{(3)} \subset \mathcal{K}_6.$$

The extensions are minimal except for $\mathcal{D}_5^{(3)} \subset \mathcal{D}_6^{(3)} \subset \mathcal{K}_6$.

Proof. We may suppose that $\mathfrak{L}$ is the star-closure in $\mathbb{C}^{n+1}$ of $\mathcal{D}_n^{(3)}$ and a line ℓ with root $x = (\alpha_1, \alpha_2, \ldots, \alpha_{n+1})$. Furthermore, by Lemma 7.20, we may suppose $\alpha_h \neq 0$ for all $h \leq n$.

If, for $h, j \leq n$, none of $\alpha_h - \alpha_j$, $\alpha_h - \omega \alpha_j$, $\alpha_h - \omega^2 \alpha_j$ are 0, then

$$|\alpha_h|^2 - \alpha_h \bar{\alpha}_j - \bar{\alpha}_h \alpha_j + |\alpha_j|^2 = 1,$$
$$|\alpha_h|^2 - \omega^2 \alpha_h \bar{\alpha}_j - \omega \bar{\alpha}_h \alpha_j + |\alpha_j|^2 = 1, \text{ and}$$
$$|\alpha_h|^2 - \omega \alpha_h \bar{\alpha}_j - \omega^2 \bar{\alpha}_h \alpha_j + |\alpha_j|^2 = 1.$$

Therefore $|\alpha_h|^2 + |\alpha_j|^2 = 1$ and hence $\alpha_h \bar{\alpha}_j + \bar{\alpha}_h \alpha_j = \omega^2 \alpha_h \bar{\alpha}_j + \omega \bar{\alpha}_h \alpha_j = 0$. Consequently $\alpha_h \bar{\alpha}_j = 0$, contrary to assumption. Thus for $1 \leq j \leq n$ we find that α_j is equal to α_1, $\omega \alpha_1$ or $\omega^2 \alpha_1$. We can now transform x so that $\alpha_1 = \alpha_2 = \cdots = \alpha_n$. Taking the inner product with $e_1 - \omega e_2$ we find that $|\alpha_1 - \omega^2 \alpha_1| = 1$ and hence $|\alpha_j| = \frac{1}{\sqrt{3}}$ for all $j \leq n$.

We have $(x, x) = \frac{1}{3}n + |\alpha_{n+1}|^2 = 2$ and therefore $n \le 6$. In the following, let $\theta = i\sqrt{3} = \omega - \omega^2$.

The case $n = 2$. It follows from Lemma 7.16 with $a = (\omega, -\omega^2, 0)$, $b = (1, -1, 0)$ and $v = x = -\theta^{-1}(1, 1, 2)$ that $\mathfrak{L}$ is equivalent to $\mathcal{D}_3^{(2)}$.

The case $n = 3$. We have $|\alpha_{n+1}| = 1$ and we may scale x so that $x = -\theta^{-1}(1, 1, 1, \theta)$. The group $G(3, 3, 3)$ acts on $\mathfrak{L}$ and so $y := -\theta^{-1}(1, 1, 1, \omega\theta)$ is a root of $\mathfrak{L}$ such that $(x, y) = -\omega$. Thus from Lemma 7.19 with $a = (1, -1, 0, 0)$ and $b = (\omega, -\omega^2, 0, 0)$ it follows that $\mathfrak{L}$ is equivalent to $\mathcal{D}_4^{(3)}$.

The case $n = 4$. We have $|\alpha_{n+1}|^2 = \frac{2}{3}$ and therefore we may suppose that $x = \theta^{-1}(1, 1, 1, 1, \sqrt{2})$. The linear transformation $\varphi : \mathbb{C}^5 \to \mathbb{C}^6$ given by $e_h\varphi = e_h$ ($1 \le h \le 4$) and $e_5\varphi = \frac{1}{\sqrt{2}}(e_5 + e_6)$ defines an equivalence between $\mathfrak{L}$ and the $\mathcal{K}_5$ subsystem of $\mathcal{K}_6$ orthogonal to $e_5 - e_6$ (see §6).

The case $n = 5$. In this case $|\alpha_{n+1}|^2 = \frac{1}{3}$ and therefore we may suppose that $x = \theta^{-1}(1, 1, 1, 1, 1, 1)$. The group $G(3, 3, 5)$ acts on $\mathfrak{L}$ and therefore the element $y := \theta^{-1}(\omega, \omega, \omega, \omega, \omega^2, 1)$ is a root of $\mathfrak{L}$. The third line of the 3-star of x and y is spanned by $z := (0, 0, 0, 0, 1, -\omega^2)$. The star-closure of $\mathcal{D}_5^{(3)}$ and z is $\mathcal{D}_6^{(3)}$. Thus $\mathfrak{L}$ is the line system $\mathcal{K}_6$ defined in §6 and the extension is simple but not minimal.

The case $n = 6$. In this case $\alpha_{n+1} = 0$. Therefore we may suppose that $\mathfrak{L}$ is a line system in $\mathbb{C}^6$ and that $x = \theta^{-1}(1, 1, 1, 1, 1, 1)$. As in the previous case we have $\mathfrak{L} = \mathcal{K}_6$.

All extensions except $\mathcal{D}_5^{(3)} \subset \mathcal{D}_6^{(3)} \subset \mathcal{K}_6$ are minimal. □

If $a := e_1 - e_2$, $b := e_2 - e_3$ and $d := e_1 + e_2$, then in the Goethals–Seidel decomposition of $\mathcal{D}_n^{(2)}$ determined by a and b we have $d \in \Gamma_a$ and $(x, d) = 0$ for all $x \in \Gamma_a \setminus \{d\}$. We leave it as an exercise to show that this property characterises $\mathcal{D}_n^{(2)}$.

9. Further structure of line systems in $\mathbb{C}^n$

Throughout this section suppose that $\mathfrak{L}$ is an indecomposable 3-system in $\mathbb{C}^n$ and that Σ is a set of roots for $\mathfrak{L}$. As in §7 choose roots a and b corresponding to a pair of lines at $60°$ and scale b so that $(a, b) = -1$. Let $c = -a - b$ and define Γ_a, Γ_b, Γ_c, Δ and Λ as before.

Lemma 7.23. *If $\Gamma_a \ne \emptyset$, then $\{ \langle x - y \rangle \mid x, y \in \Gamma_a$ and $(x, y) = 1 \}$ is the set of lines spanned by the elements of Δ.*

Proof. If $x, y \in \Gamma_a$ and $(x, y) = 1$, then by Lemma 7.19 (*ii*) $x - y$ is a root of $\mathfrak{L}$ and a multiple of $x - y$ belongs to Δ.

To prove the converse we extend the argument of [**45**]. Suppose that $z \in \Delta$ and that there exists $x \in \Gamma_a$ such that $(z, x) \ne 0$. We may scale z so that $(z, x) = -1$.

Then $x + z$ is a root of $\mathfrak{L}$ and $z + x \in \Gamma_a$ because $(a, x + z) = 0$ and $(b, x + z) = 1$. Furthermore $(x + z, x) = 1$ and $z = (x + z) - x$. To complete the proof suppose that $\Delta' \neq \emptyset$, where

$$\Delta' = \{\, z \in \Delta \mid (z, x) = 0 \text{ for all } x \in \Gamma_a \,\}.$$

We shall show that this leads to a contradiction. The roots a, b and c and, by Lemma 7.15, the elements of Γ_a, Γ_b and Γ_c are orthogonal to every element of Δ'. Moreover, we have just shown that every element of $\Delta \setminus \Delta'$ is a multiple of the difference of two elements of Γ_a and therefore orthogonal to every element of Δ'. Since $\mathfrak{L}$ is indecomposable it follows that there exists $s \in \Lambda$ and $t \in \Delta'$ such that $(s, t) \neq 0$.

It follows from Lemma 7.16 (*ii*) that the star-closure of a, b and s is equivalent to $\mathcal{D}_3^{(3)}$ and then from Theorem 7.22 the star-closure $\mathfrak{M}$ of a, b, s and t is equivalent to $\mathcal{D}_4^{(3)}$. Since t is orthogonal to a and b we may suppose that a and b correspond to $(1, -1, 0, 0)$ and $(1, -\omega, 0, 0)$, respectively. If $u \in \Gamma_a$, it follows from Theorem 7.22 that the star-closure of $\mathfrak{M}$ and u is $\mathcal{D}_5^{(3)}$ or $\mathcal{K}_5$. But in $\mathcal{D}_5^{(3)}$ we have $\Gamma_a = \emptyset$ and in $\mathcal{K}_5$ there is an element of Γ_a not orthogonal to t. This contradiction completes the proof.   □

Corollary 7.24. *If* $\Gamma_a \neq \emptyset$, *we may normalise the roots* Σ *of* $\mathfrak{L}$ *so that every root is a linear combination of elements of* $\{a, b, c\} \cup \Gamma_a$ *with coefficients from* $\{\pm 1, \pm \omega, \pm \omega^2\}$.

Proof. From Lemma 7.15 and the previous lemma the result holds for the elements of Γ_b, Γ_c and Δ. If $s \in \Lambda$ and $x \in \Gamma_a$ it follows from Lemma 7.16 (*ii*) and Theorem 7.22 that the star-closure $\mathfrak{M}$ of a, b, x and s is $\mathcal{D}_4^{(3)}$. In $\mathcal{D}_4^{(3)}$ we may suppose that $a = (1, -1, 0, 0)$, $b = (0, 1, -1, 0)$ and $x = (0, 0, -1, 1)$. If $y = (0, 0, -1, \omega)$, then $\mathfrak{M}$ is the star-closure of a, b, x and y. We have $y \in \Gamma_a$ and every element of $\mathfrak{M}$ is a linear combination of a, b, x and y with coefficients from $\{\pm 1, \pm \omega, \pm \omega^2\}$.   □

Theorem 7.25. *If* $\Gamma_a = \emptyset$, *then* $\mathfrak{L}$ *is* $\mathcal{D}_n^{(3)}$ *for some* n.

Proof. By assumption $\mathfrak{L}$ is indecomposable and therefore $\Lambda \neq \emptyset$, otherwise $\Sigma = \{a, b, c\} \perp \Delta$. If $s \in \Lambda$, then the star-closure of a, b and s is $\mathcal{D}_3^{(3)}$ and it follows from Theorems 7.13 and 7.22 that $\mathfrak{L}$ is $\mathcal{D}_n^{(3)}$ or that $\mathfrak{L}$ contains $\mathcal{K}_5$ or $\mathcal{K}_6$. However, if $\mathfrak{L}$ contains $\mathcal{K}_5$ or $\mathcal{K}_6$ we have $\Gamma_a \neq \emptyset$.   □

10. Extensions of Euclidean line systems

This section includes an account of some of the results of D. Cvetković, P. Rowlinson and S. Simić on Euclidean line systems presented in [68]. Our classification (Theorem 7.31) of the indecomposable complex 3-systems will depend on the corresponding result for real systems, which we give in this section.

We continue to use the notation introduced in the previous section. In particular, $\mathfrak{L}$ denotes an indecomposable 3-system in $\mathbb{C}^n$ with a 3-star σ whose roots a, b and c have been chosen so that $a + b + c = 0$ and $(a, b) = (b, c) = (c, a) = -1$. Furthermore, $\Sigma := \sigma \cup \Gamma_a \cup \Gamma_b \cup \Gamma_c \cup \Delta \cup \Lambda$ is the Goethals–Seidel decomposition (7.14) for a set Σ of representatives of the roots of $\mathfrak{L}$.

Theorem 7.26. *The 3-system $\mathfrak{L}$ is the complexification of a Euclidean line system if and only if $\Lambda = \emptyset$.*

Proof. If $\mathfrak{L}$ is Euclidean, it follows from Lemma 7.16 *(ii)* that $\Lambda = \emptyset$.

Conversely, if $\Lambda = \emptyset$, then $(x, y) \in \{0, 1\}$ for all $x \neq y \in \Gamma_a$. Let R denote the vector space of linear combinations of the elements of $\{a, b\} \cup \Gamma_a$ with real coefficients. It follows from Corollary 7.24 and Lemma 7.23 that $\Sigma \subset R$ and hence $\mathfrak{L}$ is the complexification of a Euclidean line system. $\qquad\square$

Theorem 7.27. *If $\mathfrak{L}$ is an indecomposable 3-system such that $(x, y) = 1$ for all $x, y \in \Gamma_a$, whenever $x \neq y$, then $\mathfrak{L}$ is equivalent to $\mathcal{A}_n$.*

Proof. We shall prove that $\{b, c\} \cup \Gamma_a$ is a basis for the space spanned by the line system $\mathfrak{L}$. It follows from Corollary 7.24 that it is sufficient to prove that the set $\{b, c\} \cup \Gamma_a$ is linearly independent. To this end, suppose that

$$\beta b + \gamma c + \sum_{x \in \Gamma_a} \alpha_x x = 0.$$

On taking inner products with b, c and $y \in \Gamma_a$, we have

$$2\beta - \gamma + \sum_{x \in \Gamma_a} \alpha_x = 0,$$

$$-\beta + 2\gamma - \sum_{x \in \Gamma_a} \alpha_x = 0, \quad \text{and}$$

$$\beta - \gamma + \alpha_y + \sum_{x \in \Gamma_a} \alpha_x = 0 \quad \text{for all } y \in \Gamma_a.$$

Thus $\beta + \gamma = 0$ and then $\alpha_y = \beta$ for all y. It follows that $\beta = \gamma = \alpha_y = 0$, as required. In particular, $|\Gamma_a| = n - 2$. The inner products between the basis elements are uniquely determined and therefore $\mathfrak{L}$ must be equivalent to $\mathcal{A}_n$.

An explicit description of the equivalence between $\mathfrak{L}$ and $\mathcal{A}_n$ can be obtained as follows. Choose a labelling $x_1, x_2, \ldots, x_{n-2}$ for the elements of Γ_a and observe that the map $\varphi : \mathcal{A}_n \rightarrow \mathfrak{L}$ defined by $\varphi(-e_2 + e_3) = b$, $\varphi(e_1 - e_3) = c$ and $\varphi(e_3 - e_h) = x_{h-3}$ (for $4 \leq h \leq n + 1$) is an isometry. $\qquad\square$

Theorem 7.28. *Suppose that $\mathfrak{L}$ is an indecomposable 3-system in a real vector space and a simple extension of $\mathcal{A}_n$ $(n \geq 2)$. Then the extension is equivalent to one of*

$$\mathcal{A}_n \subset \mathcal{A}_{n+1}, \qquad \mathcal{A}_n \subset \mathcal{D}_{n+1}^{(2)}, \qquad \mathcal{A}_5 \subset \mathcal{E}_6, \qquad \mathcal{A}_6 \subset \mathcal{E}_7,$$
$$\mathcal{A}_7 \subset \mathcal{E}_7, \qquad \mathcal{A}_7 \subset \mathcal{E}_8 \quad or \quad \mathcal{A}_8 \subset \mathcal{E}_8.$$

The extensions are minimal except for

$$\mathcal{A}_5 \subset \mathcal{A}_5 \perp \mathcal{A}_1 \subset \mathcal{E}_6, \qquad \mathcal{A}_6 \subset \mathcal{A}_7 \subset \mathcal{E}_7,$$
$$\mathcal{A}_7 \subset \mathcal{A}_8 \subset \mathcal{E}_8 \quad and \quad \mathcal{A}_7 \subset \mathcal{D}_8^{(2)} \subset \mathcal{E}_8.$$

Proof. We may take $\mathfrak{L}$ to be a line system in $\mathbb{R}^{n+1}$ and choose an orthonormal basis $e_1, e_2, \ldots, e_{n+1}$ of $\mathbb{R}^{n+1}$.

The line system $\mathfrak{L}$ is the star-closure of $\mathcal{A}_n$ and a line ℓ with root x. As roots of $\mathcal{A}_n$ we take the vectors $e_i - e_j$, where $1 \leq i < j \leq n + 1$. We may write $x = (\alpha_1, \alpha_2, \ldots, \alpha_{n+1})$, using coordinates with respect to the basis $e_1, e_2, \ldots, e_{n+1}$. Since all inner products are real, only 0, 1 and -1 occur as possible values for (u, v), where $u \neq v$ are roots of $\mathfrak{L}$. Furthermore, the group $W(\mathcal{A}_n)$ acts as $\mathrm{Sym}(n + 1)$ on the coordinate values.

If $\alpha_i \neq \alpha_j$, then $\alpha_j = \alpha_i \pm 1$ and therefore there exists α such that for all i we have $\alpha_i = \alpha$ or $\alpha_i = \alpha + 1$. Suppose that there are k values of i such that $\alpha_i = \alpha + 1$. Then

$$2 = (x, x) = k(\alpha + 1)^2 + (n + 1 - k)\alpha^2 = (n + 1)\alpha^2 + 2k\alpha + k.$$

This quadratic in α has real roots and therefore $k^2 \geq (n+1)(k-2)$. Replacing x by its negative, if necessary, we may suppose that $k \leq (n+1)/2$ and so $k^2 \geq 2k(k-2)$, whence $k \leq 4$. If $k = 0$, then x would be orthogonal to every root of $\mathcal{A}_n$ and $\mathfrak{L}$ would be reducible; thus $1 \leq k \leq 4$.

We may take $x = \alpha \mathbf{j} + e_1 + e_2 + \cdots + e_k$, where $\mathbf{j} = \mathbf{j}_9 = (1, 1, \ldots, 1)$ and where α is a root of $f(X) = (n+1)X^2 + 2kX + (k-2)$. The orthogonal matrix $I - \frac{2}{n+1}J$ fixes every line of $\mathcal{A}_n$ and interchanges the two extensions corresponding to the roots of $f(X)$. We shall see that these extensions coincide when (k, n) is $(2, 3)$, $(4, 7)$ or $(3, 8)$.

The case $k = 1$. In this case $\alpha = (-1 \pm \sqrt{n + 2})/(n + 1)$. For each choice of α, the star-closure of $\mathcal{A}_n$ and x consists of the roots of $\mathcal{A}_n$ together with the $n + 1$ roots $\alpha \mathbf{j} + e_i$. This line system satisfies the hypotheses of Theorem 7.27, where $a = -e_1 + e_2$, $b = -e_2 + e_3$ and Γ_a consists of the $n - 2$ vectors $e_3 - e_i$ $(4 \leq i \leq n + 1)$ together with $\alpha \mathbf{j} + e_3$. Thus $\mathfrak{L}$ is equivalent to $\mathcal{A}_{n+1}$.

The case $k = 2$. In this case $\alpha = 0$ or $\alpha = -4/(n + 1)$ and when $n = 3$ the two extensions coincide. If $\alpha = 0$, then $x = e_1 + e_2$ and so $\mathfrak{L} = \mathcal{D}_{n+1}^{(2)}$.

The case $k = 4$. In this case $\alpha = (-4 \pm \sqrt{14 - 2n})/(n + 1)$. Since we have chosen $k \leq (n + 1)/2$ it follows that $n = 7$ and $\alpha = -\frac{1}{2}$. Thus $x = \frac{1}{2}(-1, -1, -1, -1, 1, 1, 1, 1)$ and $\mathfrak{L}$ consists of the 63 lines of $\mathcal{E}_8$ orthogonal to $\mathbf{j}$. That is, $\mathfrak{L} = \mathcal{E}_7$ and this is the unique extension of $\mathcal{A}_7$ to $\mathcal{E}_7$ in $\mathbb{R}^8$.

The case $k = 3$. In this case $\alpha = (-3 \pm \sqrt{8 - n})/(n + 1)$ and therefore n is 5, 6, 7 or 8.

The orbit of x under the action of $W(\mathcal{A}_n)$ is the set of all x_L, where L is a 3-element subset of $\{1, 2, \ldots, n + 1\}$ and $x_L := \alpha \mathbf{j} + \sum_{i \in L} e_i$. If $L \cap M = \emptyset$, then $(x_L, x_M) = -1$ and therefore $x_L + x_M$ spans a line of $\mathfrak{L}$. Except when $n = 8$ the sets $\{\langle x_L + x_M \rangle \mid L \cap M = \emptyset\}$ and $\{\langle x_N \rangle \mid |N| = 3\}$ are disjoint. If $|L \cap M| = 1$, then $(x_L, x_M) = 0$ whereas if $|L \cap M| = 2$, then $(x_L, x_M) = 1$ and $x_L - x_M$ spans a line of $\mathcal{A}_n$.

If $n \neq 8$, $\mathfrak{L}$ contains the $\binom{n+1}{2}$ lines of $\mathcal{A}_n$, the $\binom{n+1}{3}$ lines $\langle x_L \rangle$ and the $\binom{n+1}{6}$ lines $\langle x_L + x_M \rangle$, where L and M are disjoint.

If $n = 5$, then $\mathfrak{L}$ contains $\binom{6}{2} + \binom{6}{3} + 1 = 36$ lines. The linear transformation $\varphi : \mathbb{R}^6 \to \mathbb{R}^8$ with matrix

$$\begin{bmatrix} I - \frac{1}{6} J_6 \\ \frac{1}{2\sqrt{3}} \mathbf{j}_6 \\ -\frac{1}{2\sqrt{3}} \mathbf{j}_6 \end{bmatrix}$$

is an isometry that fixes every root of $\mathcal{A}_5$ and sends $\frac{1}{\sqrt{3}} \mathbf{j}_6$ to $e_7 - e_8$. Thus $\varphi(S)$ is the set of 36 lines of $\mathcal{E}_8$ orthogonal to $\frac{1}{2} \mathbf{j}_8$ and $e_7 + e_8$. Therefore $\mathfrak{L} \simeq \mathcal{E}_6$ and since $\frac{1}{\sqrt{3}} \mathbf{j}_6$ is orthogonal to $\mathcal{A}_5$ we have $\mathcal{A}_5 \subset \mathcal{A}_5 \perp \mathcal{A}_1 \subset \mathcal{E}_6$.

If $n = 6$ and L and M are disjoint, then $x_L + x_M = (2\alpha + 1)\mathbf{j} - e_i$ for some i. Thus, by the case $k = 1$ handled above, the line system formed by these lines together with $\mathcal{A}_6$ is equivalent to $\mathcal{A}_7$. Furthermore $\mathfrak{L}$ contains $\binom{7}{2} + \binom{7}{3} + \binom{7}{6} = 63$ lines and it is an extension of $\mathcal{A}_7$ in $\mathbb{R}^7$. This is equivalent to the case $k = 4$ above. Therefore $\mathfrak{L} \simeq \mathcal{E}_7$ and so $\mathcal{A}_6 \subset \mathcal{A}_7 \subset \mathcal{E}_7$.

If $n = 7$, then by the case $k = 2$ above, the lines of $\mathcal{A}_7$ together with the 28 lines spanned by the roots $x_L + x_M$ where L and M are disjoint is equivalent to $\mathcal{D}_8^{(2)}$. It follows from Theorem 7.21 that $\mathfrak{L} \simeq \mathcal{E}_8$ and therefore $\mathcal{A}_7 \subset \mathcal{D}_8^{(2)} \subset \mathcal{E}_8$.

If $n = 8$, then $\alpha = -\frac{1}{3}$ and the $\binom{9}{3} = 84$ lines spanned by the roots x_L are orthogonal to $\mathbf{j}_9$. This is the line system $\overline{\mathcal{E}}_8$ described in §6 and therefore $\mathfrak{L} \simeq \mathcal{E}_8$. This is the unique extension of $\mathcal{A}_8$ to a line system of type $\mathcal{E}_8$ in $\mathbb{R}^9$. $\square$

This completes the determination of all Euclidean extensions of line systems of type $\mathcal{A}_n$ and we now turn our attention to the line systems $\mathcal{E}_6, \mathcal{E}_7$ and $\mathcal{E}_8$.

Theorem 7.29. *Suppose that $\mathfrak{L}$ is an indecomposable 3-system in a real vector space and a simple extension of $\mathfrak{M}$, where $\mathfrak{M}$ is one of $\mathcal{E}_n$ ($n = 6, 7, 8$). Then the extension*

$\mathfrak{M} \subset \mathfrak{L}$ *is equivalent to* $\mathcal{E}_6 \subset \mathcal{E}_7$ *or* $\mathcal{E}_7 \subset \mathcal{E}_8$. *The extension* $\mathcal{E}_6 \subset \mathcal{E}_7$ *is minimal but* $\mathcal{E}_7 \subset \mathcal{E}_7 \perp \mathcal{A}_1 \subset \mathcal{E}_8$.

Proof. Suppose first that $\mathfrak{L}$ is the star-closure in $\mathbb{R}^9$ of $\mathcal{E}_8$ and a line $\ell \notin \mathcal{E}_8$. We have $\mathcal{A}_8 \subset \mathcal{E}_8$ and by Theorem 7.28 the star-closure of $\mathcal{A}_8$ and ℓ is equivalent to $\mathcal{A}_9$, $\mathcal{D}_9^{(2)}$ or $\mathcal{E}_8$. We have shown that in $\mathbb{R}^9$ there is a unique extension of $\mathcal{A}_8$ to $\mathcal{E}_8$ and therefore $\mathfrak{L}$ contains $\mathcal{A}_9$ or $\mathcal{D}_9^{(2)}$. However, by Theorem 7.21 and Theorem 7.28 there are no proper extensions of $\mathcal{A}_9$ or $\mathcal{D}_9^{(2)}$ in $\mathbb{R}^9$ and since these line systems contain only 45 and 72 lines, respectively, we reach a contradiction. That is, there are no simple extensions of $\mathcal{E}_8$.

Next suppose that $\mathfrak{L}$ is the star-closure in $\mathbb{R}^8$ of $\mathcal{E}_7$ and a line $\ell \notin \mathcal{E}_7$. The argument is completely analogous to that for $\mathcal{E}_8$. We have $\mathcal{A}_7 \subset \mathcal{E}_7$ and by Theorem 7.28 the star-closure of $\mathcal{A}_7$ and ℓ is equivalent to $\mathcal{A}_8$, $\mathcal{D}_8^{(2)}$, $\mathcal{E}_7$ or $\mathcal{E}_8$. In $\mathbb{R}^8$ there is a unique extension of $\mathcal{A}_7$ to $\mathcal{E}_7$ and by Theorem 7.21 and Theorem 7.28 there are no proper extensions of $\mathcal{A}_8$ or $\mathcal{D}_8^{(2)}$ in $\mathbb{R}^8$. As above, a counting argument shows that $\mathfrak{L}$ is not equivalent to $\mathcal{A}_8$ or $\mathcal{D}_8^{(2)}$. The only possibility is that $\mathfrak{L} \simeq \mathcal{E}_8$.

Finally, we consider the extensions of $\mathcal{E}_6$ and to this end we represent the line system $\mathcal{E}_6$ by the 36 roots of $\mathcal{E}_8$ orthogonal to $\mathbf{j}_8$ and $e_7 + e_8$. This version of $\mathcal{E}_6$ consists of a copy of $\mathcal{A}_5$ on the first 6 coordinates, the line spanned by $w := e_7 - e_8$ and the 20 images of the line spanned by $y := \frac{1}{2}(1, 1, 1, -1, -1, -1, 1, -1)$ under the action of $W(\mathcal{A}_5)$.

Suppose that $\mathfrak{L}$ is the star-closure in $\mathbb{R}^8$ of $\mathcal{E}_6$ and a root $x = (\alpha_1, \alpha_2, \ldots, \alpha_8)$. If $\alpha_i = 0$ for some $i \leq 6$ then for $j \leq 6$ we have $\alpha_j = 0$ or $\alpha_j = \pm 1$. Suppose that $\alpha_j \neq 0$ for just one value of $j \leq 6$. On taking inner products with w and y we see that, up to equivalence, we may take $x = e_6 - e_7$ or $x = e_6 + e_7$. The orthogonal matrix $\mathrm{diag}(1, 1, 1, 1, 1, 1, -1, -1)$ fixes $\mathcal{E}_6$ and interchanges these two possibilities; thus we may suppose that $x = e_6 - e_7$. Then the star-closure of $\mathcal{A}_5 \perp \mathcal{A}_1$ and x is $\mathcal{A}_7$. The lines of this system are orthogonal to $\mathbf{j}_8$ and, as in the case $k = 4$ above, we have $\mathfrak{L} = \mathcal{E}_7$. If $\alpha_j \neq 0$ and $\alpha_k \neq 0$ for $j \neq k \leq 6$, then we may take $x = e_5 + e_6$. In this case the star-closure of $\mathcal{A}_5$ and x is $\mathcal{D}_6^{(2)}$. Thus $\mathfrak{L}$ contains at least 51 lines and from Theorem 7.21 it follows that $\mathfrak{L}$ is the line system $\mathcal{E}_7$ orthogonal to $e_7 + e_8$.

From now on we may suppose that $\alpha_i \neq 0$ for $i \leq 6$. The argument used at the beginning of the proof of Theorem 7.28 shows that there exists α such that for $i \leq 6$ we have $\alpha_i = \alpha$ or $\alpha_i = \alpha + 1$. Suppose that there are k values of $i \leq 6$ such that $\alpha_i = \alpha + 1$. Replacing x be its negative, if necessary, we may assume that $k \leq 3$ and that the first k coordinate values equal $\alpha + 1$. If $\alpha_7 \neq \alpha_8$, then $\varepsilon = \alpha_7 - \alpha_8 = \pm 1$. If $q := \frac{1}{2}(1, 1, 1, -1, -1, -1, \varepsilon, -\varepsilon)$, then $(x, q) = \frac{1}{2}(k+1)$. Therefore $k = 1$ and $x - q$ is a root of $\mathfrak{L}$. Thus, we may replace x by $x - q$ and henceforth assume that $\alpha_7 = \alpha_8$. Since $(x, y) = \frac{1}{2}k$ and x is not orthogonal to all the lines of $\mathcal{E}_6$ we have $k = 2$ and we may write $x = (\alpha + 1, \alpha + 1, \alpha, \alpha, \alpha, \alpha, \beta, \beta)$. Since $3\alpha^2 + 2\alpha + \beta^2 = 0$, the matrix

$$I + \frac{1}{2}\begin{bmatrix} \alpha J_6 & \beta\,\mathbf{j}_6{}^t & \beta\,\mathbf{j}_6{}^t \\ \beta\,\mathbf{j}_6 & & \\ & -(3\alpha+2)J_2 \\ \beta\,\mathbf{j}_6 & & \end{bmatrix}$$

is orthogonal. It fixes every line of $\mathcal{E}_6$ and sends x to $e_1 + e_2$. This reduces the situation to a previous case and therefore $\mathfrak{L}$ is equivalent to $\mathcal{E}_7$. This completes the proof. $\qquad\square$

11. Extensions of $\mathcal{A}_n$, $\mathcal{E}_n$ and $\mathcal{K}_n$ in $\mathbb{C}^n$

Theorem 7.30. *Suppose that $\mathfrak{L}$ is an indecomposable 3-system and a simple extension of $\mathfrak{M}$, where $\mathfrak{M}$ is one of $\mathcal{A}_n$ $(n \geq 2)$, $\mathcal{E}_n$ $(n = 6,7,8)$ or $\mathcal{K}_n$ $(n = 5,6)$. Then the extension $\mathfrak{M} \subset \mathfrak{L}$ is equivalent to one of*

$$\mathcal{A}_n \subset \mathcal{A}_{n+1}, \qquad \mathcal{A}_n \subset \mathcal{D}_{n+1}^{(2)}, \qquad \mathcal{A}_n \subset \mathcal{D}_{n+1}^{(3)},$$
$$\mathcal{A}_4 \subset \mathcal{K}_5, \qquad \mathcal{A}_5 \subset \mathcal{K}_5, \qquad \mathcal{A}_5 \subset \mathcal{E}_6, \qquad \mathcal{A}_5 \subset \mathcal{K}_6,$$
$$\mathcal{A}_6 \subset \mathcal{K}_6, \qquad \mathcal{A}_6 \subset \mathcal{E}_7, \qquad \mathcal{A}_7 \subset \mathcal{E}_7, \qquad \mathcal{A}_7 \subset \mathcal{E}_8, \qquad \mathcal{A}_8 \subset \mathcal{E}_8,$$
$$\mathcal{E}_6 \subset \mathcal{E}_7, \qquad \mathcal{E}_6 \subset \mathcal{K}_6, \qquad \mathcal{E}_7 \subset \mathcal{E}_8 \quad \text{or} \quad \mathcal{K}_5 \subset \mathcal{K}_6.$$

The extensions are minimal except that for suitable choices of $\mathcal{A}_5$ subsystems we have

$$\mathcal{A}_4 \subset \mathcal{A}_5 \subset \mathcal{K}_5,$$
$$\mathcal{A}_5 \subset \mathcal{A}_5 \perp \mathcal{A}_1 \subset \mathcal{E}_6 \subset \mathcal{K}_6, \qquad \mathcal{A}_5 \subset \mathcal{K}_5 \subset \mathcal{K}_5 \perp \mathcal{A}_1 \subset \mathcal{K}_6,$$
$$\mathcal{A}_5 \subset \mathcal{A}_6 \subset \mathcal{K}_6, \qquad \mathcal{A}_5 \subset \mathcal{D}_6^{(2)} \subset \mathcal{K}_6, \qquad \mathcal{A}_5 \subset \mathcal{D}_6^{(3)} \subset \mathcal{K}_6,$$
$$\mathcal{A}_6 \subset \mathcal{A}_7 \subset \mathcal{E}_7, \qquad \mathcal{A}_7 \subset \mathcal{A}_8 \subset \mathcal{E}_8 \quad \text{and} \quad \mathcal{E}_7 \subset \mathcal{E}_7 \perp \mathcal{A}_1 \subset \mathcal{E}_8.$$

Proof. We take $\mathfrak{L}$ to be a line system in $\mathbb{C}^{n+1}$, where $n = \dim \mathfrak{M}$. In all cases $\mathcal{A}_2 \subset \mathfrak{L}$ and therefore, by Theorem 7.18, either $\mathcal{D}_3^{(2)} \subseteq \mathfrak{L}$ or $\mathcal{D}_3^{(3)} \subseteq \mathfrak{L}$.

If there is no subsystem of $\mathfrak{L}$ equivalent to $\mathcal{D}_3^{(3)}$ then, by Theorem 7.26, $\mathfrak{L}$ is Euclidean and the result follows from Theorems 7.28 and 7.29. Thus we may suppose that $\mathcal{D}_3^{(3)} \subseteq \mathfrak{L}$ and therefore, by Theorems 7.13 and 7.22, $\mathfrak{L}$ is equivalent to $\mathcal{D}_n^{(3)}$, $\mathcal{D}_{n+1}^{(3)}$, $\mathcal{K}_5$, $\mathcal{K}_6$ or to a simple extension of $\mathcal{K}_5$ or $\mathcal{K}_6$.

If $\mathfrak{L}$ is equivalent to $\mathcal{D}_m^{(3)}$ for some m, then each element of $\mathfrak{L}$ is in a unique 3-star with roots a, b and c such that $\Gamma_a = \emptyset$. It follows from Theorem 7.25 that 3-stars of this type cannot wholly lie in $\mathfrak{M}$ and therefore $|\mathfrak{M}| \leq \frac{1}{3}|\mathcal{D}_m^{(3)}| = \binom{m}{2}$. This eliminates all possibilities for $\mathfrak{M}$ except $\mathcal{A}_n$ and therefore the extension is $\mathcal{A}_n \subset \mathcal{D}_{n+1}^{(3)}$.

If $\mathfrak{L}$ is $\mathcal{K}_5$ or $\mathcal{K}_6$, then from §6 we see that all possible simple extensions exist, namely

$$\mathcal{A}_4 \subset \mathcal{K}_5, \qquad \mathcal{A}_5 \subset \mathcal{K}_5, \qquad \mathcal{A}_5 \subset \mathcal{K}_6,$$
$$\mathcal{A}_6 \subset \mathcal{K}_6, \qquad \mathcal{K}_5 \subset \mathcal{K}_6 \quad \text{and} \quad \mathcal{E}_6 \subset \mathcal{K}_6.$$

If $\mathfrak{L}$ is a simple extension of $\mathcal{K}_5$, then we may suppose, in the notation of §6, that $\overline{\mathcal{K}}_5 \subset \mathfrak{L} \subset \mathbb{C}^6$. In particular, $\mathfrak{L}$ contains the 12 lines of $\mathcal{D}_4^{(2)}$, the line spanned by the vector $u = (0,0,0,0,\sqrt{2},0)$ and the 32 lines spanned by the images of $v = \frac{1}{2}(1,1,1,i\sqrt{3},\sqrt{2},0)$ under the action of $W(\mathcal{D}_4^{(2)})$.

Suppose that $\mathfrak{L}$ is the star-closure of $\mathcal{K}_5$ and $x = (\alpha_1,\alpha_2,\ldots,\alpha_6)$. If $\alpha_1 = \alpha_2 = \alpha_3 = \alpha_4 = 0$, then $(x,u) = \sqrt{2}\,\alpha_5$ and $(x,v) = \frac{1}{\sqrt{2}}\alpha_5$, whence $\alpha_5 = 0$ and $\mathfrak{L}$ is equivalent to $\mathcal{K}_5 \perp \mathcal{A}_1$. Therefore, if $\alpha_i = 0$ for some $i \le 4$, the argument of Lemma 7.20 shows that we may take x to be $(1,0,0,0,\alpha_5,\alpha_6)$. Then $(x,u) = \sqrt{2}\alpha_5$ and $(x,v) = \frac{1}{2}+\frac{1}{\sqrt{2}}\alpha_5$, whence $\alpha_5 = \pm\frac{1}{\sqrt{2}}$. Thus, up to equivalence, $x = (1,0,0,0,\frac{1}{\sqrt{2}},\frac{1}{\sqrt{2}})$. The unitary transformation $\mathbb{C}^6 \to \mathbb{C}^6$ given by $e_i \mapsto e_i$ $(1 \le i \le 4)$, $e_5 \mapsto \frac{1}{\sqrt{2}}(e_5 + e_6)$, and $e_6 \mapsto \frac{1}{\sqrt{2}}(e_5 - e_6)$ defines an equivalence between $\mathfrak{L}$ and $\overline{\mathcal{K}}_6$.

We may now suppose that $\alpha_i \ne 0$ for $i \le 4$. A variation of the argument of Theorem 7.21 shows that we may assume that $\alpha_1 = \alpha_2 = \alpha_3 = \frac{1}{2}$ and that α_4 is either $\frac{1}{2}\varepsilon$ or $\frac{1}{2}\varepsilon i\sqrt{3}$, where $\varepsilon = \pm 1$. If $x = \frac{1}{2}(1,1,1,\varepsilon,0,2\alpha_6)$, then $(x,v) = \frac{1}{4}(3 - \varepsilon i\sqrt{3})$, which is a contradiction. If $x = \frac{1}{2}(1,1,1,\varepsilon i\sqrt{3},0,2\alpha_6)$, then $(x,v) = \frac{3}{4}(1 + \varepsilon)$ and so $\varepsilon = -1$ in this case. But $w = \frac{1}{2}(-1,1,1,-i\sqrt{3},\sqrt{2},0)$ is a root of $\mathfrak{L}$ and $(x,w) = 1$, hence $x - w = (1,0,0,0,-\frac{1}{\sqrt{2}},\alpha_6)$ is also a root of $\mathfrak{L}$, reducing us to the previous case. Thus from now on we may suppose that $\alpha_5 \ne 0$ and hence $|\alpha_5| = \frac{1}{\sqrt{2}}$. If $\alpha_4 = \varepsilon i\sqrt{3}$, then $\alpha_6 = 0$ and $(x,v) = \frac{1}{4}(3 + 3\varepsilon + 2\sqrt{2}\alpha_5)$. It follows that $\varepsilon = -1$ and $\alpha_5 = \pm\frac{1}{\sqrt{2}}$. But then $x \in \mathcal{K}_5$, which is a contradiction.

Thus we may suppose that $x = \frac{1}{2}(1,1,1,\varepsilon,2\alpha_5,2\alpha_6)$, whence $|\alpha_5| = |\alpha_6| = \frac{1}{\sqrt{2}}$ and so $(x,v) = \frac{1}{4}(3 - \varepsilon i\sqrt{3} + 2\sqrt{2}\alpha_5)$. Consideration of the 3-star of x and u shows that we are free to chose the sign of α_5. Thus if $\varepsilon = 1$ we may suppose $\alpha_5 = -\omega^2/\sqrt{2}$ and if $\varepsilon = -1$, we may take $\alpha_5 = -\omega/\sqrt{2}$. Then $(x,v) = 1$ in both cases. But then $x - v$ is a root whose first 3 coordinates are 0 and again we reduce to a previous case. Thus, up to equivalence, $\mathcal{K}_6$ is the only indecomposable simple extension of $\mathcal{K}_5$.

Finally, suppose that $\mathfrak{L}$ is the star-closure of $\mathcal{K}_6$ and $x = (\alpha_1,\alpha_2,\ldots,\alpha_7)$ in $\mathbb{C}^7$. We have $\mathcal{D}_6^{(3)} \subset \mathcal{K}_6$ and the star-closure of $\mathcal{D}_6^{(3)}$ and x is indecomposable. From Theorem 7.22 the only indecomposable simple extensions of $\mathcal{D}_6^{(3)}$ are $\mathcal{D}_7^{(3)}$ and $\mathcal{K}_6$. Furthermore, if $\mathcal{D}_7^{(3)} \subseteq \mathfrak{L}$, then equality holds, since by Theorem 7.22 there are no indecomposable extensions of $\mathcal{D}_7^{(3)}$ in $\mathbb{C}^7$. But $|\mathcal{K}_6| \leq |\mathcal{D}_7^{(3)}|$, which is a contradiction.

From the proof of case $n = 6$ of Theorem 7.22 we see that $\alpha_7 = 0$, hence $\mathcal{L}$ is contained in $\mathbb{C}^6$. In $\mathbb{C}^6$ there are just three extensions of $\mathcal{D}_6^{(3)}$ equivalent to $\mathcal{K}_6$: the star-closures of $\mathcal{D}_6^{(3)}$ with $\theta^{-1}(1,1,1,1,1,1)$, $\theta^{-1}(1,1,1,1,1,\omega)$ and $\theta^{-1}(1,1,1,1,1,\omega^2)$, respectively, where $\theta = \omega - \omega^2$. On the other hand, no two of the roots just given are at $60°$ or $90°$. Therefore, $\mathcal{K}_6$ has no indecomposable simple extensions.

The list of simple extensions that are not minimal follows from the descriptions of the line systems given in sections 5 and 6. $\qquad\square$

Theorem 7.31. *If $\mathcal{L}$ is an indecomposable 3-system, then for some n, $\mathcal{L}$ is equivalent to $\mathcal{A}_n$, $\mathcal{D}_n^{(2)}$, or $\mathcal{D}_n^{(3)}$, or to $\mathcal{E}_6$, $\mathcal{E}_7$, $\mathcal{E}_8$, $\mathcal{K}_5$ or $\mathcal{K}_6$.*

Proof. Since $\mathcal{L}$ is a 3-system it contains a pair of lines at $60°$ and hence a 3-star σ. By Theorem 7.13 $\mathcal{L}$ is the end of a chain of simple indecomposable extensions, beginning at σ. The theorem follows from the combined results of Theorems 7.18, 7.21, 7.22 and 7.30. $\qquad\square$

12. Extensions of 4-systems

Suppose that $\mathcal{L}$ is an indecomposable 4-system in $\mathbb{C}^n$. Then $\mathcal{L}$ contains a pair of lines at $45°$. If a and b are long roots corresponding to these lines we may scale b so that $(a, b) = -\sqrt{2}$. Then $c := r_a(b) = \sqrt{2}\,a + b$ and $d := r_b(a) = a + \sqrt{2}\,b$ are roots that span the other two lines of the 4-star of a and b.

A 4-star can be represented either as the line system $\mathcal{B}_2^{(2)}$ with roots $(1, -1)$, $(1, 1)$, $(\sqrt{2}, 0)$ and $(0, \sqrt{2})$ or as the line system $\mathcal{D}_2^{(4)}$ with roots $(1, -1)$, $(1, 1)$, $(1, i)$ and $(1, -i)$. The equivalence is given by the unitary transformation with matrix $\frac{1}{\sqrt{2}} \begin{bmatrix} 1 & i \\ i & 1 \end{bmatrix}$.

Lemma 7.32. *Suppose that σ is a 4-star and that $\mathcal{L}$ is an indecomposable 4-system which is the star-closure of σ and a line $\ell \notin \sigma$.*

(i)   *If ℓ is orthogonal to a line of σ then $\mathcal{L}$ is equivalent to $\mathcal{B}_3^{(2)}$.*
(ii)   *If ℓ is not orthogonal to any line of σ, then $\mathcal{L}$ is equivalent to one of*

$$\mathcal{B}_2^{(4)}, \quad \mathcal{D}_3^{(4)} \quad or \quad \mathcal{J}_3^{(4)}.$$

The extensions are minimal except for $\sigma \subset \mathcal{B}_3^{(2)} \subset \mathcal{J}_3^{(4)}$.

Proof. We may suppose that $\mathcal{L}$ is a line system in $\mathbb{C}^3$ and that $\sigma = \{a, b, c, d\}$, where $(a, b) = -\sqrt{2}$, $c := r_a(b)$ and $d := r_b(a)$. Let x be a root of ℓ. We may choose a basis for $\mathbb{C}^3$ so that, as row vectors, we have $a = (1, -1, 0)$, $b = (0, \sqrt{2}, 0)$, $c = (\sqrt{2}, 0, 0)$, $d = (1, 1, 0)$ and $x = (\alpha, \beta, \gamma)$.

To prove (i), choose the notation so that $(b, x) = 0$ and $(c, x) = \sqrt{2}$. It follows immediately that $\alpha = 1$, $\beta = 0$ and hence $|\gamma| = 1$. Thus, up to equivalence, we may suppose that $\gamma = 1$. Hence $x = (1, 0, 1)$ and $\mathfrak{L}$ is equivalent to $\mathcal{B}_3^{(2)}$.

For (ii), suppose first that ℓ is at 45° to a line of σ. Then we may choose the notation so that $(a, x) = \sqrt{2}$. If $|(b, x)| = |(c, x)| = \sqrt{2}$, then $|\alpha| = |\alpha - \sqrt{2}| = 1$ so that $\alpha = (1 \pm i)/\sqrt{2}$, $\beta = (-1 \pm i)/\sqrt{2}$ and consequently $\gamma = 0$. Thus, up to a scalar multiple, x is $(1, i, 0)$ or $(1, -i, 0)$. Therefore $\mathfrak{L} = \mathcal{B}_2^{(4)}$ and $|(c, x)| = \sqrt{2}$.

If $(a, x) = \sqrt{2}$ and $|(b, x)| = |(c, x)| = 1$, then $|\alpha| = |\alpha - \sqrt{2}| = \frac{1}{\sqrt{2}}$ and so $\alpha = \frac{1}{\sqrt{2}}$, and $\beta = -\frac{1}{\sqrt{2}}$, whence $(d, x) = 0$, contrary to the choice of ℓ in this case.

If $(a, x) = |(b, x)| = \sqrt{2}$ and $|(c, x)| = 1$, then $|\alpha| = 1/\sqrt{2}$, $|\beta| = 1$ and hence $|\gamma| = 1/\sqrt{2}$. Therefore we may choose $\gamma = \alpha$. In this case $r_x(b)$ is a scalar multiple of $(1, 0, 1)$ and it follows that $\mathfrak{L}$ contains $\mathcal{B}_3^{(2)}$. Let $\lambda = -\frac{1}{2}(1 + i\sqrt{7})$ and replace b by $(0, \bar{\lambda}, 0)$. Then $(a, b) = -\lambda$ and $c = (\bar{\lambda}, 0, 0)$. Scale x so that $(a, x) = -\bar{\lambda}$. Then $\beta = \alpha + \lambda$ and we have $2\bar{\lambda}\alpha^2 + 3\alpha + \lambda = 0$, hence $\alpha = \frac{1}{2}\bar{\lambda}$ or $\alpha = \frac{1}{4}\lambda^3$.

If $\alpha = \frac{1}{2}\bar{\lambda}$, then $\beta = \frac{1}{2}\bar{\lambda}^2$. In this case, replace x by $(\lambda/\bar{\lambda})x$ so that $x = \frac{1}{2}(\lambda, 2, \lambda)$ and hence $\mathfrak{L} = \mathcal{J}_3^{(4)}$. Similarly, if $\alpha = \frac{1}{4}\lambda^3$, then $\beta = \frac{1}{4}\lambda^4$. On replacing x by $(2\bar{\lambda}/\lambda^3)x$ we find that $x = \frac{1}{2}(\bar{\lambda}, 2, \bar{\lambda})$ and therefore $\mathfrak{L} = \overline{\mathcal{J}}_3^{(4)}$, the complex conjugate of $\mathcal{J}_3^{(4)}$.

If $(a, x) = |(c, x)| = \sqrt{2}$ and $|(b, x)| = 1$, then replacing x by $r_d(x)$ reduces this case to the previous one.

The remaining possibility is that ℓ is not at 45° to any line of σ; that is, we may scale x so that $(a, x) = |(b, x)| = |(c, x)| = 1$. In this case $\beta = \alpha - 1$, $|\alpha| = |\alpha - 1| = \frac{1}{\sqrt{2}}$ and therefore $|\gamma| = 1$. It follows that $\alpha^2 - \alpha + \frac{1}{2} = 0$ and therefore $\alpha = \frac{1}{2}(1 \pm i)$. Thus $x = \frac{1}{2}(1 + i, -1 + i, 2\gamma)$ or $x = \frac{1}{2}(1 - i, -1 - i, 2\gamma)$. If we take $\gamma = \sqrt{2}\beta$, then in both cases the unitary transformation $\mathbb{C}^3 \to \mathbb{C}^3$ with matrix

$$\frac{1}{\sqrt{2}} \begin{bmatrix} 1 & i & 0 \\ i & 1 & 0 \\ 0 & 0 & \sqrt{2} \end{bmatrix}$$

sends σ to $\mathcal{D}_2^{(4)}$ and x to a multiple of $(0, 1, 1)$. Thus $\mathfrak{L} \simeq \mathcal{D}_3^{(4)}$. □

The lines of the 4-systems $\mathcal{B}_n^{(2)}$, $\mathcal{D}_n^{(4)}$ and $\mathcal{B}_n^{(4)}$ are spanned by the roots of the reflections in the imprimitive reflection groups $G(2, 1, n)$, $G(4, 4, n)$ and $G(4, 2, n)$ respectively. In §6 we described four other indecomposable 4-systems: $\mathcal{J}_3^{(4)}$, $\mathcal{F}_4$, $\mathcal{N}_4$ and $\mathcal{O}_4$. In this section we show that there are no other indecomposable 4-systems.

We first consider extensions of $\mathcal{B}_n^{(2)}$. In some cases it is possible to determine some simple extensions of $\mathcal{B}_n^{(2)}$ exactly – not just up to equivalence. These cases are dealt with in the next three lemmas.

Lemma 7.33. *Let $\mathfrak{L}$ be an indecomposable 4-system and a simple extension of $\mathcal{B}_n^{(2)}$, where $n \geq 2$. Then $\mathfrak{L}$ may be represented as the star-closure in $\mathbb{C}^{n+1}$ of $\mathcal{B}_n^{(2)}$ and a line ℓ with root $x = (\alpha_1, \alpha_2, \ldots, \alpha_{n+1})$. Suppose that $\alpha_j = 0$ for some $j \leq n$. If $\alpha_{n+1} = 0$, then $\mathfrak{L} = \mathcal{B}_n^{(4)}$, otherwise $\mathfrak{L}$ is equivalent $\mathcal{B}_{n+1}^{(2)}$.*

Proof. We may suppose that $\alpha_1 = 0$ and since $\mathfrak{L}$ is indecomposable $\alpha_h \neq 0$ for some $h \leq n$. On taking inner products with $\sqrt{2}\,e_h$ we see that for $h \leq n$, either $\alpha_h = 0$ or $|\alpha_h| = 1$. Since $(x, x) = 2$ it follows that $\alpha_h \neq 0$ for exactly two values of $h \leq n + 1$. If these values are h and k with $h < k$, we scale x so that $\alpha_h = 1$.

If $\alpha_{n+1} = 0$, then $k \leq n$ and $\alpha_k \neq 1$. Thus $|1 - \alpha_k|$ is 1 or $\sqrt{2}$. If $|1 - \alpha_k| = 1$, then α_k is $-\omega$ or $-\omega^2$, which is a contradiction. If $|1 - \alpha_k| = \sqrt{2}$, then $\alpha_k = \pm i$. In this case we have $\mathfrak{L} = \mathcal{B}_n^{(4)}$.

If $\alpha_{n+1} \neq 0$, then $k = n + 1$ and the $(n + 1) \times (n + 1)$ diagonal matrix $\mathrm{diag}(1, 1, \ldots, 1, \alpha_{n+1}^{-1})$ is unitary; it fixes every root of $\mathcal{B}_n^{(2)}$ and sends x to $e_i + e_{n+1}$. Thus $\mathfrak{L}$ is equivalent to $\mathcal{B}_{n+1}^{(2)}$. $\square$

Lemma 7.34. *Suppose that $\mathfrak{L}$ is an indecomposable 4-system in $\mathbb{C}^3$ and a simple extension of $\mathcal{B}_3^{(2)}$. Then $\mathfrak{L}$ is $\mathcal{B}_3^{(4)}$, $\mathcal{J}_3^{(4)}$ or $\overline{\mathcal{J}}_3^{(4)}$.*

Proof. We may suppose that $\mathfrak{L}$ is the star-closure of $\mathcal{B}_3^{(2)}$ and a line ℓ with root $x = (\alpha_1, \alpha_2, \alpha_3)$. If $\alpha_h = 0$ for some h, then by Lemma 7.33 we have $\mathfrak{L} = \mathcal{B}_3^{(4)}$. Therefore we assume $\alpha_h \neq 0$ for $1 \leq h \leq 3$.

On taking inner products with the roots $\sqrt{2}\,e_h$ we find that for $h \leq 3$, $|\alpha_h|$ is 1 or $\frac{1}{\sqrt{2}}$. Since $\alpha_h \neq 0$ for $h \leq 3$, the only possibility is that $|\alpha_h| = 1$ for exactly one value of h. Thus we may suppose that $|\alpha_1| = |\alpha_2| = \frac{1}{\sqrt{2}}$ and $|\alpha_3| = 1$, then scale x so that $x = (\frac{1}{2}\lambda, \frac{1}{2}\lambda, \alpha_3)$. Then $|\frac{1}{2}\lambda - \alpha_3|$ is 1 or 2; that is, $\lambda\bar{\alpha}_3 + \bar{\lambda}\alpha_3 = \pm 1$. If $c = (0, 0, \sqrt{2})$, then $r_c(x) = (\frac{1}{2}\lambda, \frac{1}{2}\lambda, -\alpha_3)$ and so we may suppose that $\lambda\bar{\alpha}_3 + \bar{\lambda}\alpha_3 = -1$. Thus $\alpha_3 = 1$ or $\alpha_3 = \lambda/\bar{\lambda}$. In the first case $x = (\frac{1}{2}\lambda, \frac{1}{2}\lambda, 1)$ and $\mathfrak{L} = \mathcal{J}_3^{(4)}$. In the second case the scalar multiple $(\bar{\lambda}/\lambda)\,x$ of x is $(\frac{1}{2}\bar{\lambda}, \frac{1}{2}\bar{\lambda}, 1)$ and therefore $\mathfrak{L} = \overline{\mathcal{J}}_3^{(4)}$. $\square$

Lemma 7.35. *Suppose that $\mathfrak{L}$ is an indecomposable 4-system in $\mathbb{C}^4$ and a simple extension of $\mathcal{B}_4^{(2)}$. Then $\mathfrak{L}$ is $\mathcal{B}_4^{(4)}$, $\mathcal{F}_4$, $\mathcal{N}_4$ or $\overline{\mathcal{O}}_4$. Furthermore, $\mathcal{O}_4$ is the union of $\mathcal{B}_4^{(4)}$, $\mathcal{F}_4$ and $\mathcal{N}_4$.*

Proof. We may suppose that $\mathcal{L}$ is the star-closure of $\mathcal{B}_4^{(2)}$ and a line ℓ with root $x = (\alpha_1, \alpha_2, \alpha_3, \alpha_4)$. If $\alpha_h = 0$ for some h, then by Lemma 7.33 we have $\mathcal{L} = \mathcal{B}_4^{(4)}$. Therefore we assume $\alpha_h \neq 0$ for $h \leq 4$.

On taking inner products with the roots $\sqrt{2}\, e_h$ we find that $|\alpha_h|$ is 1 or $\frac{1}{\sqrt{2}}$. Since $(x, x) = 2$, the only possibility is $|\alpha_h| = \frac{1}{\sqrt{2}}$ for all $h \leq 4$.

Suppose that $\alpha_2 \neq \pm\alpha_3$. On taking inner products with the roots $e_2 \pm e_3$, we have

$$p := |\alpha_2 + \alpha_3|^2 = 1 + \bar{\alpha}_2\alpha_3 + \alpha_2\bar{\alpha}_3 \in \{1, 2\} \quad \text{and}$$

$$m := |\alpha_2 - \alpha_3|^2 = 1 - \bar{\alpha}_2\alpha_3 - \alpha_2\bar{\alpha}_3 \in \{1, 2\}.$$

Thus $p + m = 2$ and consequently $p = m = 1$. From this we deduce that $\alpha_3 = \pm i\alpha_2$. If in addition $\alpha_1 \neq \pm\alpha_2$ and $\alpha_1 \neq \pm\alpha_3$ we reach a contradiction. The group $G(2, 1, 4)$ acts on $\mathbb{C}^4$ by permuting the coordinates and negating their values. Therefore we may suppose that $\alpha_1 = \alpha_2 = \frac{1}{\sqrt{2}}$. Furthermore, taking into account the action of $G(2, 1, 4)$, $\alpha_3 = \frac{1}{\sqrt{2}}$ or $\alpha_3 = \frac{i}{\sqrt{2}}$ and similarly $\alpha_4 = \frac{1}{\sqrt{2}}$ or $\alpha_4 = \frac{i}{\sqrt{2}}$. Thus there are essentially three possibilities for x, namely $x = \frac{1}{\sqrt{2}}(1, 1, 1, 1)$, $x = \frac{1}{\sqrt{2}}(1, 1, i, i)$ and $x = \frac{1}{\sqrt{2}}(1, 1, 1, i)$. The corresponding extensions are $\mathcal{F}_4, \mathcal{N}_4$ and $\overline{\mathcal{O}}_4$.

Since $\mathcal{B}_4^{(2)} \subset \mathcal{O}_4$ and since the roots $(0, 0, 1, 1)$, $\frac{1}{\sqrt{2}}(1, 1, 1, 1)$ and $\frac{1}{\sqrt{2}}(1, 1, i, i)$ span lines of $\mathcal{O}_4$ a simple counting argument shows that $\mathcal{O}_4 = \mathcal{B}_4^{(4)} \cup \mathcal{F}_4 \cup \mathcal{N}_4$. $\square$

Theorem 7.36. *Suppose that $\mathcal{L}$ is an indecomposable 4-system and a simple extension of $\mathcal{B}_n^{(2)}$, where $n \geq 2$. Then the extension is equivalent to one of*

$$\mathcal{B}_n^{(2)} \subset \mathcal{B}_n^{(4)}, \quad \mathcal{B}_n^{(2)} \subset \mathcal{B}_{n+1}^{(2)}, \quad \mathcal{B}_2^{(2)} \subset \mathcal{D}_3^{(4)}, \quad \mathcal{B}_2^{(2)} \subset \mathcal{J}_3^{(4)},$$

$$\mathcal{B}_3^{(2)} \subset \mathcal{J}_3^{(4)}, \quad \mathcal{B}_3^{(2)} \subset \mathcal{F}_4, \quad \mathcal{B}_3^{(2)} \subset \mathcal{N}_4,$$

$$\mathcal{B}_4^{(2)} \subset \mathcal{F}_4, \quad \mathcal{B}_4^{(2)} \subset \mathcal{N}_4 \quad \text{or} \quad \mathcal{B}_4^{(2)} \subset \mathcal{O}_4.$$

Proof. We may suppose that $\mathcal{L}$ is a line system in $\mathbb{C}^{n+1}$ and the star-closure of $\mathcal{B}_n^{(2)}$ and a line ℓ with root $x = (\alpha_1, \alpha_2, \ldots, \alpha_{n+1})$. By Lemma 7.32 we may suppose that $n \geq 3$ and by Lemma 7.33 we can assume that $\alpha_h \neq 0$ for all $h \leq n$.

On taking inner products with the roots $\sqrt{2}\, e_h$ we find that for $h \leq n$, $|\alpha_h|$ is 1 or $\frac{1}{\sqrt{2}}$. In particular, $n \leq 4$ and $|\alpha_h| = 1$ for at most one value of h. Thus we may suppose that $|\alpha_1| = |\alpha_2| = \frac{1}{\sqrt{2}}$. If $|\alpha_3| = 1$, then n is 2 or 3. If $n = 3$, then $\alpha_4 = 0$ and in both cases we may suppose that $\mathcal{L}$ is contained in $\mathbb{C}^3$. By Lemma 7.34, $\mathcal{L} = \mathcal{J}_3^{(4)}$ or $\mathcal{L} = \overline{\mathcal{J}}_3^{(4)}$.

We may now assume that $|\alpha_h| = \frac{1}{\sqrt{2}}$ for $1 \leq h \leq n$, where n is 3 or 4. If $n = 3$, then $|\alpha_4| = \frac{1}{\sqrt{2}}$ whereas if $n = 4$, then $\alpha_5 = 0$. In both cases we may suppose that

$\mathcal{L}$ is contained in $\mathbb{C}^4$. The calculations of Lemma 7.35 show that we may assume that $\alpha_1 = \alpha_2 = \frac{1}{\sqrt{2}}$. If $a = (1, 1, 0, 0)$, then $r_x(a) = (0, 0, -\sqrt{2}\,\alpha_3, -\sqrt{2}\,\alpha_4)$ and therefore $\mathcal{L}$ contains a line system equivalent to $\mathcal{B}_4^{(2)}$. The result now follows from Lemma 7.35. $\qquad\square$

Corollary 7.37. *No indecomposable 4-system is a simple extension of $\mathcal{J}_3^{(4)}$.*

Proof. Suppose that $\mathcal{L}$ is an indecomposable 4-system and is the star-closure of $\mathcal{J}_3^{(4)}$ and a line ℓ with root x. We may suppose that $\mathcal{L}$ is contained in $\mathbb{C}^4$ and that $\mathcal{J}_3^{(4)}$ is the star-closure in $\mathbb{C}^4$ of $\mathcal{B}_3^{(2)}$ and the line spanned by $e := (\frac{1}{2}\lambda, \frac{1}{2}\lambda, 1, 0)$, where $\lambda = -\frac{1}{2}(1 + i\sqrt{7})$. From the theorem, the star-closure of $\mathcal{B}_3^{(2)}$ and ℓ is equivalent to $\mathcal{B}_3^{(4)}$, $\mathcal{B}_4^{(2)}$, $\mathcal{J}_3^{(4)}$, $\mathcal{F}_4$ or $\mathcal{N}_4$. If $\mathcal{L}$ is equivalent to $\mathcal{F}_4$ or $\mathcal{N}_4$ then $\mathcal{L}$ contains a line system equivalent to $\mathcal{B}_4^{(2)}$. On taking the inner product of e with $(1, 0, i, 0)$ and $(1, 0, 0, \alpha_4)$ we see that neither $\mathcal{B}_3^{(4)}$ nor $\mathcal{B}_4^{(2)}$ can occur. Thus the only possibility is that $\mathcal{L} \subset \mathbb{C}^3$ and $\mathcal{L}$ contains both $\mathcal{J}_3^{(4)}$ and $\overline{\mathcal{J}}_3^{(4)}$. But the lines spanned by e and $(\frac{1}{2}\bar{\lambda}, \frac{1}{2}\bar{\lambda}, -1, 0)$ are not part of a 4-system. This contradiction completes the proof. $\qquad\square$

Corollary 7.38. *If $\mathcal{L}$ is an indecomposable 4-system and a simple extension of $\mathcal{F}_4$ or $\mathcal{N}_4$, then $\mathcal{L} \simeq \mathcal{O}_4$.*

Proof. Suppose that $\mathcal{L}$ an indecomposable 4-system and the star-closure of $\mathcal{F}_4$ or $\mathcal{N}_4$ and a line ℓ with root x. We may suppose that $\mathcal{L}$ is contained in $\mathbb{C}^5$, that $\mathcal{F}_4$ is the star-closure of $\mathcal{B}_4^{(2)}$ with $e := \frac{1}{\sqrt{2}}(1, 1, 1, 1, 0)$ and that $\mathcal{N}_4$ is the star-closure of $\mathcal{B}_4^{(2)}$ with $f := \frac{1}{\sqrt{2}}(1, 1, i, i, 0)$. If the star-closure of $\mathcal{B}_4^{(2)}$ and ℓ is equivalent to $\mathcal{B}_5^{(2)}$, then $\mathcal{L}$ contains a line with root $g := (1, 0, 0, 0, \gamma)$ for some γ. But then $(e, g) = (f, g) = \frac{1}{\sqrt{2}}$, which is a contradiction. It now follows from the theorem that $\dim \mathcal{L} = 4$ and we may suppose that $\mathcal{L} \subset \mathbb{C}^4$. By Lemma 7.35, $\mathcal{L} = \mathcal{O}_4$. $\qquad\square$

Theorem 7.39. *Suppose that $\mathcal{L}$ is an indecomposable 4-system and a simple extension of $\mathcal{D}_n^{(4)}$, where $n \geq 3$. Then the extension is equivalent to one of*

$$\mathcal{D}_n^{(4)} \subset \mathcal{B}_n^{(4)}, \quad \mathcal{D}_n^{(4)} \subset \mathcal{D}_{n+1}^{(4)}, \quad \mathcal{D}_3^{(4)} \subset \mathcal{N}_4 \quad or \quad \mathcal{D}_4^{(4)} \subset \mathcal{N}_4.$$

Furthermore, there are exactly four extensions of $\mathcal{D}_4^{(4)}$ in $\mathbb{C}^4$ equivalent to $\mathcal{N}_4$.

Proof. We may suppose that $\mathcal{L}$ is a line system in $\mathbb{C}^{n+1}$ which is the star-closure of $\mathcal{D}_n^{(4)}$ and a line ℓ with root $x = (\alpha_1, \alpha_2, \ldots, \alpha_{n+1})$. By Lemma 7.20 we can assume that $\alpha_h \neq 0$ for all $h \leq n$.

For $1 \leq h < j \leq 3$ set

$$p_{hj} := |\alpha_h + \alpha_j|^2, \quad q_{hj} := |\alpha_h - \alpha_j|^2,$$
$$\hat{p}_{hj} := |\alpha_h + i\alpha_j|^2, \quad \hat{q}_{hj} := |\alpha_h - i\alpha_j|^2$$

and suppose that none of these quantities vanish. If $r_{hj} := p_{hj} + q_{hj}$, then

$$6 \leq r_{12} + r_{23} + r_{13} = 4(|\alpha_1|^2 + |\alpha_2|^2 + |\alpha_3|^2) \leq 8$$

and therefore we may choose the notation so that $r_{12} = 2$. It follows that $p_{12} = q_{12} = 1$ and as $\hat{p}_{hj} + \hat{q}_{hj} = p_{hj} + q_{hj} = 2$ we also have $\hat{p}_{12} = \hat{q}_{12} = 1$. But now $\alpha_1 \bar{\alpha}_2 = 0$, contrary to assumption.

The group $G(4, 1, n)$ fixes $\mathcal{D}_n^{(4)}$ and acts on $\mathbb{C}^n$ by permuting the coordinates and multiplying their values by 4th roots of unity. Therefore, up to equivalence, there exist α and β such that for $1 \leq h \leq n$, α_h is equal to either α or β. Suppose that there are k values of h such that $\alpha_h = \alpha$ and choose the notation so that $k \geq n - k$. Then $k \geq 2$ and it follows that $|\alpha| = \frac{1}{\sqrt{2}}$. If $k \neq n$, the calculation just carried out shows that $r := 2(|\alpha|^2 + |\beta|^2) \in \{2, 3, 4\}$. It follows that $r = 3$ and therefore $|\beta| = 1$. But now $k = 2$ and $n = 3$. Taking inner products with $(0, 1, \pm 1, 0)$ and $(0, 1, \pm i, 0)$ leads to a contradiction.

Thus $k = n$ and therefore $n \leq 4$. If $n = 3$ we may choose $\alpha_4 = \frac{1}{\sqrt{2}}$ so that $x = \frac{1}{\sqrt{2}}(1, 1, 1, 1)$. Then $a = (1, 1, 0, 0) \in \mathcal{D}_3^{(4)}$ and $r_x(a) = (0, 0, -1, -1)$, hence $\mathcal{D}_4^{(4)} \subset \mathcal{L}$. If $n = 4$, then $\alpha_5 = 0$ and so in both cases we may suppose that $\mathcal{L} \subset \mathbb{C}^4$, $\mathcal{D}_4^{(4)} \subset \mathcal{L}$ and $x = \frac{1}{\sqrt{2}}(1, 1, 1, 1)$. It follows that $\mathcal{L} = \mathcal{N}_4$. There are four images of $\mathcal{N}_4$ under the action of $G(4, 1, 4)$ and these are the only extensions of $\mathcal{D}_4^{(4)}$ in $\mathbb{C}^4$ equivalent to $\mathcal{N}_4$. $\qquad \square$

Theorem 7.40. *Suppose that $\mathcal{L}$ is an indecomposable 4-system and a simple extension of $\mathcal{B}_n^{(4)}$, where $n \geq 3$. Then the extension is equivalent to one of*

$$\mathcal{B}_n^{(4)} \subset \mathcal{B}_{n+1}^{(4)}, \quad \mathcal{B}_3^{(4)} \subset \mathcal{O}_4 \quad or \quad \mathcal{B}_4^{(4)} \subset \mathcal{O}_4.$$

Furthermore, there are exactly two extensions of $\mathcal{B}_4^{(4)}$ in $\mathbb{C}^4$ equivalent to $\mathcal{O}_4$.

Proof. We may suppose that $\mathcal{L}$ is a line system in $\mathbb{C}^{n+1}$ which is the star-closure of $\mathcal{B}_n^{(4)}$ and a line ℓ with root $x = (\alpha_1, \alpha_2, \ldots, \alpha_{n+1})$. From Lemma 7.33 we can assume that $\alpha_h \neq 0$ for all $h \leq n$. As in the proof of Theorem 7.36, α_h is 1 or $\frac{1}{\sqrt{2}}$ and we may suppose that $\alpha_1 = \alpha_2 = \frac{1}{\sqrt{2}}$. If $|\alpha_3| = 1$, then taking inner products with $(0, 1, \pm 1, \ldots)$ and $(0, 1, \pm i, \ldots)$ leads to a contradiction. Thus $|\alpha_h| = \frac{1}{\sqrt{2}}$ for all $h \leq n + 1$ and therefore $n = 3$ or $n = 4$.

If $n = 3$ and $a = (1, 1, 0, 0)$, then the star-closure of $r_x(a)$ and $\mathcal{B}_3^{(4)}$ is $\mathcal{B}_4^{(4)}$. If $n = 4$ we have $\alpha_5 = 0$ and therefore in both cases we may suppose that $\mathcal{L}$

is an extension of $\mathcal{B}_4^{(4)}$ in $\mathbb{C}^4$. But now $x = \frac{1}{\sqrt{2}}(1,1,1,1)$, $x = \frac{1}{\sqrt{2}}(1,1,1,i)$ or $x = \frac{1}{\sqrt{2}}(1,1,i,i)$ and it follows that $\mathcal{L} = \mathcal{O}_4$ or $\mathcal{L} = \overline{\mathcal{O}}_4$. $\qquad\square$

Corollary 7.41. *No indecomposable 4-system is a simple extension of $\mathcal{O}_4$.*

Proof. Suppose that $\mathcal{L}$ an indecomposable 4-system and the star-closure of $\mathcal{O}_4$ and a line ℓ with root x. We may suppose that $\mathcal{L}$ is contained in $\mathbb{C}^5$. From the theorem, the star-closure of $\mathcal{B}_4^{(4)}$ and ℓ is equivalent to $\mathcal{B}_5^{(4)}$ or $\mathcal{O}_4$. If the star-closure is $\mathcal{B}_5^{(4)}$, then $\mathcal{L} = \mathcal{B}_5^{(4)}$ because there are no extensions of $\mathcal{B}_5^{(4)}$ in $\mathbb{C}^5$. But $|\mathcal{B}_5^{(4)}| = 45$, which is a contradiction. Therefore we may suppose that $\mathcal{L} \subset \mathbb{C}^4$. In this case $\mathcal{L}$ contains both $\mathcal{O}_4$ and $\overline{\mathcal{O}}_4$ and hence contains lines with roots $\frac{1}{\sqrt{2}}(1,1,1,1)$ and $\frac{1}{\sqrt{2}}(1,1,1,i)$, which is a contradiction. $\qquad\square$

Theorem 7.42. *If $\mathcal{L}$ is an indecomposable 4-system, then for some n, $\mathcal{L}$ is equivalent to $\mathcal{B}_n^{(2)}$, $\mathcal{B}_n^{(4)}$, or $\mathcal{D}_n^{(4)}$, or to $\mathcal{F}_4$, $\mathcal{J}_3^{(4)}$, $\mathcal{N}_4$ or $\mathcal{O}_4$.*

Proof. Since $\mathcal{L}$ is a 4-system it contains a pair of lines at $45°$ and hence a 4-star σ. By Theorem 7.13, $\mathcal{L}$ is the end of a chain of simple indecomposable extensions beginning at σ. The theorem follows from the combined results of Theorems 7.36, 7.39, 7.40 and their corollaries. $\qquad\square$

Exercises

1. Let Σ be a set of unit vectors that span distinct one-dimensional subspaces in $\mathbb{R}^n$. If $A := \{\,|(u,v)|^2 \mid u, v \in \Sigma, u \neq v\,\}$ and $s := |A|$, then

$$|\Sigma| \leq \begin{cases} \dbinom{n + 2s - 2}{2s - 1} & \text{if } 0 \in A \\[2ex] \dbinom{n + 2s - 1}{2s} & \text{if } 0 \notin A. \end{cases}$$

   Show that the line system $\mathcal{E}_8$ meets the bound.

2. Which of the indecomposable line systems described in this chapter meet the bounds of Theorem 7.2 or the previous exercise?

3. Suppose that $\mathcal{L}$ and $\mathfrak{M}$ are indecomposable star-closed line systems such that $\mathfrak{M}$ is a simple extension of $\mathcal{L}$. If U is a decomposable line system such that $\mathcal{L} \subset U \subset \mathfrak{M}$, show that for some $\ell \in \mathfrak{M}$, $U = U_1 \perp \{\ell\}$, where $\mathcal{L} \subseteq U_1$, $\dim \mathcal{L} = \dim U_1$ and $\dim \mathfrak{M} = \dim \mathcal{L} + 1$.

4. Verify the inclusions:

$$3\mathcal{A}_2 \subset \mathcal{D}_4^{(3)} \perp \mathcal{A}_2 \subset \mathcal{D}_6^{(3)} \subset \mathcal{K}_6.$$

5. Using the description of $\mathcal{E}_8$ given in §6, let H be the stabiliser in $W(\mathcal{D}_8)$ of the line ℓ spanned by $z = \frac{1}{2}(e_1 + e_2 + \cdots + e_8)$. Show that H is isomorphic to $C_2 \times W(\mathcal{A}_7)$ and that H has three orbits on the lines spanned by the elements of X; namely $\{\ell\}$, the 35 lines orthogonal to ℓ and the 28 lines at $60°$ to ℓ.

6. Show that the orbit of $\frac{1}{2}(1,1,1,1,1,i\sqrt{3})$ under the action of $W(\mathcal{D}_6^{(2)})$ contains 192 vectors, which span 96 lines. Show that these lines together with the 30 lines of $\mathcal{D}_6^{(2)}$ form a 3-system in $\mathbb{C}^6$.

7. Let $\mathcal{L}$ be an indecomposable 3-system in $\mathbb{C}^n$ and choose a star $\sigma = \{a, b, c\}$ such that $(a,b) = (b,c) = (c,a) = -1$. Let Γ_a be the roots x of $\mathcal{L}$ such that $(a,x) = 0$ and $(b,x) = 1$, as in §7. Define a graph on Γ_a by declaring x to be adjacent to y whenever $(x,y) = 0$. Define a triangle to be three pairwise adjacent elements of Γ_a. Prove the following.

   (i)   Each edge in Γ_a is in a unique triangle.

   (ii)   Given a triangle Δ_1 and an element u of Γ_a such that u is not joined to any vertex of Δ_1, there is a unique triangle Δ_2 such that $u \in \Delta_2$ and no vertex of Δ_1 is joined to a vertex of Δ_2.

   (iii) Given triangles Δ_1 and Δ_2 in Γ_a such that there are no edges between vertices of Δ_1 and Δ_2, there is a unique triangle Δ_3 in Γ_a such that no vertex of Δ_3 is joined to a vertex of Δ_1 or Δ_2.

   (iv)   Given $x \in \Delta_1$ show that there is a unique element $y \in \Delta_2$ such that $(x,y) = 1$ and a unique element $z \in \Delta_3$ such that $(y,z) = 1$. Furthermore the line spanned by the 9 elements of $\Delta_1 \cup \Delta_2 \cup \Delta_3$ together with the 3 elements $x - y$, $y - z$ and $z - x$ is a star-closed line system of type $\mathcal{D}_4^{(2)}$.

8. Let $\mathcal{L}$ be an indecomposable 3-system. For $\ell \in \mathcal{L}$ define

$$\sigma(\ell) = \{\, m \in \mathcal{L} \mid \mathcal{L} \cap \ell^\perp = \mathcal{L} \cap m^\perp \,\} \text{ and}$$
$$\tau(\ell) = \{\, m \in \mathcal{L} \mid \{\ell\} \cup (\mathcal{L} \cap \ell^\perp) = \{m\} \cup (\mathcal{L} \cap m^\perp) \,\}.$$

Prove that

   (i) $\sigma(\ell)$ and $\tau(\ell)$ are blocks of imprimitivity for the permutation action of $W(\mathcal{L})$ on $\mathcal{L}$.

   (ii)   If $\sigma(\ell) \neq \{\ell\}$, then either $|\mathcal{L}| = |\sigma(\ell)| = 9$ and $\mathcal{L}$ is $\mathcal{D}_3^{(3)}$ or $\sigma(\ell)$ is a star and $\mathcal{L}$ is $\mathcal{D}_n^{(3)}$ for some $n \neq 3$.

   (iii) If $\tau(\ell) \neq \{\ell\}$, then $\tau(\ell) = \{\ell, m\}$, where m is orthogonal to ℓ, and $\mathcal{L}$ is $\mathcal{D}_n^{(2)}$ for some n.

This is a special case of a theorem of Fischer [97]. Further details can be found in the book by Aschbacher [7].

9. Show that the line system $\mathcal{K}_5$ has the following proper subsystems and in each case show that $W(\mathcal{K}_5)$ has a single orbit on the subsystems of a given type.
   (i)   Rank 1: $\mathcal{A}_1$.
   (ii)   Rank 2: $\mathcal{A}_2$ and $2\mathcal{A}_1$.
   (iii)   Rank 3: $\mathcal{A}_3$, $\mathcal{A}_2 \perp \mathcal{A}_1$, $3\mathcal{A}_1$ and $\mathcal{D}_3^{(3)}$.
   (iv)   Rank 4: $\mathcal{A}_4$, $\mathcal{A}_3 \perp \mathcal{A}_1$, $2\mathcal{A}_2$, $4\mathcal{A}_1$, $\mathcal{D}_4^{(2)}$ and $\mathcal{D}_4^{(3)}$.
   (v)   Rank 5: $\mathcal{A}_5$, $5\mathcal{A}_1$ and $\mathcal{D}_4^{(2)} \perp \mathcal{A}_1$.

10. Define a subsystem $\mathfrak{M}$ of a line system $\mathfrak{L}$ in a space V to be *parabolic* if it consists of all the lines of $\mathfrak{L}$ orthogonal to some subspace of V. Show that every proper subsystem of $\mathcal{K}_5$ is parabolic except for $2\mathcal{A}_2$, $4\mathcal{A}_1$ and those of rank 5.

11. Show that the line system $\mathcal{K}_6$ has the following proper subsystems.
   (i)   Rank 1: $\mathcal{A}_1$.
   (ii)   Rank 2: $\mathcal{A}_2$ and $2\mathcal{A}_1$.
   (iii)   Rank 3: $\mathcal{A}_3$, $\mathcal{A}_2 \perp \mathcal{A}_1$, $3\mathcal{A}_1$ and $\mathcal{D}_3^{(3)}$.
   (iv)   Rank 4: $\mathcal{A}_4$, $\mathcal{A}_3 \perp \mathcal{A}_1$, $2\mathcal{A}_2$, $\mathcal{A}_2 \perp 2\mathcal{A}_1$, $4\mathcal{A}_1$, $\mathcal{D}_4^{(2)}$, $\mathcal{D}_4^{(3)}$ and $\mathcal{D}_3^{(3)} \perp \mathcal{A}_1$.
   (v)   Rank 5: $\mathcal{A}_5$, $\mathcal{A}_4 \perp \mathcal{A}_1$, $\mathcal{A}_3 \perp \mathcal{A}_2$, $\mathcal{A}_3 \perp 2\mathcal{A}_1$, $2\mathcal{A}_2 \perp \mathcal{A}_1$, $5\mathcal{A}_1$, $\mathcal{D}_5^{(2)}$, $\mathcal{D}_4^{(2)} \perp \mathcal{A}_1$, $\mathcal{D}_5^{(3)}$, $\mathcal{D}_4^{(3)} \perp \mathcal{A}_1$, $\mathcal{D}_3^{(3)} \perp \mathcal{A}_2$ and $\mathcal{K}_5$.
   (vi)   Rank 6: $\mathcal{A}_6$, $\mathcal{A}_5 \perp \mathcal{A}_1$, $2\mathcal{A}_3$, $3\mathcal{A}_2$, $6\mathcal{A}_1$, $\mathcal{D}_6^{(2)}$, $\mathcal{D}_4^{(2)} \perp 2\mathcal{A}_1$, $\mathcal{D}_6^{(3)}$, $\mathcal{D}_4^{(3)} \perp \mathcal{A}_2$, $2\mathcal{D}_3^{(3)}$, $\mathcal{E}_6$ and $\mathcal{K}_5 \perp \mathcal{A}_1$.

   Show that $W(\mathcal{K}_6)$ has a single orbit on the subsystems of a given type except that there are two orbits on subsystems of type $2\mathcal{A}_2$, three orbits on subsystems of type $\mathcal{A}_5$ and two orbits on subsystems of type $\mathcal{A}_6$.

12. Show that every proper subsystem of $\mathcal{K}_6$ of rank less than 6 is parabolic except for $4\mathcal{A}_1$, $\mathcal{A}_3 \perp 2\mathcal{A}_1$, $2\mathcal{A}_2 \perp \mathcal{A}_1$, $5\mathcal{A}_1$, $\mathcal{D}_3^{(3)} \perp \mathcal{A}_2$, $\mathcal{D}_4^{(2)} \perp \mathcal{A}_1$, one orbit of subsystems of type $2\mathcal{A}_2$ and one orbit of subsystems of type $\mathcal{A}_5$.

13. Suppose that $\mathfrak{L}$ is an indecomposable 3-system containing a star with roots a, b and c. If there exists $d \in \Gamma_a$ such that $(x, d) = 0$ for all $x \in \Gamma_a \setminus \{d\}$, then $\mathfrak{L}$ is equivalent to $\mathcal{D}_n^{(2)}$.

14. Suppose that $\mathfrak{L}$ is an indecomposable 3-system in $\mathbb{R}^7$ and a simple extension of $\mathcal{A}_6$ (as in the proof of Theorem 7.28). Show that the linear transformation $\varphi : \mathbb{R}^7 \to \mathbb{R}^8$ with matrix $\begin{bmatrix} I_7 - \frac{4+\sqrt{2}}{28} J_7 \\ \frac{1}{2\sqrt{2}} \mathbf{j}_7 \end{bmatrix}$ is an isometry that defines an equivalence between $\mathfrak{L}$ and $\mathcal{E}_7$.

15. Define a *frame* in a line system of dimension n to be a set of n pairwise orthogonal lines. Show that there are 14 frames in $\mathcal{J}_3^{(4)}$ and that they form two orbits, each of 7 frames, for the group $W(\mathcal{J}_3^{(4)})$. Define two frames to be *incident* if they have exactly one line in common. Show that the stabiliser of a frame in $W(\mathcal{J}_3^{(4)})$ is a conjugate of $W(\mathcal{B}_3^{(2)})$ and that each frame is incident with 3 frames in the opposite orbit. Deduce that $W(\mathcal{J}_3^{(4)}) = \langle -I \rangle \times G$, where $G \simeq SL_3(\mathbb{F}_2)$ is the group of collineations of the 7-point plane.

CHAPTER 8

The Shephard and Todd classification

This chapter contains the complete classification of finite unitary reflection groups. For the primitive groups in higher dimensions this is largely due to Mitchell [**166**], who published his results in 1914. However, the successful determination of the finite groups in dimensions three and four was achieved by a number of authors: Bagnera [**9**], Blichfeldt [**22, 23, 24, 25**] following earlier work of Klein [**130**], Jordan [**123**] and Valentiner [**220**] and then reproved and generalised by Mitchell [**164, 165**] using more geometric methods. Their work was expressed in terms of projective groups and in this setting the reflection groups correspond to groups generated by homologies. The first complete list of the linear reflection groups was compiled by Shephard and Todd [**193**] in 1954. In 1976 Cohen [**54**] gave a new proof of the classification – extending the classification of finite Coxeter groups but independent of the earlier results of Mitchell and others. A brief history of the subject can be found in the article by Kantor [**125**].

1. Outline of the classification

Fix an hermitian inner product $(-, -)$ on the complex vector space V of dimension n and let $U(V)$ be the corresponding unitary group. It follows from the equivalence of any two positive definite hermitian forms on V (page 7) and Lemma 1.3 that if G is any finite subgroup of $GL(V)$, then G is conjugate to a subgroup of $U(V)$.

Hence by Theorem 1.33 any finite reflection group G in V belongs to a unique $U(V)$-conjugacy class of reflection subgroups of $U(V)$. Therefore, the classification of finite reflection subgroups of $GL(V)$, up to conjugacy under $GL(V)$, is equivalent to the classification of finite reflection subgroups of $U(V)$, up to conjugacy under $U(V)$. We now give an outline of the steps in this classification.

Suppose that G is a finite group generated by unitary reflections acting on a vector space V of dimension n and leaving the positive definite hermitian form $(-, -)$ invariant.

1. By Theorem 1.27, V is the direct sum of mutually orthogonal subspaces V_1, $V_2, \ldots, V_m$ such that the restriction G_i of G to V_i acts irreducibly on V_i and

137

$G \simeq G_1 \times G_2 \times \cdots \times G_m$. Therefore we may suppose that V is an irreducible G-module.

2. If $n = 1$, then G is cyclic and so we may suppose that $n \geq 2$.

3. If G is irreducible and imprimitive, then by Theorem 2.14 G is conjugate to $G(m, p, n)$ for some m and some divisor p of m.

4. If G is primitive and $n = 2$, then G is of tetrahedral, octahedral or icosahedral type. The results of Chapter 6 show that there are 19 possibilities: 4 of tetrahedral type, 8 of octahedral type and 7 of icosahedral type.

5. By a result of Blichfeldt (see Theorem 8.1 below), if G is primitive and $n \geq 3$, the reflections in G have orders 2 or 3. Furthermore, if G contains reflections of order 3, there are just three possibilities: the groups $W(\mathcal{L}_3)$, $W(\mathcal{L}_4)$ or $W(\mathcal{M}_3)$, defined in §5 below.

6. If $n \geq 3$ and there are no reflections of order 3 in G, then another application of Blichfeldt's Theorem shows that the order of the product of two reflections is at most 5 (Theorem 8.5).

7. Suppose that G is primitive, that $n \geq 3$, and that G is generated by reflections of order 2. Then the set $\mathcal{L}$ of lines spanned by the roots of the reflections in G is an indecomposable star-closed line system and $G = W(\mathcal{L})$.

   (a) If $\mathcal{L}$ is a 3-system, then by Theorem 7.31, $\mathcal{L}$ is equivalent to $\mathcal{A}_n$ or to one of $\mathcal{K}_5$, $\mathcal{K}_6$, $\mathcal{E}_6$, $\mathcal{E}_7$ or $\mathcal{E}_8$.

   (b) If $\mathcal{L}$ is a 4-system, then by Theorem 7.42, $\mathcal{L}$ is equivalent to one of $\mathcal{J}_3^{(4)}$, $\mathcal{F}_4$, $\mathcal{N}_4$ or $\mathcal{O}_4$.

   (c) If $\mathcal{L}$ is a 5-system, then by Theorems 8.10 and 8.15 and their corollaries $\mathcal{L}$ is equivalent to one of $\mathcal{H}_3$, $\mathcal{J}_3^{(5)}$ or $\mathcal{H}_4$.

8. Thus if G is irreducible and neither imprimitive, cyclic nor isomorphic to $\mathrm{Sym}(n)$ for some n, then it is one of 34 primitive reflection groups of rank at most 8.

2. Blichfeldt's Theorem

In this section we prove a theorem of Blichfeldt [26] that is fundamental to the classification of primitive unitary reflection groups. The proof given here follows that of Theorem B of Robinson [**186**]. Another proof can be found in [**86**].

If (u, v) denotes the usual inner product on $\mathbb{C}^n$, the *length* of $v \in \mathbb{C}^n$ is $\|v\| := \sqrt{(v, v)}$ and the *norm* of a linear transformation $t : \mathbb{C}^n \to \mathbb{C}^n$ is defined by

$$\|t\| := \sup\{ \|tv\| \mid \|v\| = 1 \}.$$

Then $\|tv\| \leq \|t\| \, \|v\|$ for all v and $\|t\|$ is the largest absolute value of an eigenvalue of t.

If $s \in U_n(\mathbb{C})$ has finite order, then its eigenvalues are roots of unity. Thus in the following theorem the condition that s has an eigenvalue λ such that $|\lambda - \mu| \leq 1$ for all eigenvalues μ of s is equivalent to saying that all eigenvalues of s lie within $\pi/3$ of λ on the unit circle.

Theorem 8.1. *Let G be a finite primitive subgroup of $U_n(\mathbb{C})$ and suppose that $s \in G$ has an eigenvalue λ such that $|\lambda - \mu| \leq 1$ for all eigenvalues μ of s. Then $s \in Z(G)$.*

Proof. Suppose, by way of contradiction, that $s \notin Z(G)$ and let $H := \langle s^G \rangle$ be the normal closure of s in G. Let $V := \mathbb{C}^n$ denote the space on which G acts and put $V_\lambda := \{ v \in V \mid sv = \lambda v \}$.

Since $s \notin Z(G)$, we have $V_\lambda \neq V$. If V_λ were H-invariant, then its images under the action of G would form a system of imprimitivity for G, contrary to hypothesis. Thus V_λ cannot be H-invariant. Let $v_1, v_2, \ldots, v_r$ be an orthonormal basis for V_λ and choose a conjugate t of s such that $tV_\lambda \neq V_\lambda$ and such that $\sum_{i=1}^r \|(t - \lambda)v_i\|$ is minimal.

Suppose first that $t^{-1}stV_\lambda = V_\lambda$. If $t^{-1}stv = \lambda v$ for all $v \in V_\lambda$, then $tv \in V_\lambda$ for all $v \in V_\lambda$, contrary to the choice of t. Hence $t^{-1}st$ has an eigenvector $v \in V_\lambda$ with eigenvalue $\mu \neq \lambda$. That is, $t^{-1}stv = \mu v$ and so $stv = \mu tv$. Since v and tv are eigenvectors of s corresponding to different eigenvalues, we have $(v, tv) = 0$ and consequently $\|(t - \lambda)v\|^2 = (tv - \lambda v, tv - \lambda v) = 2\|v\|^2$. On the other hand, for every eigenvalue θ of t we have $|\theta - \lambda| \leq 1$ and hence $\|t - \lambda\| \leq 1$, which is a contradiction. This proves that V_λ is not $t^{-1}st$-invariant.

For any $\mu_1, \mu_2, \ldots, \mu_r \in \mathbb{C}$ we have

$$\sum_{i=1}^r \|(t^{-1}st - \lambda)v_i\| = \sum_{i=1}^r \|(t^{-1}st - s)v_i\| = \sum_{i=1}^r \|(st - ts)v_i\|$$

$$= \sum_{i=1}^r \|((s - \lambda)(t - \mu_i) - (t - \mu_i)(s - \lambda))v_i\|$$

$$= \sum_{i=1}^r \|(s - \lambda)(t - \mu_i)v_i\|$$

$$\leq \sum_{i=1}^r \|(t - \mu_i)v_i\|, \quad \text{because } \|s - \lambda\| \leq 1.$$

For each i, choose μ_i so that $\|(t - \mu_i)v_i\|$ is minimal. Then $\sum_{i=1}^r \|(t - \mu_i)v_i\| \leq \sum_{i=1}^r \|(t - \lambda)v_i\|$, whence

$$\sum_{i=1}^r \|(t^{-1}st - \lambda)v_i\| \leq \sum_{i=1}^r \|(t - \mu_i)v_i\| \leq \sum_{i=1}^r \|(t - \lambda)v_i\|$$

and our initial choice of t forces equality throughout, and implies that $\mu_i = \lambda$ realises the minimum value of $\|(t - \mu_i)v_i\|$. But now

$$
\begin{aligned}
\|(t - \mu_i)v_i\|^2 &= \|tv_i\|^2 - \overline{\mu}_i(tv_i, v_i) - \mu_i\overline{(tv_i, v_i)} + |\mu_i|^2\|v_i\|^2 \\
&= 1 + |\mu_i|^2 - \overline{\mu}_i(tv_i, v_i) - \mu_i\overline{(tv_i, v_i)} \\
&= 1 + |\mu_i - (tv_i, v_i)|^2 - |(tv_i, v_i)|^2
\end{aligned}
$$

and so the minimum of this expression is 0 and occurs when $\mu_i = (tv_i, v_i) = \lambda$ for all i. Hence $tv_i = \lambda v_i$ for all i, a contradiction. This completes the proof. $\qquad\square$

The following Corollary appears as Exercise 2 in §73 of Blichfeldt [26].

Corollary 8.2. *Let G be a finite primitive subgroup of $U_n(\mathbb{C})$ and suppose that for $s \in G$, all eigenvalues of s lie on an arc of length less than $2\pi/5$ on the unit circle. Then $s \in Z(G)$.*

Proof. (Robinson [186]) The eigenvalues of s have the form $\exp(i\theta)\lambda$ for some eigenvalue λ, where $0 \le \theta < 2\pi/5$. If there is an eigenvalue $\exp(i\theta)\lambda$ such that $\pi/15 \le \theta \le \pi/3$, then all eigenvalues of s lie within $\pi/3$ of $\exp(i\theta)\lambda$ and therefore $s \in Z(G)$ by the theorem. Thus we may suppose that there is no such eigenvalue. It follows that every eigenvalue of s^5 lies within $\pi/3$ of λ^5 and applying the theorem once more we see that $s^5 \in Z(G)$, whence $s^5 = \lambda^5 I$. But all eigenvalues of s are strictly within $2\pi/5$ of λ and therefore $s = \lambda I$ and hence $s \in Z(G)$. $\qquad\square$

3. Consequences of Blichfeldt's Theorem

Blichfeldt's Theorem imposes considerable restrictions on the structure of a finite unitary reflection group. Some of these restrictions are set out in the following theorems, most of which can be found in Cohen [54]. Many of these results, as well as the classification of primitive reflection groups, were known to Bagnera [9] who obtained them by geometric methods.

Theorem 8.3. *If $n > 1$ and G is a finite primitive subgroup of $U_n(\mathbb{C})$, then every reflection in G has order 2, 3, 4 or 5.*

Proof. If $t \in G$ is a reflection of order m, then the eigenvalues of t are 1 and $\exp(2\pi ik/m)$ for some k coprime to m. Replacing t by a power we may suppose that $k = 1$. From Blichfeldt's Theorem we have $2\pi/m > \pi/3$; that is, $m < 6$. $\qquad\square$

The classification of the two-dimensional primitive reflection groups in Chapter 6 shows that all possibilities allowed by this corollary do occur. However, in dimensions beyond two the situation is more restricted.

Theorem 8.4. *Suppose that G is a finite primitive group acting on the space V and that $\dim V \ge 3$.*

(i)   If H is a reflection subgroup of rank 2 in G and if the action of H on its support is primitive, then H may be identified with the group $G_4 \simeq SL_2(\mathbb{F}_3)$ of Chapter 6.

(ii)   The reflections in G have orders 2 or 3.

Proof. Let W be the subspace of V spanned by the roots of the reflections in H. The group H cannot contain the binary tetrahedral group $\mathcal{T}$ of Chapter 5. This is because $\mathcal{T}$ contains an element of order 6 and determinant 1 which would have eigenvalues $-\omega$ and $-\omega^2$ in its action on W. These eigenvalues are within $\pi/3$ of 1 and so by Blichfeldt's Theorem the element of order 6 is central in G, and hence in $\mathcal{T}$, which is impossible since $\mathcal{T}$ has centre of order 2. From the list of 19 primitive reflection subgroups of $U_2(\mathbb{C})$ this leaves only $G_4 \simeq SL_2(\mathbb{F}_3)$ and $G_6 \simeq C_4 \circ SL_2(\mathbb{F}_3)$ as candidates for H.

Suppose that $t \in G$ is a reflection of order at least 4 and let v be a root of t. Since G is primitive, the subspaces $\mathbb{C}gv$ (for $g \in G$) cannot form a system of imprimitivity for G. Hence there exists $g \in G$ such that v and $w := gv$ are linearly independent and w is not orthogonal to v. Let K be the subgroup of G generated by the reflections with root v or w. This group acts on the subspace spanned by v and w and since any generating set of reflections of the imprimitive group $G(m, p, 2)$ contains an involution, the group K must be primitive. Consequently, as in the previous paragraph, it must be isomorphic to $SL_2(\mathbb{F}_3)$ or $C_4 \circ SL_2(\mathbb{F}_3)$. But neither of these groups contains a reflection of order greater than 3. This proves (ii).

To complete the proof of (i), choose a reflection s with root $a \notin W \cup W^\perp$. This can be done since G is primitive and hence the normal subgroup generated by the reflections in G acts irreducibly on V. Then the group $K := \langle H, s \rangle$ acts irreducibly on $W \oplus \mathbb{C}a$ and from Theorem 2.15 we see that K is primitive. If $H \simeq C_4 \circ SL_2(\mathbb{F}_3)$, then K contains an element conjugate to $\begin{bmatrix} 1 & 0 & 0 \\ 0 & i & 0 \\ 0 & 0 & i \end{bmatrix}$. But then the group $\langle K, iI \rangle$ is primitive and contains a reflection of order 4 contrary to the previous paragraph. This prove (i). $\square$

The proof of this theorem shows that a primitive group that acts on a space of dimension at least 3 and that contains reflections of order 3 must contain a reflection subgroup of type $G_4 \simeq SL_2(\mathbb{F}_3)$.

Theorem 8.5. *Suppose that G is a finite primitive group of rank at least 3. If H is a reflection subgroup of rank 2 that acts irreducibly and imprimitively on its support, then H is conjugate to one of*

$$G(3, 1, 2), \quad G(3, 3, 2), \quad G(4, 2, 2), \quad G(2, 1, 2) \simeq G(4, 4, 2) \quad or \quad G(5, 5, 2).$$

If H contains a reflection of order 3, then $H \simeq G(3, 1, 2)$.

Proof. From Theorem 2.14, H is conjugate to $G(m, p, 2)$ for some divisor p of m. Thus H contains an element with eigenvalues $\exp(2\pi i/m)$, $\exp(-2\pi i/m)$ and 1. By

Blichfeldt's Theorem we have $m \leq 5$. In addition, if $p \neq m$, H contains a reflection of order m/p and thus from Theorem 8.4 we have $m \leq 3p$. Except for $G(2,2,2)$, the groups $G(m,p,2)$ with $m \leq 5$ and $m \leq 3p$ are irreducible and this completes the proof.   □

Corollary 8.6. *Suppose that G is a finite primitive unitary reflection group of rank at least 3. If G contains no reflections of order 3, then the set of lines spanned by the roots of the reflections is an m-system for some $m \leq 5$.*

Proof. If r and s are reflections in G and if r and s do not commute, then $H :=$ $\langle r, s \rangle$ is an irreducible imprimitive reflection group of rank 2. The group $G(4,2,2)$ cannot be generated by just two of its reflections and therefore, up to conjugacy, H is $G(m,m,2)$, where $m \leq 5$.   □

Theorem 8.7. *If H is a primitive reflection subgroup of the finite primitive group G acting on the space V of dimension $n \geq 3$ and if the rank of H is less than n, then $|Z(H)| \leq 3$.*

Proof. Let m be the rank of H. If $m = 1$, the result follows from Theorem 8.4, so we assume that $m > 1$. Suppose that $m = n - 1$ and that $Z(H)$ contains an element z of order k. If λ is an eigenvalue of z, then $G\langle \lambda I \rangle$ is a primitive group which contains a reflection of order k and therefore $k \leq 3$.

In general, if H is a reflection subgroup of type A_m $(m > 1)$, then $|Z(H)| = 1$ and we are done. Otherwise, by Corollary 1.31, we may find a reflection r such that $K := \langle H, r \rangle$ is an irreducible reflection subgroup of rank $m + 1$, which by Theorem 2.15 is primitive. This reduces the situation to the previous paragraph, with G replaced by K, and the proof is complete.   □

4. Extensions of 5-systems

In this section we complete the classification of the finite primitive reflection groups generated by reflections of order 2 by classifying the indecomposable 5-systems. The 5-system $\mathcal{D}_n^{(5)}$ with reflection group $W(\mathcal{D}_n^{(5)}) = G(5,5,n)$ was defined in Chapter 7. Every 5-system contains $\mathcal{D}_2^{(5)}$ and we begin by determining its extensions.

Lemma 8.8. *Suppose that $\mathfrak{L}$ is an indecomposable 5-system and the star-closure in $\mathbb{C}^3$ of $\sigma := \mathcal{D}_2^{(5)}$ and a line ℓ. Then $\mathfrak{L}$ is equivalent to $\mathcal{H}_3$ if and only if ℓ is orthogonal to a line of σ.*

Proof. If $\mathfrak{L} = \mathcal{H}_3$, then one can check directly that every line of $\mathfrak{L}$ not in σ is orthogonal to exactly one line of σ.

Conversely, let x be a root of ℓ and suppose that ℓ is orthogonal to a line of σ spanned by the root a. Choose a root b of σ such that $(a, b) = \frac{1}{2}\tau^{-1}$ and scale x

so that (b, x) is real and positive. Then $\mathfrak{L}$ is the star-closure of a, b and x and the matrices of the reflections r_a, r_b and r_x have real entries. Therefore we may suppose that $\mathfrak{L} \subset \mathbb{R}^3$ and choose a basis of $\mathbb{R}^3$ so that $a = (0, 0, 1)$, $b = (1, \tau, \tau^{-1})$, and $x = (\alpha, \beta, 0)$ for some $\alpha, \beta \in \mathbb{R}$.

Put $d := r_b r_a(b) = \frac{1}{2}(\tau^{-1}, 1, \tau)$. Then $(x, d) = \tau^{-1}(x, b)$ and since (x, b) and (x, d) belong to $\{\frac{1}{2}, \frac{1}{\sqrt{2}}, \frac{1}{2}\tau, \frac{1}{2}\tau^{-1}\}$ it follows that $(x, b) \in \{\frac{1}{2}, \frac{1}{2}\tau\}$.

Suppose first that $(x, b) = \frac{1}{2}$. Then $\alpha^2 + \beta^2 = 1$ and $\alpha + \tau\beta = 1$, hence $\beta((\tau + 2)\beta - 2\tau) = 0$. Thus either $\beta = 0$, $x = (1, 0, 0)$ and $\mathfrak{L} = \mathcal{H}_3$ or $\beta = 2\tau/(\tau + 2) = \frac{2}{\sqrt{5}}$. In the latter case $x = \frac{1}{\sqrt{5}}(-1, 2, 0)$ and the linear transformation $\varphi : \mathbb{C}^3 \to \mathbb{C}^3$ with orthogonal matrix

$$
A := \begin{bmatrix} -\frac{1}{\sqrt{5}} & \frac{2}{\sqrt{5}} & 0 \\ \frac{2}{\sqrt{5}} & \frac{1}{\sqrt{5}} & 0 \\ 0 & 0 & 1 \end{bmatrix}
$$

interchanges $\mathfrak{L}$ and $\mathcal{H}_3$.

If $(x, b) = \frac{1}{2}\tau$, then $\tau^{-1}\alpha + \beta = 1$ and a similar calculation to that above shows that either $\alpha = 0$ or $\alpha = 2\tau^{-1}/(2 - \tau^{-1}) = \frac{2}{\sqrt{5}}$. We have the same possibilities for $\mathfrak{L}$ as before: if $\alpha = 0$, then $\mathfrak{L} = \mathcal{H}_3$ and if $\alpha = \frac{2}{\sqrt{5}}$, then $x = \frac{1}{\sqrt{5}}(2, 1, 0)$ and φ defines an equivalence between $\mathfrak{L}$ and $\mathcal{H}_3$. $\qquad\square$

If $\mathfrak{L}$ is a 5-system, then by definition $\mathfrak{L}$ contains a 5-star $\sigma \simeq \mathcal{D}_2^{(5)}$. The standard representation of $\mathcal{D}_2^{(5)}$, given in §5 of Chapter 7 has (long) roots $(1, -\zeta^i)$ where $(0 \le i \le 4)$ and where $\zeta := \exp(2\pi/5)$. In what follows we shall represent the lines of $\mathfrak{L}$ by long roots (that is, vectors of length $\sqrt{2}$) and use the embedding of $\mathcal{D}_2^{(5)}$ in $\mathbb{C}^3$ with long roots $a_i := (1, -\zeta^i, 0)$. Note that $\tau = 1 + \zeta + \zeta^4 = -\zeta^2 - \zeta^3$ and so

$$
|(a_i, a_j)| = \begin{cases} \tau & \text{if } j \equiv \pm i + 1 \pmod 5 \\ \tau^{-1} & \text{if } j \equiv \pm i + 2 \pmod 5. \end{cases}
$$

Lemma 8.9. *Let* $\sigma = \mathcal{D}_2^{(5)}$ *and suppose that* $\mathfrak{L}$ *is an indecomposable 5-system and the star-closure in* $\mathbb{C}^3$ *of* σ *and a line* ℓ *spanned by a (long) root* x. *For* $0 \le i < 5$ *put* $n_i := |(x, a_i)|^2$. *If* ℓ *is not orthogonal to any line of* σ, *then either*

(i) $n_i = 1$ *for all* i *and* $\mathfrak{L}$ *is equivalent to* $\mathcal{D}_3^{(5)}$ *or*

(ii)   *we may choose* x *so that* $n_0 = \tau^2$, $n_1 = n_4 = 2$, $n_2 = n_3 = 1$ *and* $\mathfrak{L}$ *properly contains* $\mathcal{H}_3$.

Proof. Suppose $x := (\alpha, \beta, \gamma)$. If $\beta = \zeta^i \alpha$ for some i, then $(a_i, x) = 0$, contrary to assumption. Thus we may suppose that $\beta \ne \zeta^i \alpha$ for $0 \le i < 5$. In particular, the quantities $n_i = |\alpha - \bar{\zeta}^i \beta|^2$ are non-zero and take values in $\{1, 2, \tau^2, \tau^{-2}\}$.

For $0 \leq i < 5$ we have $n_i = |\alpha|^2 - (\bar{\zeta}^i \bar{\alpha} \beta + \zeta^i \alpha \bar{\beta}) + |\beta|^2$ and on summing these equations we find that

$$5(|\alpha|^2 + |\beta|^2) = n_0 + n_1 + n_2 + n_3 + n_4.$$

Setting $n := |\alpha|^2 + |\beta|^2$ and $\mu := \alpha \bar{\beta}$, we have $\zeta^i \mu + \bar{\zeta}^i \bar{\mu} = n - n_i$ for $0 \leq i < 5$. If three of these values coincide, the only solution is $\mu = 0$. But then $\alpha = 0$ or $\beta = 0$ and from Lemma 7.20 $\mathfrak{L} \simeq \mathcal{D}_3^{(5)}$, hence $n_i = 1$ for all i.

Therefore, at least one of the n_i is τ^2 or τ^{-2}. Since $(a_j, r_{a_i}(x)) = (r_{a_i}(a_j), x)$ we may replace x by an image under the action of $W(\mathcal{D}_2^{(5)})$ and suppose that $n_0 \in \{\tau^2, \tau^{-2}\}$. A short calculation shows that in this case the 5-tuple $(n_0, n_1, n_2, n_3, n_4)$ is either $(\tau^2, 2, 1, 1, 2)$ or $(\tau^{-2}, 1, 2, 2, 1)$. If $n_0 = \tau^{-1}$, let $y := r_x(a_0) = a_0 - (a_0, x)x$. Then $(y, a_i) = (a_0, a_i) - (a_0, x)(x, a_i)$, hence $(y, a_0) = \tau$ and $(y, a_i) \neq 0$ for $1 \leq i \leq 4$. Therefore we may replace x by y and assume that $(x, a_0) = \tau$. In this case $(a_1, r_x(a_1)) = 2 - n_1 = 0$ and from Lemma 8.8 the star-closure of σ and $r_x(a_1)$ is equivalent to $\mathcal{H}_3$. By the previous lemma, $\mathfrak{L}$ properly contains $\mathcal{H}_3$. $\qquad\square$

Theorem 8.10. *If $\mathfrak{L}$ is an indecomposable 5-system in $\mathbb{C}^3$, then one of the following holds:*

(i) $\mathfrak{L}$ *is a simple extension of $\mathcal{D}_2^{(5)}$ equivalent to $\mathcal{D}_3^{(5)}$, $\mathcal{H}_3$ or $\mathcal{J}_3^{(5)}$, or*

(ii) $\mathfrak{L}$ *is a simple extension of $\mathcal{H}_3$ and $\mathfrak{L} = \mathcal{J}_3^{(5)}$ or $\mathfrak{L} = \overline{\mathcal{J}}_3^{(5)}$.*

Proof. Suppose first that $\mathfrak{L}$ is the star-closure of $\mathcal{D}_2^{(5)}$ and a line ℓ. If ℓ is orthogonal to a line of σ, then by Lemma 8.8 $\mathfrak{L} \simeq \mathcal{H}_3$. If this is not the case, then by Lemma 8.9 $\mathfrak{L}$ is either equivalent to $\mathcal{D}_3^{(5)}$ or properly contains $\mathcal{H}_3$.

Thus we may suppose that $\mathfrak{L}$ is a simple extension of $\mathcal{H}_3$. In particular, we choose $a := (0, 1, 0)$ and $b := \frac{1}{2}(1, \tau, \tau^{-1})$ in $\mathcal{D}_2^{(5)}$. Then $\mathcal{H}_3$ is the star-closure of a, b and $p := (1, 0, 0)$. Let $c := r_a(b) = \frac{1}{2}(1, -\tau, \tau^{-1})$, $d := -r_b(a) = \frac{1}{2}(\tau, \tau^{-1}, 1)$ and $e := r_c(a) = \frac{1}{2}(\tau, -\tau^{-1}, 1)$. The lines spanned by a, b, c, d and e correspond to the lines spanned by a_0, a_1, a_4, a_2 and a_3 in Lemma 8.9 and therefore we may suppose that $\mathfrak{L}$ is the star-closure of $\mathcal{D}_2^{(5)}$ and a line with root $x = (\alpha, \beta, \gamma)$ such that $(a, x) = \frac{1}{2}\tau$ and $|(b, x)| = |(c, x)| = \frac{1}{\sqrt{2}}$. Therefore

(8.11) $$\tau(\alpha + \bar{\alpha}) + \gamma + \bar{\gamma} = 0,$$

(8.12) $$\tau^2 |\alpha^2| + \tau(\alpha \bar{\gamma} + \bar{\alpha} \gamma) + |\gamma|^2 = \frac{3}{4} \quad \text{and}$$

(8.13) $$|\alpha|^2 + \frac{1}{4}\tau^2 + |\gamma|^2 = 1.$$

It follows from these equations that $\alpha \neq 0$ and therefore $|\alpha| = |(p, x)|$ is an element of $\{\frac{1}{2}, \frac{1}{\sqrt{2}}, \frac{1}{2}\tau, \frac{1}{2}\tau^{-1}\}$. From (8.13) the only possibilities for $|\alpha|$ are $\frac{1}{2}$ and $\frac{1}{2}\tau^{-1}$ and then $|\gamma|$ is $\frac{1}{2}\tau^{-1}$ or $\frac{1}{2}$, respectively. If $\theta := 4\tau\alpha\bar{\gamma}$, then $|\theta| = 1$ in both cases. If $|\alpha| = \frac{1}{2}\tau^{-1}$, then $\gamma = \bar{\theta}\tau\alpha$, $\theta + \bar{\theta} = 1$ and hence θ is $-\omega$ or $-\omega^2$. If $\theta = -\omega$, then from (8.11), $\alpha = \pm\frac{1}{2}\omega^2\tau^{-1}$ and $\mathfrak{L} = \overline{\mathcal{J}}_3^{(5)}$; if $\theta = -\omega$, then $\alpha = \pm\frac{1}{2}\omega\tau^{-1}$ and $\mathfrak{L} = \mathcal{J}_3^{(5)}$.

Suppose that $|\alpha| = \frac{1}{2}$ and $|\gamma| = \frac{1}{2}\tau^{-1}$. Then $\alpha = \theta\tau\gamma$ and from (8.12) we have $\theta = \pm i$. It follows from (8.11) that $\gamma = \pm(\bar{\theta} + \tau^{-2})/2i\sqrt{3}$ and $\alpha = \pm(\tau + \theta\tau^{-1})/2i\sqrt{3}$. Consideration of the inner products of x and $\frac{1}{2}(\tau^{-1}, 1, \tau)$ leads to a contradiction in all cases. (However, the star-closure of $\mathcal{D}_2^{(5)}$ and x is similar to $\mathcal{J}_3^{(5)}$.)

To complete the proof we show that there are no proper extensions of $\mathcal{D}_3^{(5)}$ or $\mathcal{J}_3^{(5)}$ in $\mathbb{C}^3$. If $\mathfrak{L}$ is a simple extension of $\mathcal{J}_3^{(5)}$, then $\mathfrak{L}$ is also a simple extension of $\mathcal{H}_3$ and therefore $\mathfrak{L}$ contains both $\mathcal{J}_3^{(5)}$ and $\overline{\mathcal{J}}_3^{(5)}$. But the inner product of $\frac{1}{2}(\tau\omega^2, \omega, \tau^{-1})$ and $\frac{1}{2}(\tau\omega, \omega^2, -\tau^{-1})$ is $\frac{1}{8}(5 - \tau)$, which is a contradiction.

Finally, suppose that $\mathfrak{L}$ is the star-closure of $\mathcal{D}_3^{(5)}$ and $x = (\alpha, \beta, \gamma)$. It follows from Lemma 7.20 that $\alpha\beta\gamma \neq 0$ and therefore by the first part of the proof, $\mathfrak{L}$ contains $\mathcal{H}_3$ or $\mathcal{J}_3^{(5)}$ and therefore by what has just been shown $\mathfrak{L}$ is equivalent to $\mathcal{H}_3$ or $\mathcal{J}_3^{(5)}$. Comparing sizes, the only possibility is $\mathfrak{L} \simeq \mathcal{J}_3^{(5)}$. But $\mathcal{J}_3^{(5)}$ does not contain $\mathcal{D}_3^{(5)}$ because every line of $\mathcal{D}_3^{(5)}$ not in $\sigma := \mathcal{D}_2^{(5)}$ is at $60°$ to every line of σ, whereas this is not the case in $\mathcal{J}_3^{(5)}$. $\qquad\square$

Corollary 8.14. *No indecomposable 5-system is a proper extension of $\mathcal{J}_3^{(5)}$.*

Proof. Suppose that $\mathfrak{L}$ is an indecomposable 5-system and a simple extension of $\mathcal{J}_3^{(5)}$. It follows from the theorem that $\dim \mathfrak{L} = 4$ and then from Theorem 2.15, $W(\mathfrak{L})$ is primitive. However, the centre of $W(\mathcal{J}_3^{(5)})$ contains an element of order 6, contradicting Theorem 8.7. $\qquad\square$

Theorem 8.15. *Suppose that $\mathfrak{L}$ is an indecomposable 5-system and a simple extension of $\mathcal{H}_3$. Then $\mathfrak{L}$ is equivalent to $\mathcal{J}_3^{(5)}$ or $\mathcal{H}_4$.*

Proof. If $\mathfrak{L} \subset \mathbb{C}^3$, the result follows from the previous theorem and so from now on we suppose that $\mathfrak{L} \not\subset \mathbb{C}^3$. Thus $\mathfrak{L}$ is the star-closure in $\mathbb{C}^4$ of $\mathcal{H}_3$ and a line ℓ spanned by a root $x = (\alpha_1, \alpha_2, \alpha_3, \alpha_4)$, where $\alpha_4 \neq 0$.

The line system $\mathcal{H}_3$ contains six subsystems equivalent to $\mathcal{D}_2^{(5)}$, any two of which have just one line in common. If σ is any such subsystem and if ℓ is not orthogonal to any line of σ, then the star-closure of σ and ℓ is equivalent to $\mathcal{J}_3^{(5)}$. By Corollary 8.14 and our assumptions this cannot happen and so ℓ is orthogonal to at least one line of σ.

Therefore ℓ is orthogonal to at least two lines, say m_1 and m_2, of $\mathcal{H}_3$. Furthermore $\mathcal{H}_3$ is the star-closure of m_1, m_2 and a line m_3. We may choose roots r_1, r_2 and r_3 for m_1, m_2 and m_3 so that their pairwise inner products are real. We have $(r_1, x) = (r_2, x) = 0$ and we may scale x so that $(r_3, x) \in \mathbb{R}$. It follows that $\mathfrak{L}$ is equivalent to a line system in $\mathbb{R}^4$.

We may choose coordinates so that $\mathcal{H}_3$ is the star-closure of $p := (1,0,0,0)$, $q := (0,1,0,0)$ and $b := \frac{1}{2}(1,\tau,\tau^{-1},0)$. If $s := (0,0,1,0)$ and $t := (0,0,0,1)$, the reflections r_p, r_q and r_s belong to $W(\mathcal{H}_3)$ and their product is $-r_t$. Thus the group generated by $W(\mathcal{L})$ and $-I$ is a finite reflection group and by Theorems 7.9 and 8.5, the roots of its reflections span the lines of a 5-system that contains $\mathcal{L}$ and the line spanned by t. It follows that, up to equivalence, we may suppose that $\alpha_i \in \{0, \frac{1}{2}, \frac{1}{\sqrt{2}}, \frac{1}{2}\tau, \frac{1}{2}\tau^{-1}\}$, for $1 \le i \le 4$.

If two of the α_i are zero, we may suppose that $\alpha_1 = \alpha_2 = 0$. Then $\alpha_3 = \alpha_4 = \frac{1}{\sqrt{2}}$ and $(x, b) = \frac{1}{2\sqrt{2}}\tau^{-1}$, which is a contradiction.

If $\alpha_1 = 0$, then α_2, α_3 and α_4 take either the values $\frac{1}{2}, \frac{1}{\sqrt{2}}, \frac{1}{\sqrt{2}}$ or else the values $\frac{1}{2}, \frac{1}{2}\tau, \frac{1}{2}\tau^{-1}$ in some order. Consideration of the inner products (b, x) and (c, x), where $c := \frac{1}{2}(\tau^{-1}, 1, \tau, 0)$ shows that x is $\frac{1}{2}(0, \tau^{-1}, \tau, 1)$, $\frac{1}{2}(0, \tau, 1, \tau^{-1})$ or $\frac{1}{2}(0, 1, \tau^{-1}, \tau)$. In all cases $\mathcal{L} = \mathcal{H}_4$.

If $\alpha_i \ne 0$ for $1 \le i \le 4$, then $x = \frac{1}{2}(1,1,1,1)$ and again $\mathcal{L} = \mathcal{H}_4$. $\square$

Corollary 8.16. *No indecomposable 5-system is a proper extension of $\mathcal{H}_4$.*

Proof. Suppose that $\mathcal{L}$ is indecomposable and the star-closure in $\mathbb{C}^5$ of $\mathcal{H}_4$ and the line spanned by $x := (\alpha_1, \alpha_2, \ldots, \alpha_5)$, where $(x, x) = 1$. As in the proof of the theorem we may suppose that $\mathcal{L}$ is contained in a 5-system that also contains $(0,0,0,0,1)$. It follows that $|\alpha_i| \in \{0, \frac{1}{2}, \frac{1}{\sqrt{2}}, \frac{1}{2}\tau, \frac{1}{2}\tau^{-1}\}$. We may suppose that x is not orthogonal to the copy of $\mathcal{H}_3$ that is the star-closure of $a := (0,1,0,0,0)$, $b := \frac{1}{2}(1, \tau, \tau^{-1}, 0, 0)$ and $c := (1,0,0,0,0)$. Then by Corollary 8.14 and the theorem, the star-closure of a, b, c and x is equivalent to $\mathcal{H}_4$. We may replace x by an image under the action of $W(\mathcal{H}_4)$ and assume that $(a, x) = (b, x) = 0$ and $(c, x) = \frac{1}{2}$. It follows that $x = (\frac{1}{2}, 0, -\frac{1}{2}\tau, \alpha_4, \alpha_5)$. Since $(x, x) = 1$ and $\alpha_5 \ne 0$, the only possibility is $\alpha_4 = 0$ and $|\alpha_5| = \frac{1}{2}\tau^{-1}$. But now the inner product between x and $\frac{1}{2}(1,1,1,1,0) \in \mathcal{H}_4$ is $-\frac{1}{4}\tau^{-1}$, a contradiction. $\square$

We now have all the information we need about 5-systems for the purposes of the classification. The following result is stated for completeness of the exposition.

Theorem 8.17. *If $\mathcal{L}$ is an indecomposable 5-system, then $\mathcal{L}$ is equivalent to $\mathcal{H}_3$, $\mathcal{H}_4$, $\mathcal{J}_3^{(5)}$ or $\mathcal{D}_n^{(5)}$ for $n \ge 2$.*

Proof. See Exercise 2 at the end of this chapter. $\square$

5. Line systems and reflections of order three

The line systems we have considered hitherto are specifically adapted to the study of groups generated by reflections of order two. By Theorem 8.4, in classifying primitive groups in dimension at least three, only reflections of order two or three occur.

In this section we study line systems adapted to groups generated by reflections of order three, (the 'ternary' case) and to groups generated by reflections of order two and three (the 'mixed' case).

Let ℓ be a line through the origin of $\mathbb{C}^n$ spanned by a vector a. Recall that a root a of ℓ is *short*, *long* or *tall* according to whether (a, a) is 1, 2 or 3.

Definition 8.18. Given a tall root a, the reflection t_a of order three is defined by

$$t_a(v) := v - \tfrac{1}{3}(1 - w)(v, a)a = v + \theta^{-1}w^2(v, a)a,$$

where $w = \tfrac{1}{2}(-1 + i\sqrt{3})$ and $\theta := w - w^2 = i\sqrt{3}$.

Thus $t_a(a) = wa$ and $t_a^2(v) = v - \theta^{-1}w(v, a)a$. Since $t_a = t_b$ if and only if a and b span the same line ℓ, this reflection may be denoted unambiguously by t_ℓ.

Lemma 7.7 gives the possible values for $|(a, b)|$, where a and b are long roots of reflections of order two. We extend this result to include the case where at least one of a or b is a tall root of a reflection of order three in a primitive reflection group of rank at least 3. (If the rank is 2, other values are possible.)

Lemma 8.19. *Suppose that s and t are non-commuting reflections in a finite primitive reflection group of rank at least 3 and suppose that the order of s is 3. Let a be a tall root of s.*

(i) *If the order of t is 3 and if b is a tall root of t, then $\langle s, t \rangle \simeq SL_2(\mathbb{F}_3)$ and $|(a, b)| = \sqrt{3}$.*

(ii) *If the order of t is 2 and if b is a long root of t, then $\langle s, t \rangle \simeq G(3, 1, 2)$ and $|(a, b)| = \sqrt{3}$.*

Proof. (i) The reflections s and t fix the subspace $W := \langle a, b \rangle$ and by replacing s by s^2 and t by t^2, if necessary, the matrices of s and t with respect to the basis a, b are

$$A := \begin{bmatrix} w & \tfrac{1}{3}(w - 1)(b, a) \\ 0 & 1 \end{bmatrix} \quad \text{and} \quad B := \begin{bmatrix} 1 & 0 \\ \tfrac{1}{3}(w^2 - 1)(a, b) & w^2 \end{bmatrix}.$$

The the action of the group $\langle s, t \rangle$ on W is primitive because the imprimitive groups $G(m, p, n)$ cannot be generated by reflections of order 3. Thus by Theorem 8.4, $\langle s, t \rangle \simeq SL_2(\mathbb{F}_3)$ and s is a conjugate of t^{-1}. That is, $s = g^{-1}t^{-1}g$ for some g and therefore $st = g^{-1}t^{-1}gt$ has order 4. It follows that the trace of AB is 0 and since

$$AB = \begin{bmatrix} w + \tfrac{1}{3}|(a, b)|^2 & \tfrac{1}{3}(1 - w^2)(b, a) \\ \tfrac{1}{3}(w^2 - 1)(a, b) & w^2 \end{bmatrix},$$

we see that $|(a, b)|^2 = 3$, as claimed.

(ii) In this case it follows from Theorem 8.4 (i) that $\langle s, t \rangle$ is imprimitive and from Lemma 2.8 and Theorem 8.5 $\langle s, t \rangle \simeq G(3, 1, 2)$ and $|(a, b)| = \sqrt{3}$. $\square$

Corresponding to each line ℓ in a line system $\mathfrak{L}$ we have defined a canonical reflection t_ℓ of order three. Therefore, in analogy with Definition 7.6, we define a *ternary k-system* as follows.

Definition 8.20. A line system $\mathfrak{L}$ is a *ternary k-system* if

(i)   for all lines $\ell \in \mathfrak{L}$, $t_\ell(\mathfrak{L}) \subseteq \mathfrak{L}$,

(ii)   for all $\ell, m \in \mathfrak{L}$ the order of $t_\ell t_m$ is at most k, and

(iii)   there exist $\ell, m \in \mathfrak{L}$ such that the order of $t_\ell t_m$ equals k.

Given a ternary k-system $\mathfrak{L}$, $W(\mathfrak{L})$ denotes the group generated by the reflections t_ℓ, for all $\ell \in \mathfrak{L}$. The star-closure X^* of a subset X of $\mathfrak{L}$ is the intersection of all ternary k-systems $\mathfrak{K}$ such that $X \subseteq \mathfrak{K}$. Evidently $X^* \subseteq \mathfrak{L}$.

If $a := (1,1,1)$ and $b := (\omega, 1, 1)$, then a and b are tall roots, $|(a,b)| = \sqrt{3}$ and $G := \langle t_a, t_b \rangle$ is the primitive reflection group $G_4 \simeq SL_2(\mathbb{F}_3)$. A reflection of order 3 is not conjugate to its inverse and so the 8 reflections in G form two conjugacy classes, each of size 4. If $r, s \in G$ are reflections whose roots are linearly independent, then the order of rs is 6 or 4 according to whether or not r and s are conjugate in G.

Let $\mathcal{L}_1$ denote the unique ternary k-system of dimension 1. Then $W(\mathcal{L}_1) = C_3$ is a cyclic group of order 3. If $\mathfrak{L}$ is the line system of a primitive reflection group generated by reflections of order 3 and of rank $n \geq 3$, it follows from Lemma 8.19 (i) that either $\mathfrak{L} = n\mathcal{L}_1$ or $\mathfrak{L}$ is a ternary 6-system. Furthermore, if a and b are tall roots that span lines of $\mathfrak{L}$, then $(a,b) = 0$ or $|(a,b)| = \sqrt{3}$.

Definition 8.21. If $\mathfrak{K}$ is a k-system and $\mathfrak{M}$ is a ternary h-system disjoint from $\mathfrak{K}$ such that $r_\ell(\mathfrak{M}) \subseteq \mathfrak{M}$ and $t_m(\mathfrak{K}) \subseteq \mathfrak{K}$ for all $\ell \in \mathfrak{K}$ and all $m \in \mathfrak{M}$, the pair $\mathfrak{L} = (\mathfrak{K}, \mathfrak{M})$ is called a *mixed (k,h)-system*. The set of lines of $\mathfrak{L}$ is defined to be the union $\mathfrak{K} \cup \mathfrak{M}$. We put $W(\mathfrak{L}) := W(\mathfrak{K})W(\mathfrak{M})$ and call $W(\mathfrak{L})$ the *Weyl group* of $\mathfrak{L}$. Note that $W(\mathfrak{K})$ and $W(\mathfrak{M})$ are normal subgroups of $W(\mathfrak{L})$.

The imprimitive reflection groups provide infinitely many examples of mixed systems. However, if the rank is at least 3, we shall see that there is just one finite primitive reflection group that is the Weyl group of a mixed system.

5.1. The mixed line systems $\mathcal{B}_n^{(3)}$.

If G is an imprimitive irreducible reflection group that contains reflections of order 3, then $G = G(3p, p, n)$ for some n and some p. The reflections of order 2 in G generate the subgroup $G(3p, 3p, n)$, whose line system is $\mathcal{D}_n^{(3p)}$. The reflections of order 3 commute and their line system is the ternary 3-system $n\mathcal{L}_1$, namely the coordinate axes. If G is a subgroup of a primitive reflection group, it follows from Theorem 8.5 that $p = 1$. In this case, we denote the mixed $(3,6)$-system $(\mathcal{D}_n^{(3)}, n\mathcal{L}_1)$ of $G(3,1,n)$ by $\mathcal{B}_n^{(3)}$.

5.2. The line system $\mathcal{L}_2$. Suppose that a and b are tall roots such that $|(a, b)| = \sqrt{3}$. If we scale b so that $(a, b) = \theta$, where $\theta = i\sqrt{3}$, then $t_a(b) = -\omega^2 a + b$ and $t_a^2(b) = \omega t_b(a) = \omega a + b$. The ternary 6-system $\mathcal{L}_2$ is defined to be the set of four lines spanned by a, b, $t_a(b)$ and $t_a^2(b)$. Thus $W(\mathcal{L}_2) = \langle t_a, t_b \rangle$ is $G_4 \simeq SL_2(\mathbb{F}_3)$ and it acts on the lines of $\mathcal{L}_2$ as $\mathrm{Sym}(4)$.

5.3. The line system $\mathcal{L}_3$. The group $G(3, 1, 3)$ acts on $\mathbb{C}^3$ by permuting the coordinates and multiplying their values by ω and ω^2. Thus there are 9 lines in the $G(3, 1, 3)$ orbit of $\langle (1, 1, 1) \rangle$. The ternary 6-system $\mathcal{L}_3$ is the union of these 9 lines with the coordinate axes.

5.4. The line system $\mathcal{L}_4$. We may regard $\mathcal{L}_3$ as a subset of $\mathbb{C}^4$ via the embedding that sends the vector (x_1, x_2, x_3) to $(x_1, x_2, x_3, 0)$. The $W(\mathcal{L}_3)$ orbit Γ of $\langle (0, 1, -1, -1) \rangle$ contains 27 lines and the roots of these lines are obtained by cyclically permuting the first three coordinates and multiplying the coordinate values by ω or ω^2. The ternary 6-system $\mathcal{L}_4$ has 40 lines and consists of the 27 lines of Γ, the 12 lines of $\mathcal{L}_3$ and the line $\langle (0, 0, 0, \theta) \rangle$. The 240 vectors obtained by multiplying these tall roots by 6th roots unity is a $\mathbb{Z}[\omega]$-root system.

5.5. The mixed line system $\mathcal{M}_3$. The group $G(3, 1, 3)$ acts on the line system $\mathcal{L}_3$ and normalises $W(\mathcal{L}_3)$. It follows that the product $W(\mathcal{L}_3)G(3, 1, 3)$ is a finite reflection group, which contains reflections of order 2 as well as reflections of order 3.

The 3 reflections of order 3 in $G(3, 1, 3)$ belong to $W(\mathcal{L}_3)$ and the 9 reflections of order 2 in $G(3, 1, 3)$ generate the group $G(3, 3, 3)$ whose line system is $\mathcal{D}_3^{(3)}$. The line system $\mathcal{M}_3$ is the mixed $(3, 6)$-system $(\mathcal{D}_3^{(3)}, \mathcal{L}_3)$.

It follows from Theorem 8.5 that if we choose tall roots for the lines of $\mathcal{L}_3$ and long roots for the lines of $\mathcal{D}_3^{(3)}$, then $|(a, b)|$ is 0, 1 or $\sqrt{3}$ for all roots a and b spanning distinct lines of $\mathcal{M}_3$.

6. Extensions of ternary 6-systems

In this section we provide the final step in the classification of the finite primitive reflection groups G of rank greater than 2 by showing that if G contains reflections of order three, then G is a subgroup of $U_3(\mathbb{C})$ or $U_4(\mathbb{C})$. We shall see that there are just three possibilities for G, namely $W(\mathcal{L}_3)$, $W(\mathcal{M}_3)$ and $W(\mathcal{L}_4)$.

If G is a finite primitive group generated by reflections of order three, it follows from Lemma 8.19 that the lines spanned by the roots of the reflections is a ternary 6-system. Conversely, if $\mathfrak{L}$ is an indecomposable ternary 6-system, then $W(\mathfrak{L})$ contains only reflections of order three. Therefore $W(\mathfrak{L})$ is primitive because, by Lemma 2.7, the imprimitive groups $G(m, p, n)$ contain reflections of order two.

Lemma 8.22. *Suppose that r and s are non-commuting reflections of order 3 in a primitive reflection group G. If $t \in G$ is a reflection of order 2 or if t is a reflection of order 3 such that $t \notin \langle r, s \rangle$, then t commutes with a reflection in $\langle r, s \rangle$.*

Proof. Let a and b be tall roots of r and s, respectively, and let c be a root of t, which is long or tall according to whether the order of t is 2 or 3. By Lemma 8.19 we may scale b and c so that $(a, b) = (b, c) = \theta$ and $(a, c) = \eta\theta$ for some η such that $|\eta| = 1$. We may suppose that $s = t_b$ and then $s(a) = a + \omega^2 b$ and $s^2(a) = a - \omega b$. The assumption $t \notin \langle r, s \rangle$ means that the line spanned by c is not one of the lines spanned by a, b, $s(a)$ or $s^2(a)$. Therefore, if both $(s(a), c)$ and $(s^2(a), c)$ are non-zero, we have $|(s(a), c)| = |(s^2(a), c)| = \sqrt{3}$ and hence $|\eta + \omega^2| = |\eta - \omega| = 1$, which has no solution for η. Therefore, one of $(s(a), c)$ or $(s^2(a), c)$ is zero. $\square$

Lemma 8.23. *If $\mathfrak{L}$ is an indecomposable ternary 6-system and a simple extension of $\mathcal{L}_2$, then $\mathfrak{L}$ is equivalent to $\mathcal{L}_3$.*

Proof. From Lemma 8.22 we may suppose that $\mathfrak{L}$ is the star-closure of tall roots a, b and c such that $(a, b) = (b, c) = \theta$ and $(a, c) = 0$, where $\theta = i\sqrt{3}$. Then $t_b(a) = a + \omega^2 b$, $t_b(c) = -\omega^2 b + c$ and $t_c^2(b) = b - \omega c$. Thus $d := t_b t_c^2 t_b(a) = a - \theta b - c$ and $(a, d) = (c, d) = 0$. That is, $\{\theta^{-1} a, -\theta^{-1} c, -\theta^{-1} d\}$ is an orthonormal basis for $\mathbb{C}^3$ such that, with respect to this basis, we have $a = (\theta, 0, 0)$, $b = (1, 1, 1)$, $c = (0, -\theta, 0)$ and $d = (0, 0, -\theta)$. It follows that $\mathfrak{L}$ is equivalent to $\mathcal{L}_3$. $\square$

Lemma 8.24. *If $\mathfrak{L}$ is an indecomposable ternary 6-system and a simple extension of $\mathcal{L}_3$, then $\mathfrak{L}$ is equivalent to $\mathcal{L}_4$.*

Proof. Suppose that $\mathfrak{L}$ is indecomposable and is the star-closure in $\mathbb{C}^4$ of $\mathcal{L}_3$ and the line spanned by $x := (\alpha_1, \alpha_2, \alpha_3, \alpha_4)$, where $(x, x) = 3$. We may suppose that $\mathfrak{L}$ contains the lines spanned by the tall roots $(\theta, 0, 0, 0)$, $(0, \theta, 0, 0)$, $(0, 0, \theta, 0)$ and $b := (1, 1, 1, 0)$. If $1 \le i \le 3$ and $\alpha_i \neq 0$, then $|\alpha_i| = 1$. Thus there is just one value of i for which $\alpha_i = 0$.

If $\alpha_4 = 0$, we may suppose that $\alpha_1 = 1$ and from Lemma 8.22 $(b, x) = 0$. But then $1 + \alpha_2 + \alpha_3 = 0$ and α_2 is ω or ω^2, hence $x \in \mathcal{L}_3$. It follows that we may assume that $\alpha_1 = 0$ and $\alpha_2 = |\alpha_3| = |\alpha_4| = 1$. Once again, from Lemma 8.22 we may suppose that $(b, x) = 0$. Thus $\alpha_3 = -1$ and, up to equivalence, we may take $\alpha_4 = -1$. It follows that $\mathfrak{L} \simeq \mathcal{L}_4$. $\square$

Lemma 8.25. *No indecomposable ternary 6-system is a proper extension of $\mathcal{L}_4$.*

Proof. Suppose that $\mathfrak{L}$ is indecomposable and is the star-closure in $\mathbb{C}^5$ of $\mathcal{L}_4$ and the line spanned by $x := (\alpha_1, \alpha_2, \alpha_3, \alpha_4, \alpha_5)$, where $(x, x) = 3$. If $e_1, e_2, \ldots, e_5$ is an orthonormal basis for $\mathbb{C}^5$, then $\mathcal{L}_4$ contains the lines spanned by the vectors θe_i, $1 \le i \le 4$. It follows that for $1 \le i \le 4$ we have $\alpha_i = 0$ or $|\alpha_i| = 1$. If $\alpha_5 \neq 0$, then up to equivalence, we may suppose that $\alpha_1 = \alpha_2 = 0$. Consideration of the inner product between x and $(1, 1, 1, 0, 0)$ leads to a contradiction. Thus $\alpha_5 = 0$ and $\mathfrak{L}$ is contained in $\mathbb{C}^4$. The previous lemma shows that we may take x to be $(0, 1, -1, \alpha_4, 0)$. Since $|\alpha_4| = 1$, the inner product between x and $d := (0, 1, -1, -1, 0)$ cannot be 0, therefore $|(d, x)| = \sqrt{3}$ and hence α_4 is $-\omega$ or $-\omega^2$. But then x is a tall root of $\mathcal{L}_4$ and so $\mathfrak{L}$ is not a proper extension of $\mathcal{L}_4$. $\square$

Theorem 8.26. *Suppose that G is a finite primitive reflection group of rank at least 3.*

(i)   *If G is generated by reflections of order 3, then G is conjugate to $W(\mathcal{L}_3)$ or $W(\mathcal{L}_4)$.*

(ii)   *If G contains reflections of orders 2 and 3, then G is conjugate to $W(\mathcal{M}_3)$.*

Proof. (i) In this case G contains reflections r and s of order 3 that do not commute. If a and b are tall roots of r and s, then by Lemma 8.19 the star-closure of a and b is $\mathcal{L}_2$. Then Lemmas 8.23, 8.24 and 8.25 show that the ternary 6-system defined by the reflections of order 3 is $\mathcal{L}_3$ or $\mathcal{L}_4$. Therefore G is conjugate to $W(\mathcal{L}_3)$ or $W(\mathcal{L}_4)$.

(ii) We may suppose that G contains a reflection of order 3 and a reflection of order 2 that do not commute. The subgroup N generated by the reflections of order 3 in G is normal and by Theorem 2.4 it acts irreducibly on V. Since no irreducible imprimitive group is generated by reflections of order 3 it follows that N is primitive and therefore, by (i), N is conjugate to $W(\mathcal{L}_3)$ or $W(\mathcal{L}_4)$.

Suppose first that $N = W(\mathcal{L}_3)$ and that $\mathcal{L}_3$ contains the lines spanned by the tall roots $(\theta, 0, 0)$, $(0, \theta, 0)$, $(0, 0, \theta)$ and $b := (1, 1, 1)$. If $x := (\alpha_1, \alpha_2, \alpha_3)$ is a long root such that $r_x \in G$, then $(x, x) = 2$ and it follows from Lemma 8.19 (ii) that $|\alpha_i| \in \{0, 1\}$. Therefore we may suppose that $x = (0, 1, \alpha_3)$. If $(x, b) = 0$, then $\alpha_3 = -1$, whereas if $|(x, b)| = \sqrt{3}$, then α_3 is $-\omega$ or $-\omega^2$. In both cases the line spanned by x belongs to $\mathcal{D}_3^{(3)}$ and it follows that $N\langle t \rangle \simeq W(\mathcal{M}_3)$. Furthermore we have shown that, up to conjugacy, $W(\mathcal{M}_3)$ is the unique finite primitive reflection subgroup of $U_3(\mathbb{C})$.

Finally, suppose that $N = W(\mathcal{L}_4)$ and that $x := (\alpha_1, \alpha_2, \alpha_3, \alpha_4)$ is a long root such that $r_x \in G$. As above, it follows that $|\alpha_i| \in \{0, 1\}$ for $1 \leq i \leq 4$. Since $(x, x) = 2$ it follows that $\alpha_i = 0$ for exactly two values of i. But then the inner product with either $(1, 1, 1, 0)$ or $(0, 1, -1, -1)$ leads to a contradiction. $\square$

Corollary 8.27. *A finite primitive reflection group of rank at least 4 does not contain both a reflection of order 2 and a reflection of order 3.*

Corollary 8.28. *If G is a finite primitive reflection group of rank at least 5, then all reflections in G have order 2.*

7. The classification

By Theorem 1.27 every finite unitary reflection group G is the direct product of irreducible reflection groups. Furthermore, if G is irreducible, then by Theorem 7.9,

$G = W(\mathfrak{L})$ for some indecomposable k-system $\mathfrak{L}$ and by the results of this chapter and Chapter 7 all indecomposable k-systems for $k \le 6$ are known.

Theorem 8.29 (Shephard and Todd). *Suppose that G is a finite irreducible unitary reflection group of rank n. Then, up to conjugacy in $U_n(\mathbb{C})$, G belongs to precisely one of the following ten classes:*

(i) $n = 1$ *and G is a cyclic group;*

(ii) $n \ge 2$ *and G is the imprimitive group $G(m, p, n)$ for some $m > 1$ and some divisor p of m;*

(iii) $n = 2$ *and G is one of the 19 primitive unitary reflection groups listed in Table D.1 of Appendix D;*

(iv) $n = 3$ *and G is $W(\mathcal{H}_3)$, $W(\mathcal{J}_3^{(4)})$, $W(\mathcal{J}_3^{(5)})$, $W(\mathcal{L}_3)$ or $W(\mathcal{M}_3)$;*

(v) $n = 4$ *and G is $W(\mathcal{F}_4)$, $W(\mathcal{H}_4)$, $W(\mathcal{L}_4)$, $W(\mathcal{N}_4)$ or $W(\mathcal{O}_4)$;*

(vi) $n = 5$ *and G is $W(\mathcal{K}_5)$;*

(vii) $n = 6$ *and G is $W(\mathcal{E}_6)$ or $W(\mathcal{K}_6)$;*

(viii) $n = 7$ *and G is $W(\mathcal{E}_7)$;*

(ix) $n = 8$ *and G is $W(\mathcal{E}_8)$;*

(x) $n \ge 4$ *and G is $W(\mathcal{A}_n) \simeq \mathrm{Sym}(n+1)$.*

In the list above, the group $W(\mathfrak{L})$ is the reflection group with corresponding line system $\mathfrak{L}$. These are explicitly described in §6 of Chapter 7.

Proof. If G is imprimitive, then by Theorem 2.14, G is conjugate to $G(m, p, n)$ for some $m > 1$ and some divisor p of m. If $n = 1$ then G is cyclic. If $n = 2$ and G is primitive, there are 19 possibilities for G and they are listed in Chapter 6. Therefore, from now on we may suppose that G is primitive and $n \ge 3$.

If G contains reflections of order three, it follows from Theorems 8.26 that G is $W(\mathcal{L}_3)$, $W(\mathcal{L}_4)$ or $W(\mathcal{M}_3)$. Thus we may suppose that the order of every reflection in G is two.

By Theorem 7.9 the roots of the reflections in G span the lines of an indecomposable k-system $\mathfrak{L}$ such that $G = W(\mathfrak{L})$. From Corollary 8.6, $k \le 5$ and $\mathfrak{L}$ contains at least one of the line systems $\mathcal{A}_2 \simeq \mathcal{D}_2^{(3)}$, $\mathcal{B}_2^{(4)}$, $\mathcal{B}_2^{(2)} \simeq \mathcal{D}_2^{(4)}$ or $\mathcal{D}_2^{(5)}$.

If $\mathfrak{L}$ is a 3-system, then $\mathfrak{L}$ is an extension of a 3-star $\mathcal{A}_2$. From Theorem 7.31, $\mathfrak{L}$ is equivalent to one of the line systems $\mathcal{A}_n$, $\mathcal{E}_6$, $\mathcal{E}_7$, $\mathcal{E}_8$, $\mathcal{K}_5$ or $\mathcal{K}_6$. The group $W(\mathcal{A}_1) \simeq \mathrm{Sym}(2)$ is cyclic, $W(\mathcal{A}_2) \simeq G(3, 3, 2)$ and $W(\mathcal{A}_3) \simeq G(2, 2, 3)$, therefore (to avoid duplication) we have $n \ge 4$ in (x).

If $\mathfrak{L}$ is a 4-system, then $\mathfrak{L}$ is an extension of a 4-star $\mathcal{D}_2^{(4)}$. From Theorem 7.42, $\mathfrak{L}$ is equivalent to $\mathcal{J}_3^{(4)}$, $\mathcal{F}_4$, $\mathcal{N}_4$ or $\mathcal{O}_4$.

If $\mathfrak{L}$ is a 5-system, then $\mathfrak{L}$ is an extension of a 5-star $\mathcal{D}_2^{(5)}$ and G contains $G(5, 5, 2)$. If $\dim \mathfrak{L} = 3$, then from Theorem 8.10, $\mathfrak{L}$ is equivalent to $\mathcal{H}_3$ or $\mathcal{J}_3^{(5)}$. If $\dim \mathfrak{L} > 3$, then from Theorem 2.19, G contains a primitive reflection subgroup of rank 3. Thus

$\mathfrak{L}$ is an extension of $\mathcal{H}_3$ or $\mathcal{J}_3^{(5)}$ and it follows from Theorem 8.15 and Corollaries 8.14 and 8.16 that $\mathfrak{L}$ is equivalent to $\mathcal{H}_4$.   □

The case $n = 3$ of this theorem was attempted by Jordan [123] in 1878 but he missed the second and third groups in the list. The second group in the list (modulo its centre) was described by Klein [130] in 1879 and the third group (modulo its centre) was described by Valentiner [220] in 1889 and shown by Wiman [224] in 1896 to be isomorphic to $\mathrm{Alt}(6)$.

The idea of a reflection acting on $\mathbb{C}^n$ was introduced by Shephard [192] in 1953 and the first complete list of the finite unitary reflection groups was given by Shephard and Todd [193] in 1954. However, the equivalent concept of an homology acting on complex projective space had been studied since the nineteenth century (see Wiman [225]). Shephard and Todd's determination of the primitive reflection groups was based on previous work by Blichfeldt, Mitchell and others on groups generated by homologies.

7.1. The Shephard–Todd numbering. Shephard and Todd number the irreducible unitary reflection groups from 1 to 37. The first three entries in their list refer to infinite series.

1	the symmetric groups $\mathrm{Sym}(n)$
2	the imprimitive groups $G(m, p, n)$, for $n \geq 2$
3	the cyclic groups $\mathcal{C}_m \simeq G(m, 1, 1)$
$4 - 22$	19 primitive groups of rank 2
$23 - 37$	15 primitive groups of rank ≥ 3

Tables D.1 and D.2 of Appendix D describe the correspondence with the line system notation used throughout this and previous chapters.

8. Root systems and the ring of definition

In Chapter 6 we showed that every finite primitive reflection group G of rank two can be represented by matrices over the ring of definition $\mathbb{Z}(G)$ (see Definition 1.42) of G. In this section we extend this result to all finite primitive reflection groups G by showing that in each case there is a root system (see Definition 1.43) for G defined over $\mathbb{Z}(G)$. The rings that occur are the rational integers $\mathbb{Z}$, the Gaussian integers $\mathbb{Z}[i]$, the Eisenstein integers $\mathbb{Z}[\omega]$, the golden integers $\mathbb{Z}[\tau]$, the Kleinian integers $\mathbb{Z}[\lambda]$ and, in the case of $\mathcal{J}_3^{(5)}$, the golden Eisenstein integers $\mathbb{Z}[\tau, \omega]$. (See [59] for the origins of this terminology.)

Theorem 8.29 shows that, other than the groups $W(\mathcal{A}_n)$, there are just 15 possibilities for a finite primitive reflection group of rank at least 3. Furthermore, except for $W(\mathcal{L}_3)$, $W(\mathcal{L}_4)$ and $W(\mathcal{M}_3)$, the order of every reflection in these groups is two.

For each short or long root a there is a unique reflection r_a of order two with root a. Thus if a and b are short or long roots, the Cartan coefficient of a and b is $\langle a \mid b \rangle = 2(a, b)/(b, b)$.

On the other hand, to each tall root b we associate the reflection t_b of order three such that $t_b(b) = \omega b$. Therefore, if b is a tall root, the Cartan coefficient of a and b is $\langle a \mid b \rangle = \frac{1}{3}(1 - \omega)(a, b)$. In all cases the image of a under the action of the reflection r_b or t_b is $a - \langle a \mid b \rangle b$.

Theorem 8.30. *If G is a finite primitive reflection group of rank at least 3, the ring of definition $\mathbb{Z}(G)$ is a principal ideal domain and there exists a $\mathbb{Z}(G)$-root system (Σ, f) whose roots span the line system $\mathfrak{L}$ of G and where $f(a) = -1$ for all long and short roots and $f(a) = \omega$ for all tall roots. The details are as follows.*

(i)   *If $\mathfrak{L}$ is $\mathcal{A}_n$, $\mathcal{E}_6$, $\mathcal{E}_7$ or $\mathcal{E}_8$, then $\mathbb{Z}(G) = \mathbb{Z}$, $\mathfrak{L}$ is a 3-system and Σ consists of long roots.*

(ii)   *If $\mathfrak{L}$ is $\mathcal{K}_5$ or $\mathcal{K}_6$, then $\mathbb{Z}(G) = \mathbb{Z}[\omega]$, $\mathfrak{L}$ is a 3-system and Σ consists of long roots.*

(iii) *If $\mathfrak{L} = \mathcal{F}_4$, then $\mathbb{Z}(G) = \mathbb{Z}$, $\mathfrak{L}$ is a 4-system and Σ has 24 short roots and 24 long roots.*

(iv)   *If $\mathfrak{L} = \mathcal{J}_3^{(4)}$, then $\mathbb{Z}(G) = \mathbb{Z}[\lambda]$, where $\lambda^2 + \lambda + 2 = 0$, $\mathfrak{L}$ is a 4-system and Σ consists of long roots.*

(v)   *If $\mathfrak{L}$ is $\mathcal{N}_4$ or $\mathcal{O}_4$, then $\mathbb{Z}(G) = \mathbb{Z}[i]$, $\mathfrak{L}$ is a 4-system and Σ consists of long roots.*

(vi)   *If $\mathfrak{L}$ is $\mathcal{H}_3$ or $\mathcal{H}_4$, then $\mathbb{Z}(G) = \mathbb{Z}[\tau]$, $\mathfrak{L}$ is a 5-system and Σ consists of short roots.*

(vii) *If $\mathfrak{L} = \mathcal{J}_3^{(5)}$, then $\mathbb{Z}(G) = \mathbb{Z}[\tau, \omega]$, $\mathfrak{L}$ is a 5-system and Σ consists of short roots.*

(viii) *If $\mathfrak{L}$ is $\mathcal{L}_3$ or $\mathcal{L}_4$, then $\mathbb{Z}(G) = \mathbb{Z}[\omega]$, $\mathfrak{L}$ is a ternary 6-system and Σ consists of tall roots.*

(ix)   *If $\mathfrak{L}$ is $\mathcal{M}_3$, then $\mathbb{Z}(G) = \mathbb{Z}[\omega]$ and Σ contains both long and tall roots.*

Proof. For each group G we derive the root system Σ from the line systems described in §6 of Chapter 7 and §5 of the present chapter. For each line $\ell \in \mathfrak{L}$ we have a root a_ℓ that spans ℓ and the coordinates of the roots lie in a field F. The fields that occur are $\mathbb{Q}$, $\mathbb{Q}[i]$, $\mathbb{Q}[\omega]$, $\mathbb{Q}[\tau]$, $\mathbb{Q}[\lambda]$ and $\mathbb{Q}[\tau, \omega]$. Let A be the ring of integers of F and put $\Sigma := \{ \alpha a_\ell \mid \ell \in S, \alpha \in \mu(A) \}$. Define $f : \Sigma \to \mu(A)$ by setting $f(a) = -1$ for all long or short roots a and $f(a) = \omega$ for all tall roots a. Then (Σ, f) is an A-root system for G.

Except for $\mathcal{F}_4$ we obtain the descriptions given in (i) to (ix) above and to complete the proof we must show that $A = \mathbb{Z}(G)$. In the case of $\mathcal{F}_4$ the construction produces a $\mathbb{Z}[i]$-root system. However, if e_1, e_2, e_3, e_4 is an orthonormal basis we may use the 24 long roots $\pm e_i \pm e_j$ $(1 \le i < j \le 4)$ and the 24 short roots $\pm e_i$ $(1 \le i \le 4)$ and $\frac{1}{2}(\pm e_1 \pm e_2 \pm e_3 \pm e_4)$ to obtain a $\mathbb{Z}$-root system for $\mathcal{F}_4$.

The vector space V over F generated by Σ has a basis consisting of roots. Therefore the reflections of G can be represented by matrices with entries in F and hence G itself is represented by matrices whose entries lie in F. It follows that $\mathbb{Q}(G) \subseteq F$.

In every case the field F is either $\mathbb{Q}$, an extension of $\mathbb{Q}$ of degree 2 or, in the case of $\mathcal{J}_3^{(5)}$, an extension of degree 4.

Suppose that $\mathbb{Q}(G) = \mathbb{Q}$. If $\mathcal{L}$ is a 3-system it follows from Theorem 7.28 that we have case (*i*). If $\mathcal{L}$ is a 4-system it follows from Theorem 7.42 that we have case (*iii*) because $\mathcal{N}_4$ and $\mathcal{O}_4$ contain $\mathcal{D}_4^{(4)}$ and from the proof of Lemma 7.32, $\mathcal{J}_3^{(4)}$ is not a Euclidean line system. If $\mathcal{L}$ is a Euclidean 5-system, then from Theorems 8.10 and 8.15, the only possibility for $\mathcal{L}$ is $\mathcal{H}_3$ or $\mathcal{H}_4$. However, from the description of their root systems in §6 of Chapter 5 we see that in this case $\mathbb{Q}(G) = \mathbb{Q}[\tau]$. Therefore, if $F = \mathbb{Q}$ or if F is an extension of $\mathbb{Q}$ of degree 2, then $F = \mathbb{Q}(G)$. The line system $\mathcal{J}_3^{(5)}$ contains both $\mathcal{D}_3^{(3)}$ and $\mathcal{H}_3$, hence $F = \mathbb{Q}(G)$ in this case as well. $\qquad\square$

The next result is equivalent to [**171**, Corollary 13].

Theorem 8.31. *If G is a finite unitary reflection group of rank n, then G is definable over its ring of definition, i.e. G can be represented by $n \times n$ matrices with entries in $\mathbb{Z}(G)$.*

Proof. We may suppose that G is irreducible. If G is imprimitive, the result follows from Theorem 2.20 whereas, if the rank of G is two, it follows from Theorem 6.2. Thus we may now suppose that G is primitive and that its rank is ≥ 3.

By Theorem 8.30 the ring of definition A of G is a principal ideal domain and $G = W(\Sigma, f)$, where (Σ, f) is a A-root system.

The set L of A-linear combinations of the elements of Σ is a G-invariant finitely generated torsion-free A-module, hence L is a free A-module of rank n. (That is, L is a *lattice* in V.) It follows that the entries of the matrices of the elements of G with respect to a basis of L are in A. $\qquad\square$

We close this section with a statement which is a consequence of our treatment of the classification.

Theorem 8.32. *Let $\sigma \in \mathrm{Gal}(\overline{\mathbb{Q}}, \mathbb{Q})$ be any Galois automorphism. Then $\sigma(G)$ is a reflection group in V which is $GL(V)$-conjugate to G.*

An example of such an automorphism is complex conjugation.

9. Reduction modulo p

In order to prepare for the identification of a primitive reflection group G as a finite linear group, we consider the reduction of G modulo p, where p is a prime (more precisely, modulo $\mathfrak{p}$, where $\mathfrak{p}$ is a prime ideal dividing p in the ring of integers of a number field). To some extent this is the approach to the classification taken by Blichfeldt [**26**] and has its origins in the work of Kronecker [**141**] as used by Gordon [**102**] in his determination of the finite subgroups of $PSU_2(\mathbb{C})$.

The elements of G are matrices and the basic idea is to replace their entries by their reductions modulo $\mathfrak{p}$. For this to be meaningful we first ensure that the matrix entries are integers in some algebraic number field and choose a prime ideal $\mathfrak{p}$ that contains the rational prime p.

Suppose that G acts irreducibly on a vector space V of dimension n over $\mathbb{Q}(G)$, where n is the rank of G. We shall suppose that the ring of definition $A := \mathbb{Z}(G)$ is a principal ideal domain since, by Theorem 8.30, this is the case for the primitive groups of rank at least 3. If $v \neq 0$ is an element of V, then

$$L := \sum_{g \in G} Ag(v).$$

is G-invariant and a finitely generated torsion-free A-module. Hence L is a free A-module of rank n (see Lang [**142**, Chapter III, Theorem 7.8]). In particular, $V = \mathbb{Q}(G) \otimes_A L$ and we may represent G by matrices with entries in A. If $\mathfrak{p}$ is a prime ideal of A that contains the rational prime p, then $\mathbb{F}_q := A/\mathfrak{p}$ is a finite field of order $q = p^k$, for some k. If $L_q := L/\mathfrak{p}L$, then the natural map $\mu : A \to \mathbb{F}_q$ extends to a homomorphism $\hat{\mu} : G \to GL(L_q) \simeq GL_n(\mathbb{F}_q)$.

Lemma 8.33. Ker $\hat{\mu} \subseteq O_p(G)$, *where* $O_p(G)$ *denotes the largest normal* p-*subgroup of* G.

Proof. (Brauer [**34**]) The kernel of $\hat{\mu}$ is certainly a normal subgroup of G and so all that needs to be shown is that the order of every element $h \in$ Ker $\hat{\mu}$ is a power of p. To this end, suppose that $h \in G$ and $h \equiv I \pmod{\mathfrak{p}}$. By Corollary A.6 we may choose k sufficiently large so that for all $g \in G$, no entry in $g - I$ belongs to $\mathfrak{p}^k$. Then $h^{p^k} = I$ because $h^{p^k} - I \equiv 0 \pmod{\mathfrak{p}}$. □

Lemma 8.34. *Suppose that* G *is a finite primitive reflection group of rank* $n \geq 3$ *and that* p *is a prime.*

(i)   *If* n *is not a power of* p, *then* $O_p(G) \subseteq Z(G)$.
(ii)   *If* $p > 3$, *then* $O_p(G) = 1$.

Proof. (i) If $O_p(G) \not\subseteq Z(G)$, then by Theorem 2.4 $O_p(G)$ is irreducible. The degree of an irreducible representation divides the order of the group (Lang [**142**, Chapter XVIII, Corollary 4.8]) and therefore n is a power of p, contrary to our assumption. Thus $O_p(G) \subseteq Z(G)$.

(ii) If $r \in G$ is a reflection and if $x \in O_p(G)$, then by Theorems 8.4 and 8.5 $r^{-1}x^{-1}rx \in O_p(G)$ has order at most 6. If $xr = rx$ for all reflections r, then $x \in Z(G)$. Thus we may suppose that $p = 5$ and that r and s are reflections of order 2 such that rs is an element of order 5 in $O_5(G)$. Then, by Theorem 2.19, there exists a reflection t such that $H := \langle r, s, t \rangle$ is primitive. But now $rs \in O_5(G) \cap H \subseteq Z(H)$, by (i). This contradiction means that $x \in Z(G)$ and hence $O_p(G) \subseteq Z(G)$.

If the order of $x \in O_p(G)$ is p, then $x = \lambda I$, where λ is a pth root of unity. We have $\det x = \lambda^n$ and since G is generated by reflections of orders 2 or 3 it follows that $\lambda^{6n} = 1$ and hence p divides n. If $G = W(\mathcal{A}_n)$, then $Z(G) = 1$ and therefore, by Theorem 8.29, we may suppose that $n \leq 8$. It follows that n is 5 or 7 and $p = n$. But in theses cases it follows from Theorem 8.7 that $|Z(G)| \leq 3$, which is a contradiction. Hence $O_p(G) = 1$. $\qquad\square$

If $p = 2$ and $n = 4$, or if $p = n = 3$ we shall see that there are finite primitive reflection groups G such that $O_p(G) \not\subseteq Z(G)$.

10. Identification of the primitive reflection groups

The reflection group $W(\mathcal{A}_n)$ is the symmetric group $\text{Sym}(n + 1)$ and the groups $W(\mathcal{D}_n^{(k)}) = G(k, k, n)$ and $W(\mathcal{B}_n^{(k)}) = G(k, \frac{1}{2}k, n)$ have been described in Chapter 2 as subgroups of the wreath product $C_k \wr \text{Sym}(n)$.

If G is one of the 19 primitive groups of rank 2, then $G/Z(G)$ is $\text{Alt}(4)$, $\text{Sym}(4)$ or $\text{Alt}(5)$ and explicit generators and descriptions have been given in Chapter 6.

In this section we construct homomorphisms from the 15 'exceptional' primitive reflection groups of rank at least 3 to linear groups over finite fields.

10.1. Quadratic, alternating and hermitian forms. Suppose that G is a finite primitive unitary reflection group of rank n, let $A := \mathbb{Z}(G)$ be its ring of definition and let Σ be the A-root system for G, as in Theorem 8.30. Since $\mathbb{Q}(G)$ is a finite abelian extension of $\mathbb{Q}$, there is a well-defined operation of complex conjugation on $\mathbb{Q}(G)$. If $\mathbb{Q}(G)$ is $\mathbb{Q}$ or $\mathbb{Q}[\tau]$ this automorphism is the identity.

Let V be the n-dimensional vector space over $\mathbb{Q}(G)$ spanned by Σ and let L be the A-module spanned by Σ. Then L is free of rank n and $V = \mathbb{Q}(G) \otimes L$.

As in §9 suppose that $\mathfrak{p}$ is a prime ideal of A that contains the rational prime p. Then $A/\mathfrak{p}$ is a finite field $\mathbb{F}_q$, where q is a power of p. The natural map $\mu : A \to \mathbb{F}_q$ extends to a homomorphism $\hat{\mu} : G \to GL_n(\mathbb{F}_q)$. In general, we shall see that $\text{Ker}\, \hat{\mu} = O_p(G)$ and that $\hat{\mu}(G)$ is an orthogonal, symplectic or unitary group. This approach to the identification of the groups $W(\mathcal{E}_6)$, $W(\mathcal{E}_7)$ and $W(\mathcal{E}_8)$ is the basis of several exercises in Bourbaki [33, Ch. VI §4].

If Σ consists of short roots, define $h(u, v) := 2(u, v)$ for all $u.v \in L$. If Σ consists of long roots or tall roots, define $h(u, v) := (u, v)$. In both cases $h(u, v)$ is an hermitian form on L with values in A.

Suppose that $\mathfrak{p}$ is fixed by complex conjugation. Then the map $\sigma : \mathbb{F}_q \to \mathbb{F}_q$ defined by $\sigma(a + \mathfrak{p}) = \bar{a} + \mathfrak{p}$ is an automorphism of $\mathbb{F}_q$. If σ is not the identity, then q is a square and there is an induced hermitian form h_q on the vector space $L_q = L/\mathfrak{p}L$ over the field $\mathbb{F}_q$. If $\hat{\mu} : G \to GL_n(\mathbb{F}_q)$ is induced by the natural map $A \to A/\mathfrak{p}$, then $\hat{\mu}(G)$ preserves this form and therefore, if h_q is not identically zero, $\hat{\mu}(G)$ is a subgroup of the corresponding finite unitary group $U_n(\mathbb{F}_q)$.

For the remainder of this section suppose that $\mathfrak{p}$ is fixed by complex conjugation but that the induced automorphism of $\mathbb{F}_q$ is the identity. For example, this is always the case when $q = p$. In some cases the form induced on L_q by the hermitian form on L vanishes identically. In other cases there is an alternating or quadratic form on L_q. Appendix B contains a summary of notation and relevant results on quadratic forms and orthogonal groups over finite fields.

For $v \in L$, let $\bar{v}$ denote the image of v in L_q. If the A-module L is spanned by short roots, then the map $Q_q : L_q \to A$ defined by $Q_q(\bar{v}) := (v, v) + \mathfrak{p}$ is a quadratic form with polar form $\beta_q(\bar{u}, \bar{v}) := 2(u, v) + \mathfrak{p}$. Similarly, if $p \neq 2$ and L is spanned by long roots, then the map $Q_q : L_q \to A$ defined by $Q_q(\bar{v}) := \frac{1}{2}(v, v) + \mathfrak{p}$ is a quadratic form Q_q with polar form $\beta_q(\bar{u}, \bar{v}) = (u, v) + \mathfrak{p}$. In both cases $\hat{\mu}(G)$ preserves Q_q and therefore $\hat{\mu}(G)$ is a subgroup of the finite orthogonal group $O(L_q, Q_q)$.

If $a \in \Sigma$, then $Q_q(\bar{a}) = 1$; that is, $\bar{a}$ is non-singular. Thus $\bar{a}$ defines a reflection $t_{\bar{a}} \in \widehat{\Omega}(L_q, Q_q)$, where $\widehat{\Omega}(L_q, Q_q)$ is the kernel of the spinor norm (see Appendix B, §3.2). If q is a power of 2, $\widehat{\Omega}(L_q, Q_q) = O(L_q, Q_q)$.

It can happen that $t_{\bar{a}} = t_{\bar{b}}$ even though $a \neq b$ and furthermore, it is not always true that every reflection in $\widehat{\Omega}(L_q, Q_q)$ has the form $t_{\bar{a}}$. However, we shall see that, in general, for small primes p, the homomorphism $\hat{\mu} : G \to \widehat{\Omega}(L_q, Q_q)$ is onto.

If $p = 2$ and if L is spanned by long roots, then the bilinear form $\beta_q(\bar{u}, \bar{v}) := 2(u, v) + \mathfrak{p}$ is alternating and hence $\hat{\mu}(G)$ is a subgroup of the finite symplectic group $Sp_n(L_q)$.

10.2. The groups $W(\mathcal{H}_3)$ and $W(\mathcal{H}_4)$. The 15 lines of $\mathcal{H}_3$ and the 60 lines of $\mathcal{H}_4$ are spanned by roots whose coordinates belong to $\mathbb{Z}[\tau]$. Therefore $W(\mathcal{H}_3)$ is a finite subgroup of $O_3(\mathbb{R})$ and $W(\mathcal{H}_4)$ is a finite subgroup of $O_4(\mathbb{R})$. This means that we may use the results of §2 of Chapter 6 and regard the roots of $\mathcal{H}_3$ and $\mathcal{H}_4$ as quaternions of norm 1.

Theorem 8.35. $W(\mathcal{H}_4)$ *is a group of order* $2^6 \cdot 3^2 \cdot 5^2 = 14\,400$. *It is isomorphic to*

(i)   *the central product $\mathcal{I} \circ \mathcal{I}$ extended by an element of order 2 that interchanges the two factors, where $\mathcal{I} \simeq SL_2(\mathbb{F}_5)$ is the binary icosahedral group,*

(ii)   *a non-split extension of C_2 by the orthogonal group $O_4^+(\mathbb{F}_4)$, and*

(iii)   *the group $\widehat{\Omega}_4^+(\mathbb{F}_5)$.*

Proof. The order of the binary icosahedral group $\mathcal{I}$ is 120 and $\mathcal{I}$ is a subgroup of $\mathbb{H}$, the algebra of quaternions. Bearing this in mind, and using the description of $\mathcal{I}$ in §6 of Chapter 5 as well as the description of $\mathcal{H}_4$ in §6.3 of Chapter 7, we see that the elements of $\mathcal{I}$ span the 60 lines of $\mathcal{H}_4$. It follows from Theorem 5.17 that $W(\mathcal{H}_4) \simeq (\mathcal{I} \circ \mathcal{I})\langle \rho \rangle$. In particular, $W(\mathcal{H}_4)$ is transitive on the 60 lines of its line system.

(*ii*) If η is an element of order 3 in the Galois field $\mathbb{F}_4$, there is a homomorphism $\mu_4 : \mathbb{Z}[\tau] \to \mathbb{F}_4$ such that $\mu_4(\tau) = \eta$. The map $Q(v) = (v, v)$ on the $\mathbb{Z}[\tau]$-module L spanned by the roots of $\mathcal{H}_4$ defines a non-degenerate quadratic form Q_4 on $L_4 = \mathbb{F}_4^4$. The Witt index of Q_4 is 2 because the images of $(1, 1, 0, 0)$ and $(0, 0, 1, 1)$ in L_4 span a 2-dimensional subspace on which Q_4 vanishes identically. Therefore μ_4 extends to a homomorphism $\hat{\mu}_4 : W(\mathcal{H}_4) \to O_4^+(\mathbb{F}_4)$. By Lemma 8.33 its kernel is $\{\pm I\}$. By Lemma B.4, $O_4^+(\mathbb{F}_4)$ contains 60 reflections. Thus $\hat{\mu}_4$ induces a one-to-one correspondence between the reflections of $W(\mathcal{H}_4)$ and the reflections of $O_4^+(\mathbb{F}_4)$. Since $O_4^+(\mathbb{F}_4)$ is generated by its reflections, it follows that $\hat{\mu}_4$ is onto and hence $W(\mathcal{H}_4)/\{\pm I\} \simeq O_4^+(\mathbb{F}_4)$. The extension does not split because $W(\mathcal{H}_4)$ contains elements whose square is $-I$.

(*iii*) Similarly there is a homomorphism $\mu_5 : \mathbb{Z}[\tau] \to \mathbb{F}_5$ such that $\mu_5(\tau) = 3$. The quadratic form Q_5 that is induced by Q on $L_5 = \mathbb{F}_5^4$ is non-degenerate and its Witt index is 2 because the images of $(1, 2, 0, 0)$ and $(0, 0, 1, 2)$ in L_5 span a 2-dimensional subspace on which Q_5 vanishes identically. By Lemma 8.33 the induced homomorphism $\hat{\mu}_5 : W(\mathcal{H}_4) \to \widehat{\Omega}_4^+(\mathbb{F}_5)$ is one-to-one. By Theorems B.5 and B.7, $\widehat{\Omega}_4^+(\mathbb{F}_5)$ is generated by its 60 reflections and so $\hat{\mu}_5$ is onto, whence $W(\mathcal{H}_4) \simeq \widehat{\Omega}_4^+(\mathbb{F}_5)$. $\qquad\square$

Theorem 8.36. $W(\mathcal{H}_3)$ *is a group of order* $2^3 \cdot 3 \cdot 5 = 120$. *It is isomorphic to* $C_2 \times \mathrm{Alt}(5)$, $C_2 \times SL_2(\mathbb{F}_4)$ *and* $C_2 \times PSL_2(\mathbb{F}_5)$.

Proof. (*i*) The 15 lines of $\mathcal{H}_3$ are spanned by the pure quaternions i, j, k and the cyclic shifts of $\frac{1}{2}(i \pm \tau j \pm \tau^{-1} k)$. The corresponding reflections fix $1 \in \mathbb{H}$ and therefore, in the notation of the previous theorem, $W(\mathcal{H}_3)$ is a normal subgroup of $\langle \rho \rangle \times B(\mathcal{I})$. Furthermore, $r_i r_j r_k$ is a central element of order 2 in $W(\mathcal{H}_3)$ and $B(\mathcal{I}) \simeq \mathrm{Alt}(5)$ has no non-trivial normal subgroups. Therefore $W(\mathcal{H}_3) = \langle \rho \rangle \times B(\mathcal{I})$.

(*ii*) Let N be the $\mathbb{Z}[\tau]$-module spanned by the roots of $W(\mathcal{H}_3)$. As in the previous theorem, there is a non-degenerate quadratic form Q_4 on $N_4 = \mathbb{F}_4^3$ induced by $Q(v) = (v, v)$. Therefore μ_4 extends to a homomorphism $\hat{\mu}_4 : W(\mathcal{H}_3) \to O_3(\mathbb{F}_4) \simeq SL_2(\mathbb{F}_4)$. By Lemma 8.34 the kernel is $\{\pm I\}$. By Lemma B.4 and Theorem B.7, the group $O_3(\mathbb{F}_4)$ is generated by its 15 reflections and hence $\hat{\mu}_4$ is onto. Consequently $W(\mathcal{H}_3) \simeq C_2 \times SL_2(\mathbb{F}_4)$ and $\mathrm{Alt}(5) \simeq SL_2(\mathbb{F}_4)$.

(*iii*) The restriction of the homomorphism $\hat{\mu}_5$ of Theorem 8.35 (*iii*) to $W(\mathcal{H}_3)$ is an isomorphism $\hat{\mu}_5 : W(\mathcal{H}_3) \to C_2 \times \Omega_3(\mathbb{F}_5) \simeq C_2 \times PSL_2(\mathbb{F}_5)$. $\qquad\square$

Corollary 8.37. $\mathrm{Alt}(5) \simeq SL_2(\mathbb{F}_4) \simeq PSL_2(\mathbb{F}_5)$.

Corollary 8.38. $PSL_2(\mathbb{F}_9) \simeq \mathrm{Alt}(6)$.

Proof. The polynomial $X^2 - X - 1 \in \mathbb{F}_3[X]$ is irreducible and therefore $\mathbb{F}_9 = \mathbb{F}_3[\xi]$, where $\xi^2 = \xi + 1$. The homomorphism $\mathbb{Z}[\tau] \to \mathbb{F}_9$ that sends τ to ξ extends to an embedding $\hat{\mu}_9 : W(\mathcal{H}_3) \to \mathcal{C}_2 \times \Omega_3(\mathbb{F}_9) \simeq \mathcal{C}_2 \times PSL_2(\mathbb{F}_9)$ and thus $PSL_2(\mathbb{F}_9)$ contains Alt(5) as a subgroup of index 6. Therefore $PSL_2(\mathbb{F}_9)$ is isomorphic to Alt(6). $\qquad\square$

10.3. The groups $W(\mathcal{J}_3^{(4)})$ and $W(\mathcal{J}_3^{(5)})$. A *frame* in $\mathbb{C}^n$ is a set of n mutually perpendicular lines. Let $\mathfrak{L}$ be the line system $\mathcal{J}_3^{(4)}$ or $\mathcal{J}_3^{(5)}$ and set $G := W(\mathfrak{L})$. Then $-I \in G$ since in both cases $\mathfrak{L}$ contains a frame of $\mathbb{C}^3$ and the product of the reflections of order 2 whose roots are the lines of the frame is the scalar matrix $-I$.

Theorem 8.39. $W(\mathcal{J}_3^{(4)})$ *is a group of order* $2^4 \cdot 3 \cdot 7 = 336$. *It is isomorphic to* $\mathcal{C}_2 \times PSL_2(\mathbb{F}_7)$ *and to* $\mathcal{C}_2 \times SL_3(\mathbb{F}_2)$.

Proof. The 21 lines of $\mathcal{J}_3^{(4)}$ are spanned by $(\lambda, 0, 0)$, $\frac{1}{2}(\lambda^2, \lambda^2, 0)$, $\frac{1}{2}(\lambda, \lambda, 2)$ and their images under the action of $W(\mathcal{B}_3^{(2)})$, where $\lambda^2 + \lambda + 2 = 0$. The Cartan coefficients belong to $\mathbb{Z}[\lambda]$ and therefore the entries of the matrices of the reflections (with respect to a basis of roots) also belong to $\mathbb{Z}[\lambda]$.

The homomorphism $\mathbb{Z}[\lambda] \to \mathbb{F}_7$ that sends λ to 3 extends to a homomorphism $\hat{\mu}_7 : G \to \mathcal{C}_2 \times \Omega_3(\mathbb{F}_7) \simeq \mathcal{C}_2 \times PSL_2(\mathbb{F}_7)$. From Lemma B.4, $\mathcal{C}_2 \times \Omega_3(\mathbb{F}_7)$ is generated by its 21 reflections and hence $\hat{\mu}_7$ is onto. From Lemma 8.33, $\hat{\mu}_7$ is one-to-one and therefore $\hat{\mu}_7$ is an isomorphism. Thus $|Z(G)| = 2$ and $|G| = 336$.

The kernel $\mathfrak{p}$ of the homomorphism $\mu_2 : \mathbb{Z}[\lambda] \to \mathbb{F}_2$ that sends λ to 0 is the ideal $\lambda\mathbb{Z}[\lambda]$, which is not fixed by complex conjugation. Thus in this case there is no G-invariant form on $L/\mathfrak{p}L$. However, μ_2 extends to a group homomorphism $\hat{\mu}_2 : G \to GL_3(\mathbb{F}_2) = SL_3(\mathbb{F}_2)$ with kernel $Z(G)$. On comparing orders we see that $\hat{\mu}_2$ is onto. $\qquad\square$

Corollary 8.40. $PSL_2(\mathbb{F}_7) \simeq SL_3(\mathbb{F}_2)$.

Theorem 8.41. $W(\mathcal{J}_3^{(5)})$ *is a group of order* $2^4 \cdot 3^3 \cdot 5 = 2160$. *It is isomorphic to* $\mathcal{C}_2 \times (\mathcal{C}_3 \cdot \mathrm{Alt}(6))$, *where* $\mathcal{C}_3 \cdot \mathrm{Alt}(6)$ *denotes the non-split extension of the cyclic group* $\mathcal{C}_3$ *by the alternating group* $\mathrm{Alt}(6)$.

Proof. The 45 lines of $\mathfrak{L} := \mathcal{J}_3^{(5)}$ are the 15 lines of $\mathcal{H}_3$ and the 30 lines spanned by the images of $\frac{1}{2}(\tau + \omega, \tau^{-1}\omega - 1, 0)$ under the action of $W(\mathcal{H}_3)$. Representatives for the roots of $\mathcal{J}_3^{(5)}$ not in $\mathcal{H}_3$ can be obtained by applying cyclic shifts and sign changes to the vectors $\frac{1}{2}(\tau + \omega, \tau^{-1}\omega - 1, 0)$, $\frac{1}{2}(\tau\omega, 1, \tau^{-1}\omega^2)$ and $\frac{1}{2}(\omega^2, 1, \tau\omega + 1)$. The Cartan coefficients belong to $\mathbb{Z}[\tau, \omega]$.

We have $\mathbb{F}_9 = \mathbb{F}_3[\xi]$, where $\xi^2 = \xi + 1$ and hence $\xi^4 = -1$. The kernel of the homomorphism $\mu_9 : \mathbb{Z}[\tau, \omega] \to \mathbb{F}_9$ that sends τ to ξ and ω to 1 is fixed by complex conjugation and the induced automorphism of $\mathbb{F}_9$ is the identity. Thus there is a group homomorphism $\hat{\mu}_9 : W(\mathcal{J}_3^{(5)}) \to \mathcal{C}_2 \times \Omega_3(\mathbb{F}_9) \simeq \mathcal{C}_2 \times PSL_2(\mathbb{F}_9)$. The group $\mathcal{C}_2 \times \Omega_3(\mathbb{F}_9)$ is generated by its 45 reflections and therefore $\hat{\mu}_9$ is onto. We know from its construction that $W(\mathcal{J}_3^{(5)})$ contains a central element of order 6 and so from Lemma 8.33 the kernel of $\hat{\mu}_9$ is a non-trivial 3-group.

We have $\mathbb{F}_4 := \mathbb{F}_2[\eta]$, where $\eta^2 + \eta + 1 = 0$. The kernel of the homomorphism $\mu_4 : \mathbb{Z}[\omega, \tau] \to \mathbb{F}_4$ that sends both ω and τ to η is $\mathfrak{p} = (\tau + \omega)\mathbb{Z}[\tau, \omega]$. The ideal $\mathfrak{p}$ is not fixed by complex conjugation and there is no $W(\mathcal{J}_3^{(5)})$-invariant form on $L/\mathfrak{p}L$. On the other hand, the ring homomorphism induces a group homomorphism $\hat{\mu}_4 : W(\mathcal{J}_3^{(5)}) \to SL_3(\mathbb{F}_4)$. From Lemma 8.33 and the previous paragraph, $\mathrm{Ker}\,\hat{\mu}_4 = \{\pm I\}$, therefore $|\mathrm{Ker}\,\hat{\mu}_9| = 3$ and the centre of $W(\mathcal{J}_3^{(5)})$ has order 6. Thus $|W(\mathcal{J}_3^{(5)})| = 2160$ and the image of $\hat{\mu}_4$ is a group $\hat{A}$ of index 56 in $SL_3(\mathbb{F}_4)$. The line system $\mathcal{J}_3^{(5)}$ contains $\mathcal{D}_3^{(3)}$ and therefore a Sylow 3-subgroup of $G(3,3,3)$ is also a Sylow 3-subgroup of $W(\mathcal{J}_3^{(5)})$, namely an extraspecial group of order 27 and exponent 3. It follows that $\hat{A}$ is a non-split extension of $\mathcal{C}_3$ by $\mathrm{Alt}(6) \simeq PSL_2(\mathbb{F}_9)$. $\square$

10.4. The groups $W(\mathcal{L}_3)$, $W(\mathcal{L}_4)$ and $W(\mathcal{M}_3)$. The group $W(\mathcal{M}_3)$ is the only primitive reflection group of rank at least 3 that contains reflections of orders 2 and 3. The following theorem shows that its structure is completely determined by that of $W(\mathcal{L}_3)$.

Theorem 8.42.

(i) $W(\mathcal{M}_3)$ is a group of order $2^4 \cdot 3^4 = 1296$ and equal to $\{\pm I\} \times W(\mathcal{L}_3)$.

(ii) $W(\mathcal{L}_3)$ is a group of order $2^3 \cdot 3^4 = 648$. It is isomorphic to $U_3(\mathbb{F}_4)$ and to the semidirect product $E \cdot W(\mathcal{L}_2)$, where $W(\mathcal{L}_2) \simeq SL_2(\mathbb{F}_3)$ and where $E := O_3(W(\mathcal{D}_3^{(3)}))$ is an extraspecial group of order 27 and exponent 3.

Proof. (i) The group $W(\mathcal{L}_2) \simeq SL_2(\mathbb{F}_3)$ is a subgroup of $W(\mathcal{L}_3)$ and therefore $W(\mathcal{L}_3)$ contains an element t of order 2 whose eigenvalues are $-1, -1$ and 1. Thus $-r$ is a reflection and therefore $\{\pm I\} \times W(\mathcal{L}_3)$ is generated by reflections of orders 2 and 3. It follows from Theorem 8.26 that $\{\pm I\} \times W(\mathcal{L}_3) \simeq W(\mathcal{M}_3)$.

The imprimitive reflection group $G(3,3,3)$ is a normal subgroup of index 3 in $H := G(3,1,3)$. Furthermore, H fixes one of the four systems of imprimitivity of $G(3,3,3)$ and permutes the others in a cycle of length three. By symmetry we may choose a conjugate K of H in $U_3(\mathbb{C})$ that normalises $G(3,3,3)$ and fixes a system of imprimitivity other than the one fixed by H. Then the group $G := \langle H, K \rangle$ normalises $G(3,3,3)$ and is a primitive group generated by reflections of orders 2 and 3; thus, from Theorem 8.26, $G \simeq W(\mathcal{M}_3)$.

By construction, $W(\mathcal{M}_3)$ acts as the alternating group $\mathrm{Alt}(4)$ on the 4 systems of imprimitivity of $G(3,3,3)$. The kernel of this action is $Z(W(\mathcal{M}_3))$. The group $G(3,3,3)$ contains the central element ωI and this element belongs to $W(\mathcal{L}_3)$. Thus

$|Z(W(\mathcal{L}_3))| = 3$ and $|Z(W(\mathcal{M}_3))| = 6$. It follows that $|W(\mathcal{M}_3)| = 2^4 \cdot 3^4 = 1296$ and $|W(\mathcal{L}_3)| = 648$.

(*ii*) If $\theta := \omega - \omega^2$, the 12 lines of $\mathcal{L}_3$ are spanned by the tall roots $(\theta, 0, 0)$, $(0, \theta, 0)$, $(0, 0, \theta)$ and $(1, \omega^i, \omega^j)$, where $i, j \in \{0, 1, 2\}$ and where the coordinates belong to $\mathbb{Z}[\omega]$. Let Σ be the root system and let L be the $\mathbb{Z}[\omega]$-module spanned by Σ.

If η is an element of order 3 in the Galois field $\mathbb{F}_4$, there is a homomorphism $\mu_4 : \mathbb{Z}[\omega] \to \mathbb{F}_4$ such that $\mu_4(\omega) = \eta$. The kernel $\mathfrak{p}$ of μ_4 is fixed by complex conjugation and therefore the hermitian form on $\mathbb{C}^3$ induces an hermitian form on $L_4 = L/\mathfrak{p}L$. Thus μ_4 extends to a homomorphism $\hat{\mu}_4 : W(\mathcal{L}_3) \to U_3(\mathbb{F}_4)$ whose kernel is trivial. On comparing orders we see that $\hat{\mu}_4$ is an isomorphism.

Let $E := K \cap W(\mathcal{L}_3)$, where $K := G(3, 3, 3)$. Then E is a subgroup of index 2 in K and therefore $E = O_3(K)$, which is an extraspecial group of exponent 3 (see Remark 2.17). We have $E \cap W(\mathcal{L}_2) = 1$ and therefore $|E W(\mathcal{L}_2)| = 648$, whence $W(\mathcal{L}_3) = E W(\mathcal{L}_2)$. □

Theorem 8.43. $W(\mathcal{L}_4)$ *is a group of order* $2^7 \cdot 3^5 \cdot 5 = 155\,520$. *It is isomorphic to* $C_3 \times Sp_4(\mathbb{F}_3)$ *and to a non-split extension of* C_2 *by* $U_4(\mathbb{F}_4)$. *Its centre has order* 6.

Proof. Let Σ be the $\mathbb{Z}[\omega]$-root system for $\mathcal{L}_4$ defined in §5.4. For $u, v \in L$ define $h(u, v) := \theta^{-1}\omega^2(u, v)$, where $\theta = \omega - \omega^2$. Then for $a, b \in \Sigma$, $h(a, b)$ is the Cartan coefficient $\langle a \,|\, b \rangle$; it is either 0 or a 6th root of unity. Thus $h(a, b) + h(b, a)$ is either 0 or a multiple of θ. The kernel of the homomorphism $\mu_3 : \mathbb{Z}[\omega] \to \mathbb{F}_3$ that sends ω to 1 is generated by θ and therefore h induces an alternating form on $L_3 := L/\theta L$.

Consequently there is a homomorphism $\hat{\mu}_3 : W(\mathcal{L}_4) \to Sp_4(\mathbb{F}_3)$ that maps each reflection to a symplectic transvection. Furthermore, every transvection is the image of a reflection. By Theorem B.7 $Sp_4(\mathbb{F}_3)$ is generated by its transvections and therefore $\hat{\mu}_3$ is onto. Since $O_3(Sp_4(\mathbb{F}_3)) = 1$ it follows from Lemma 8.34 that $\operatorname{Ker} \hat{\mu}_3 = O_3(W(\mathcal{L}_4)) \subseteq Z(W(\mathcal{L}_4))$. Since $W(\mathcal{L}_4)$ can be represented by matrices over $\mathbb{Z}[\omega]$, the order of the centre of $W(\mathcal{L}_4)$ divides 6. On the other hand, $\mathcal{L}_4$ contains a frame and therefore the centre contains an element of order 3. This proves that $|O_3(W(\mathcal{L}_4))| = 3$ and hence $W(\mathcal{L}_4) \simeq C_3 \times Sp_4(\mathbb{F}_3)$. It follows that $|Z(W(\mathcal{L}_4))| = 6$ and $|W(\mathcal{L}_4)| = 2^7 \cdot 3^5 \cdot 5$.

If $\mathbb{F}_4 := \mathbb{F}_2[\eta]$, the homomorphism $\mu_4 : \mathbb{Z}[\omega] \to \mathbb{F}_4 = \mathbb{F}_2[\eta]$ that sends ω to η extends to a group homomorphism $\hat{\mu}_4 : W(\mathcal{L}_4) \to U_4(\mathbb{F}_4)$. The kernel of μ_4 is the central subgroup of order 2. On comparing orders we have $U_4(\mathbb{F}_4) \simeq C_3 \times PSp_4(\mathbb{F}_3)$. The group $W(\mathcal{L}_4)$ contains elements of order 4 whose square is central and hence the extension is non-split. □

Corollary 8.44. $PSU_4(\mathbb{F}_4) \simeq PSp_4(\mathbb{F}_3)$.

Proof. We obtain the result from the isomorphism $U_4(\mathbb{F}_4) \simeq C_3 \times PSp_4(\mathbb{F}_3)$ on taking the quotient by the centre. (This isomorphism was obtained by Kneser [**132**] using similar methods.) $\qquad\qquad\square$

10.5. The groups $W(\mathcal{F}_4)$, $W(\mathcal{N}_4)$ and $W(\mathcal{O}_4)$. The line systems $\mathcal{A}_4$, $\mathcal{D}_4^{(2)}$, $\mathcal{D}_4^{(4)}$, $\mathcal{B}_4^{(2)}$, $\mathcal{B}_4^{(4)}$, $\mathcal{F}_4$ and $\mathcal{N}_4$ are subsystems of $\mathcal{O}_4$. We elucidate the structure of $\mathcal{O}_4$ and the interplay between these subsystems in a series of lemmas.

Recall, from §6.2 of Chapter 7, that the line system $\mathcal{O}_4$ is the union of $\mathcal{B}_4^{(4)}$ with the 32 images, under the action of $G(4,2,4) = W(\mathcal{B}_4^{(4)})$, of the line ℓ spanned by $x := \frac{1+i}{2}(1,1,1,1)$. The 16 images of $(1+i,0,0,0)$ and the 96 images of $(1,1,0,0)$, under the action of $G(4,2,4)$, form a $\mathbb{Z}[i]$-root system for $\mathcal{B}_4^{(4)}$. The union of this root system with the 128 images of x is the $\mathbb{Z}[i]$-root system Σ for $\mathcal{O}_4$.

Let m, n, p and q be the lines spanned by

$$(1+i,0,0,0), \quad (1,i,0,0), \quad \tfrac{1+i}{2}(-1,1,1,1) \quad \text{and} \quad \tfrac{1+i}{2}(i,-i,1,1),$$

respectively.

Lemma 8.45. *The group $W(\mathcal{O}_4)$ acts transitively on $\mathcal{O}_4$ and its subgroup $W(\mathcal{N}_4)$ has orbits $\mathcal{N}_4$ and $\mathcal{O}_4 \setminus \mathcal{N}_4$ of lengths 40 and 20 respectively.*

Proof. The group $W(\mathcal{B}_4^{(4)})$ has three orbits on $\mathcal{O}_4$ (of lengths 32, 4 and 24) with representatives ℓ, m and n. The lines ℓ, p and q are in the same orbit of $W(\mathcal{B}_4^{(4)})$ and the sets $\{\ell, m, p\}$ and $\{\ell, n, q\}$ are 3-stars. Therefore the reflection r_ℓ interchanges m and p and also n and q and so $W(\mathcal{O}_4)$ has a single orbit on $\mathcal{O}_4$.

The lines q and n are representatives for the orbits (of lengths of 16 and 24) of $W(\mathcal{D}_4^{(4)})$ on $\mathcal{N}_4$ and m and p represent the orbits (of lengths 4 and 16) of $W(\mathcal{D}_4^{(4)})$ on $\mathcal{O}_4 \setminus \mathcal{N}_4$. Therefore $W(\mathcal{N}_4)$ is transitive on $\mathcal{N}_4$ and on $\mathcal{O}_4 \setminus \mathcal{N}_4$. $\qquad\square$

If Δ is the set of lines of $\mathcal{O}_4$ at $60°$ to m, then $|\Delta| = 32$ and Δ is an orbit of $W(\mathcal{B}_4^{(4)})$. Therefore, there are exactly four lines which are equal or orthogonal to m and at $60°$ to every line of Δ, namely the frame F whose lines are spanned by the roots $(1+i,0,0,0)$, $(0,1+i,0,0)$, $(0,0,1+i,0)$ and $(0,0,0,1+i)$. This is the (unique) system of imprimitivity for $W(\mathcal{B}_4^{(4)})$. Call a frame obtained in this manner a *special frame* and let $\mathfrak{F}$ be the set of all special frames.

Every line of $\mathcal{O}_4$ is in a unique special frame and two special frame are either equal or disjoint, hence $|\mathfrak{F}| = 15$. Furthermore, by Lemma 8.45, $W(\mathcal{O}_4)$ acts transitively on $\mathfrak{F}$. (Thus the permutation action of $W(\mathcal{O}_4)$ on $\mathcal{O}_4$ is imprimitive and the special frames are 'blocks of imprimitivity'.)

Lemma 8.46. *The stabiliser of F in $W(\mathcal{O}_4)$ is $W(\mathcal{B}_4^{(4)})$, its stabiliser in $W(\mathcal{N}_4)$ is $W(\mathcal{D}_4^{(4)})$ and its stabiliser in $W(\mathcal{F}_4)$ is $W(\mathcal{B}_4^{(2)})$.*

Proof. Let H be the stabiliser of F in $W(\mathcal{O}_4)$. Then H is imprimitive and contains $W(\mathcal{B}_4^{(4)})$. If $h \in H$ fixes a line of F spanned by a root $a \in \Sigma$, then $ha = \alpha a$, where $\alpha \in \{\pm 1, \pm i\}$. Thus H is a subgroup of $G(4, 1, 4)$. However, $W(\mathcal{O}_4)$ does not contain reflections of order 4 and therefore $H = W(\mathcal{B}_4^{(4)})$.

The index of $W(\mathcal{D}_4^{(4)})$ in $W(\mathcal{B}_4^{(4)})$ is 2 and therefore the stabiliser of F in $W(\mathcal{N}_4)$ is $W(\mathcal{D}_4^{(4)}) = W(\mathcal{N}_4) \cap W(\mathcal{B}_4^{(4)})$.

The stabiliser of F in $W(\mathcal{F}_4)$ contains $W(\mathcal{B}_4^{(2)})$. By Theorem 8.30 we may scale the roots of $\mathcal{F}_4$ so that their Cartan coefficients belong to $\mathbb{Z}$. Therefore, if $h \in W(\mathcal{F}_4)$ fixes a root a, then $ha = \pm a$. The argument of the first paragraph shows that $W(\mathcal{B}_4^{(2)})$ is the stabiliser of F in $W(\mathcal{F}_4)$.   □

Corollary 8.47.

(i) $|W(\mathcal{O}_4)| = 2^{10} \cdot 3^2 \cdot 5 = 46\,080,$

(ii) $|W(\mathcal{N}_4)| = 2^9 \cdot 3 \cdot 5 = 7680,$ *and*

(iii) $|W(\mathcal{F}_4)| = 2^7 \cdot 3^2 = 1152.$

Proof. Consider the action of these groups on the set $\mathfrak{F}$ of special frames. The group $W(\mathcal{O}_4)$ is transitive on the 15 frames and the stabiliser of F has order $2^{10} \cdot 3$, hence $|W(\mathcal{O}_4)| = 2^{10} \cdot 3^2 \cdot 5$. The frame F is in an orbit of length 5 for $W(\mathcal{N}_4)$ and its stabiliser has order $2^9 \cdot 3$, hence $|W(\mathcal{N}_4)| = 2^9 \cdot 3 \cdot 5$.

The group $W(\mathcal{F}_4)$ has two orbits of length 3 on $\mathfrak{F}$ and the stabiliser of F has order $2^7 \cdot 3$, therefore $|W(\mathcal{F}_4)| = 2^7 \cdot 3^2$.   □

Let L be the $\mathbb{Z}[i]$-module spanned by the root system Σ for $\mathcal{O}_4$ and let $\mu_2 : \mathbb{Z}[i] \to \mathbb{F}_2$ be the homomorphism such that $\mu_2(i) = 1$. The kernel of μ_2 is the ideal $\mathfrak{p} = (i + 1)\mathbb{Z}[i]$ and we put $L_2 := L/\mathfrak{p}L$.

The elements of Σ are long roots and therefore the hermitian inner product $(-, -)$ on L induces an alternating form on L_2. Thus μ_2 extends to a homomorphism $\hat{\mu}_2 : W(\mathcal{O}_4) \to Sp_4(\mathbb{F}_2)$.

The group $E := O_2(W(\mathcal{D}_4^{(2)}))$ is an extraspecial 2-group isomorphic to the central product $\mathcal{Q} \circ \mathcal{Q}$ of two quaternion groups. (See Remark 2.17 and the description of $W(\mathcal{D}_4^{(2)}) = W(\mathcal{T})$ following Theorem 5.17.) The centre of $W(\mathcal{O}_4)$ is the cyclic group Z generated by the scalar matrix iI and we define $D := Z \circ E$.

Lemma 8.48. *The homomorphism $\hat{\mu}_2 : W(\mathcal{O}_4) \to Sp_4(\mathbb{F}_2)$ is surjective and its kernel is $O_2(W(\mathcal{O}_4)) = D$.*

Proof. If a and b are roots that span distinct lines of the frame F, it is clear that $a - b \in \mathfrak{p}L$. Thus $\hat{\mu}_2$ maps the four reflections with roots in a given frame to the same element (a transvection) of $Sp_4(\mathbb{F}_2)$ and hence the 60 reflections of $W(\mathcal{O}_4)$ are mapped to the 15 transvections of $Sp_4(\mathbb{F}_2)$. The group $Sp_4(\mathbb{F}_2)$ is generated by its transvections and therefore $\hat{\mu}_2$ is onto. Since $O_2(Sp_4(\mathbb{F}_2)) = 1$ it follows from Lemma 8.33 that $\mathrm{Ker}\,\hat{\mu}_2 = O_2(W(\mathcal{O}_4))$.

The elements of Z and of E fix every special frame and therefore D is contained in $\mathrm{Ker}\,\hat{\mu}_2$. Both groups have order 64 and therefore $D = \mathrm{Ker}\,\hat{\mu}_2$.   □

Theorem 8.49. *The group $W(\mathcal{N}_4)$ is the semidirect product of D by $\mathrm{Sym}(5)$, where $D = Z \circ E$ as above.*

Proof. The description of $\mathcal{N}_4$ in §6.2 of Chapter 7 includes an explicit construction of a subsystem of type $\mathcal{A}_4$. Thus $H := W(\mathcal{A}_4) \simeq \mathrm{Sym}(5)$ is a subgroup of $W(\mathcal{N}_4)$. We have $H \cap D = 1$ and on comparing orders it follows that $W(\mathcal{N}_4) = DH$.   □

Note that $W(\mathcal{N}_4)$ acts as $\mathrm{Sym}(5)$ on the 5 special frames not in $\mathcal{N}_4$ and that D is the kernel of this action.

Theorem 8.50. *The group $W(\mathcal{O}_4)$ is a non-split extension of D by $Sp_4(\mathbb{F}_2) \simeq O_5(\mathbb{F}_2) \simeq \mathrm{Sym}(6)$.*

Proof. It follows from the previous lemmas that $W(\mathcal{O}_4)/D \simeq Sp_4(\mathbb{F}_2)$. This group has a subgroup of index 6, namely $W(\mathcal{N}_4)/D \simeq \mathrm{Sym}(5)$ and therefore $Sp_4(\mathbb{F}_2) \simeq \mathrm{Sym}(6)$. The group $\mathrm{Sym}(6)$ does not have a faithful representation of degree 4 and therefore the extension does not split.

Regard $V := D/D'$ as a vector space of dimension 5 over $\mathbb{F}_2$ and identify D' with $\mathbb{F}_2$. Then the map $Q : V \to \mathbb{F}_2$ induced by $D \to D' : g \mapsto g^2$ defines a quadratic form on V (see Aschbacher [**6**, 23.10]). The group $W(\mathcal{O}_4)$ acts on D/D' by conjugation and D is in the kernel of this action. Thus we have an embedding of $Sp_4(\mathbb{F}_2) \simeq W(\mathcal{O}_4)/D$ in the orthogonal group $O_5(\mathbb{F}_2)$ of Q. On comparing orders we see that $Sp_4(\mathbb{F}_2) \simeq O_5(\mathbb{F}_2)$.   □

The group $W(\mathcal{F}_4)$ has been described, following Theorem 5.17, as a subgroup of index 2 in the central product of two copies of the binary octahedral group $\mathcal{O}$ extended by a reflection ρ interchanging the factors. The binary octahedral group is a non-split extension of the quaternion group $\mathcal{Q}$ by $\mathrm{Sym}(3)$; the central product $\mathcal{Q} \circ \mathcal{Q}$ is the group E defined above.

Theorem 8.51.

(i)   *The kernel of the restriction of $\hat{\mu}_2$ to $W(\mathcal{F}_4)$ is E and $W(\mathcal{F}_4)$ is the semidirect product of E by $W(\mathcal{A}_2) \times W(\mathcal{A}_2)$.*

(ii)   *The normaliser N of $W(\mathcal{F}_4)$ in $W(\mathcal{O}_4)$ is a group of order $2^9 \cdot 3^2 = 4608$. It is a non-split extension of $Z \circ W(\mathcal{F}_4)$ by C_2 and a non-split extension of D by $O_4^+(\mathbb{F}_2) \simeq \mathrm{Sym}(3) \wr C_2$.*

Proof. (i) We have $E \subseteq W(\mathcal{F}_4)$ but $W(\mathcal{F}_4)$ does not contain Z, hence $E = O_2(W(\mathcal{F}_4))$ is the kernel of $\hat{\mu}_2$. The lines spanned by $(1, -1, 0, 0)$, $(0, 1, -1, 0)$

and $(1, 0, -1, 0)$ is a 3-star σ in $\mathcal{F}_4$. The three lines of $\mathcal{F}_4$ orthogonal to σ also form a 3-star and so $W(\mathcal{F}_4)$ contains $H := W(\mathcal{A}_2) \times W(\mathcal{A}_2)$. Since $W(\mathcal{A}_2) \simeq \mathrm{Sym}(3)$ it follows that $E \cap H = 1$ and therefore $W(\mathcal{F}_4) = EH$.

(*ii*) As in Theorem 8.50, the map $x \mapsto x^2$ defines a quadratic form on E/E' of Witt index 2 and since $E = O_2(W(\mathcal{F}_4))$, conjugation by $g \in N$ preserves this quadratic form. Thus N/D is isomorphic to a subgroup of $O_4^+(\mathbb{F}_2)$. But $W(\mathcal{F}_4)/E$ is a subgroup of index 2 in $O_4^+(\mathbb{F}_2)$ and therefore $N/D \simeq O_4^+(\mathbb{F}_2)$. Thus $N \simeq \mathcal{C}_4 \circ (\mathcal{O} \circ \mathcal{O}) \langle \rho \rangle$ and the extensions are non-split. $\qquad \square$

10.6. The groups $W(\mathcal{K}_5)$ and $W(\mathcal{K}_6)$. Let $e_1, e_2, \ldots, e_6$ be the standard orthonormal basis of $\mathbb{C}^6$. The 126 lines of $\mathcal{K}_6$ are the images of the lines spanned by $e_1 - e_2$ and $\frac{1}{3}(\omega^2 - \omega)(e_1 + e_2 + \cdots + e_6)$ under the action of $W(\mathcal{D}_6^{(3)})$. The Cartan coefficients belong to the ring $\mathbb{Z}[\omega]$.

Theorem 8.52. $W(\mathcal{K}_6)$ *is a group of order* $2^9 \cdot 3^7 \cdot 5 \cdot 7 = 39\,191\,040$ *isomorphic to the non-split extension of the cyclic group $\mathcal{C}_3$ by $\widehat{\Omega}_6^-(\mathbb{F}_3)$.*

Proof. Let L be the $\mathbb{Z}[\omega]$-module spanned by the set Σ of roots. There is a homomorphism $\mu_3 : \mathbb{Z}[\omega] \to \mathbb{F}_3$ that sends ω to 1. The kernel $\mathfrak{p} = \mathrm{Ker}\, \mu_3$ is fixed by complex conjugation and therefore the map $Q(u) := (u, u)$ induces a non-degenerate quadratic form Q_3 on $L_3 := L/\mathfrak{p}L$. The Witt index of Q_3 is 2 and therefore we have a homomorphism $\hat{\mu}_3 : W(\mathcal{K}_6) \to \widehat{\Omega}_6^-(\mathbb{F}_3)$. It follows from Theorem B.5 that $\widehat{\Omega}_6^-(\mathbb{F}_3)$ contains 126 reflections and they generate $\widehat{\Omega}_6^-(\mathbb{F}_3)$. Thus $\hat{\mu}_3$ is onto.

We have $\mathbb{F}_4 := \mathbb{F}_2[\eta]$, where $\eta^2 + \eta + 1 = 0$. The kernel $\mathfrak{q}$ of the homomorphism $\mu_4 : \mathbb{Z}[\omega] \to \mathbb{F}_4$ that sends ω to η is fixed by complex conjugation and therefore the hermitian form on L induces a non-degenerate hermitian form on $L_4 := L/\mathfrak{q}L$. The reflections in $W(\mathcal{K}_6)$ act as elements of determinant 1 on L_4 and therefore we have a homomorphism $\hat{\mu}_4 : W(\mathcal{K}_6) \to SU_6(\mathbb{F}_4)$. From Theorem B.3, the order of $SU_6(\mathbb{F}_4)$ is $2^{15} \cdot 3^7 \cdot 5 \cdot 7 \cdot 11$ and the order of $\widehat{\Omega}_6^-(\mathbb{F}_3)$ is $2^9 \cdot 3^6 \cdot 5 \cdot 7$. Therefore $|\mathrm{Ker}\, \hat{\mu}_3| = 3$ and the order of $W(\mathcal{K}_6)$ is $2^9 \cdot 3^7 \cdot 5 \cdot 7 = 39\,191\,040$. $\qquad \square$

Theorem 8.53. $W(\mathcal{K}_5)$ *is a group of order* $2^7 \cdot 3^4 \cdot 5 = 51\,840$ *isomorphic to $\mathcal{C}_2 \times \Omega_5(\mathbb{F}_3)$.*

Proof. We continue the notation of the previous proof. Given a line $\ell \in \mathcal{K}_6$, the 45 lines of $\mathcal{K}_6$ orthogonal to ℓ form a line system of type $\mathcal{K}_5$ and so $W(\mathcal{K}_5)$ acts on the submodule of L_3 orthogonal to ℓ.

Let ψ be the restriction of $\hat{\mu}_3$ to $W(\mathcal{K}_5)$. Then $\psi : W(\mathcal{K}_5) \to \mathcal{C}_2 \times \Omega_5(\mathbb{F}_3)$ is one-to-one because the central element of order 3 in $W(\mathcal{K}_6)$ does not belong to $W(\mathcal{K}_5)$. The 45 reflections of $W(\mathcal{K}_5)$ are mapped to the 45 reflections of $\mathcal{C}_2 \times \Omega_5(\mathbb{F}_3)$ and therefore ψ is an isomorphism. $\qquad \square$

10.7. The groups $W(\mathcal{E}_6)$, $W(\mathcal{E}_7)$ and $W(\mathcal{E}_8)$. Let $e_1, e_2, \ldots, e_8$ be the standard orthonormal basis of $\mathbb{R}^8$. The line system $\mathcal{E}_8$ is contained in $\mathbb{R}^8$ and as its set Σ of long roots we choose the 56 vectors $e_i \pm e_j$ $(1 \le i < j \le 8)$, and the 64 vectors $\frac{1}{2}(e_1 \pm e_2 \pm \cdots \pm e_8)$ where the number of positive coefficients is even.

Theorem 8.54.

(i) $W(\mathcal{E}_8)$ is a group of order $2^{14} \cdot 3^5 \cdot 5^2 \cdot 7 = 696\,729\,600$ isomorphic to the non-split extension of the cyclic group C_2 by $O_8^+(\mathbb{F}_2)$.

(ii) $W(\mathcal{E}_7)$ is a group of order $2^{10} \cdot 3^4 \cdot 5 \cdot 7 = 2\,902\,040$ isomorphic to $C_2 \times O_7(\mathbb{F}_2)$ and to $C_2 \times Sp_6(\mathbb{F}_2)$.

(iii) $W(\mathcal{E}_6)$ is a group of order $2^7 \cdot 3^4 \cdot 5 = 51\,840$ isomorphic to $O_6^-(\mathbb{F}_2)$ and to $\widehat{\Omega}_5(\mathbb{F}_3) \simeq SO_5(\mathbb{F}_3)$.

Proof. (i) The restriction of the quadratic form $Q(v) := \frac{1}{2}(v, v)$ to the $\mathbb{Z}$-module $L := \mathbb{Z}\Sigma$ takes integer values and we define $Q_2(v)$ to be the induced form on $L_2 = L/2L$. For $i = 1, \ldots 7$, let $a_i := e_i - e_{i+1}$ and $b_i := e_i + e_{i+1}$. Then the vectors $a_1 + b_1$, $a_3 + b_3$, $a_5 + b_5$ and $a_7 + b_7$ belong to L and their images in L_2 span a 4-dimensional subspace on which Q_2 vanishes identically. Thus the Witt index of Q_2 is 4.

The elements of $G := W(\mathcal{E}_8)$ act on L and preserve Q. Their reductions modulo 2 act on L_2 and preserve Q_2. Thus there is a homomorphism $\hat{\mu}_2 : G \to O_8^+(\mathbb{F}_2)$. If a is a root and $\bar{a}$ the image of a in L_2, then $Q_2(\bar{a}) = 1$; that is, $\bar{a}$ is a non-singular vector. From Lemma B.4 the group $O_8^+(\mathbb{F}_2)$ contains 120 reflections and these reflections generate $O_8^+(\mathbb{F}_2)$; thus $\hat{\mu}_2$ is onto.

If $g \in \operatorname{Ker} \hat{\mu}_2$ and if a is a root, then $g(a) = \varepsilon_a a$, where $\varepsilon_a = \pm 1$. If a and b are roots such that $(a, b) \ne 0$, then $\varepsilon_a = \varepsilon_b$. Since $\mathcal{E}_8$ is indecomposable it follows that $\varepsilon_a = \varepsilon_b$ for all roots a and b; thus $\operatorname{Ker} \hat{\mu}_2 = \{\pm I\}$. From Theorem B.3 the order of $O_8^+(\mathbb{F}_2)$ is $2^{13} \cdot 3^5 \cdot 5^2 \cdot 7 = 3\,483\,648\,000$ and hence the order of $W(\mathcal{E}_8)$ is $2^{14} \cdot 3^5 \cdot 5^2 \cdot 7 = 696\,729\,600$.

In §6.2 we described an embedding of $W(\mathcal{O}_4)$ in $W(\mathcal{E}_8)$. The central element iI of $W(\mathcal{O}_4)$ corresponds to an element of $W(\mathcal{E}_8)$ whose square is $-I$. Thus the extension of $\langle -I \rangle$ by $O_8^+(\mathbb{F}_2)$ does not split.

(ii) The line system $\mathcal{E}_7$ consists of the 63 lines of $\mathcal{E}_8$ orthogonal to the root $z := \frac{1}{2}(e_1 + e_2 + \cdots + e_8)$. Since $Q(z) = 1$, the homomorphism $\hat{\mu}_2 : W(\mathcal{E}_8) \to O_8^+(\mathbb{F}_2)$ sends $W(\mathcal{E}_7)$ to a subgroup of the orthogonal group $O_7(\mathbb{F}_2)$ acting on the hyperplane H orthogonal to the image $\bar{z}$ of z in L_2. From Lemma B.4 the number of number of non-singular vectors in H is 63 and the corresponding reflections generate $O_7(\mathbb{F}_2)$. The kernel of the restriction of $\hat{\mu}_2$ to $W(\mathcal{E}_7)$ is $\langle -r_z \rangle$ and since $\det(-r_z) = -1$ we have $W(\mathcal{E}_7) \simeq C_2 \times O_7(\mathbb{F}_2)$. From Theorem B.3, the order of $O_7(\mathbb{F}_2)$ is $2^9 \cdot 3^4 \cdot 5 \cdot 7$ and therefore the order of $W(\mathcal{E}_7)$ is $2^{10} \cdot 3^4 \cdot 5 \cdot 7 = 2\,902\,040$.

The polar form of the restriction of q to H is an alternating form whose radical $H^\perp$ is one-dimensional. The 63 lines of $\mathcal{E}_7$ correspond to the 63 non-zero vectors of $H/H^\perp$. Therefore the 63 reflections of $W(\mathcal{E}_7)$ are mapped to the 63 transvections of $Sp_6(\mathbb{F}_2)$. Since $Sp_6(\mathbb{F}_2)$ is generated by its transvections it follows that $W(\mathcal{E}_7) \simeq C_2 \times Sp_6(\mathbb{F}_2)$.

(*iii*) For $\mathcal{E}_6$ we take the 36 lines of $\mathcal{E}_8$ orthogonal to $a_6 := e_6 - e_7$ and $a_7 := e_7 - e_8$. The image of $W(\mathcal{E}_6)$ under $\hat{\mu}_2$ acts on the 6-dimensional subspace L of L_2 orthogonal to the subspace spanned by the images of a_6 and a_7. Since $|\mathcal{E}_6| = 36$ it follows from (*i*) that L has 36 non-singular vectors and therefore, by Lemma B.4, the restriction of Q_2 to L has Witt index 2. Thus μ_2 induces a homomorphism $\hat{\mu}_2 : W(\mathcal{E}_6) \rightarrow O_6^-(\mathbb{F}_2)$ and since $O_6^-(\mathbb{F}_2)$ is generated by its reflections, this homomorphism is onto. If $g \in W(\mathcal{E}_6)$ and $\hat{\mu}_2(g)$ fixes every element of L, then $g = 1$ and therefore $W(\mathcal{E}_6) \simeq O_6^-(\mathbb{F}_2)$. From Theorem B.3, the order of $O_6^-(\mathbb{F}_2)$ is $2^7 \cdot 3^4 \cdot 5 = 51\,840$.

To complete the proof we consider the reduction $L_3 := L/3L$ of L modulo 3. Let Q_3 be the quadratic form on L_3 induced by the from $Q(v) := \frac{1}{2}(v, v)$ on L. The vector $v := \frac{1}{2}(e_1 - e_2 - e_3 - e_4 - e_5 + e_6 + e_7 + e_8)$ is a root of $\mathcal{E}_6$ and for each root a of $\mathcal{E}_6$ the vector $z := v + a_1 - a_3 + a_4$ is either orthogonal to a or $(a, z) = \pm 3$. Thus $W(\mathcal{E}_6)$ acts on the 5-dimensional subspace of L_3 orthogonal to the subspace spanned by the images of a_6, a_7 and z and there is an embedding $\hat{\mu}_3 : W(\mathcal{E}_6) \rightarrow \widehat{\Omega}_5(\mathbb{F}_3)$. From Lemma B.4 there are 36 reflections in $\widehat{\Omega}_5(\mathbb{F}_3)$ and they generate $\widehat{\Omega}_5(\mathbb{F}_3)$. Thus $\hat{\mu}_3$ is an isomorphism. $\qquad\square$

Exercises

1. Show that $W(\mathcal{J}_3^{(5)})$ is transitive on the 45 lines of $\mathcal{J}_3^{(5)}$ and for $\ell \in \mathcal{J}_3^{(5)}$ there are 4 lines orthogonal to ℓ, 8 lines at $45°$ to ℓ, 16 lines at $60°$ to ℓ, 8 lines at $32°$ to ℓ and 8 lines at $72°$ to ℓ. Furthermore, if $\ell, m \in \mathcal{J}_3^{(5)}$ are at $60°$, show that there is a unique subsystem $\mathfrak{K} \simeq \mathcal{D}_3^{(3)}$ containing ℓ and m, a unique subsystem $\mathfrak{M} \simeq \mathcal{B}_3^{(2)}$ containing ℓ and m, and $\mathfrak{K} \cap \mathfrak{M}$ is a 3-star.

2. Prove that if $\mathfrak{L}$ is an indecomposable 5-system, then $\mathfrak{L}$ is equivalent to $\mathcal{D}_n^{(5)}$, for some n, or to $\mathcal{H}_3$, $\mathcal{H}_4$ or $\mathcal{J}_3^{(5)}$.

3. Given linear transformations x and y of a finite dimensional complex vector space V show that if x and y have quadratic minimal polynomials, then there is a 2-dimensional subspace of V which is $\langle x, y \rangle$-invariant.

4. If $\lambda = -\frac{1}{2}(1 + i\sqrt{7})$, show that there is a homomorphism $\mu : \mathbb{Z}[\lambda] \rightarrow \mathbb{F}_9 = \mathbb{F}_3[\xi]$ such that $\mu(\lambda) = \xi$, where $\xi^2 + \xi + 2 = 0$. Show that $\mu(\bar{\lambda}) = \xi^3$ and deduce that there is an embedding of $W(\mathcal{J}_3^{(4)})$ in the unitary group $U_3(\mathbb{F}_9)$.

5. Using the construction of $\mathcal{H}_4$ given in §6.3, prove that the group $C_2 \wr \mathrm{Alt}(4)$ is a subgroup of $W(\mathcal{H}_4)$.

6. Show that there is a homomorphism μ from $W(\mathcal{M}_3)$ to $GL_3(\mathbb{F}_3)$ such that $|\operatorname{Ker}\mu| = 3$ and such the image of μ is a subgroup of $GL_3(\mathbb{F}_3)$ fixing a line.

7. Suppose that G is a finite unitary reflection group generated by reflections of order 2 such that $-1 \in Z(G)$. Show that $\psi : G \to G : g \mapsto (-1)^{\det(g)} g$ is an automorphism of G. If the rank of G is at least 3, show that ψ is an outer automorphism. For which groups of rank 2 is ψ an outer automorphism? State and prove a corresponding result for the groups $W(\mathcal{L}_3)$ and $W(\mathcal{L}_4)$.

8. Complex conjugation fixes the line systems $\mathcal{N}_4$ and $\mathcal{O}_4$ and thereby defines automorphisms of $W(\mathcal{N}_4)$ and $W(\mathcal{O}_4)$. Show that these automorphisms are outer.

The next three exercises construct embeddings of the primitive unitary reflection groups of rank four in $W(\mathcal{E}_8)$. Throughout these exercises, G is a finite reflection subgroup of $U_4(\mathbb{C})$, Σ is a $\mathbb{Z}(G)$-root system for G (see Theorem 8.30) and $\mathfrak{L}$ is the set of lines spanned by the elements of Σ. Thus G acts on a vector space V of dimension four over $\mathbb{Q}(G)$, where $\mathbb{Q}(G)$ is $\mathbb{Q}[i]$, $\mathbb{Q}[\omega]$ or $\mathbb{Q}[\tau]$. By restricting the scalars, V may be regarded as a vector space $V_{\mathbb{Q}}$ of dimension eight over $\mathbb{Q}$. In each case there is a 'natural' Euclidean inner product on $V_{\mathbb{Q}}$. For $a \in V_{\mathbb{Q}}$, let ρ_a be the reflection of order two, with root a, acting on $V_{\mathbb{Q}}$.

9. Suppose that $\mathfrak{L} = \mathcal{O}_4$. Then $\mathbb{Z}(G) = \mathbb{Z}[i]$ and every element of Σ is a long root. For $u, v \in V$, define $[u, v]$ to be the real part of (u, v) and show that $[-, -]$ is a Euclidean inner product on $V_{\mathbb{Q}}$. If $a, b \in \Sigma$ and a is not a multiple of b, show that $(a, b) \in \{0, \pm 1, \pm i, \pm 1 \pm i\}$. Deduce that the set $\mathfrak{L}_{\mathbb{Q}}$ of lines in $V_{\mathbb{Q}}$ spanned by the elements of Σ is the 3-system $\mathcal{E}_8$ and if $a \in \Sigma$, then $r_a = \rho_a \rho_{ia} = \rho_{ia} \rho_a$. Thus $W(\mathcal{O}_4) \leq W(\mathcal{E}_8)$.

10. Suppose that $\mathfrak{L} = \mathcal{L}_4$. Then $\mathbb{Z}(G) = \mathbb{Z}[\omega]$, and every element of Σ is a tall root. For $u, v \in V$, define $[u, v]$ to be the real part of (u, v) and show that $[-, -]$ is a Euclidean inner product on $V_{\mathbb{Q}}$. If $a, b \in \Sigma$ and a is not a multiple of b, show that $(a, b) \in \{0, \pm(1 - \omega), \pm(1 + 2\omega), \pm(2 + \omega)\}$. If ℓ is the line spanned by $a \in \Sigma$, show that ℓ is spanned by six elements of Σ and the corresponding lines of $V_{\mathbb{Q}}$ form a 3-star. Deduce that the set $\mathfrak{L}_{\mathbb{Q}}$ of lines in $V_{\mathbb{Q}}$ spanned by the elements of Σ is the 3-system $\mathcal{E}_8$ and that $t_a = \rho_a \rho_{\omega a}$ and $t_a^2 = \rho_{\omega a} \rho_a$ are the reflections of order 3 in $W(\mathcal{L}_4)$ with root a. Thus $W(\mathcal{L}_4) \leq W(\mathcal{E}_8)$. Similarly, show that $\mathcal{L}_{2,\mathbb{Q}} = \mathcal{D}_4^{(2)}$ and $\mathcal{L}_{3,\mathbb{Q}} = \mathcal{E}_6$.

11. Suppose that $\mathfrak{L} = \mathcal{H}_4$. Then $\mathbb{Z}(G) = \mathbb{Z}[\tau]$, and every element of Σ is a short root. Define $\kappa : \mathbb{Q}[\tau] \to \mathbb{Q}$ by $\kappa(a + b\tau) = a$ and show that $[u, v] := \kappa(u, v)$ is a Euclidean inner product on $V_{\mathbb{Q}}$ such that for $a \in \Sigma$, both a and τa are short roots in $V_{\mathbb{Q}}$. Deduce that the set $\mathfrak{L}_{\mathbb{Q}}$ of lines in $V_{\mathbb{Q}}$, spanned by the

elements a and τa, where $a \in \Sigma$, is the 3-system $\mathcal{E}_8$ and that $r_a = \rho_a \rho_{\tau a} = \rho_{\tau a} \rho_a$. Thus $W(\mathcal{H}_4) \leq W(\mathcal{E}_8)$. Similarly, show that $\mathcal{H}_{3,\mathbb{Q}} = \mathcal{D}_6^{(2)}$. (For more information see [**159**, §3.9], [**190**] and [**91**].)

12. Extend the previous exercise to $\mathbb{Z}[\omega, \tau]$-root systems and hence obtain an embedding of $W(\mathcal{J}_3^{(5)})$ in $W(\mathcal{K}_6)$. That is, let Σ be the root system of $\mathcal{J}_3^{(5)}$ and let $\kappa : \mathbb{Q}[\omega, \tau] \to \mathbb{Q}[\omega]$ be the $\mathbb{Q}[\omega]$-linear map such that $\kappa(1) = 1$ and $\kappa(\tau) = 0$. Show that $[u, v] := \kappa(u, v)$ is an hermitian inner product on $V_{\mathbb{Q}[\omega]}$ and that the star-closure of the lines in $V_{\mathbb{Q}[\omega]}$ spanned by a and τa, where $a \in \Sigma$, is the 3-system $\mathcal{K}_6$. Show that, under this embedding, $W(\mathcal{J}_3^{(5)})$ has two orbits of length 45 and one orbit of length 36 on $\mathcal{K}_6$.

13. Suppose that $\lambda = -\frac{1}{2}(1 + i\sqrt{7})$ and let $\kappa : \mathbb{Q}[\omega, \lambda] \to \mathbb{Q}[\omega]$ be the $\mathbb{Q}[\omega]$-linear map such that $\kappa(1) = 1$ and $\kappa(\lambda) = \omega$. Let V be the vector space of dimension three over $\mathbb{Q}[\omega, \lambda]$ spanned by the $\mathbb{Q}[\lambda]$-root system Σ of $\mathcal{J}_3^{(4)}$. Show that $[u, v] := \kappa(u, v)$ is an hermitian inner product on $V_{\mathbb{Q}[\omega]}$ and that the star-closure of the lines in $V_{\mathbb{Q}[\omega]}$ spanned by a and $(\lambda - \omega)a$, where $a \in \Sigma$, is the 3-system $\mathcal{K}_6$. Show that, under this embedding, $W(\mathcal{J}_3^{(4)})$ has two orbits of length 21 and one orbit of length 84 on $\mathcal{K}_6$.

CHAPTER 9

The orbit map, harmonic polynomials and semi-invariants

In this chapter we make a detailed study of the Jacobian of the orbit map $\mathbb{C}^n \to \mathbb{C}^n$, which is defined by the invariants of a unitary reflection group G. Properties of this map are then applied to prove the fundamental result that subspace stabilisers in reflection groups are again reflection groups. These are the 'parabolic subgroups' of a reflection group.

Definition 9.1. A *parabolic subgroup* of the reflection group G in V is the pointwise stabiliser G_A of some subset $A \subseteq V$. Since G acts linearly on V, $G_A = G_{\langle A \rangle}$ and so A may be taken to be a subspace of V.

The chapter also includes a study of the space of G-harmonic polynomials. Basic references for this material are Steinberg [**207**], Flatto [**98**], and Lehrer [**148**]. (See also Lehrer–Michel [**151**].)

1. The orbit map

Suppose that G is a finite subgroup of $U(V)$, where $n := \dim V$, and that G is generated by reflections. Let $S := \mathbb{C}[V]$ be the coordinate ring of V and let $J := S^G$ be the ring of invariants. Then J is generated as an algebra by *basic invariants*; that is, algebraically independent homogeneous polynomials $I_1, I_2, \ldots, I_n$, where the degree of I_i is d_i.

Definition 9.2. The *orbit map* $\omega_G : V \to \mathbb{C}^n$ is defined by

$$\omega_G(v) = (I_1(v), I_2(v), \ldots, I_n(v)).$$

It is clear that ω_G is constant on the orbits of G and thus ω_G defines a map $\overline{\omega}_G : V/G \to \mathbb{C}^n$, where V/G is the set of orbits of G on V.

Proposition 9.3. *The map* $\overline{\omega}_G : V/G \to \mathbb{C}^n$ *is a bijection.*

Proof. The map $\overline{\omega}_G$ is surjective by Theorem 3.15, and injective by Theorem 3.5. $\square$

2. Skew invariants and the Jacobian

Definition 9.4. The *Jacobian matrix* of the map ω_G is the matrix of partial derivatives:

$$\mathrm{Jac}(\omega_G) := \left(\frac{\partial I_i}{\partial X_j}\right)_{1 \le i, j \le n},$$

and the polynomial $\Pi \in S$ is defined to be the determinant of this matrix, namely the *Jacobian*

$$\Pi := \frac{\partial(I_1, I_2, \ldots, I_n)}{\partial(X_1, X_2, \ldots, X_n)} = \det \mathrm{Jac}(\omega_G).$$

Lemma 9.5. *The polynomials $f_1, f_2, \ldots, f_n \in \mathbb{C}[x_1, x_2, \ldots, x_n]$ are algebraically independent if and only if $\dfrac{\partial(f_1, f_2, \ldots, f_n)}{\partial(x_1, x_2, \ldots, x_n)} \ne 0.$*

Proof. (Flatto [98]) Suppose that $f_1, f_2, \ldots, f_n$ are algebraically dependent polynomials. By definition there is a non-constant polynomial $P(y_1, y_2, \ldots, y_n)$ such that $P(f_1, f_2, \ldots, f_n) = 0$. On differentiating with respect to x_j we find that

$$\sum_{i=1}^{n} \frac{\partial P}{\partial y_i} \frac{\partial f_i}{\partial x_j} = 0.$$

If $\dfrac{\partial(f_1, f_2, \ldots, f_n)}{\partial(x_1, x_2, \ldots, x_n)} \ne 0$ then $\dfrac{\partial P}{\partial y_i} = 0$ for all i and hence P is a constant, contradicting the choice of P.

Conversely, suppose that $f_1, f_2, \ldots, f_n$ are algebraically independent. The transcendence degree of $\mathbb{C}(x_1, x_2, \ldots, x_n)$ is n and therefore there are non-zero polynomials $Q_i(y_0, y_1, \ldots, y_n)$ such that $Q_i(x_i, f_1, \ldots, f_n) = 0$ for $1 \le i \le n$. On differentiating with respect to x_j we have

$$\delta_{ij} \frac{\partial Q_i}{\partial y_0} + \sum_{k=0}^{n} \frac{\partial Q_i}{\partial y_k} \frac{\partial f_k}{\partial x_j} = 0 \quad \text{for } 1 \le i \le n.$$

Interpreting this as a matrix equation we see that the matrix $\left(\dfrac{\partial f_k}{\partial x_j}\right)$ is invertible because each Q_i contains a term involving y_0 and hence $\dfrac{\partial Q_i}{\partial y_0} \ne 0$ for all i. On taking determinants it follows that $\dfrac{\partial(f_1, f_2, \ldots, f_n)}{\partial(x_1, x_2, \ldots, x_n)} \ne 0.$ $\qquad\square$

Corollary 9.6. *The degree of Π is the number of reflections in G.*

Proof. Since Π is non-zero it is a homogeneous polynomial of degree $\sum_{i=1}^{n}(d_i - 1)$. By Lemma 4.14 this is the number of reflections in G. $\qquad\square$

Let $H_1, H_2, \ldots, H_\ell$ be the hyperplanes Fix r, where r runs through the reflections of G and call these the *reflecting hyperplanes* of G. For each i, let e_i be the order of the cyclic group G_{H_i} fixing H_i pointwise and let $L_1, L_2, \ldots, L_\ell$ be linear forms such that $H_i = \text{Ker } L_i$.

Lemma 9.7. $\prod_{i=1}^{\ell} L_i^{e_i-1}$ *divides* $\dfrac{\partial(g_1, g_2, \ldots, g_n)}{\partial(X_1, X_2, \ldots, X_n)}$ *for all* $g_1, g_2, \ldots, g_n \in J$.

Proof. We may suppose that V is provided with a positive definite hermitian form and that $G \subseteq U(V)$. Let $b_1, b_2, \ldots, b_n$ be an orthonormal basis of V such that $H_1 = b_1^\perp$ and let $X_1, X_2, \ldots, X_n$ be the dual basis of V^*. Then L_1 is a multiple of X_1.

Let r_1 be a generator of the group G_{H_1}. Then $r_1 X_1 = \varepsilon_1 X_1$, where ε_1 is a primitive e_1^{th} root of unity. We also have $r_1 X_i = X_i$ for $i > 1$. Thus

$$r_1(X_1^{m_1} X_2^{m_2} \cdots X_n^{m_n}) = \varepsilon_1^{m_1} X_1^{m_1} X_2^{m_2} \cdots X_n^{m_n}$$

and this implies that any polynomial $f \in S$ such that $r_1 f = f$ is a polynomial in $X_1^{e_1}$ with coefficients in $\mathbb{C}[X_2, \ldots, X_n]$. In particular, since $g_j \in J$, the derivative $\dfrac{\partial g_j}{\partial X_1}$ is divisible by $X_1^{e_1-1}$. Thus $\Delta := \dfrac{\partial(g_1, g_2, \ldots, g_n)}{\partial(X_1, X_2, \ldots, X_n)}$ is divisible by $L_1^{e_1-1}$. Similarly Δ is divisible by $L_i^{e_i-1}$ for all i and therefore $\prod_{i=1}^{\ell} L_i^{e_i-1}$ divides Δ. □

Theorem 9.8. *For some non-zero constant c we have* $\Pi = c \prod_{i=1}^{\ell} L_i^{e_i-1}$.

Proof. The degree of $\prod_{i=1}^{\ell} L_i^{e_i-1}$ is $\sum_{i=1}^{\ell}(e_i-1)$, which is the number of reflections in G. On the other hand, by Corollary 9.6, the number of reflections in G is the degree of Π. From Lemma 9.7 $\prod_{i=1}^{\ell} L_i^{e_i-1}$ divides Π, thus Π and $\prod_{i=1}^{\ell} L_i^{e_i-1}$ are equal up to a non-zero multiplicative constant. □

Definition 9.9. An element $P \in S$ is a *skew polynomial* if for all $g \in G$ we have $gP = (\det g)P$.

Lemma 9.10. *The polynomial Π satisfies*

(i) Π *is a skew polynomial,*

(ii) Π *divides every skew polynomial.*

Proof. Suppose that r is a reflection in G. If $H := \text{Fix } r$ and if L is the corresponding linear form, then $rL = (\det r)^{-1}L$ and $(\det r)^{e_H} = 1$, where e_H is the order of G_H. Thus $r(L^{e_H-1}) = (\det r)^{-e_H+1}L^{e_H-1} = (\det r)L^{e_H-1}$.

(i) Let K be a reflecting hyperplane other than H. Then there exists a smallest integer m such that $r^m L_K = cL_K$ for some $c \in \mathbb{C}$. It follows from Lemma 3.17 that $r^m L_K - L_K = (1-c)L_K$ is a multiple of L_H; hence $c = 1$. This means that r fixes the product

$$L_K (rL_K) \cdots (r^{m-1}L_K).$$

The product $\prod_{K \neq H} L_K^{e_K - 1}$ may be written as a product of terms of the above type, which therefore are r-invariant. Thus

$$r \prod_K L_K^{e_K - 1} = r(L^{e_H - 1}) \, r \prod_{K \neq H} L_K^{e_K - 1} = (\det r) \prod_K L_K^{e_K - 1}$$

and, since G is generated by reflections, this proves (i).

(ii) Suppose that P is a skew polynomial and let r_1 be a generator of the group G_{H_1}. As in Lemma 9.8 we may choose a basis $X_1, X_2, \ldots, X_n$ for V^* such that $r_1 X_1 = \varepsilon_1 X_1$ and $r_1 X_i = X_i$ for $i > 1$, where $\varepsilon_1 := (\det r_1)^{-1}$ is a primitive e_1^{th} root of unity. Thus

$$r_1(X_1^{m_1} X_2^{m_2} \cdots X_n^{m_n}) = \varepsilon_1^{m_1} X_1^{m_1} X_2^{m_2} \cdots X_n^{m_n}$$

and if $X_1^{m_1} X_2^{m_2} \cdots X_n^{m_n}$ occurs in P, then $\varepsilon_1^{m_1} = \varepsilon_1^{-1}$. Hence $\varepsilon_1^{m_1 + 1} = 1$ and so e_1 divides $m_1 + 1$. Thus $m_1 \geq e_1 - 1$ and therefore $X_1^{e_H - 1}$ divides P. This argument applies to all hyperplanes H_i and so Π divides P. $\qquad \square$

Corollary 9.11. *If P is a skew polynomial, then $P = \Pi I$, where $I \in J$.*

Proof. From the lemma we may write $P = \Pi I$ for some polynomial I. Then for $g \in G$ we have $(\det g)P = gP = (g\Pi)(gI) = (\det g)\Pi(gI)$, whence $gI = I$. $\qquad \square$

The next statement may be thought of as an explicit version of Corollary 4.23 (iii).

Corollary 9.12. $(S/F)_N = \mathbb{C}\Pi$.

Proof. We know from Corollary 4.23 (iii) that the top degree component $(S/F)_N$ of S/F yields the determinant representation of G; hence it is spanned by (the residue class of) a homogeneous polynomial P which is skew and of degree N. That is, P is a multiple of Π. $\qquad \square$

3. The rank of the Jacobian

In this section we present a result of Steinberg [**207**], which gives the local rank of the Jacobian matrix of the orbit map ω_G.

Theorem 9.13. *Let G be a finite reflection group in V and suppose that $p \in V$. Then the following numbers are equal:*

(i) $\dim X(p)$, *where $X(p)$ is the intersection of all the reflecting hyperplanes containing p;*

(ii) $\min\{\, \dim(\text{Fix}\, x) \mid x \in G_p \,\}$;

(iii) $\text{rank}_p \, \text{Jac}(\omega_G)$.

Proof. Let $H_1, H_2, \ldots, H_\ell$ be the reflecting hyperplanes of G, labelled so that H_1, $H_2, \ldots, H_m$ are those that contain p. For $1 \leq i \leq m$ let r_i be a generator of the group G_{H_i}, let a_i be a unit vector orthogonal to H_i, and define L_i by $L_i(v) = (v, a_i)$. Then $k = n - \dim X(p)$ is the dimension of the subspace spanned by the a_i.

Choose an orthonormal basis $b_1, \ldots, b_n$ of V such that $b_1, \ldots, b_k$ is a basis for the subspace spanned by the a_i. Let $X_1, \ldots, X_n$ be the dual basis of V^*.

Let G_1 be the subgroup of G generated by the reflections r_i, where $1 \leq i \leq m$. Then G_1 is a finite reflection group and $G_1 \subseteq G_p$. Moreover, G_1 acts trivially on $\langle a_1, a_2, \ldots, a_m \rangle^{\perp}$ and hence it fixes X_i for $i > k$. That is, $X_{k+1}, \ldots, X_n \in S^{G_1}$.

Suppose that $g_1, g_2, \ldots, g_k \in J$ and for $1 < k \leq m$ define $g_i := X_i$. The g_i are G_1-invariants and since $\dfrac{\partial g_i}{\partial X_j} = \delta_{ij}$ for $i > k$ it follows from Lemma 9.7 that

$$\frac{\partial(g_1, g_2, \ldots, g_n)}{\partial(X_1, X_2, \ldots, X_n)} = \frac{\partial(g_1, g_2, \ldots, g_k)}{\partial(X_1, X_2, \ldots, X_k)}$$

is divisible by $\prod_{i=1}^{m} L_i^{e_i - 1}$.

Now consider the Laplace expansion of $\Pi = \dfrac{\partial(I_1, I_2, \ldots, I_n)}{\partial(X_1, X_2, \ldots, X_n)}$ by the first k columns:

$$\Pi = \sum \pm \frac{\partial(I_{i_1}, \ldots, I_{i_k})}{\partial(X_1, \ldots, X_k)} \frac{\partial(I_{i_{k+1}}, \ldots, I_{i_n})}{\partial(X_{k+1}, \ldots, X_n)},$$

where the sum is over all permutations $i_1, \ldots, i_n$ of $1, \ldots, n$ for which $i_1 < \cdots < i_k$ and $i_{k+1} < \cdots < i_n$. As shown above $\dfrac{\partial(I_{i_1}, \ldots, I_{i_k})}{\partial(X_1, \ldots, X_k)}$ is divisible by $\prod_{i=1}^{m} L_i^{e_i - 1}$ and, by Theorem 9.8, $\Pi = c \prod_{i=1}^{\ell} L_i^{e_i - 1}$ for some $c \neq 0$. Therefore there are polynomials $M_{i_1, \ldots, i_k}$ such that

$$\prod_{i=k+1}^{\ell} L_i^{e_i - 1} = \sum \pm M_{i_1, \ldots, i_k} \frac{\partial(I_{i_{k+1}}, \ldots, I_{i_n})}{\partial(X_{k+1}, \ldots, X_n)}.$$

Since the left side of this equality does not vanish at p there are indices i_{k+1}, $i_{k+2}, \ldots, i_n$ such that $\dfrac{\partial(I_{i_{k+1}}, \ldots, I_{i_n})}{\partial(X_{k+1}, \ldots, X_n)} \neq 0$. That is, $\operatorname{rank}_p \operatorname{Jac}(\omega_G) \geq n - k = \dim X(p)$.

Now suppose that the H_i are ordered so that $H_1, H_2, \ldots, H_k$ contain p and the corresponding linear forms $L_1, L_2, \ldots, L_k$ are linearly independent.

If $v \in \operatorname{Fix}(r_1 r_2 \cdots r_k)$, then $r_2 \cdots r_k v = r_1^{-1} v$ and therefore $v + \sum_{i=2}^{k} \alpha_i a_i = v + \alpha_1 a_1$, for some $\alpha_i \in \mathbb{C}$. It follows that $\alpha_1 = 0$ and hence $r_1 v = v$. On repeating the argument we have $r_i v = v$ for all i and so $v \in H_1 \cap \cdots \cap H_k$. Thus $\operatorname{Fix}(r_1 r_2 \cdots r_k) \subseteq H_1 \cap \cdots \cap H_k$ and therefore $\min\{\dim(\operatorname{Fix} x) \mid x \in G_p\} \leq \dim X(p)$.

Finally, if $x \in G_p$, let $v_1, v_2, \ldots, v_n$ be an orthonormal basis of V consisting of eigenvectors of x. Choose the notation so that $\operatorname{Fix} x$ is spanned by $v_{h+1}, \ldots, v_n$, where $h := n - \dim(\operatorname{Fix} x)$ and let $X_1, X_2, \ldots, X_n$ be the basis of V^* dual to v_1, $v_2, \ldots, v_n$. Then $X_i(p) = 0$ for $1 \leq i \leq h$.

If $f \in J$, then $xf = f$ and so no monomial occurring in f has total degree 1 in $X_1, X_2, \ldots, X_h$. Therefore $\dfrac{\partial f}{\partial X_i}(p) = 0$ for $1 \leq i \leq h$. That is, the first h columns of $\mathrm{Jac}(\omega_G)$ vanish at p and therefore $\mathrm{rank}_p \, \mathrm{Jac}(\omega_G) \leq n - h = \dim(\mathrm{Fix}\, x)$.

We have shown that

$$\mathrm{rank}_p \, \mathrm{Jac}(\omega_G) \geq \dim X(p) \geq \min\{\, \dim(\mathrm{Fix}\, x) \mid x \in G_p \,\} \geq \mathrm{rank}_p \, \mathrm{Jac}(\omega_G),$$

hence equality holds throughout and the proof is complete. $\qquad\square$

Corollary 9.14. *If X is an intersection of reflecting hyperplanes, then there exists $x \in G$ such that $\mathrm{Fix}\, x = X$.*

Proof. We may write $X = H_1 \cap \cdots \cap H_k$, where the H_i are linearly independent reflecting hyperplanes. For all i let r_i be a reflection such that $H_i = \mathrm{Fix}\, r_i$. From the proof of the theorem, with $x = r_1 \cdots r_k$, we have $\mathrm{Fix}(x) \subseteq H_1 \cap \cdots \cap H_k = X$ and hence $\dim(\mathrm{Fix}\, x) \leq \dim X$. But there exists $p \in X$ such that the only reflecting hyperplanes that contain p are those containing X. Now $X = X(p)$ and from the theorem we have $\dim(\mathrm{Fix}\, x) \geq \dim X$. Therefore $\dim(\mathrm{Fix}\, x) = \dim X$ and thus $\mathrm{Fix}\, x = X$. $\qquad\square$

Corollary 9.15. *If p does not lie on any reflecting hyperplane, then $G_p = 1$.*

Proof. For $x \in G_p$ we have $\dim(\mathrm{Fix}\, x) \geq \dim \bigcap_{p \in H} H = n$ and so $x = 1$. $\qquad\square$

4. Semi-invariants

In this section we shall characterise all homomorphisms $\lambda : G \to \mathbb{C}$, where G is a reflection group. These homomorphisms are known as the *linear characters* of G. Our treatment is in the spirit of Springer [**202**, §4.3].

Recall from Definition 6.3 that the polynomial $f \in S$ is a semi-invariant of G if for each element $g \in G$, we have $gf = \lambda(g)f$ for some scalar $\lambda(g) \in \mathbb{C}$. In this situation, the map $g \mapsto \lambda(g)$ is a group homomorphism, so that semi-invariants give rise to linear characters.

We begin by describing some semi-invariants of G, which will provide some linear characters, and then show that all semi-invariants, and linear characters, arise in the way we describe. Let $\mathcal{A}(G)$ be the set of reflecting hyperplanes (i.e. the fixed-point subspaces of the reflections) of G; we shall write $\mathcal{A}$ when there is no risk of confusion. It is clear from Proposition 1.19 (*ii*) that G permutes the hyperplanes; write $\mathcal{O}_1, \mathcal{O}_2, \ldots, \mathcal{O}_\ell$ for the orbits of G on $\mathcal{A}$. If $H \in \mathcal{A}$, write $L_H \in V^*$ for a linear form whose kernel is H.

Lemma 9.16. *Let $H \in \mathcal{A}$ and let r be a reflection in G with $\mathrm{Fix}\, r = H$. Then we have the following.*

(i)   If L is any non-zero element of V^* such that $rL = aL$ for some non-zero
        scalar a, then either L is a multiple of L_H and $a = \zeta_r^{-1}$, or $a = 1$, where ζ_r
        is the non-trivial eigenvalue of r on V.

(ii)   Suppose that $L_1, L_2, \ldots, L_c$ are non-zero elements of V^* such that for $i = 1, 2 \ldots, c - 1$, $rL_i = a_i L_{i+1}$, and $rL_c = a_c L_1$ for non-zero scalars $a_1, \ldots, a_c$. If no L_i is a multiple of L_H, then $r(L_1 L_2 \cdots L_c) = L_1 L_2 \cdots L_c$.

Proof. For (i), observe that it follows from Lemma 3.17, that $rL = L + bL_H$ for
some scalar b, whence $(a - 1)L = bL_H$. The statement (i) is immediate.

Now note that $r^c L_1 = a_1 \cdots a_c L_1$ and $r(L_1 L_2 \cdots L_c) = a_1 \cdots a_c L_1 L_2 \cdots L_c$. If
$r^c \neq 1$ we may apply (i) with r replaced by r^c. Since L_1 is not a multiple of L_H, it
follows that $a_1 \cdots a_c = 1$, which proves (ii).   □

Let $\mathcal{O}$ be a G-orbit in $\mathcal{A}$, and define

$$f_{\mathcal{O}} = \prod_{H \in \mathcal{O}} L_H.$$

Then $f_{\mathcal{O}}$ is determined up to a non-zero scalar multiple.

Corollary 9.17. *The polynomial $f_{\mathcal{O}}$ is a semi-invariant of G with corresponding
linear character $\lambda_{\mathcal{O}}$ defined on a reflection r with $\mathrm{Fix}\, r = H \in \mathcal{A}$ by*

$$\lambda_{\mathcal{O}}(r) = \begin{cases} 1 & \text{if } H \notin \mathcal{O} \\ \zeta_r^{-1} & \text{otherwise,} \end{cases}$$

where ζ_r is the non-trivial eigenvalue of r on V.

Note that all values of any linear character of G are determined by its values on
the reflections, since these generate G.

Proof. The G-orbit $\mathcal{O}$ is a union of orbits of the cyclic group generated by r. One
of these orbits is $\{H\}$ and we have $rL_H = \zeta_r^{-1} L_H$. If $\{H_1, \ldots, H_c\}$ is any other
orbit, then it follows from Lemma 9.16 (ii) that r fixes $L_{H_1} \cdots L_{H_c}$. The statement
is now clear.   □

The following corollary is the analogue of Lemma (4.11) of Cohen [54] for root
systems that satisfy Definition 1.43.

Corollary 9.18. *Suppose that F is a finite abelian extension of $\mathbb{Q}$ and let A be the
ring of integers of F. Let (Σ, f) be an A-root system and let $G := W(\Sigma, f)$ be its
Weyl group.*

(i)   *If $r \in G$ is a reflection, then r has a root $a \in \Sigma$ such that $r = r_{a,f(a)}^m$ for some
        integer m.*

(ii)   *Suppose that G is generated by the reflections $\{r_{a,f(a)} \mid a \in \Delta\}$, where $\Delta \subseteq \Sigma$. Then $\Sigma = \{\alpha g(a) \mid \alpha \in \mu(A), g \in G, a \in \Delta\}$ and every reflection in G is
        conjugate to a reflection of the form $r_{a,f(a)}^m$ for some $a \in \Delta$ and some integer
        m.*

Proof. (i) Let $\mathcal{O}$ be the G-orbit of $\operatorname{Fix} r$. By Corollary 9.17 there is a linear character χ of G such that for all reflections $s \in G$ we have

$$\chi(s) = \begin{cases} \det(s) & \text{if } \operatorname{Fix} s \in \mathcal{O}, \\ 1 & \text{if } \operatorname{Fix} s \notin \mathcal{O}. \end{cases}$$

Since $\chi \neq 1$ and since G is generated by the reflections $r_{b,f(b)}\,(b \in \Sigma)$, there exists $b \in \Sigma$ such that $\chi(r_{b,f(b)}) \neq 1$. It follows that $b^{\perp} = \operatorname{Fix} r_{b,f(b)} \in \mathcal{O}$. Thus $g(b^{\perp}) = \operatorname{Fix} r$ for some $g \in G$ and so $a = g(b) \in \Sigma$ is a root of r.

Furthermore, $r = r_{a_1,f(a_1)} \cdots r_{a_k,f(a_k)}$ for some $a_1, \dots, a_k \in \Sigma$. If $a_i^{\perp} \in \mathcal{O}$, then $\chi(r_{a_i,f(a_i)}) = f(a_i)$, otherwise $\chi(r_{a_i,f(a_i)}) = 1$. By definition, f is constant on $\mu(A)G$-orbits and therefore

$$\det(r) = \prod_{a_i^{\perp} \in \mathcal{O}} f(a_i) = f(a)^m \quad \text{for some } m.$$

It follows that $r = r_{a,f(a)}^m$.

(ii) Let $\Sigma_1 = \mu(A)G\Delta$ and let f_1 be the restriction of f to Σ_1. Then (Σ_1, f_1) is an A-root system such that $\Sigma_1 \subseteq \Sigma$ and $W(\Sigma_1, f_1) = G$. For all $a \in \Sigma$, it follows from (i) that $\alpha a \in \Sigma_1$ for some $\alpha \in \mu(A)$; thus $a \in \Sigma_1$ and hence $\Sigma_1 = \Sigma$. Therefore, if $b \in \Sigma$, then $b = \alpha g(a)$ for some $\alpha \in \mu(A), a \in \Delta$ and $g \in G$. If r is a reflection with root b, then from (i) we have $r = g r_{a,f(a)}^m g^{-1}$ for some m. $\qquad \square$

If the reflecting hyperplanes H and H' are in the same G-orbit, then the (cyclic) groups G_H and $G_{H'}$ are conjugate and therefore $|G_H| = |G_{H'}|$. Hence for any G-orbit $\mathcal{O} \subseteq \mathcal{A}$, we write $e(\mathcal{O})$ for the common value of $|G_H|$ for $H \in \mathcal{O}$. The main result of this section is as follows.

Theorem 9.19.

(i) *Using the above notation, any homogeneous semi-invariant $f \in S$ of G may be written uniquely in the form*

$$f = \left(\prod_{\mathcal{O}} f_{\mathcal{O}}^{m(\mathcal{O})} \right) h,$$

where $0 \leq m(\mathcal{O}) \leq e(\mathcal{O}) - 1$ and $h \in J\,(= S^G)$ is a homogeneous invariant polynomial.

(ii) *The linear character corresponding to the semi-invariant in (i) is*

$$\prod_{\mathcal{O}} \lambda_{\mathcal{O}}^{m(\mathcal{O})}.$$

These characters are all distinct and constitute a complete set of linear characters of G.

Proof. To see (i), we use induction on the degree of f. The result is certainly true if $\deg(f) = 0$ or if f is invariant. If f is not invariant, then there is a reflection r such that $rf = \lambda f$ with $\lambda \neq 1$. If $H := \operatorname{Fix}(r)$, then by Lemma 3.17 $rf - f$ is divisible by L_H. By semi-invariance, it follows that for each hyperplane H' in the G-orbit $\mathcal{O}$ of H, $L_{H'}$ divides f, whence $f = f_1 f_{\mathcal{O}}$, where f_1 is a homogeneous semi-invariant of smaller degree than f. This completes the proof of (i).

For (ii), note that there is a G-stable homogeneous complement $\mathcal{C}$ (i.e. one which is spanned by homogeneous polynomials) of F in S. We shall explicitly construct such a complement below (the harmonic polynomials), but here we require only the existence of $\mathcal{C}$, which can be seen as follows. Since F is a homogeneous subspace (ideal) of finite codimension in S, there is certainly some homogeneous complement C, which may not be G-stable. Let $P_C : S \to F$ be the projection with kernel C, and write $\overline{P}_C = |G|^{-1} \sum_{g \in G} g P_C g^{-1}$. It is evident that $\overline{P}_C$ acts as the identity on F, and hence is also a projection from S to J. Since $\overline{P}_C$ commutes with G, $\mathcal{C} := \operatorname{Ker} \overline{P}_C$ is a G-stable complement of F in S.

Now as G-module, $\mathcal{C}$ is isomorphic to S/F, which by Proposition 3.32 is isomorphic to the regular representation. It follows that for each linear character λ of G, up to multiplication by a non-zero scalar, there is a unique polynomial $f_\lambda \in \mathcal{C}$ which is semi-invariant, with corresponding linear character λ. Since G preserves degree, f_λ is homogeneous. The statement (ii) now follows from (i). □

Corollary 9.20. *If G' is the derived group of G, we have $|G/G'| = \prod_{\mathcal{O}} e(\mathcal{O})$, the product being over the G-orbits in $\mathcal{A}$.*

Proof. This is immediate from Theorem 9.19, since $|G/G'|$ is the number of linear characters of G. □

The last few results may be summarised as follows.

Corollary 9.21. *Let $m : \mathcal{A}/G \to \mathbb{N}$ be an integer valued function on the G-orbits of reflecting hyperplanes of G which satisfies, for each orbit $\mathcal{O}$, $0 \leq m(\mathcal{O}) \leq e(\mathcal{O}) - 1$, where $e(\mathcal{O})$ is the order of the cyclic group G_H for any $H \in \mathcal{O}$. Define $f_m = \prod_{\mathcal{O} \in \mathcal{A}/G} f_{\mathcal{O}}^{m(\mathcal{O})}$. Then f_m is semi-invariant, with corresponding linear character λ_m, and the λ_m form a complete set of distinct linear characters of G. We shall sometimes refer to the polynomial f_m as f_λ, where $\lambda = \lambda_m$ is the corresponding linear character.*

There are two noteworthy special cases of this result. One is when for each orbit $\mathcal{O}$, $m(\mathcal{O}) = e(\mathcal{O}) - 1$, in which case $f_m = c\Pi$ for some $c \neq 0$ and the corresponding linear character is det; the other is when $m(\mathcal{O}) = 1$, in which case the corresponding linear character is $\det^{-1}$.

5. Differential operators

In the proof of Theorem 9.19 it was shown that the ideal F of S which is generated by the invariant polynomials of positive degree has a homogeneous G-stable complement. In this section we develop tools for the explicit construction of such complements. It will turn out that these tools will also be useful for further analysis of the orbit map, and in the study of eigenspace theory for elements of G.

Definition 9.22. If A is a commutative algebra over $\mathbb{C}$, a *derivation* of A is a $\mathbb{C}$-linear transformation $D : A \to A$ such that

$$D(ab) = D(a)b + aD(b)$$

for all $a, b \in A$. The set $\mathrm{Der}(A)$ of all derivations of A is a vector space over $\mathbb{C}$.

A *differential operator* on V is a polynomial in derivations of $S(V^*)$.

Example 9.23. If $S = \mathbb{C}[X_1, \ldots, X_n]$ is the algebra of polynomials, then partial differentiation with respect to X_i defines a derivation of S and so the map D defined by $Df = \dfrac{\partial^2 f}{\partial X_i^2} \dfrac{\partial f}{\partial X_j}$ is a differential operator.

For $v \in V$ we define $D_v \in \mathrm{Der}(S)$ to be the *directional derivative* in direction v, namely

$$D_v f(x) = \lim_{t \to 0} \frac{f(x + tv) - f(x)}{t}.$$

If $b_1, b_2, \ldots, b_n$ is a basis of V and if $X_1, X_2, \ldots, X_n$ is its dual basis, then $D_{b_i} = \dfrac{\partial}{\partial X_i}$. The map $V \to \mathrm{Der}(S) : v \mapsto D_v$ is linear and extends to an algebra homomorphism $S(V) \to \mathrm{Diff}(V) : f \mapsto D_f$, where $\mathrm{Diff}(V)$ is the algebra of differential operators on V.

For example, if $f = b_1^3 b_2 + 3b_3^2$, then $D_f = \dfrac{\partial^3}{\partial X_1^3} \dfrac{\partial}{\partial X_2} + 3\dfrac{\partial^2}{\partial X_3^2}$.

Definition 9.24. We shall denote by $\mathcal{D}_S$ the algebra of differential operators on V which is generated by the derivations D_v (for $v \in V$). Equivalently, $\mathcal{D}_S$ is the image of the map $f \mapsto D_f$ above.

The map $f \mapsto X_j \dfrac{\partial f}{\partial X_i}$ is a derivation which is not in $\mathcal{D}_S$.

Definition 9.25. If U is a complex vector space, a *real structure* on U is an involution $\sigma : U \to U$ which is semi-linear, i.e. $\sigma(\alpha u) = \bar{\alpha}\sigma(u)$ for $\alpha \in \mathbb{C}$ and $u \in U$, and is such that $U = U^\sigma \otimes_\mathbb{R} \mathbb{C}$. Thus U^σ is a 'real form' of U.

For the remainder of this section, we shall fix a basis $b_1, b_2, \ldots, b_n$ of V, which we assume is orthonormal for a positive definite hermitian form which is respected by G. Given this data, we may define a real structure on V by

$$v = \sum_i \alpha_i b_i \mapsto \bar{v} = \sum_i \bar{\alpha}_i b_i,$$

where $\overline{\alpha}_i$ is the usual complex conjugate of α_i. In a similar fashion we define a real structure on V^* with respect to the dual basis $X_1, X_2, \ldots, X_n$. That is,

$$\sum_i \alpha_i X_i \mapsto \sum_i \overline{\alpha}_i X_i.$$

Since V and V^* respectively freely generate $S(V)$ and $S(V^*)$ as commutative $\mathbb{C}$-algebras, there are unique real structures on $S(V)$ and $S(V^*)$ which satisfy the additional requirement of being ring homomorphisms. Moreover, we have a (non-canonical) isomorphism $\varphi : V \to V^*$ such that $\varphi(b_i) = X_i$, which extends uniquely to an algebra isomorphism $\varphi : S(V) \to S(V^*)$.

Lemma 9.26.

(i)   For $g \in G$, the matrix $M(g)$ of g with respect to the basis $b_1, b_2, \ldots, b_n$ of V satisfies $M(g)^t \, \overline{M(g)} = I$.

(ii)   For $g \in G$, the matrix $M^*(g)$ of g acting on V^* with respect to the basis $X_1, X_2, \ldots, X_n$ satisfies $M^*(g) = \overline{M(g)}$.

(iii) For $f \in S(V)$ and $g \in G$ we have $\overline{\varphi(gf)} = g\varphi(\overline{f})$.

(iv) $f \in S(V)^G$ if and only if $\overline{\varphi(f)} \in S(V^*)^G$.

Proof. Statement (i) says simply that g is unitary, and (ii) is an easy consequence. Notice that if (iii) is true for a set B of elements $f \in S(V)$, it is easily verified that (iii) holds for all f in the algebra generated by B. Since (iii) holds for $f \in V$, it is true for all $f \in S(V)$. The statement (iv) follows directly from (iii).   □

Definition 9.27. Define the bilinear pairing

$$[-, -] : S(V) \times S(V^*) \to \mathbb{C}$$

by $[f, P] = (D_f P)(0)$.

Notice that this pairing is a canonical extension of the pairing $V \times V^* \to \mathbb{C}$.

Lemma 9.28.

(i)   Suppose that $H \in S$ is homogeneous of degree m and that $v \in V$. Then $[v^m, H] = m! H(v)$.

(ii)   For all non-zero elements $f \in S(V)$, $[f, \overline{\varphi(f)}]$ is a positive real number.

Proof. We may write $v = \sum \alpha_i b_i$, and then $D_v = \sum \alpha_i \dfrac{\partial}{\partial X_i} = \sum X_i(v) \dfrac{\partial}{\partial X_i}$. As H is homogeneous of degree m we have Euler's formula:

$$\left(\sum_i X_i \frac{\partial}{\partial X_i} \right) H = mH,$$

whence $D_v H(v) = mH(v)$, $D_v^2 H(v) = D_v(D_v(H))(v) = m(m-1)H(v)$, and so on. Since $D_v^m H$ is a constant we have $[v^m, H] = m! H(v)$, proving (i).

For (ii), if $g \in S(V)_k$ and $P \in S(V^*)_\ell$, then $[g, P] = 0$ unless $k = \ell$, since $D_g P$ is homogenous of degree $\ell - k$; it therefore suffices to prove (ii) for f homogeneous.

Now suppose that f_1 is a monomial of degree m in the b_i and that P_1 is a monomial of degree m in the X_i. Then $[f_1, P_1] = 0$ unless $\varphi(f_1)$ and P_1 are proportional. That is, if $f_1 = cb_1^{m_1} \cdots b_n^{m_n}$ and $P_1 = c' X_1^{k_1} \cdots X_n^{k_n}$, then $D_{f_1} P_1 = 0$ unless $k_i = m_i$ for all i. Hence it is sufficient to take $f = cb_1^{m_1} \cdots b_n^{m_n}$ with $\sum m_i = m$ and $c \in \mathbb{C}$. In this case we have

$$[f, \overline{\varphi(f)}] = c \frac{\partial^{m_1}}{\partial X_1^{m_1}} \cdots \frac{\partial^{m_n}}{\partial X_n^{m_n}} (\bar{c} X_1^{m_1} \cdots X_n^{m_n}) = |c|^2 \prod_i m_i! > 0.$$

$\square$

Corollary 9.29. *The pairing* $[-, -] : S(V) \times S(V^*) \to \mathbb{C}$, *defined by* $[f, P] := (D_f P)(0)$, *is non-degenerate in both variables.*

Proof. Suppose that $[f, P] = 0$ for all $P \in S$. Then by the previous lemma, $[f, \overline{\varphi(f)}] = 0$ implies that $f = 0$, hence $[-, -]$ is non-degenerate in the first variable.

Suppose P is such that $[f, P] = 0$ for all $f \in S(V)$. Since this is true if and only if it is true for all homogeneous f, we may take P to be homogeneous of degree m. If $P \neq 0$, then for some $v \in V$ we have $P(v) \neq 0$ and hence $[v^m, P] = m! P(v) \neq 0$, a contradiction. Thus $P = 0$ and $[-, -]$ is non-degenerate in the second variable. $\square$

Note that although we have used the (non-canonical) real structures and isomorphism $\varphi : S(V) \to S(V^*)$ in proving the above lemma, the pairing $[-, -]$ and Corollary 9.29, are independent of any choices.

Corollary 9.30. *The map* $D : f \mapsto D_f$ *is an algebra isomorphism:* $S(V) \to \mathcal{D}_S$.

Proof. The map is an algebra homomorphism by construction, and is onto by definition. If $f \in S(V)$ and $D_f = 0$, then the previous corollary shows that $f = 0$. $\square$

This corollary shows that $S(V)$ may be identified with an algebra of differential operators on V. The action of G on $S(V)$ may by transferred to $\mathcal{D}_S$ using this isomorphism: $gD_f := D_{gf}$ for $g \in G$ and $f \in S(V)$.

We shall now use the perfect pairing $[-, -] : S(V) \times S(V^*) \to \mathbb{C}$ to study these polynomial algebras. One of the fundamental properties we shall require is the G-invariant nature of the pairing, which is proved in Corollary 9.34 below.

Definition 9.31. If U is a subspace of $S(V)$, define

$$U^\perp := \{ P \in S(V^*) \mid [u, P] = 0 \text{ for all } u \in U \},$$

and similarly for any subspace U^* of $S(V^*)$.

The next statement is easily verified.

Lemma 9.32.

(i)   *For any degree k, the pairing* $S(V)_k \times S(V^*)_k \to \mathbb{C}$ *given by* $(f, P) \mapsto [f, P]$ *is non-degenerate.*

(ii)   *We have $U^{\perp\perp} = U$ for all homogeneous subspaces U of $S(V)$.*

A simple consequence of this lemma is that $S(V)_k^{\perp} = \bigoplus_{\ell \neq k} S(V^*)_\ell$.

Lemma 9.33. *Let g be any invertible linear transformation of V. For $f \in S(V)$ and $P \in S(V^*)$ we have*

$$g(D_f P) = D_{gf}(gP).$$

Proof. If $f = f_1 f_2$ and we have the result for f_1 and f_2, then

$$g(D_f P) = g(D_{f_1}(D_{f_2} P)) = D_{gf_1}(g(D_{f_2} P))$$
$$= D_{gf_1}(D_{gf_2}(gP)) = D_{(gf_1)(gf_2)}(gP)$$
$$= D_{gf}(gP).$$

As the statement is linear in f it is enough to check it on any set $\{f\}$ of algebra generators of $S(V)$. In particular, it suffices to consider the case $f = v \in V$. In this case, for all $x \in V$,

$$g(D_v P)(x) = D_v P(g^{-1}x)$$
$$= \lim_{t \to 0} \frac{P(g^{-1}x + tv) - P(g^{-1}x))}{t}$$
$$= \lim_{t \to 0} \frac{(gP)(x + tgv) - (gP)(x))}{t}$$
$$= D_{gv}(gP)(x)$$

and so $g(D_v P) = D_{gv}(gP)$.   □

Corollary 9.34. *The pairing $[-, -]$ is invariant under the group $GL(V)$ of invertible linear transformations of V. That is, for any $g \in GL(V)$, we have*

$$[gf, gP] = [f, P] \quad \text{for all } g \in G.$$

In particular, if $G \subset GL(V)$ is a reflection group, $[-, -]$ is invariant under G.

Proof. We have $[gf, gP] = D_{gf}(gP)(0) = (D_f P)(g^{-1}0) = [f, P]$.   □

6. The space of G-harmonic polynomials

In this section we shall use the differential operators to produce a natural G-stable homogeneous complement of the ideal $F := SJ^+$ in S, where J^+ is the set of polynomials of J of positive degree. Of course as G-module, this complement will be isomorphic to the coinvariant algebra.

Definition 9.35.

(i)   We say that $D_f \in \mathcal{D}_S$ is an *invariant differential operator* if $gD_f = D_f$ for all $g \in G$; i.e. if $f \in S(V)^G$.

(ii)   The set of elements of $S(V)^G$ of positive degree is a maximal ideal of $S(V)^G$, denoted by $J(V)^+$.

(iii)   A polynomial function $P \in S$ is G-harmonic if $D_f P = 0$ for all $f \in J(V)^+$. Let $\mathcal{H}$ be the space of G-harmonic polynomials. If there is no risk of confusion we shall refer simply to 'harmonic' polynomials.

The usual harmonic functions are obtained by replacing G by $O(V)$, in which case $S(V)^G$ is generated by $\sum_i b_i^2$ and so a polynomial function P is harmonic if

$$\sum \frac{\partial^2 P}{\partial X_i^2} = 0.$$

Lemma 9.36. *If $F(V)$ is the ideal of $S(V)$ generated by $J(V)^+$, then $\mathcal{H} = F(V)^\perp$.*

Proof. We have

$$
\begin{aligned}
P \in F(V)^\perp &\iff [\xi f, P] = 0 \quad \text{for all } \xi \in J(V)^+ \text{ and } f \in S(V) \\
&\iff [f, D_\xi P] = 0 \quad \text{for all } \xi \in J(V)^+ \text{ and } f \in S(V) \\
&\iff D_\xi P = 0 \quad \text{for all } \xi \in J(V)^+ \\
&\iff P \in \mathcal{H}. \qquad \qquad \qquad \square
\end{aligned}
$$

Corollary 9.37. *The set $\mathcal{H}$ of harmonic polynomials is a G-invariant complement of F in S.*

Proof. The ideal $F(V)$ is G-invariant and hence so is $\mathcal{H} = F(V)^\perp$.

Suppose that $P \in \mathcal{H} \cap F$. We may write $P = \sum_j P_j I_j$, where $I_j \in J^+$. Then $\overline{\varphi^{-1}(P)} = \sum_j p_j i_j \in F(V)$, where $p_j = \overline{\varphi^{-1}(P_j)}$ and $i_j = \overline{\varphi^{-1}(I_j)} \in J(V)^+$. By Lemma 9.28 (ii), we have $P = 0$. Thus $\mathcal{H}$ is G-invariant and we have $\mathcal{H} \cap F = 0$.

Now $(F + \mathcal{H})/F \simeq \mathcal{H}/F \cap \mathcal{H} = \mathcal{H}$ and $\dim \mathcal{H} = \operatorname{codim}_{S(V)} F(V) = |G| = \dim S/F$. Therefore $\dim(F + \mathcal{H})/F = \dim S/F$ and hence $F + \mathcal{H} = S$. $\qquad \square$

The next result shows that the G-harmonic polynomials are precisely the derivatives of the skew polynomial Π of minimal degree. This generalises Corollary 9.12.

Theorem 9.38. *We have*

(i) $f \in F(V)$ *if and only if* $D_f \Pi = 0$, *and*

(ii) $P \in \mathcal{H}$ *if and only if* $P = D_f \Pi$ *for some* $f \in S(V)$.

Proof. (i) If $f \in J(V)^+$, then $D_f \Pi$ is skew because

$$g(D_f \Pi) = D_{gf}(g\Pi) = (\det g) D_f \Pi.$$

But every skew polynomial is divisible by Π and so $D_f \Pi = 0$; that is, Π is harmonic. Since $F(V)$ is generated by $J(V)^+$ it follows that $D_f \Pi = 0$ for all $f \in F(V)$.

Conversely, suppose that $D_f\Pi = 0$ for some $f \in S(V)$. In order to show that $f \in F(V)$ it is sufficient to take f to be homogeneous. If the degree of f is greater than N, the degree of Π, then by Corollary 4.23 (i), $f \in F(V)$. Thus we assume the result for all non-constant multiples of f and prove the result by downward induction on the degree of f. Let H be any reflecting hyperplane of G and let v be a unit vector perpendicular to H. Then $vf \in F(V)$ and so $vf = u_1 f_1 + u_2 f_2 + \cdots + u_k f_k$ for some $f_j \in J(V)^+$. If r is a reflection such that $H = \text{Fix}\, r$, then

(9.38) $(\det r)vr(f) = r(u_1)f_1 + r(u_2)f_2 + \cdots + r(u_k)f_k.$

From Lemma 3.17 (applied to $S(V)$), for $u \in S(V)$ there exists $u' \in S(V)$ such that $ru - u = vu'$. On subtracting vf from (9.38) and cancelling v we find that

$$(\det r)r(f) - f = u_1' f_1 + u_2' f_2 + \cdots + u_k' f_k \in F(V).$$

Therefore $f \equiv (\det r)r(f) \pmod{F(V)}$ for all reflections $r \in G$, and consequently $f \equiv (\det g)g(f) \pmod{F(V)}$ for all $g \in G$.

If $\hat{f} = |G|^{-1} \sum_g (\det g)gf$, then $f \equiv \hat{f} \pmod{F(V)}$. Furthermore, for $g \in G$, $g\hat{f} = (\det g)^{-1}\hat{f}$ and therefore, by the dual of Lemma 9.10, $\hat{f} = \pi\xi$ where ξ is G-invariant and π is the canonical skew element of $S(V)$. So $f \equiv \pi\xi \pmod{F(V)}$.

Again applying the dual of Lemma 9.10, $\bar{\pi}$ is a non-zero scalar multiple of π, and so $D_\pi\Pi$ is a non-zero constant by Lemma 9.28 (ii). Thus $0 = D_f\Pi = D_\xi(D_\pi\Pi)$ implies that $\xi \in J(V)^+$; that is, ξ has zero constant term. It follows that $f \equiv \pi\xi \equiv 0 \pmod{F(V)}$; i.e. $f \in F(V)$.

(ii) Let $\mathcal{D}\Pi$ be the space of derivatives of Π. We want to show that $\mathcal{D}\Pi = \mathcal{H}$. These are both graded subspaces of S and we shall show that $\mathcal{D}\Pi^\perp = \mathcal{H}^\perp$, which by Lemma 9.32 will suffice. Indeed, for $h \in S(V)$ we have $h \in \mathcal{D}\Pi^\perp$ if and only if $[h, D_f\Pi] = 0$ for all $f \in S(V)$ and this holds if and only if $[f, D_h\Pi] = 0$ in which case $D_h\Pi = 0$. This in turn is equivalent to $h \in F(V) = \mathcal{H}^\perp$. $\square$

Corollary 9.39. $S = J \otimes \mathcal{H}$. That is, if $H_1, \ldots, H_N$ is a basis of $\mathcal{H}$, then every polynomial function on V may be written uniquely in the form $P = \sum_j F_j H_j$ where F_j is G-invariant.

Proof. This is immediate from Corollaries 3.31 and 9.37. $\square$

We observe next that we have already met some harmonic polynomials.

Proposition 9.40. *For any linear character λ_m of G let f_m be the corresponding polynomial defined in Corollary 9.21. Then f_m is harmonic. Moreover any non-zero harmonic polynomial P such that $\mathbb{C}P$ is stable under G is a multiple of some f_m.*

Proof. Let i be an invariant of positive degree in $S(V)$. Then $D_i f_m$ is a semi-invariant with corresponding character λ_m. It follows that $D_i f_m = f_m P$, where

$P \in S(V^*)$ is invariant. By degree, we conclude that $D_i f_m = 0$, i.e. that f_m is annihilated by all invariant differential operators in $\mathcal{D}_S$. Hence $f_m \in \mathcal{H}$.

Since $\mathcal{H}$ realises the regular representation of G, each one-dimensional representation occurs just once. Since the λ_m exhaust all the linear characters of G, the last statement follows.   □

Corollary 9.41. *Suppose G is a cyclic group acting linearly on V via reflections in a hyperplane H. If $|G| = e$ and $L_H \in S$ is a linear form with kernel H, then the space $\mathcal{H}$ of G-harmonic polynomials is given by $\mathcal{H} = \mathbb{C} \oplus \mathbb{C}L_H \oplus \mathbb{C}L_H^2 \oplus \ldots \oplus \mathbb{C}L_H^{e-1}$.*

This is immediate from the previous proposition. Note that the powers of L_H are just the polynomials f_m in this case.

We conclude this section with a criterion for a set of harmonic polynomials to be a basis.

Definition 9.42. An element $v \in V$ is *regular* if it does not lie on any reflecting hyperplane.

Proposition 9.43.

(i)   Given $v \in V$, let S_v be the space of polynomial functions on the G-orbit $Gv = \{gv \mid g \in G\}$. Then $\dim S_v = |G : G_v|$ and the restriction map $\mathrm{Res}_v : S \to S_v$ is surjective.

(ii)   The restriction map $\mathrm{Res}_v : \mathcal{H} \to S_v$ is also surjective. If v is regular, then $\mathrm{Res}_v : \mathcal{H} \to S_v$ is an isomorphism.

(iii)   The set $\{A_1, A_2, \ldots, A_{|G|}\} \subset \mathcal{H}$ is a basis for $\mathcal{H}$ if and only if for some regular element $v \in V$ we have $\det \big(A_i(gv)\big)_{i,g} \neq 0$.

Proof. (i) By Lagrange interpolation (Lemma 3.3), given any finite set $\{v_1, \ldots, v_k\}$ in $\mathbb{C}^n$ and scalars $a_1, \ldots, a_k$, there exists a polynomial $f(X_1, \ldots, X_n)$ such that $f(v_i) = a_i$. Thus Res_v is surjective and it is clear that $\dim S_v = |G : G_v|$.

(ii) The map $\mathrm{Res}_v : S = J \otimes \mathcal{H} \to S_v$ is surjective and all elements of J restrict to constants. It follows that $\mathrm{Res}_v(S) = \mathrm{Res}_v(\mathcal{H})$. If v is regular, then $G_v = 1$ so that $\dim S_v = |G| = \dim \mathcal{H}$.

(iii) The set $\{A_1, A_2, \ldots, A_{|G|}\}$ forms a basis for $\mathcal{H}$ if and only if it maps to a basis for S_v (when v is regular). This is equivalent to the nonsingularity of the matrix.   □

7. Steinberg's fixed point theorem

In this section we prove a fundamental result of Steinberg [**208**, Theorem 1.5]. The proof below is due to Lehrer [**148**].

Theorem 9.44. *Let G be a finite reflection group in the space $V \cong \mathbb{C}^n$. If $v \in V$, the stabiliser $G_v := \{g \in G \mid gv = v\}$ is the reflection group generated by the reflections which fix v.*

Proof. Let $K = G_v$ and write K_0 for the subgroup of K which is generated by the reflections in hyperplanes which contain v. If r_H is a reflection in the hyperplane H and $g \in G$, then $g r_H g^{-1}$ is a reflection in the hyperplane gH. Hence if $r_H \in K_0$ and $g \in K$, $g r_H g^{-1} \in K_0$, and so K_0 is a normal subgroup of K. We need to show that $K = K_0$.

Since K_0 is (by definition) a reflection group, the algebra of invariants S^{K_0} is generated by a set $\{P_1, P_2, \ldots, P_n\}$ of algebraically independent homogeneous polynomials, by Theorem 3.20. To prove the theorem, we claim that it suffices to prove that

(9.45)   *each polynomial P_i above is invariant under K.*

For if we assume (9.45), it would follow, since the P_i generate S^{K_0}, that $S^{K_0} \subseteq S^K$, from which we deduce equality since $K \supseteq K_0$ implies the reverse inequality. But then it is immediate from Theorem 4.19 that $|K| = |K_0|$, whence the result. Thus we are reduced to proving (9.45).

Take an element $g \in K$. We shall show that g acts trivially on the P_i. First observe that since g normalises K_0, g transforms S^{K_0} into itself, for if $f \in S^{K_0}$ and $h \in K_0$, then $h(gf) = g(g^{-1}hg)f = gf$. We shall show that,

(9.46)   given $g \in K$, the generators P_i may be chosen so that
             $gP_i = \varepsilon_i P_i$ for each i (for some $\varepsilon_i \in \mathbb{C}$).

To prove (9.46), assume without loss of generality that, if we write $d_i = \deg(P_i)$, then $d_1 \leq d_2 \leq \cdots \leq d_n$. Now $S^{K_0} = \bigoplus_{i \geq 0} S_i^{K_0}$ is the direct sum of finite dimensional spaces which are invariant under g, and on which g acts semisimply. Let $A(r)$ be the subalgebra of S^{K_0} generated by the P_i of degree $< r$. It follows from Lemma 4.18 that, if B_r is the linear span of the P_i of degree r, then $S_r^{K_0} = A(r)_r \oplus B_r$. Since $A(r)_r$ is clearly g-invariant, by semisimplicity, we may choose a basis of $S_r^{K_0}$, consisting of g-eigenfunctions, which extends some basis of $A(r)$. The basic invariants of degree r may then be replaced by the eigenfunctions of this basis which are not in $A(r)$. Since the case $r = 1$ is trivial (as $A(1) = \mathbb{C}$), this proves that the P_i may recursively be replaced by eigenfunctions of g.

Now let $\omega_G : V \to \mathbb{C}^n$ be the orbit map (see §1), given by

$$\omega_G(v) = (I_1(v), \ldots, I_n(v)),$$

where $I_1, \ldots, I_n$ is a set of basic invariants of G, and let ω_{K_0} be the corresponding map for (the reflection group) K_0. Since each polynomial $I_i \in S$ is K_0-invariant, there are unique polynomials $Q_1, \ldots, Q_n \in \mathbb{C}[y_1, \ldots, y_n]$ such that for each i, $I_i(X_1, \ldots, X_n) = Q_i(P_1, \ldots, P_n)$. The polynomials Q_i may be thought of as the coordinate functions of the (polynomial) map $\omega_{G, K_0} : \mathbb{C}^n \to \mathbb{C}^n$ in the diagram below.

$$\mathbb{C}^n \, (= V) \xrightarrow{\ \omega_{K_0}\ } \mathbb{C}^n \, (= V/K_0)$$

(9.47)

with ω_G going diagonally and ω_{G,K_0} going down to $\mathbb{C}^n \, (= V/G)$.

Now since $\omega_G = \omega_{G,K_0} \circ \omega_{K_0}$, it follows from the chain rule that

(9.48)
$$\frac{\partial(I_1,\ldots,I_n)}{\partial(X_1,\ldots,X_n)} = \frac{\partial(Q_1,\ldots,Q_n)}{\partial(P_1,\ldots,P_n)}\frac{\partial(P_1,\ldots,P_n)}{\partial(X_1,\ldots,X_n)} \in S = S(V^*).$$

Applying Theorem 9.8 to G and K_0, we see that

(9.49)
$$\frac{\partial(Q_1,\ldots,Q_n)}{\partial(P_1,\ldots,P_n)}(X_1,\ldots,X_n) = \prod_{H\in\mathcal{A}, v\notin H} L_H^{e_H-1},$$

from which it is evident that $\dfrac{\partial(Q_1,\ldots,Q_n)}{\partial(P_1,\ldots,P_n)}(v) \neq 0$. Hence, by expanding the determinant, there is a permutation π of $\{1,2,\ldots,n\}$ such that

(9.50)
$$\prod_{i=1}^{n} \frac{\partial Q_{\pi i}}{\partial P_i}(v) \neq 0,$$

whence $\dfrac{\partial Q_{\pi i}}{\partial P_i}(v) \neq 0$ for each i.

Now regard $P_1, P_2, \ldots, P_n$ as the coordinate functions on the affine space $\mathbb{C}^n$, identified as V/K_0. The element $g \in K$ is an automorphism (which is diagonal with respect to the basis of coordinates defined by the P_i) of this affine space since g normalises K_0, and g fixes $v \in V/K_0$. Moreover the argument of Lemma 9.33 shows that $g\dfrac{\partial Q_i}{\partial P_j} = \dfrac{\partial(gQ_i)}{\partial(gP_j)} = \varepsilon_j^{-1}\dfrac{\partial Q_i}{\partial P_j}$. It follows that

$$g\frac{\partial Q_i}{\partial P_j}(v) = \varepsilon_j^{-1}\frac{\partial Q_i}{\partial P_j}(v) = \frac{\partial Q_i}{\partial P_j}(g^{-1}v) = \frac{\partial Q_i}{\partial P_j}(v).$$

Hence for $i = 1,\ldots,n$, we have from (9.50)

$$g\frac{\partial Q_{\pi i}}{\partial P_i}(v) = \varepsilon_i^{-1}\frac{\partial Q_{\pi i}}{\partial P_i}(v) = \frac{\partial Q_{\pi i}}{\partial P_i}(v) \neq 0.$$

It follows that $\varepsilon_i = 1$ for all i; that is, each polynomial P_i is invariant under g. Since g was an arbitrary element of K, this proves (9.45), and hence the theorem. □

Corollary 9.51. *Let U be any subset of V. The parabolic subgroup G_U of G which fixes U pointwise is the reflection group generated by the reflections in G whose fixed point hyperplanes contain U.*

Proof. Since G acts linearly, an element $g \in G$ fixes U pointwise if and only if g fixes the linear span of U pointwise. Thus we may assume that U is a subspace of V. Again by linearity, g fixes U pointwise if and only if g fixes a basis of U pointwise. Hence we may suppose that U is a finite set. The result now follows from Theorem 9.44 by an easy induction of the number of elements in U.   □

Remark 9.52. The proof of the assertion (9.46) in the proof of Theorem 9.44 above proves the following general fact, which will be useful later.

Let G be a reflection group in $V \cong \mathbb{C}^n$ and suppose $g \in GL(V)$ is a semisimple transformation which normalises G. Then G has a set $\{F_1, \ldots, F_n\}$ of homogeneous basic invariants such that for each i, $gF_i = \varepsilon_i F_i$ for some $\varepsilon_i \in \mathbb{C}$.

Exercises

In all exercises below, G is a unitary reflection group in $V = \mathbb{C}^n$. The exercises explore some properties of operators on S which are analogous to the 'Demazure operators' of [**76**].

1. Let $r \in G$ be a reflection with $\mathrm{Fix}(r) = H$. Let $L_H \in V^*$ be a linear form with kernel H, and define $\delta_{r^i} : S \to S$ by

$$\delta_{r^i} P = \frac{r^i P - P}{L_H}$$

for $P \in S$. This makes sense by Lemma 1.15. Show that $\delta_{r^i} S_d \subseteq S_{d-1}$, and that $r\delta_{r^i} = \zeta \delta_{r^i} r$, where $\zeta = \det_V(r)$.

2. Maintaining the above notation, show that δ_{r^i} is a twisted derivation; i.e. $\delta_{r^i}(PQ) = \delta_{r^i}(P)r^i(Q) + P\delta_{r^i}(Q)$ for $P, Q \in S$.

3. Deduce from Exercise 1 above that if r generates the subgroup of G which fixes H pointwise, and e is the order of this group, then for any j,

$$\sum_{i=0}^{e-1}(\zeta^{-1}r)^i \delta_{r^j} = 0.$$

4. Deduce from the previous exercise that for any integer j,

$$\sum_{i=1}^{e-1} \zeta^{-i}\delta_{r^i}\delta_{r^j} = 0.$$

In particular, if $e = 2$, show that $\delta_r^2 = 0$.

5. (*i*)   Use a similar argument to the one above to show that for $k = 1, 2, \ldots,$

$$\Big(\sum_{i=0}^{e-1}(\zeta^{-k}r)^i\Big)\delta_r^k = 0.$$

(*ii*)   Deduce that for $k = 0, 1, 2, \ldots, e - 1$, $(\sum_{i=0}^{e-1}(\zeta^{-k}r)^i)\delta_r^e = 0$, and hence that $\delta_r^e = 0$.

6. Show that δ_{r^i} is an S^G-module endomorphism of S. Let $\mathcal{A}$ be the algebra of S^G-endomorphisms of S generated by all δ_r together with S, acting by multiplication on itself. Show that $\mathcal{A} \supset \mathbb{C}G$.

The following exercises are intended for the reader with a knowledge of finite Coxeter groups. We take G to be a finite Coxeter group, with a given set $R = \{s_1, \ldots, s_n\}$ of simple reflections and associated length function $\ell(g)$, which correspond to a chosen positive system in the root system of G. For $g \in G$, $N(g)$ denotes the set of hyperplanes orthogonal to the positive roots which are made negative by g. For $i = 1, \ldots, n$, H_i is the reflecting hyperplane of s_i and $L_i = L_{H_i}$. In general, there is a bijection between reflections and hyperplanes, and L_H denotes a linear form corresponding to the hyperplane H; these may be thought of as the coroots.

7. If $w = s_{i_1} \cdots s_{i_p}$ is a reduced expression for $w \in G$, show that if we write $\delta_i = \delta_{s_i}$, then
$$\left(\prod_{H \in N(w)} L_H \right) \delta_{i_1} \cdots \delta_{i_p} = \det(w)w + \sum_{\substack{w' \in G \\ \ell(w') < \ell(w)}} a_{w'} w',$$
where $a_{w'} \in S$ (see Exercise 6 above).

8. Let $w_0 = s_{i_1} \cdots s_{i_N}$ be a reduced expression for the longest element w_0 of G. Prove that
$$\delta_{i_1} \cdots \delta_{i_N} = \Pi^{-1} \circ \sum_{w \in G} \det(w)w,$$
where $\Pi = \prod_H L_H$ is the (unique) skew polynomial of minimal degree.

9. Prove that if $w = s_{i_1} \cdots s_{i_p}$ is a reduced expression for $w \in G$, then $\delta_w := \delta_{i_1} \cdots \delta_{i_p}$ is independent of the reduced expression. Note that this is equivalent to saying that the δ_i satisfy the braid relations.

10. Show that if $w = s_{i_1} \cdots s_{i_p}$ is not reduced, then $\delta_{i_1} \cdots \delta_{i_p} = 0$. Deduce that if $w, w' \in G$, then
$$\delta_w \delta_{w'} = \begin{cases} \delta_{ww'} & \text{if } \ell(ww') = \ell(w) + \ell(w') \\ 0 & \text{otherwise.} \end{cases}$$

11. Show that the residues modulo F of $\{\delta_w \Pi \mid w \in G\}$ form a basis of S/F.

12. Deduce from the last exercise that the Poincaré polynomial of S/F is equal to $\sum_{w \in G} t^{\ell(w)}$.

This statement is sometimes referred to as the Bott–Solomon–Tits formula.

CHAPTER 10

Covariants and related polynomial identities

In this chapter we shall study the invariants of a reflection group G on modules of the form $S \otimes_{\mathbb{C}} M$, where M is generally a finite dimensional G-module. These will be referred to as *covariants*, and for certain modules M, the structure of the space of covariants will lead to interesting polynomial identities, which will be used to relate the structure of G to the geometry of V.

Throughout this chapter G will be a finite reflection group acting on the vector space V of dimension n over $\mathbb{C}$. Recall that for each reflecting hyperplane H we have a linear function L_H with kernel H and then $\Pi = \prod_H L_H^{e_H - 1}$ is a skew polynomial of minimal degree, where e_H is the order of the cyclic group generated by the reflections with hyperplane H. If N is the number of reflections in G and $\mathcal{H}$ is the space of harmonic polynomials, then $\mathcal{H}_N = \mathbb{C}\Pi$, where we regard $\mathcal{H}$ as a graded module: $\mathcal{H} = \bigoplus_{k=0}^{N} \mathcal{H}_k$, and $\mathcal{H}$ is the space of all derivatives of Π.

1. The space of covariants

We return now to the theme of §3.6.

Definition 10.1.

(i)   If M_1 and M_2 are two finite dimensional G-modules, we define the *intertwining number* of M_1 and M_2 to be

$$(M_1, M_2) := \dim \mathrm{Hom}_{\mathbb{C}G}(M_1, M_2).$$

By Schur's Lemma, if M_1 and M_2 are irreducible, then (M_1, M_2) is 1 or 0 according to whether or not M_1 and M_2 are isomorphic.

(ii)   Given a finite dimensional G-module M, the polynomial

$$f_M(t) = \sum_i (\mathcal{H}_i, M) t^i$$

is called the *fake degree* of M. We shall write this as

$$f_M(t) = t^{q_1(M)} + t^{q_2(M)} + \cdots + t^{q_r(M)}$$

where $q_i(M) \leq q_{i+1}(M)$ for all i and we call $(q_1(M), q_2(M), \ldots, q_r(M))$ the *sequence of M-exponents* of G. These are the same polynomials as those defined for characters in §3.6.

This terminology comes from the fact that when G is the Weyl group of a reductive group $\mathbf{G}$ over $\mathbb{F}_q$, the polynomials $f_M(q)$ are the degrees of certain 'almost characters' (see, for example, [198]) of the finite reductive group $\mathbf{G}(\mathbb{F}_q)$. The almost-characters are actual unipotent characters in some cases, but are generally rational linear combinations of irreducible characters. We shall now prove some results about the structure of modules of covariants, and deduce consequences for the fake degrees.

In the case $M = V$ we shall see that $f_V(t) = \sum_i t^{m_i}$, where the $m_i = d_i - 1$ are the exponents cf. Definition 10.24 of G. And in fact, more generally, in Corollary 10.25 we shall see that $f_{\Lambda^j V}(t) = e_j(t^{m_1}, \ldots, t^{m_n})$, where e_j is the jth elementary symmetric function (see Chapter 2, page 36).

Lemma 10.2. *Given finite dimensional G-modules M_1 and M_2 we have*

(i) $M_1 \otimes M_2^* \simeq \operatorname{Hom}(M_2, M_1)$,

(ii) $(M_1 \otimes M_2^*)^G \simeq \operatorname{Hom}_{\mathbb{C}G}(M_2, M_1)$,

(iii) $\dim_{\mathbb{C}}(M_1 \otimes M_2^*)^G = (M_1, M_2)$.

The isomorphisms (i) and (ii) are canonical.

Proof. See Lang [142, Chapter XVI, §5].   □

If M is a finite-dimensional G-module we define a grading on the G-module $S \otimes M$ by defining the degree j component of $S \otimes M$ to be $S_j \otimes M$. The G-action is given (as usual) by $g(Q \otimes y) = gQ \otimes gy$, and hence preserves degree. The space $S \otimes M$ is also a graded S-module with respect to the action $P(Q \otimes y) = PQ \otimes y$.

We know that as a G-space, $\mathcal{H}$ affords the regular representation and hence that each irreducible G-module M occurs in $\mathcal{H}$ with multiplicity equal to its degree. That is, $(\mathcal{H}, M) = \dim M$. It follows that $(\mathcal{H} \otimes M^*)^G$ has a basis $u_1, u_2, \ldots, u_r$, where $r = \dim M$ and where, taking into account Definition 10.1 above, the u_i are homogeneous of degree $q_i(M)$. Since $\mathcal{H} \otimes (M_1 \oplus M_2) \cong \mathcal{H} \otimes M_1 \oplus \mathcal{H} \otimes M_2$, the same is true for any G-module.

Proposition 10.3.

(i)   *For any finite dimensional G-module M, both $S \otimes M$ and $(S \otimes M^*)^G$ are free as modules over $J = S^G$.*

(ii)   *If $u_1, u_2, \ldots, u_r$ is a $(\mathbb{C})$-basis of $(\mathcal{H} \otimes M^*)^G$, then $u_1, u_2, \ldots, u_r$ is also a J-basis of $(S \otimes M^*)^G$.*

(iii)   *If the u_i in (ii) are homogeneous, then their degrees are the M-exponents of G, i.e. the u_j may be ordered so that $\deg u_j = q_j(M)$.*

Proof. We have from Corollary 9.39 that $S \simeq J \otimes \mathcal{H}$, i.e. that S is free as J-module, with basis any linear basis of $\mathcal{H}$. Hence $S \otimes M \simeq (J \otimes \mathcal{H}) \otimes M \simeq J \otimes (\mathcal{H} \otimes M)$, which proves that $S \otimes M$ is free as J-module.

Moreover, we have

$$\begin{aligned}
(S \otimes M^*)^G &\simeq ((J \otimes \mathcal{H}) \otimes M^*)^G \\
&\simeq (J \otimes (\mathcal{H} \otimes M^*))^G \\
&\simeq J \otimes (\mathcal{H} \otimes M^*)^G \\
&\simeq Ju_1 \oplus \cdots \oplus Ju_r.
\end{aligned}$$

This proves that $(S \otimes M^*)^G$ is free as J-module, and also proves (ii).

The statement (iii) is clear from the isomorphism $\mathrm{Hom}_{\mathbb{C}G}(M, \mathcal{H}) \simeq (\mathcal{H} \otimes M^*)^G$ (see Lemma 10.2 (i)). $\qquad\square$

The following statement is now clear.

Lemma 10.4. *We have, maintaining the above notation,*

(i) $f_M(t) = \sum_{i=1}^r t^{\deg u_i}$.

(ii) $P_{(S \otimes M^*)^G}(t) = P_J(t)f_M(t) = \prod(1 - t^{d_i})^{-1} f_M(t)$.

We have seen that the space $(\mathcal{H} \otimes M)^G$ has dimension equal to $\dim M$. The next result provides a natural class of isomorphisms between these spaces.

Lemma 10.5. *Let v be any regular element of V (that is, v is not fixed by any reflection of G) and let M be a G-module. Define a map $\psi_v : S \otimes M \to M$ by*

$$\psi_v(P \otimes y) = P(v)y.$$

Then ψ_v defines an isomorphism of vector spaces $(\mathcal{H} \otimes M)^G \simeq M$.

Proof. Since $\mathcal{H}$ affords the regular representation of G there is a basis $\{\, A_g \mid g \in G \,\}$ of $\mathcal{H}$ such that $hA_g = A_{hg}$ for all $g, h \in G$. We have $(\mathcal{H} \otimes (M_1 \oplus M_2))^G \simeq (M_1 \oplus M_2)$ if and only if $(\mathcal{H} \otimes M_1)^G \simeq M_1$ and $(\mathcal{H} \otimes M_2)^G \simeq M_2$, hence to prove the result it suffices to take M to be the regular representation with basis $\{\, [g] \mid g \in G \,\}$ and action given by $x[g] = [xg]$ for $x, g \in G$. For $g \in G$, define $u_g := \sum_{h \in G} A_{hg} \otimes [h] \in \mathcal{H} \otimes M$. For $s \in G$ we have $su_g = \sum_h A_{shg} \otimes [sh] = u_g$ and so $u_g \in (\mathcal{H} \otimes M)^G$.

We shall show that

$$\psi_v\Big(\sum_g \alpha_g u_g\Big) = 0 \quad \text{if and only if} \quad \alpha_g = 0 \quad \text{for all } g.$$

To this end, suppose that $\psi_v(\sum_g \alpha_g u_g) = 0$. Then $\sum_g \alpha_g \psi_v(A_{hg} \otimes [h]) = 0$, that is $\sum_g \alpha_g A_{hg}(v) \otimes [h] = 0$ and therefore $\sum_g \alpha_g A_{hg} = 0$ for all h. But v is a regular element and so the matrix $(A_{hg}(v))$ is non-singular. Thus $\alpha_g = 0$ for all g, as required.

This proves that ψ_v is injective. But $\dim(\mathcal{H} \otimes M)^G = \dim M = |G|$ and therefore ψ_v is an isomorphism. □

2. Gutkin's Theorem

Definition 10.6. Given a G-module M, let $\{y_1, y_2, \ldots, y_r\}$ be a basis for M^*. Then $(\mathcal{H} \otimes M^*)^G$ has a homogeneous basis $u_1, u_2, \ldots, u_r$ such that

$$u_i = \sum_{j=1}^{r} A_{ij} \otimes y_j$$

where each A_{ij} is a uniquely determined homogeneous element of $\mathcal{H}$ of degree $q_i(M)$. We define $\Pi_M := \det(A_{ij})$. It is clear from its definition that the degree of Π_M is $\sum_i q_i(M)$.

Proposition 10.7. *The polynomial Π_M is a non-zero homogeneous element of S, which is uniquely determined by M up to a non-zero scalar multiple.*

Proof. Clearly Π_M is homogeneous, of degree $\sum_j q_j(M)$, or zero. To show that $\Pi_M \neq 0$, suppose that v is a regular element of V. In the notation of Lemma 10.5 we have

$$\psi_v(u_i) = \sum A_{ij}(v) y_j$$

and by *loc. cit.*, $\{\psi_v(u_i)\}$ is a basis of M^*, whence $\det A_{ij}(v) \neq 0$. In particular, $\Pi_M \neq 0$.

Changing the (linear) bases $\{u_i\}$ and $\{y_j\}$ simply multiplies $\det(A_{ij})$ by a non-zero constant and so Π_M is determined by M up to a non-zero constant multiple. □

Lemma 10.8. *For any G-module M we have*

(i) $g\Pi_M = \det(g|_M)\Pi_M$,

(ii)   *If $M \simeq \mathbb{C}A$, where $A \in S$, then there exists $B \in J$ such that $A = B\Pi_M$.*

Proof. Let ΛM^* be the exterior algebra of M^* [**142**, Chap. XIX], which has basis $\{y_{i_1} \wedge \cdots \wedge y_{i_p}\}$, where $1 \leq i_1 < \cdots < i_p \leq r$, and $p = 0, \ldots, r$. Recall that for any complex vector space U of finite dimension r, the exterior algebra $\Lambda U = \bigoplus_{i=0}^{r} \Lambda^i U$ is an associative algebra with multiplication given by

$$(u_{i_1} \wedge \cdots \wedge u_{i_p})(u_{j_1} \wedge \cdots \wedge u_{j_q}) = u_{i_1} \wedge \cdots \wedge u_{i_p} \wedge u_{j_1} \wedge \cdots \wedge u_{j_q}.$$

Consider the action of G on the associative algebra $S \otimes \Lambda M^*$.

The space $\Lambda^r M^*$ has basis consisting of the single element $y_M := y_1 \wedge y_2 \wedge \cdots \wedge y_r$, and we have $g y_M = (\det g|_M)^{-1} y_M$. Now if we write $u_M := u_1 u_2 \cdots u_r \in (S \otimes \Lambda M^*)^G$, then $u_M = (\sum A_{1j} \otimes y_j)(\sum A_{2j} \otimes y_j) \cdots (\sum A_{rj} \otimes y_j) = \det A_{ij} \otimes y_M = \Pi_M \otimes y_M$. But by definition, each u_i, and hence u_M, is G-invariant and so

$\Pi_M \otimes y_M = g(\Pi_M \otimes y_M) = g\Pi_M \otimes \det(g|_M)^{-1} y_M$ whence $g\Pi_M = \det(g|_M)\Pi_M$, proving (i).

Next suppose that $M = \mathbb{C}A$ for some $A \in S$. Then by Proposition 9.40, $M^* \simeq \mathbb{C}f_m$ for some $f_m \in \mathcal{H}$. Thus in this case we have $u_1 = f_{m'} \otimes f_m$ and $\Pi_M = f_{m'} \in \mathcal{H}$, where $m' : \mathcal{A}/G \to \mathbb{Z}_{>0}$ is the harmonic function corresponding to the character of M. On the other hand, we have $S = J \otimes \mathcal{H}$ and so $A = B\Pi_M$ for some $B \in J$. $\qquad \square$

Corollary 10.9. *If* $\dim M = 1$, *then* $\Pi_M \in \mathcal{H}$.

Corollary 10.10. *If* M *is a G-module of dimension r, then* $\Pi_{\Lambda^r M}$ *divides* Π_M, *and* $\Pi_M = B\Pi_{\Lambda^r M}$ *for some invariant polynomial $B \in J$.*

Proof. The dimension of the module $\Lambda^r M \simeq \mathbb{C}\Pi_M$ is one, and therefore by Lemma 10.8 (ii), $\Pi_M = B\Pi_{\Lambda^r M}$ for some $B \in J$. $\qquad \square$

Recall that $\mathcal{A}$ is the set of hyperplanes $\operatorname{Fix} r$, where r is a reflection in G. For $H \in \mathcal{A}$ the subgroup G_H fixing H pointwise is cyclic with generator r_H of order e_H.

Definition 10.11.

(i) For any unitary reflection group G and G-module M, $C(G, M)$ denotes the sum of the M-exponents of G; that is, in the notation of Definition 10.1 (ii), $C(G, M) = \sum_{j=1}^{\dim M} q_j(M)$.

(ii) If $H \in \mathcal{A}$ is a reflecting hyperplane of G, then $C(H, M)$ is defined as $C(G_H, \operatorname{Res}_{G_H}^G M)$ in the sense of (i).

Remark 10.12. The integers $C(H, M)$ may also be explicitly defined in terms of the representation of G_H on M as follows. If G_H is the cyclic group of reflections in H as above, write $\varepsilon_H := \det(r_H)$. The irreducible characters of G_H are of the form λ^j, $0 \leq j \leq e_H - 1$, where $\lambda(r_H) = \varepsilon_H$. By Corollary 9.41, the space $\mathcal{H}_H$ of G_H-harmonic polynomials is $\bigoplus_{i=0}^{e_H-1} \mathbb{C}L_H^i$. The coexponent $q(\lambda^j) = j$, since $(\mathcal{H}_H \otimes \lambda^j)^G$ has basis $\mathbb{C}L_H^j \otimes y$, where y is a basis of the representation λ^j. Thus if $M^* \simeq \lambda^{i_1} \oplus \cdots \oplus \lambda^{i_r}$ as G_H-modules, then

$$C(H, M) = \sum_{j=1}^{r} i_j.$$

Moreover the set $\{i_j\}$ is defined uniquely as the set of integers such that r_H acts on M^* with eigenvalues $\varepsilon_H^{i_1}, \ldots, \varepsilon_H^{i_r}$, and $0 \leq i_j \leq e_H - 1$.

Theorem 10.13 (Gutkin [107]). *For any G-module M there is a non-zero constant c such that* $\Pi_M = c \prod_{H \in \mathcal{A}} L_H^{C(H,M)}$.

Proof. From Proposition 10.7, we know that Π_M is non-zero, and that it has degree $C(G, M)$. We show next that if we write $\Lambda_M := \prod_{H \in \mathcal{A}} L_H^{C(H,M)}$, then Λ_M divides Π_M.

Fix $H \in \mathcal{A}$. It suffices to show that $L_H^{C(H,M)}$ divides Π_M. Let $\{y_1, \ldots, y_r\}$ be a basis of M^* such that for each j, $r_H y_j = \varepsilon_H^{i_j} y_j$, where the i_j are the M^*-exponents of G_H (see Remark 10.12).

Take an element $\sum_{j=1}^r A_j \otimes y_j \in (\mathcal{H} \otimes M^*)^G$; then clearly $r_H A_j = \varepsilon_H^{-i_j} A_j$ for each j. Now choose a basis $X_1, X_2, \ldots, X_n$ for V^* such that $X_1 = L_H$ and $r_H X_i = X_i$ for $i > 1$. Then $r_H X_1 = \varepsilon_H^{-1} X_1$ and therefore

$$r_H(X_1^{k_1} X_2^{k_2} \cdots X_n^{k_n}) = \varepsilon_H^{-k_1} X_1^{k_1} X_2^{k_2} \cdots X_n^{k_n}.$$

It follows that if the monomial $X_1^{k_1} X_2^{k_2} \cdots X_n^{k_n}$ occurs with non-zero coefficient in A_j, then $\varepsilon_H^{-k_1} = \varepsilon_H^{-i_j}$, i.e. $k_1 \equiv i_j \pmod{e_H}$. Hence $k_1 \geq i_j$, whence $L_H^{i_j}$ divides A_j. Thus if $\{u_i = \sum_{j=1}^r A_{ij} \otimes y_j \mid i = 1, \ldots, r\}$ is *any* set of r elements of $(\mathcal{H} \otimes M^*)^G$, Λ_M divides $\det A_{ij}$. It follows in particular that Λ_M divides Π_M. It remains to prove that Π_M and Λ_M have the same degree.

Observe that both Λ_M and Π_M are multiplicative in M in the sense that

$$\Lambda_{M_1 \oplus M_2} = \Lambda_{M_1} \otimes \Lambda_{M_2} \quad \text{and} \quad \Pi_{M_1 \oplus M_2} = \Pi_{M_1} \otimes \Pi_{M_2}.$$

Thus it suffices to prove the theorem when M is the regular representation. Since $M^* \simeq M$ we have to compute $(S \otimes M)^G$.

Choose coset representatives g_i for the right cosets $\langle r_H \rangle g$ of $\langle r_H \rangle$. The eigenvalues of r_H on M are ε_H^i for $i = 0, 1, \ldots, e_H - 1$, each with multiplicity 1 and so

$$C(H, M) = \frac{|G|}{e_H} \sum_{i=1}^{e_H - 1} i = \frac{1}{2} |G| (e_H - 1).$$

Therefore, the degree of Λ_M is $\frac{1}{2} |G| \sum_H (e_H - 1) = \frac{1}{2} |G| N$, where N is the number of reflections.

To compute the degree of Π_M we begin with the fact that if $\{[g] \mid g \in G\}$ is a basis for M, then $\sum P_g \otimes [g]$ is G-invariant if and only if $h \sum P_g \otimes [g] = \sum P_g \otimes [g]$. This is the case if and only if $\sum h P_g \otimes [hg] = \sum P_g \otimes [g]$, namely $h P_g = P_{hg}$ for all $g, h \in G$. Thus $\sum P_g \otimes [g]$ is G-invariant if and only if $P_g = g P_1$ for all $g \in G$ and so the elements of $(S \otimes M)^G$ have the form $\sum g(P_1 \otimes [1])$, where $P_1 \in \mathcal{H}$.

If the set $\{B_i\}$ is a basis for $\mathcal{H}$, then $\{\sum_g g B_i \otimes [g]\}$ is a basis for $(\mathcal{H} \otimes M)^G$ and so $\deg \Pi_M = \sum_i \deg B_i = \sum_k k \dim \mathcal{H}_k$. To compute this, note that

$$\sum_{k=0}^N \dim \mathcal{H}_k t^k = \prod_i \frac{1 - t^{d_i}}{1 - t} = \prod_i (1 + t + \cdots + t^{d_i - 1}).$$

On differentiating and setting $t = 1$ we find that

$$\sum_k k \dim \mathcal{H}_k = \sum_i (1 + 2 + \cdots + (d_i - 1)) \prod_{j \neq i} d_j$$

$$= \sum_i \frac{1}{2}(d_i - 1) \prod_j d_j$$

$$= |G| \frac{1}{2} \sum_i (d_i - 1)$$

$$= \frac{1}{2}|G|N.$$

On comparing the two degree calculations we see that $\deg \Lambda_M = \deg \Pi_M$ and hence Λ_M is a constant multiple of Π_M. $\qquad \square$

Corollary 10.10 shows that in general Π_M is a multiple of $\Pi_{\Lambda^r M}$ by a homogeneous invariant polynomial B. In many important cases we shall see that B may be taken to be constant. We therefore give a name to such representations M.

Definition 10.14. The G-module M of dimension r is said to be *amenable* if Π_M is a constant multiple of $\Pi_{\Lambda^r M}$.

Lemma 10.15. *Suppose that M is a G-module of dimension r. Then*
(i) $C(G, M) = \sum_{H \in \mathcal{A}} C(H, M)$;
(ii) M is amenable as G-module if and only if $C(H, M) \leq e_H - 1$ for all $H \in \mathcal{A}$.

Proof. The two sides of (i) are both expressions for the degree of Π_M.

Now in the notation of the proof of Theorem 10.13, since $\Lambda^r M^*$ has basis $y_1 \wedge \cdots \wedge y_r$, it is clear that r_H acts on $\Lambda^r M^*$ as $\varepsilon_H^{C(H,M)}$ and hence the character $\det_M$ of G satisfies $0 \leq q(\det_M) \leq \sum_{H \in \mathcal{A}} C(H, M)$, with equality if and only if $0 \leq C(H, M) \leq e_H - 1$ for each $H \in \mathcal{A}$. To see this last assertion, observe that in the notation of Corollary 9.21), the harmonic polynomial $\Pi_{\det_M} = f_{\det_M} = \prod_{H \in \mathcal{A}} L_H^{i_H}$, where for each $H \in \mathcal{A}$, $0 \leq i_H \leq e_H - 1$ and $i_H \equiv C(H, M)(\bmod e_H)$.

But the degree of $\Pi_{\det_M}$ is, by Remark 10.12, equal to $q(\det_M)$, and (ii) is immediate. $\qquad \square$

This leads to the following simple sufficient condition for amenability.

Corollary 10.16. *Let M be a G-module in which each reflection r_H acts as a reflection in M. Then M is amenable as a G-module.*

Proof. In this case the criterion of Lemma 10.15 (ii) is evidently satisfied. $\qquad \square$

Let $\sigma \in \mathrm{Gal}(\overline{\mathbb{Q}}/\mathbb{Q})$, where $\overline{\mathbb{Q}}$ is the algebraic closure of $\mathbb{Q}$. By applying σ to the matrices representing the elements $g \in G$ as a linear transformation of V with respect to a given basis, we obtain a new representation V^σ, whose isomorphism class is clearly independent of the basis. The representations V^σ are the *Galois twists* of V. Of course since the matrix entries of the elements of G lie in a finite extension of $\mathbb{Q}$, there are only finitely many inequivalent Galois twists.

Corollary 10.17. *Any Galois twist V^σ of V is amenable as G-module*

Proof. In V^σ, r_H acts as a reflection. $\qquad\qquad\qquad\qquad\qquad\qquad\qquad\qquad$ $\square$

3. Differential invariants

In this section we consider the module of covariants $(S \otimes \Lambda M^*)^G$ of $S \otimes \Lambda M^*$, where M is a G-module. If $M = V$, then $S \otimes \Lambda V^*$ can be identified with the ring of differential forms on V and so we call $S \otimes \Lambda M^*$ the algebra of differential M-forms. In general, we have seen (see the proof of Lemma 10.8 above) that $S \otimes \Lambda M^*$ is an associative algebra, and G acts as a group of automorphisms of $S \otimes \Lambda M^*$, preserving the grading. The space $(S \otimes \Lambda M^*)^G$ is a subalgebra, and we shall now study its structure.

Let $y_1, y_2, \ldots, y_r$ be a basis for M^* and put $y_M := y_1 \wedge y_2 \wedge \cdots \wedge y_r$. We have seen that there is a basis $u_1, u_2, \ldots, u_r$ of $(\mathcal{H} \otimes M^*)^G$, where u_i is homogeneous of degree $q_i(M)$. Then $u_1 u_2 \cdots u_r$ is a non-zero multiple of $\Pi_M \otimes y_M$.

Theorem 10.18. *Let M be an amenable G-module (see Definition 10.14), so that Π_M is a non-zero multiple of $\Pi_{\Lambda^r M}$. Then $(S \otimes \Lambda M^*)^G$ is an exterior algebra over J. That is, $(S \otimes \Lambda M^*)^G = J \otimes \Lambda W$ for some vector subspace W of $(S \otimes \Lambda M^*)^G$.*

More explicitly, if $u_1, u_2, \ldots, u_r$ are homogeneous elements of $(S \otimes \Lambda M^*)^G$, which satisfy $u_1 u_2 \cdots u_r = \Pi_M \otimes y_M$, then

$$(S \otimes \Lambda M^*)^G = \bigoplus_{i_1 \leq i_2 \leq \cdots \leq i_p} J u_{i_1} u_{i_2} \cdots u_{i_p}.$$

Proof. Take $u_1, u_2, \ldots, u_r$ as given above, and denote by K the quotient field of S. We know by Proposition 10.3 (i) that $(S \otimes \Lambda M^*)^G$ is free as J-module, and we wish to show that the elements

$$\{ u_{i_1} u_{i_2} \ldots u_{i_p} \mid 1 \leq i_1 < \cdots < i_p \leq r, \ 0 \leq p \leq r \}$$

form a J-basis of $(S \otimes \Lambda M^*)^G$. Denote sequences $i_1, \ldots, i_p$ as above by $\underline{i}$, and the corresponding product of u_{i_j} by $u_{\underline{i}}$. Let $\mathcal{S}$ be the set of all such sequences; clearly $|\mathcal{S}| = 2^r$.

Since $\Lambda M^* = \bigoplus_{p=0}^r \Lambda^p M^*$, we have $S \otimes \Lambda M^* = \bigoplus_{p=0}^r S \otimes \Lambda^p M^*$, and this decomposition is respected by the G-action. Moreover there is a unique projection $\pi_r : S \otimes \Lambda M^* \to S \otimes \Lambda^r M^*$ corresponding to this decomposition, and this projection respects the G-action.

Now observe that if $\underline{i}'$ denotes the sequence corresponding to the complement of $\underline{i}$ in $\{1, 2, \ldots, r\}$, then evidently

$$(10.19) \qquad \pi_r\left(u_{\underline{i}'} u_{\underline{j}}\right) = \begin{cases} c\Pi_M \otimes y_M & \text{for some non-zero constant } c \text{ if } \underline{j} = \underline{i} \\ 0 & \text{otherwise.} \end{cases}$$

To show that the $u_{\underline{i}}$ are independent over J, suppose that we have a relation $\sum_{j=1}^{k} f_{\underline{i}_j} u_{\underline{i}_j} = 0$, where $f_{\underline{i}_j} \in J$. Then for any sequence $\underline{i}$, applying the map $\pi_r \circ u_{\underline{i}'}$ and using the relation (10.19), we obtain $f_{\underline{i}}\Pi_M \otimes y_M = 0$. Since $(S \otimes \Lambda M^*)^G$ is free as J-module, this implies that $f_{\underline{i}} = 0$, whence the independence of the $u_{\underline{i}}$ over J.

Now $S \otimes \Lambda M^*$ is naturally a subalgebra of

$$K \otimes_S (S \otimes \Lambda M^*) \simeq (K \otimes_S S) \otimes \Lambda M^* \simeq K \otimes \Lambda M^*,$$

and hence $(S \otimes \Lambda M^*)^G$ is a subalgebra of $(K \otimes \Lambda M^*)^G$. Further, $K \otimes_{\mathbb{C}} \Lambda M^* = \bigoplus_{p=0}^{r} K \otimes \Lambda^p M^*$ is a K-vector space of dimension 2^r, with basis $\{u_{\underline{i}} \mid \underline{i} \in \mathcal{S}\}$. We wish to show that each element of $(S \otimes \Lambda M^*)^G$ is a J-linear combination of the $u_{\underline{i}}$. To this end, take $\psi \in (S \otimes \Lambda M^*)^G$. Then as noted above, there are unique elements $k_{\underline{i}} \in K$ ($\underline{i} \in \mathcal{S}$) such that $\psi = \sum_{\underline{i} \in \mathcal{S}} k_{\underline{i}} u_{\underline{i}}$.

Since $\psi \in (S \otimes \Lambda M^*)^G$, the same argument as above shows that for each $\underline{i} \in \mathcal{S}$, $k_{\underline{i}}\Pi_M \otimes y_M \in (S \otimes \Lambda^r M^*)^G$, which we know to be equal to $J\Pi_{\Lambda^r M} \otimes y_M$. Using the hypothesis of amenability, it follows that $k_{\underline{i}}\Pi_M \otimes y_M \in J\Pi_M \otimes y_M$, and hence, by the K-free nature of $K \otimes \Lambda M^*$, that $k_{\underline{i}} \in J$. Thus $(S \otimes \Lambda^r M^*)^G = \sum_{\underline{i} \in \mathcal{S}} J u_{\underline{i}}$, and the proof is complete. $\qquad \square$

The next result arises from the grading on $S \otimes \Lambda M^*$.

Corollary 10.20. *Let M be an amenable G-module.*

(i) $P_{(S \otimes \Lambda M^*)^G}(t) = P_J(t)(1 + t^{q_1(M)}) \cdots (1 + t^{q_r(M)})$.

(ii)   *The fake degree $f_{\Lambda M}(t) = \prod_i (1 + t^{q_i(M)})$.*

(iii) *The fake degree $f_{\Lambda^p M}(t) = \sum_{i_1 < \cdots < i_p} t^{q_{i_1}(M) + \cdots + q_{i_p}(M)}$.*

Proof. The formula in (i) is immediate from Theorem 10.18 given the generalities about Poincaré series in Lemma 4.7, and (ii) follows similarly, given that we have shown that $f_{\Lambda M}(t)$ is the Poincaré series of ΛW, where W is the (graded) vector space with basis $u_1, \ldots, u_r$. But the degrees of the u_i are $q_i(M)$, whence the result. The statement (iii) is obtained by selecting the basis elements $u_{\underline{i}}$ of $(S \otimes \Lambda M^*)^G$ which lie in $(S \otimes \Lambda^p M^*)^G$. $\qquad \square$

4. Some special cases of covariants

4.1. Exponents. From Theorem 10.18, it is clear that when M is an amenable G-module, the M-covariants of G may be rather explicitly described. The criterion of Corollary 10.16 provides many interesting examples. We begin by taking $M = V$, the defining (reflection) representation of G.

Lemma 10.21. *Let $v_1, v_2, \ldots, v_n$ be a basis of V and let $X_1, X_2, \ldots, X_n$ be the dual basis of V^*. Define $d : S \otimes \Lambda V^* \to S \otimes \Lambda V^*$ by*

$$d(P \otimes y) := \sum_i D_{v_i} P \otimes (X_i \wedge y).$$

Then

(i) d is independent of the basis;
(ii)   for $P, Q \in S$ and $y \in \Lambda V^*$, we have $d(PQ \otimes y) = P\, d(Q \otimes y) + Q\, d(P \otimes y)$;
(iii)   for all $g \in G$ we have $gd = dg$.

Proof. The element $\sum_i v_i \otimes X_i \in V \otimes V^*$ is independent of the basis, proving *(i)*. The other statements are easily verified.   □

Proposition 10.22. *Suppose that $I_1, I_2, \ldots, I_n$ is a set of basic invariants for G acting on V. Then $dI_1, dI_2, \ldots, dI_n$ is a J-basis of $(S \otimes V^*)^G$.*

Proof. Note first that if $I \in J$, then $dI \in (S \otimes \Lambda V)^G$, and so $u_j := dI_j \in (S \otimes \Lambda V^*)^G$. Next we note that by Theorem 10.18, it suffices to show that $u_1 u_2 \cdots u_n = \Pi_V X_1 \wedge X_2 \wedge \cdots \wedge X_n$. Let us compute Π_V. A reflection $r_H \in G$ has eigenvalues $\varepsilon_H, 1, \ldots, 1$ on V, so that its eigenvalues on V^* are $\varepsilon_H^{e_H - 1}, 1, \ldots, 1$. Hence by Remark 10.12, $C(H, V) = e_H - 1$, and it follows from Theorem 10.13 that $\Pi_V = \prod_{H \in \mathcal{A}} L_H^{e_H - 1}$.

On the other hand, we have

$$u_1 u_2 \cdots u_n = \prod_j \left(\sum_i \frac{\partial I_j}{\partial X_i} \otimes X_i \right)$$

$$= \frac{\partial(I_1, \ldots, I_n)}{\partial(X_1, \ldots, X_n)} \otimes (X_1 \wedge \cdots \wedge X_n)$$

$$= \prod L_H^{e_H - 1} \otimes (X_1 \wedge \cdots \wedge X_n) \quad \text{by Theorem 9.8}$$

$$= \Pi_V \otimes y_V,$$

which completes the proof.   □

Corollary 10.23. *The V-exponents $q_i(V)$ are given by $q_i(V) = d_i - 1$, where $d_1, \ldots, d_n$ are the degrees of the basic invariants of G.*

Proof. The V-exponents of G are the degrees of the generators $u_1, \ldots, u_n$ of Proposition 10.22. Since the degree of dI_j is $d_j - 1$, the result is clear.   □

Definition 10.24. We call the integers $q_j(V) = m_j := d_j - 1$ the *exponents* of G.

When G is a real reflection group, i.e. the matrices representing G in V may be taken to have real entries, then G is a finite Coxeter group, and the integers m_i are related to the eigenvalues of the Coxeter elements of G.

Corollary 10.25. [199] *For $0 \le p \le n$, the fake degree*

$$f_{\Lambda^p V}(t) = e_p(t^{m_1}, t^{m_2}, \ldots, t^{m_n}) = \sum_{i_1 < \cdots < i_p} t^{m_{i_1} + \cdots + m_{i_p}},$$

where e_p is the pth elementary symmetric function.

Proof. This is immediate from Corollary 10.20 (*iii*). □

4.2. Coexponents. From the criterion for amenability (Corollary 10.16), the dual V^* of V is amenable, and we next take $M = V^*$ in Theorem 10.18. To compute Π_{V^*}, observe that a reflection $r_H \in G$ has eigenvalues $\varepsilon_H, 1, \ldots, 1$ on $(V^*)^* = V$. Hence by Remark 10.12, $C(H, V^*) = 1$, and it follows from Theorem 10.13 that $\Pi_V = \prod_{H \in \mathcal{A}} L_H$. The next proposition therefore follows from Theorem 10.18.

Proposition 10.26. *There are homogeneous elements $u_1, \ldots, u_n \in (S \otimes V)^G$ such that if we write $u_i = \sum_{j=1}^n A_{ij} \otimes v_j$, where $v_1, \ldots, v_n$ is a basis of V, then*

$$\det(A_{ij}) = \prod_{H \in \mathcal{A}} L_H.$$

Definition 10.27. The V^*-exponents $q_j(V^*) = \deg u_j = \deg A_{jk}$ (for any k such that $A_{jk} \ne 0$) are called the *coexponents* of G. They are usually denoted m_j^*, and we write $d_j^* := m_j^* - 1$ for the corresponding *codegree*.

We shall see that the coexponents and codegrees are intricately involved in the geometry of G.

5. Two-variable Poincaré series and specialisations

5.1. Poincaré series. Hitherto we have used only the grading on $S \otimes \Lambda M^*$ which comes from the grading of S. In this section we shall exploit the fact that $S \otimes \Lambda M^*$ is actually *bigraded*, since the module ΛM^* is itself graded. This bigrading leads to equivariant Poincaré series which may be exploited to obtain information about G. We shall follow in spirit the exposition in [**151**].

Definition 10.28. Let M be a G-module. Define a bigrading on G-module $S \otimes \Lambda M^*$ by $(S \otimes \Lambda M^*)_{i,p} := S_i \otimes \Lambda^p M^*$ ($i = 0, 1, 2, \ldots, p = 0, 1, 2, \ldots, r = \dim M$).

The reflection group G acts on $S \otimes \Lambda M^*$, preserving the bigrading, from which it follows that $(S \otimes \Lambda M^*)^G \simeq \bigoplus_{i,p} (S \otimes \Lambda M^*)^G_{i,p}$, i.e. that $(S \otimes \Lambda M^*)^G$ is also bigraded. The following result is essentially that of Orlik–Solomon [**178**, Theorem 3.7]. It is a refinement of a formula originally due to Solomon [**199**].

Theorem 10.29. *Let G be a finite unitary reflection group in V and let M be an amenable G-module (see Definition 10.14) of dimension r. Let t and u be indeterminates. Then*

$$|G|^{-1} \sum_{g \in G} \frac{\det_M(1 - ug)}{\det_V(1 - tg)} = \frac{\prod_{j=1}^{r}(1 - ut^{q_j(M)})}{\prod_{i=1}^{n}(1 - t^{d_i})},$$

where the $q_j(M)$ are the M-exponents of G (see Definition 10.1), and the d_i are the basic degrees of G.

Proof. We have seen above that $(S \otimes \Lambda M^*)^G$ is bigraded. We shall compute the Poincaré series $P_{(S \otimes \Lambda M^*)^G}(t, u) := \sum_{i,p \geq 0} \dim(S \otimes \Lambda M^*)^G_{i,p} t^i u^p$ in two different ways to yield the stated equality.

First, let us apply the principle of Theorem 4.13 (Molien) to $(S \otimes \Lambda M^*)^G$. We have

$$P_{(S \otimes \Lambda M^*)^G}(t, u) = |G|^{-1} \sum_{i,p \geq 0} \left(\sum_{g \in G} \text{trace}(g, (S \otimes \Lambda M^*)_{i,p}) t^i u^p \right.$$

$$= |G|^{-1} \sum_{i,p \geq 0} \left(\sum_{g \in G} \text{trace}(g, (S_i \otimes \Lambda^p M^*)) t^i u^p \right.$$

(10.30)
$$= |G|^{-1} \sum_{i,p \geq 0} \left(\sum_{g \in G} \text{trace}(g, S_i) \text{trace}(g, \Lambda^p M^*)) t^i u^p \right.$$

$$= |G|^{-1} \sum_{g \in G} \left(\sum_{i \geq 0} \text{trace}(g, S_i) t^i \right) \left(\sum_{p \geq 0} \text{trace}(g, \Lambda^p M^*) u^p \right)$$

$$= |G|^{-1} \sum_{g \in G} \frac{\det_{M^*}(1 + ug)}{\det_{V^*}(1 - tg)} \quad \text{by Lemmas 4.6 and 4.10.}$$

On the other hand, the tensor decomposition

$$(S \otimes \Lambda M^*)^G \simeq J \otimes (\mathcal{H} \otimes \Lambda M^*)^G = \bigoplus_{p=0}^{r} (J \otimes (\mathcal{H} \otimes \Lambda^p M^*)^G)$$

leads to the following two-variable analogue of Lemma 10.4 (*ii*):

$$P_{(S \otimes \Lambda M^*)^G}(t, u) = P_J(t) \left(\sum_{i,p} (\mathcal{H}_i, \Lambda^p M)_G t^i u^p \right)$$

$$= P_J(t) \sum_{p} \left(\sum_{i} (\mathcal{H}_i, \Lambda^p M)_G t^i \right) u^p$$

$$= P_J(t) \sum_{p} f_{\Lambda^p M}(t) u^p \quad \text{where } f \text{ denotes fake degree}$$

(10.31) $= P_J(t) \displaystyle\sum_p \sum_{i_1 < \cdots < i_p} t^{q_{i_1}(M) + \cdots + q_{i_p}(M)} u^p$ by Cor. 10.20 (iii)

$$= P_J(t) \prod_{j=1}^r (1 + u t^{q_j(M)})$$

$$= \frac{\prod_{j=1}^r (1 + u t^{q_j(M)})}{\prod_{i=1}^n (1 - t^{d_i})}.$$

Comparing the final expressions in (10.30) and (10.31), we obtain

$$|G|^{-1} \sum_{g \in G} \frac{\det_{M^*}(1 + ug)}{\det_{V^*}(1 - tg)} = \frac{\prod_{j=1}^r (1 + u t^{q_j(M)})}{\prod_{i=1}^n (1 - t^{d_i})}.$$

If we replace u by $-u$ in this last equation, we obtain the statement of the theorem, with V and M replaced by V^* and M^*. But the right side is evidently stable under complex conjugation, whence we may take the complex conjugate of the left side, and the proof is complete.   □

5.2. Specialisations. Theorem 10.29 may be thought of as an equation in the subring $\mathbb{C}(t)[u] \subseteq \mathbb{C}[\![t, u]\!]$ of the ring of power series in the variables t, u. We shall now specialise the variables to obtain polynomial identities which will be used to illuminate the structure of G. We shall require the following notation.

Definition 10.32.

(*i*)   If $g \in GL(V)$ and $\zeta \in \mathbb{C}^\times$ we define $V(g, \zeta)$ to be the ζ-eigenspace of g, i.e.
$V(g, \zeta) = \{ v \in V \mid gv = \zeta v \}.$

(*ii*)   In the notation of (*i*), $d(g, \zeta) = \dim V(g, \zeta)$.

The next result is [**151**, Theorem 2.3].

Theorem 10.33. *Maintain the notation above. For* $\sigma \in \mathrm{Gal}(\overline{\mathbb{Q}}/\mathbb{Q})$*, let* V^σ *be the corresponding Galois twist of* V*. Fix a primitive* d^{th} *root of unity* $\zeta \in \mathbb{C}$ *for some integer* $d \geq 1$*, and let*

$$A(d) = \{ i \mid 1 \leq i \leq n, \ \zeta^{d_i} = 1 \} \quad and$$
$$B_{\hat{\sigma}}(d) = \{ j \mid 1 \leq j \leq n, \ \zeta^\sigma \zeta^{q_j(V^\sigma)} = 1 \}.$$

Let $a(d) = |A(d)|$ *and* $b_\sigma(d) = |B_\sigma(d)|$.

Then $a(d) \leq b_\sigma(d)$*, and we have the following identity in the polynomial ring* $\mathbb{C}[T]$*. We write* $\det'(E)$ *for the product of the non-zero eigenvalues of a linear transformation* E*. If* $a(d) \neq b_\sigma(d)$*, then*

(10.34) $$\sum_{g \in G} T^{d(g, \zeta)} \det'(1 - \zeta^{-1} g)^{\sigma - 1} = 0,$$

while if $a(d) = b_\sigma(d)$, then the left side is equal to

$$(10.35) \qquad \prod_{j \in B_\sigma(d)} (T + q_j(V^\sigma)) \prod_{j \notin B_\sigma(d)} (1 - \zeta^{-\sigma - q_j(V^\sigma)}) \prod_{i \notin A(d)} \frac{d_i}{1 - \zeta^{-d_i}}.$$

Proof. Since V^σ is amenable, we may apply Theorem 10.29 with $M = V^\sigma$. Denote the eigenvalues of $g \in G$ on V by $\mu_1(g), \ldots, \mu_n(g)$. Then the eigenvalues of g on V^σ are of course $\mu_1(g)^\sigma, \ldots, \mu_n(g)^\sigma$, and we may therefore write the formula in Theorem 10.29 as

$$(10.36) \qquad |G|^{-1} \sum_{g \in G} \prod_{i=1}^n \frac{(1 - u\mu_i(g)^\sigma)}{(1 - t\mu_i(g))} = \frac{\prod_{j=1}^n (1 - ut^{q_j(V^\sigma)})}{\prod_{i=1}^n (1 - t^{d_i})}.$$

Regard this equation as an identity in $\mathbb{C}(t)[u]$. This ring has elements $A(t, u) = \sum_{i=0}^k f_i(t)u^i$, where $f_i(t) \in \mathbb{C}(t)$ is a rational function of t. We say $A(t, u)$ has a pole at $t = \alpha$ if some $f_i(t)$ has a pole at $t = \alpha$. If we put $t = \zeta^{-1}$, the term corresponding to $g \in G$ on the left side of (10.36) has a zero of order $d(g, \zeta)$ in the denominator. In order to eliminate any poles at $t = \zeta^{-1}$, introduce the change of variable

$$T = \frac{1 - u\zeta^\sigma}{1 - t\zeta}.$$

Now if we substitute $u = \zeta^{-\sigma}(1 - T + Tt\zeta)$ into (10.36), then we obtain an identity in $\mathbb{C}(t)[T]$, and by the above observation, there is no pole at $t = \zeta^{-1}$ on the left side. It follows that there is no pole at $t = \zeta^{-1}$ on the right side. But the denominator of the right side clearly has a zero of multiplicity $a(d)$ at $t = \zeta^{-1}$. Moreover the factor $1 - ut^{q_j(V^\sigma)} = 1 - t^{q_j(V^\sigma)}\zeta^{-\sigma}(1 - T + tT\zeta)$ of the numerator has a zero at $t = \zeta^{-1}$ if and only if $j \in B_\sigma(d)$. Differentiating with respect to t shows that any such zero is simple (i.e. multiplicity one), so that the multiplicity of the zero $t = \zeta^{-1}$ in the numerator is $b_\sigma(d)$. Since the right side of (10.36) has no pole at $t = \zeta^{-1}$, we have $a(d) \leq b_\sigma(d)$.

We now turn to the evaluation of both sides of (10.36) at $t = \zeta^{-1}$. This will provide the stated polynomial identity in $\mathbb{C}[T]$. Let us first evaluate the right side at $t = \zeta^{-1}$. The above argument shows that the right side has a zero of order $b_\sigma(d) - a(d)$ at $t = \zeta^{-1}$, so if $a(d) \neq b_\sigma(d)$, the right side is zero. Now assume that $a(d) = b_\sigma(d)$. Take $i \in A(d)$ and $j \in B_\sigma(d)$, and consider $\lim_{t \to \zeta^{-1}} \frac{1 - ut^{q_j(V^\sigma)}}{1 - t^{d_i}}$. Taking into account that $\lim_{t \to \zeta^{-1}} u = \zeta^{-\sigma}$, a simple application of de l'Hôpital's rule shows that the value of this limit is $\frac{T + q_j(V^\sigma)}{d_i}$. It follows that the value of the right side of (10.36) at $t = \zeta^{-1}$ is

$$(10.37) \qquad \frac{\prod_{j \in B_\sigma(d)}(T + q_j(V^\sigma)) \prod_{j \notin B_\sigma(d)}(1 - \zeta^{-(\sigma + q_j(V^\sigma))})}{\prod_{i \in A(d)} d_i \prod_{i \notin A(d)}(1 - \zeta^{-d_i})}.$$

To evaluate the left side of (10.36) at $t = \zeta^{-1}$, observe first that another easy application of de l'Hôpital's rule shows that for any $\mu \in \mathbb{C}$,

$$\lim_{t \to \zeta^{-1}} \frac{1 - u\mu^\sigma}{1 - t\mu} = \begin{cases} \dfrac{1 - (\zeta^{-1}\mu)^\sigma}{1 - \zeta^{-1}\mu} & \text{if } \mu \neq \zeta \\[2ex] T & \text{if } \mu = \zeta \end{cases}.$$

It follows that the value of the left side of (10.36) at $t = \zeta^{-1}$ is given by

$$(10.38) \qquad |G|^{-1} \sum_{g \in G} T^{d(g,\zeta)} \det'(1 - \zeta^{-1}g)^{\sigma-1}.$$

The theorem is now obtained easily by equating the expressions (10.38) and (10.37), taking into account that $|G| = d_1 d_2 \ldots d_n$. $\qquad \square$

There are many special cases of Theorem 10.33 which will be of interest to us. The first is the following result of Pianzola and Weiss [**183**].

Corollary 10.39. *We have the following identity in $\mathbb{C}[T]$.*

$$\sum_{g \in G} T^{d(g,\zeta)} = \prod_{i \text{ such that } \zeta^{d_i}=1} (T + d_i - 1) \prod_{i \text{ such that } \zeta^{d_i} \neq 1} d_i.$$

Proof. This is immediate from the case $\sigma = \text{id}$ of Theorem 10.33. $\qquad \square$

For the next result, recall Definition 10.27 that the V^*-exponents $q_j(V^*)$ are called the coexponents m_j^* of G, and the codegrees are defined as $d_j^* := m_j^* - 1$.

Definition 10.40. When the Galois automorphism σ is complex conjugation, the set $B_\sigma(d) = \{ j \mid 1 \leq j \leq n, \ \zeta^{-1}\zeta^{m_j^*} = 1 \}$ defined in the statement of Theorem 10.33 is denoted $B(d)$. Accordingly, we write $b(d)$ for its cardinality.

Corollary 10.41. [**151**, Corollary 2.6] *Let ζ be a primitive d^{th} root of unity in $\mathbb{C}$. With notation as in Definition 10.40, we have $a(d) \leq b(d)$, and*

$$(-\zeta)^n \sum_{g \in G} \det(g^{-1})(-T)^{d(g,\zeta)} =$$

$$(10.42) \qquad \begin{cases} 0 & \text{if } a(d) < b(d) \\[2ex] \displaystyle\prod_{j \in B(d)} (T + m_j^*) \prod_{j \notin B(d)} (1 - \zeta^{-d_j^*}) \prod_{i \notin A(d)} \frac{d_i}{1 - \zeta^{-d_i}} & \text{if } a(d) = b(d), \end{cases}$$

where the $d_j^* = m_j^* - 1$ are the codegrees of G.

Proof. Take σ to be complex conjugation in Theorem 10.33. First note that $a(d) \leq b(d)$ is a consequence of the general statement in Theorem 10.33.

Now for any root of unity $\lambda \in \mathbb{C}$, we have $\lambda^\sigma = \bar{\lambda}$. Using this, and the fact that $d_j^* = q_j(V^*) - 1$, it is straightforward that the expression (10.35) coincides with the right side of (10.42). It remains to verify that the left side of (10.42) coincides with (10.34) in this case.

For this, observe that for any root of unity $\lambda(\neq 1) \in \mathbb{C}$, $(1 - \lambda)^{\sigma-1} = (1-\bar{\lambda})(1-\lambda)^{-1} = -\lambda^{-1}$. It follows that if $g \in G$ has eigenvalues $\mu_1(g), \ldots, \mu_n(g)$, then, taking into account that $g \in G$ has just $d(g, \zeta)$ eigenvalues equal to ζ,

$$
\begin{aligned}
\det'(1 - \zeta^{-1}g)^{\sigma-1} &= \prod_{j:\mu_j(g)\neq\zeta} (1 - \zeta^{-1}\mu_j(g))^{\sigma-1} \\
&= \prod_{j:\mu_j(g)\neq\zeta} (-\zeta\mu_j(g)^{-1}) \\
&= (-\zeta)^{n-d(g,\zeta)} \prod_{j:\mu_j(g)\neq\zeta} \mu_j(g)^{-1} \\
&= (-1)^{n-d(g,\zeta)}\zeta^n \prod_j \mu_j(g)^{-1} \\
&= (-\zeta)^n (-1)^{d(g,\zeta)} \det(g^{-1}).
\end{aligned}
$$

It is now clear that (10.34) becomes the left side of (10.42) in this case, which completes the proof. $\square$

Exercises

Maintain the notation of this chapter.

1. Show that

$$
\sum_{g\in G} T^{\dim(\mathrm{Fix}\,g)} \det'(1 - g)^{\sigma-1} = \prod_{i=1}^{n}(T + q_i(V^\sigma)),
$$

   where $\mathrm{Fix}\,g$ is the subspace of g-fixed elements of V.

2. Write out the equation yielded by Exercise 1 in the case $\sigma = 1$. This is a formula originally proved by Orlik and Solomon.

3. Show that the case $\zeta = -1$ of Theorem 10.33 implies that

$$
\sum_{g\in G} T^{d(g,-1)} \det'(1 + g)^{\sigma-1}
$$

$$
= \begin{cases} 0 & \text{if } |\{\, j \mid q_j(V^\sigma) \text{ is odd}\,\}| \neq |\{\, j \mid d_j \text{ is even}\,\}|, \\ \prod_{j\mid q_j(V^\sigma)\text{ is odd}}(T + q_j(V^\sigma)) \prod_{j\mid d_j \text{ is odd}} d_j & \text{otherwise.} \end{cases}
$$

The next five exercises may be found in [150].

4. Fix a linear character $\lambda : G \to \mathbb{C}^\times$ as in §9.4. For any G-module M, denote by M^λ the subspace $\{ m \in M \mid gm = \lambda(g)m \text{ for all } g \in G \}$.

   (i)   Show that if $\dim M < \infty$ and if χ_M is the character of M,

   $$\dim M^\lambda = \dim(M \otimes \lambda^*)^G = |G|^{-1} \sum_{g \in G} \chi_M(g)\overline{\lambda}(g).$$

   (ii)   Show that $(S \otimes V^*)^\lambda$ has a homogeneous S^G-basis $\omega_1, \ldots, \omega_n$ with $\deg \omega_j = q_j(V \otimes \lambda)$.

5. Consider the bigraded algebra $S \otimes \Lambda V^*$.

   (i)   Show (cf. [194]) that if $\omega, \omega' \in (S \otimes \Lambda V^*)^\lambda$, then $\omega\omega' = f_\lambda \omega \wedge \omega'$ for some element $\omega \wedge \omega' \in (S \otimes \Lambda V^*)^\lambda$, where f_λ is defined in Corollary 9.21.

   (ii)   Deduce that $(S \otimes \Lambda V^*)^\lambda$ has an S^G-basis

   $$\{ \omega_{i_1} \wedge \cdots \wedge \omega_{i_p} \mid i_1 < \cdots < i_p, \ 0 \le p \le n \}.$$

6. By computing the Poincaré series of $(S \otimes \Lambda V^*)^\lambda$ in two different ways, prove that

   $$|G|^{-1} \sum_{g \in G} \frac{\det_V(1 + yg)}{\det_V(1 - xg)} \lambda(g) = x^{q(\lambda)} \frac{\prod_{j=1}^n (1 + yx^{q_j(V \otimes \lambda) - q(\lambda)})}{\prod_{i=1}^n (1 - x^{d_i})},$$

   where $q(\lambda)$ is the (unique) λ-exponent of G.

7. For $j = 1, \ldots, n$, write $r_j = r_j(\lambda) = q(\lambda) - q_j(V \otimes \lambda)$. Given a positive integer d, let $a(d)$ be the number of basic degrees d_i of G which are divisible by d, and let $c(\lambda, d) := |\{ j \mid 1 \le j \le n, \ d \text{ divides } r_j - 1 \}|$. Using a similar argument to that in the proof of Theorem 10.33, show that $a(d) \le c(\lambda, d)$.

8. Order the d_i so that $d \mid d_i$ if and only if $1 \le i \le a(d)$, and fix a primitive d^{th} root of unity ζ. Again using method of Theorem 10.33, deduce from the previous two questions the following polynomial identity in $\mathbb{C}[T]$.

   $$\sum_{g \in G} \lambda(g) T^{d(g,\zeta)} = \zeta^{-q(\lambda)} \prod_{j \le a(d)} (T - r_j) \prod_{j > a(d)} d_j \left(\frac{1 - \zeta^{r_j - 1}}{1 - \zeta^{-d_j}} \right).$$

CHAPTER 11

Eigenspace theory and reflection subquotients

Throughout this chapter, G will be a unitary reflection group in $V = \mathbb{C}^n$, and $F_1, \ldots, F_n$ will be a set of homogeneous basic invariants for G, with $d_i = \deg F_i$ for each i.

For a given root of unity ζ, the set of all ζ-eigenspaces $V(g, \zeta)$ of the elements $g \in G$ form a partially ordered set, which is in some respects analogous to the set of Sylow p-subgroups of an arbitrary finite group, for some prime p. Moreover there is a theory analogous to Sylow's theory (see also Quillen's work [185]) of p-subgroups for these eigenspaces. This has led to the theory of regular elements, and has close links with reductive groups over finite fields (see, e.g. [38, 41]). We shall develop these ideas in this chapter, using no more than the invariant theoretic results we have already proved, together with some elementary affine algebraic geometry.

In this chapter, we shall prove the basic results of the theory of eigenspaces. As preparation, we shall recall some basic facts from commutative algebra.

1. Basic affine algebraic geometry

We start with an elementary and naïve exposition of some aspects of affine algebraic geometry and commutative algebra necessary for the eigenspace theory we wish to develop. In this section, we develop the basic vocabulary necessary for the statements of the results we use. Proofs of all the algebraic and geometric results we require may be found in Appendix A.

Let $V = \mathbb{C}^n$, and $S = S(V^*)$, in accord with our exposition hitherto. The algebra S is usually known as the 'coordinate ring' or ring of polynomial functions on V. Its elements may be regarded as functions on V, as we have seen. We assume that the reader is familiar with the basic theory of Noetherian rings [142, Chap. X], of which the algebra S is an example. For any subset $A \subseteq V$, we define $\mathcal{I}(A)$ as the ideal of polynomials in S which vanish on each point of A. Similarly, for any subset T of S, we define the subset $\mathcal{V}(T)$ as the set of points of V on which all the polynomials in T vanish. Clearly if I is the ideal generated by T, we have $\mathcal{V}(T) = \mathcal{V}(I)$.

A subset $A \subseteq V$ is *algebraic* if $A = \mathcal{V}(I)$ for some ideal I of S. Similarly, an ideal I of S is *algebraic* if $I = \mathcal{I}(A)$ for some subset $A \subseteq V$. The following proposition sets out the basic properties of the maps $\mathcal{I}$ and $\mathcal{V}$.

Proposition 11.1.

(i)   Both $\mathcal{I}$ and $\mathcal{V}$ are inclusion reversing; i.e. if $A_1 \subseteq A_2$, then $\mathcal{I}(A_1) \supseteq \mathcal{I}(A_2)$ and similarly for $\mathcal{V}$.

(ii) For any subset $A \subseteq V$ and ideal $I \subseteq S$, we have $\mathcal{V}\mathcal{I}(A) \supseteq A$ and $\mathcal{I}\mathcal{V}(I) \supseteq I$.

(iii) We have $\mathcal{I}\mathcal{V}\mathcal{I} = \mathcal{I}$ and $\mathcal{V}\mathcal{I}\mathcal{V} = \mathcal{V}$.

(iv) The maps $\mathcal{I}$ and $\mathcal{A}$ are mutually inverse, inclusion reversing bijections between the set of algebraic subsets of V and the set of algebraic ideals of S.

(v)   If $\{I_\alpha\}$ is a set of ideals of S, then

$$\mathcal{V}\left(\sum_\alpha I_\alpha\right) = \bigcap_\alpha \mathcal{V}(I_\alpha).$$

(vi) For ideals I_1, I_2 of S, we have

$$\mathcal{V}(I_1 \cap I_2) = \mathcal{V}(I_1) \cup \mathcal{V}(I_2) = \mathcal{V}(I_1 I_2).$$

Proof. The statements (i) and (ii) are evident. To deduce (iii), observe that for any subset $A \subseteq V$, $\mathcal{V}\mathcal{I}(A) \supseteq A$ by (ii), whence by (i), $\mathcal{I}\mathcal{V}\mathcal{I}(A) \subseteq \mathcal{I}(A)$. But again using (ii), $\mathcal{I}\mathcal{V}(\mathcal{I}(A)) \supseteq \mathcal{I}(A)$, so we have equality, i.e. $\mathcal{I}\mathcal{V}\mathcal{I} = \mathcal{I}$, and similarly $\mathcal{V}\mathcal{I}\mathcal{V} = \mathcal{V}$. It follows that if $A \subseteq S$ is algebraic, $\mathcal{V}\mathcal{I}(A) = A$, and for an algebraic ideal $I \subseteq S$, $\mathcal{I}\mathcal{V}(I) = I$, which implies (iv).

The proof of (v) is easy. To see (vi), note that since $I_1 I_2 \subseteq I_1 \cap I_2 \subseteq I_1$ and $I_1 \cap I_2 \subseteq I_2$, we have $\mathcal{V}(I_1) \cup \mathcal{V}(I_2) \subseteq \mathcal{V}(I_1 \cap I_2) \subseteq \mathcal{V}(I_1 I_2)$. For the reverse inclusion, suppose $a \in \mathcal{V}(I_1 I_2)$, but $a \notin \mathcal{V}(I_1)$. Then there is a polynomial $f_1 \in I_1$ such that $f_1(a) \neq 0$. But for each polynomial $f \in I_2$, $f_1 f \in I_1 I_2$, whence $f_1(a) f(a) = 0$, and so $f(a) = 0$. Hence $a \in \mathcal{V}(I_2)$. $\qquad\square$

Corollary 11.2. *There is a topology on V in which the closed subsets are the algebraic subsets. This is the* Zariski topology *on V.*

Proof. Proposition 11.1 (v) and (vi) show that the algebraic subsets are closed under arbitrary intersection and finite unions. It remains only to verify that V and $\emptyset$ are algebraic. But $V = \mathcal{V}(0)$, while $\emptyset = \mathcal{V}(S)$. $\qquad\square$

Every algebraic subset of V also has a Zariski topology, which is the subspace topology induced by the topology on V.

Definition 11.3.

(i)   For an algebraic subset $A \subseteq V$, the *coordinate ring* $\mathbb{C}[A]$ is the algebra of functions $f : A \to \mathbb{C}$ obtained by restricting polynomials in S. Since $\mathcal{I}(A)$ is the kernel of the restriction homomorphism $\mathrm{Res}_A^V : S = \mathbb{C}[V] \to \mathbb{C}[A]$, by definition we have $\mathbb{C}[A] \simeq S/\mathcal{I}(A)$.

(ii) The algebraic set $A \subseteq V$ is *reducible* if $A = A_1 \cup A_2$, where A_1, A_2 are closed subsets, and $A_i \neq A$ for $i = 1, 2$. We say that A is *irreducible* if A is not reducible.

The following easy characterisation of irreducible algebraic sets will be useful below.

Lemma 11.4. *Let A be an algebraic subset of V. Then A is irreducible if and only if any two non-empty open subsets of A intersect non-trivially.*

Proof. Let U_1, U_2 be two non-empty open subsets of A. Then $(A \setminus U_1) \cup (A \setminus U_2) = A \setminus (U_1 \cap U_2)$. If A is irreducible, this implies that $U_1 \cap U_2 \neq \emptyset$. Conversely, if there are non-empty open sets U_1, U_2 such that $U_1 \cap U_2 = \emptyset$, the above shows that A is not irreducible. $\qquad\square$

We next give the following geometric application of Lemma 3.14.

Lemma 11.5. *Suppose the map $\psi : V = \mathbb{C}^n \to W = \mathbb{C}^m$ is defined by $\psi(v) = (f_1(v), \ldots, f_m(v))$ where $f_i \in \mathbb{C}[X_1, \ldots, X_n] = \mathbb{C}[V]$. Assume that $\psi(V)$ is not contained in any proper closed subset of W. If $\mathbb{C}[V]$ is finitely generated as a module over $\psi^*(\mathbb{C}[W])$, where $\psi^* : f \mapsto f \circ \psi$, then $\psi(V) = W$.*

Proof. The assumption about $\psi(V)$ is equivalent to saying that $\psi^* : \mathbb{C}[W] \to \mathbb{C}[V]$ is injective, so we shall think of $\mathbb{C}[W]$ as a subalgebra of $\mathbb{C}[V]$. Now the points of V correspond to maximal ideals of $\mathbb{C}[V]$, and the map φ may be thought of as $M \mapsto M \cap \mathbb{C}[W]$. But by Lemma 3.14, every maximal ideal of $\mathbb{C}[W]$ is realised thus. The result follows. $\qquad\square$

Proposition 11.6.

(i)   *The algebraic set $A \subseteq V$ is irreducible if and only if $\mathcal{I}(A)$ is a prime ideal of S, or equivalently $\mathbb{C}[A]$ is an integral domain.*

(ii)   *Every prime ideal of S is algebraic.*

(iii) *Every algebraic subset $A \subseteq V$ has a finite decomposition*

(11.7) $$A = A_1 \cup A_2 \cup \cdots \cup A_c,$$

where the A_i are irreducible. Any two irredundant decompositions (meaning that no component A_i may be omitted) of this form are the same, up to the order of the components A_i.

Proof. Suppose $\mathcal{I}(A)$ is not prime. Then there are polynomials $f_1, f_2 \in S$ such that $f_i \notin \mathcal{I}(A)$, but $f_1 f_2 \in \mathcal{I}(A)$. For $i = 1, 2$ let

$$A_i = \mathcal{V}(f_i, \mathcal{I}(A)) = \{\, a \in A \mid f_i(a) = 0 \,\}.$$

Then $A = A_1 \cup A_2$ is a decomposition of A, so that A is not irreducible. Hence if A is irreducible then $\mathcal{I}(A)$ is prime. Conversely, suppose that $A = A_1 \cup A_2$ is a decomposition of A. By Proposition 11.1 (vi) we have $\mathcal{V}(\mathcal{I}(A_1)\mathcal{I}(A_2)) = \mathcal{V}(\mathcal{I}(A_1)) \cup \mathcal{V}(\mathcal{I}(A_2)) = A$ and hence $\mathcal{I}(A) \subseteq \mathcal{I}(A_1)\mathcal{I}(A_2)$, so that $\mathcal{I}(A)$ is not prime. Since it is standard that I is a prime ideal of S if and only if S/I is an integral domain, this completes the proof of (i).

The assertion (*ii*) is a consequence of the 'strong form' of Hilbert's Nullstellensatz [**142**, p. 378 *et seq.*].

To prove that A has a decomposition of the type given, suppose the contrary. Then A is reducible, so that $A = B_1 \cup B_2$ where the B_i are proper algebraic subsets of A, and at least one of the B_i, say B_1, does not have a finite decomposition into irreducibles. Repeating this argument, we obtain an infinite sequence $I_1 = \mathcal{I}(A) \subsetneq I_2 = \mathcal{I}(B_1) \subsetneq I_3 \subsetneq \cdots$. But since S is Noetherian, any ascending sequence of ideals terminates, which is a contradiction. Thus A has a decomposition as a finite union of irreducibles. To prove uniqueness, suppose $A = A_1 \cup \cdots \cup A_c = B_1 \cup \cdots \cup B_d$ are two irredundant decompositions of A into irreducibles.

Then $B_1 = A \cap B_1 = (A_1 \cap B_1) \cup \cdots \cup (A_c \cap B_1)$. By the irreducible nature of B_1, there is an index j such that $A_j \cap B_1 = B_1$, i.e. $B_1 \supseteq A_j$. The same argument shows that there is an index k such that $A_j \supseteq B_k$. Thus $B_1 \supseteq A_j \supseteq B_k$, whence by irredundancy $k = 1$ and $B_1 = A_j$. The same argument shows that for each i, $B_i = A_{\pi i}$, where π is a permutation of $1, \ldots, c$. Again by irredundancy, $c = d$ and the B_i coincide with the A_j up to order. $\qquad\square$

Definition 11.8. For an algebraic set A, the canonical irreducible subsets A_i of Proposition 11.6 are known as the *irreducible components* of A.

Example 11.9. Let W be a linear subspace of V. Then W is an irreducible algebraic subset of V. For if $X_1, \ldots, X_s$ is a basis of $\{ L \in V^* \mid L(W) = 0 \}$, then $\mathcal{I}(W)$ is the ideal generated by $X_1, \ldots, X_s$. If $X_1, \ldots, X_s, \ldots, X_n$ is a basis of V^*, it is easily seen that $\mathbb{C}[W] \simeq \mathbb{C}[X_{s+1}, \ldots, X_n]$, which is a polynomial ring and an integral domain.

Define the length of the sequence

$$A = A_0 \supsetneq A_1 \supsetneq \cdots \supsetneq A_r$$

of irreducible algebraic sets to be r. For an irreducible algebraic subset $A \subseteq V$, we define the dimension dim A to be maximal length of a sequence of the above form. Note that any such sequence corresponds to the associated ascending sequence of prime ideals

$$S/\mathcal{I}(A) = \mathbb{C}[A] \supsetneq P_1 \supsetneq \cdots \supsetneq P_r,$$

of $\mathbb{C}[A]$, where $P_j = \mathcal{I}(A_j)/\mathcal{I}(A)$. Note that for each j, $\mathbb{C}[A_j] = \mathbb{C}[A]/P_j$.

It follows that for an irreducible algebraic subset A of V,

(11.10) $$\dim A = \dim \mathbb{C}[A],$$

in the sense of the definition A.1.

We note that Corollary A.10 has the following geometric interpretation.

Proposition 11.11. *Let $A \subseteq V$ be an irreducible algebraic set and let f be a non-zero element of $\mathbb{C}[A]$. Then each irreducible component of the zero locus $\mathcal{V}(f) \cap A$ of f has dimension dim $A - 1$.*

Proposition 11.12. *For an irreducible algebraic subset A of V, $\dim A$ is equal to the transcendence degree of the quotient field $\mathbb{C}(A)$ of the coordinate ring $\mathbb{C}[A]$ of A.*

Proof. If the transcendence degree of $\mathbb{C}(A)$ is d, then by Noether's normalisation lemma (Theorem A.12), there are algebraically independent elements $x_1, x_2, \ldots, x_d$ of $\mathbb{C}[A]$ such that $\mathbb{C}[A]$ is integral over the polynomial ring $\mathbb{C}[x_1, \ldots, x_d]$. But by [**142**, p. 209], $\mathbb{C}[x_1, x_2, \ldots, x_d]$ is a unique factorisation domain, and hence by [**142**, Prop. 1.7, p. 337] is integrally closed. Hence we may apply Lemma A.13 to deduce that $\dim A = \dim \mathbb{C}[x_1, x_2, \ldots, x_d]$. This reduces the problem to the case where $\mathbb{C}[A] = \mathbb{C}[x_1, x_2, \ldots, x_d]$, i.e. when $A = \mathbb{C}^d$. For this case, we need to show that $\dim \mathbb{C}[x_1, x_2, \ldots, x_d] = d$.

This is easily seen by induction on d. If $d = 0$ the result is trivial. Since $\mathbb{C}^{d-1}$ is realised as the zero locus of x_d in $\mathbb{C}^d$, it follows from Proposition 11.11 that $\dim \mathbb{C}[x_1, x_2, \ldots, x_{d-1}] = \dim \mathbb{C}[x_1, x_2, \ldots, x_d] - 1$. The result now follows by induction. $\qquad\square$

Clearly in the example above, the dimension $\dim W$ of W as an algebraic set coincides with its dimension as a vector space.

2. Eigenspaces of elements of reflection groups

Let $d > 0$ be a positive integer, and fix a primitive d^{th} root of unity $\zeta \in \mathbb{C}$. We shall assume that the basic invariants $F_1, \ldots, F_n$ are ordered so that $d \mid d_i$ if and only if $1 \leq i \leq a$, where $d_i = \deg F_i$, and $a = a(d)$ is the number of degrees d_i which are divisible by d. This notation is consistent with that in Theorem 10.33. Recall that for $g \in G$, $V(g, \zeta)$ denotes the ζ-eigenspace of g on V.

Proposition 11.13. [**201**, Proposition 3.2] *Let $\mathcal{V}(F_i)$ be the algebraic subset*

$$\mathcal{V}(F_i) := \{ v \in V \mid F_i(v) = 0 \}$$

of V for $i = 1, 2, \ldots, n$, and write

$$V(d) = \mathcal{V}(F_{a+1}) \cap \mathcal{V}(F_{a+2}) \cap \cdots \cap \mathcal{V}(F_n).$$

Then

(i)   *we have*

$$V(d) = \bigcup_{g \in G} V(g, \zeta);$$

(ii)   *the distinct maximal subspaces of the form $V(g, \zeta)$ are the irreducible components of $V(d)$.*

Proof. Let $A = \bigcup_{g \in G} V(g, \zeta)$. Clearly $v \in A$ if and only if v and ζv are in the same G-orbit on V, and by Theorem 3.5, this is true precisely when $F(v) = F(\zeta v)$ for each invariant polynomial $F \in S^G$, and in our case this is equivalent to $F_i(v) = F_i(\zeta v)$ for $i = 1, \ldots, n$. But F_i is homogeneous of degree d_i, whence $F_i(\zeta v) =$

$\zeta^{d_i} F_i(v)$. Since $\zeta^{d_i} = 1$ if and only if d divides d_i, it follows that $v \in A$ if and only if $v \in \mathcal{V}(F_i)$ for all i such that $d \nmid d_i$, which proves (i).

Statement (ii) follows directly from Proposition 11.6 (ii). $\qquad\square$

Proposition 11.14. *Let E be one of the maximal ζ-eigenspaces in Proposition 11.13. Then*

(i) $\quad \dim E = a = a(d);$

(ii) $\quad$ *the restrictions of $F_1, \ldots, F_a$ to E are algebraically independent;*

(iii) $\quad$ *any two maximal ζ-eigenspaces $V(g, \zeta)$ are conjugate by an element of G.*

Proof. First observe that $v \in \mathcal{V}(F_1) \cap \cdots \cap \mathcal{V}(F_n)$ if and only if $F_i(v) = 0 = F_i(0)$ for each i. By Theorem 3.5 this implies that v is in the same G-orbit as 0, i.e. that $v = 0$. Now if A is an irreducible component of $\mathcal{V}(F_n) \cap \cdots \cap \mathcal{V}(F_i)$, then by Proposition 11.11, any irreducible component of $A \cap \mathcal{V}(F_{i-1})$ has dimension at least $\dim A - 1$. It follows that *each* irreducible component of $\mathcal{V}(F_n) \cap \cdots \cap \mathcal{V}(F_i)$ has dimension precisely $i - 1$, which proves (i).

Consider the restriction to $V(d)$ of the orbit map $\varphi : V \to \mathbb{C}^n$. If $X_1, X_2, \ldots, X_n$ are the coordinate functions on $\mathbb{C}^n$, then $V(d) = \varphi^{-1}(\mathcal{V}(X_{a+1}, X_{a+2}, \ldots, X_n))$, and by Theorem 3.15, $\varphi : V(d) \to \mathbb{C}^{a(d)}$ is surjective, where $\mathbb{C}^a$ is identified with $\mathcal{V}(X_{a+1}, X_{a+2}, \ldots, X_n)$. Thus if $E_1, \ldots, E_c$ are the maximal subspaces $V(g, \zeta)$, $\mathbb{C}^a = \varphi(E_1) \cup \varphi(E_2) \cup \cdots \cup \varphi(E_c)$. If the restrictions of $F_1, \ldots, F_a$ to E_i are algebraically dependent, then $\varphi(E_i) \subseteq C_i$, for some proper Zariski-closed subset C_i of $\mathbb{C}^a$. Since $\mathbb{C}^a$ is irreducible, it follows that there is some $E = E_i$ such that the restrictions of $F_1, \ldots, F_a$ to E_i are algebraically independent.

The map $\varphi : E \to \mathbb{C}^a$ now satisfies the conditions of Lemma 11.5, since if $F \in \mathbb{C}[E]$, take any $F' \in \mathbb{C}[V]$ whose restriction to E is F; then $\prod_{g \in G}(t - gF') \in \mathbb{C}[t, F_1, \ldots, F_n]$, and restricting to E shows that F is integral over $\varphi^*(\mathbb{C}[\mathbb{C}^a])$. Hence by Lemma 11.5, $\varphi : E \to \mathbb{C}^a$ is surjective. Now for *any* maximal eigenspace $E' = E_j$, there is a point $e' \in E'$ such that e' is in no other maximal eigenspace E_k, since E' is not a union of proper subspaces. By the surjectivity of $\varphi : E \to \mathbb{C}^a$ just proved, we have $e \in E$ such that $\varphi(e) = \varphi(e')$, which implies that $e' = ge$ for some $g \in G$.

Since G permutes the irreducible components of $V(d)$ this shows that $gE = E'$, and hence that G acts transitively on the set of components $E_1, \ldots, E_c$, and the proof of (ii) and (iii) is complete. $\qquad\square$

3. Reflection subquotients of unitary reflection groups

In this section we shall give an elementary treatment of a theory which was originally developed by Springer [201] for regular elements of reflection groups and then more generally by Lehrer and Springer [154, 155] for arbitrary maximal eigenspaces. The treatment here is more elementary than the original work, in that it does not depend on the notion of intersection multiplicities of varieties. It is partly based on the work of Lehrer and Michel [**151**].

3.1. The main theorem.

Theorem 11.15. *Let G be a unitary reflection group in $V = \mathbb{C}^n$ and let $\zeta \in \mathbb{C}$ be a primitive d^{th} root of unity, where d is a positive integer. Let $g \in G$ be such that $E := V(g, \zeta)$ is maximal among the ζ-eigenspaces of elements of G. Let N, C respectively be the stabiliser and pointwise stabiliser of the space E; i.e. $N = \{x \in G \mid xE \subseteq E\}$, and $C = \{x \in G \mid xe = e \text{ for all } e \in E\}$. Then C is a normal subgroup of N and the subquotient N/C acts faithfully as a unitary reflection group on E. Moreover the subquotient is uniquely determined up to conjugacy by an element of G.*

Proof. It follows from Proposition 11.14 (*iii*) that any two such eigenspaces E are conjugate under G, which justifies the last sentence of the statement.

To prove that N/C is a reflection group in E, we shall use the characterisation of Shephard and Todd (Theorem 4.19). We have from Proposition 11.14 (*ii*) that (in the notation of *loc. cit.*) the basic invariants $F_1, \dots, F_{a(d)}$ of G which have degree divisible by d restrict to a set of algebraically independent polynomials on E which have degrees $d_1, \dots, d_{a(d)}$. It follows from Theorem 4.19 that

$$(11.16) \qquad\qquad |N/C| \leq \prod_{i=1}^{a(d)} d_i,$$

and we have equality if and only if N/C acts as a reflection group in E.

However, recall the equation from Corollary 10.39:

$$\sum_{g \in G} T^{d(g, \zeta)} = \prod_{i \text{ such that } \zeta^{d_i} = 1} (T + d_i - 1) \prod_{i \text{ such that } \zeta^{d_i} \neq 1} d_i.$$

The coefficient of $T^{a(d)}$ on the left side is the number of elements g with $d(g, \zeta) = \dim V(g, \zeta) = a(d)$, which is $|C|$ times the number of maximal ζ-eigenspaces; i.e. by Proposition 11.14, the number of eigenspaces xE with $x \in G$. This number is $|C||G|/|N|$, and equating it to the coefficient of $T^{a(d)}$ on the right side, we obtain

$$\frac{|C||G|}{|N|} = \prod_{i=a(d)+1}^{n} d_i,$$

and using the fact that $|G| = d_1 \cdots d_{a(d)} d_{a(d)+1} \cdots d_n$, we deduce that we have equality in (11.16), and the theorem is proved. $\qquad\square$

Note that, by Theorem 9.44, C is the reflection group generated by the reflections in hyperplanes which contain E. The following statement is an immediate consequence of the proof of the above theorem.

Corollary 11.17. *If $F_1, \ldots, F_{a(d)}$ are the homogeneous invariants of G whose degree is divisible by d, then the restrictions of $F_1, \ldots, F_{a(d)}$ to E is a set of basic homogeneous invariants for the reflection group N/C on E. In particular, the degrees of N/C are precisely those degrees of G which are divisible by d.*

Proposition 11.18. *Let G be a unitary reflection group in V and let d be a positive integer. Let ζ_1 and ζ_2 be primitive d^{th} roots of unity, and let N_1/C_1 and N_2/C_2 be subquotients arising from ζ_1 and ζ_2 respectively as in Theorem 11.15. Then there is an isomorphism $N_1/C_1 \to N_2/C_2$ which is realised by conjugation in G. That is, there is an element $y \in G$ such that $N_2 = yN_1y^{-1}$ and $C_2 = yC_1y^{-1}$.*

Proof. First observe that if $E = V(g_1, \zeta_1)$ is a maximal ζ_1-eigenspace, then since $\zeta_2 = \zeta_1^a$ for some integer a, we have $E = V(g_1^a, \zeta_2)$. Hence any maximal ζ_1-eigenspace is contained in a ζ_2-eigenspace, and by symmetry, the converse is true. It follows that the sets of maximal ζ_1- and ζ_2-eigenspaces coincide.

Let E_1 and E_2 be maximal ζ_1- and ζ_2-eigenspaces respectively. By the above remark and Proposition 11.14 *(iii)*, there is an element $y \in G$ such that $E_2 = yE_1$. It is then clear that $N_2 = yN_1y^{-1}$ and $C_2 = yC_1y^{-1}$. $\qquad\square$

From Proposition 11.18 it is apparent that the reflection group N/C arising from Theorem 11.15 depends only on the *order d* of the root of unity ζ, up to conjugacy.

We therefore make the following definition.

Definition 11.19. For any reflection group G in V and integer d, the reflection group N/C of Theorem 11.15 is denoted $G(d)$. It is a reflection group in any maximal ζ_d-eigenspace of V, whose basic degrees are those degrees of G which are divisible by d.

Lemma 11.20. *With the above notation, if e divides d then $G(d) \cong G(e)(d)$.*

Proof. Write $E(d)$, $N(d)$, $\zeta(d), \ldots$ for the objects E, N, $\zeta, \ldots$ used in defining $G(d)$. It is clear that we may take $E(d)$ and $E(e)$ to be such that $E(d) \subseteq E(e)$, since $\zeta(d)^{d/e} = \zeta(e)$ for appropriate choice of roots of unity. Then $G(e) = N(e)/C(e)$, $C(d) \subseteq C(e)$ and $E(d)$ is a maximal $\zeta(d)$-eigenspace for $G(e)$ acting on $E(e)$. A coset $nC(e) \in G(e)$ $(n \in N(e))$ fixes $E(d) \subseteq E(e)$ setwise if and only if its elements lie in $N(d) \subseteq G$. Hence $G(e)(d)$ is isomorphic to a subgroup of $G(d)$. Since the two groups have the same cardinality (equal to $\prod_{i=1}^{a(d)} d_i$) the result follows. $\qquad\square$

4. Regular elements

4.1. Basic concepts. The following important definitions of regular vectors, group elements and integers are due to Springer [**201**] (cf. Definition 9.42).

Definition 11.21. Let G be a unitary reflection group in $V = \mathbb{C}^n$.

(i)   A vector $v \in V$ is *regular* if it lies on no reflecting hyperplane of a reflection in G.

(ii)   An element $g \in G$ is *regular* if g has an eigenspace $V(g, \zeta)$ which contains a regular vector v.

(iii)   The integer d is *regular* for G if there is a primitive d^{th} root of unity ζ and an element $g \in G$ such that $V(g, \zeta)$ contains a regular vector.

In (ii) and (iii) above, the eigenspace $V(g, \zeta)$ is referred to as a *regular eigenspace*.

Note that by Theorem 9.44, v is regular if and only if v is fixed by no non-trivial element of G. The following lemma is easy but useful.

Lemma 11.22. *The following are equivalent.*

(i)   *The eigenspace $E := V(g, \zeta)$ is regular.*
(ii)   *The pointwise stabiliser $C = G_E$ of E is trivial.*
(iii)   *E is not contained in any reflecting hyperplane of G.*

Proof. It is trivial that (i) implies (ii) and that (ii) implies (iii). To see that (iii) implies (i), note that G has only finitely many reflecting hyperplanes, and if E were not regular, then E would be the union of the intersections of these hyperplanes with E, which by hypothesis (iii) are proper subspaces of E. Since this is impossible, E is regular. $\qquad\square$

Remark 11.23. If d is a regular number and $E = V(g, \zeta)$ is a regular eigenspace for a primitive d^{th} root of unity ζ, then $V(g^i, \zeta^i)$ contains E, and is therefore also regular. In particular, d is a regular number for G if and only if there is a regular ζ-eigenspace for each dth root of unity ζ.

Theorem 11.24. *Using the notation of Definition 11.21 above, suppose that $g \in G$ and $\zeta \in \mathbb{C}$ are such that $E := V(g, \zeta)$ contains a regular vector. Then*

(i)   *$V(g, \zeta)$ is a maximal ζ-eigenspace for G;*
(ii)   *the stabiliser $N(E) = \{ x \in G \mid xE \subseteq E \}$ coincides with the centraliser $C_G(g) = \{ x \in G \mid xg = gx \}$ of g in G;*
(iii)   *the centraliser $C_G(g)$ acts on E as a unitary reflection group. Its basic degrees coincide with those degrees of G which are divisible by d, where d is the order of ζ.*

Proof. Suppose that v is a regular vector in E, and let $V(h, \zeta)$ be a maximal ζ-eigenspace such that $E \subseteq V(h, \zeta)$. Then $gv = hv = \zeta v$, so $g^{-1}h$ fixes v, and by Theorem 9.44 it follows that $g = h$, and so $E = V(g, \zeta) = V(h, \zeta)$, proving (i).

Now if $x \in N(E)$, then $xv \in E$, whence $gxv = \zeta xv = xgv$. Again applying Theorem 9.44, it follows that $gx = xg$, i.e. that $x \in C_G(g)$. Conversely, if $x \in C_G(g)$, then since in general $xV(g, \zeta) = V(xgx^{-1}, \zeta)$, it follows that $xE = E$, whence (ii).

Finally, note that as already observed, the pointwise stabiliser $C = C(E)$ is trivial, since any element of C must fix the regular vector v. Thus the group N/C of Theorem 11.15 coincides in this case with $C_G(g)$, and all the assertions in (iii) follow from that Theorem. □

If $V(g, \zeta)$ is a regular eigenspace, and the order of ζ is d, we say that g is a *d-regular* element of G.

Corollary 11.25. *Any d-regular element of G has order d. If ζ is a primitive d^{th} root of unity and g, g' are such that both $V(g, \zeta)$ and $V(g', \zeta)$ are regular eigenspaces, then g and g' are conjugate in G.*

Proof. If $V(g, \zeta)$ contains a regular vector v, and ζ has order d, then g^d fixes v, and hence is trivial. Thus g clearly has order d. If g' is any element such that $\dim V(g', \zeta) = \dim V(g, \zeta)$, then by Theorem 11.24 (i), both $V(g', \zeta)$ and $V(g, \zeta)$ are maximal ζ-eigenspaces and hence by Theorem 10.33 (iii) are conjugate in G. Thus there is an element $x \in G$ such that $V(g', \zeta) = xV(g, \zeta) = V(xgx^{-1}, \zeta)$, and so $g' = xgx^{-1}$ by regularity. □

Definition 11.26. The element $g \in G$ is *pseudoregular* if there is an element $\zeta \in \mathbb{C}$ such that $V(g, \zeta)$ is a maximal ζ-eigenspace for the reflection group G.

Remark 11.27. Note that the first assertion of the previous theorem is that a regular element of G is pseudoregular; of course the converse is generally false.

4.2. A criterion for regularity. From Theorem 11.24 it is clear that an integer d is regular in the sense of Definition 11.21 precisely when a maximal ζ_d-eigenspace contains a regular vector, where ζ_d is any primitive d^{th} root of unity. We conclude this section with a necessary and sufficient criterion for d to be regular in terms of the exponents and coexponents of G (see Definitions 10.24 and 10.27).

Theorem 11.28. *Let G be a unitary reflection group in $\mathbb{C}^n$, and suppose the degrees of the basic invariants of G are $d_1, \ldots, d_n$. Let the codegrees of G be $d_1^*, \ldots, d_n^*$ (Definition 10.27). Given an integer d, let $a(d) = |\{ i \mid d$ divides $d_i \}|$, and let $b(d) = |\{ i \mid d$ divides $d_i^* \}|$.*
Then $a(d) \leq b(d)$, and d is a regular number for G if and only if $a(d) = b(d)$.

Proof. The fact that $a(d) \leq b(d)$ is immediate from Theorem 10.33.

Let ζ be a primitive d^{th} root of unity, and consider equation (10.42). The highest power of T which occurs on the left side is $T^{a(d)}$ by Proposition 11.14 (i), and by (10.42), its coefficient is 0 if and only if $a(d) \neq b(d)$. Let us compute the coefficient of $T^{a(d)}$ on the left side of (10.42).

If E is a maximal ζ-eigenspace, then the set $S(E) := \{ g \in G \mid E = V(g, \zeta) \}$ is a coset xC of the pointwise stabiliser C of E. Hence the contribution to the coefficient of $T^{a(d)}$ of the terms corresponding to $S(E)$ is

$$s(E) := (-1)^{a(d)}(-\zeta)^n \det g \sum_{x \in C} \det x,$$

where g is any element of $S(E)$. Moreover if E' is another maximal ζ-eigenspace, then by the conjugacy result (Proposition 11.14 (iii)), $S(E')$ is conjugate to $S(E)$ in G, and so $s(E') = s(E)$, so that the coefficient of $T^{a(d)}$ on the left side of (10.42) is $|G/N|s(E)$, where N is the stabiliser of E. But this coefficient is zero if and only if $\sum_{x \in C} \det x = 0$. Since C is a reflection group, this happens if and only if C is non-trivial.

It follows that C is trivial if and only if $a(d) = b(d)$, which is the required assertion, since by Lemma 11.22, d is a regular number if and only if C is trivial. $\square$

When G is a real reflection group, this criterion takes the following simple form.

Corollary 11.29. *Suppose the reflection group G is real, i.e. that there is a basis of V with respect to which the matrices representing G are real. Let $d_1, d_2, \ldots, d_n$ be the basic degrees of G. Then d is a regular number for G if and only if*

$$|\{ i \mid d_i \equiv 0 \pmod{d} \}| = |\{ i \mid d_i \equiv 2 \pmod{d} \}|.$$

Proof. Let $m_1, m_2, \ldots, m_n$ be the exponents (i.e. V-exponents) of G. Since in this case V is self-dual, i.e. $V \simeq V^*$, the V-exponents coincide with the V^*-exponents. It follows that the degrees $d_i = m_i + 1$ while the codegrees $d_i^* = m_i - 1 = d_i - 2$. The result is now clear from Theorem 11.28. $\square$

5. Properties of the reflection subquotients

In this section we shall determine the reflecting hyperplanes of the groups N/C of Theorem 11.15 and study its properties. In particular, we shall prove that if G is irreducible, then so is N/C. This makes the identification of these subquotients straightforward.

5.1. Fixed points. Let $\mathcal{A} = \mathcal{A}(G)$ be the set of reflecting hyperplanes of G, i.e. the set of hyperplanes in V which are the fixed points of reflections in G. Let $\mathcal{L} = \mathcal{L}_G$ be the lattice of intersections of the hyperplanes in $\mathcal{A}$, ordered by the reverse of inclusion. A set $H_1, \ldots, H_p$ of hyperplanes of V will be called *independent* if the corresponding linear forms $L_{H_1}, \ldots, L_{H_p}$ are linearly independent in V^*. Equivalently, if $H_i^\perp = \mathbb{C}v_i$ for vectors $v_i \in V$ ($i = 1, \ldots, p$) the v_i are linearly independent. The reader is referred to the book of Orlik and Terao [**182**] for general properties of collections of hyperplanes, and their complements. The next lemma is an easy exercise in linear algebra, whose proof is left to the reader.

Lemma 11.30. *The hyperplanes $H_1, \ldots, H_p$ are independent if and only if the codimension of $H_1 \cap \cdots \cap H_p$ in V is p.*

As usual, we denote by $\operatorname{Fix} g$ the subspace of g-fixed points of V. The next result summarises the essential properties of these subspaces which we shall require.

Proposition 11.31.

(i)   Let $X = H_1 \cap \cdots \cap H_p$, where $H_1, \ldots, H_p$ is an independent set of hyperplanes of G. For each i let r_i be a reflection in H_i. Then writing $y = r_1 r_2 \cdots r_p$, we have $\operatorname{Fix} y = X$.

(ii)   For any element $g \in G$, we have

$$\operatorname{Fix} g = \bigcap_{H \in \mathcal{A}(G),\ \operatorname{Fix} g \subseteq H} H.$$

(iii)   The set of subspaces $\{ \operatorname{Fix} g \mid g \in G \}$ coincides with $\mathcal{L}_G$.

Proof. Since each r_i fixes X pointwise, it is clear that $\operatorname{Fix} y \supseteq X$. We prove the converse by induction on p; it is clearly true for $p = 1$. Let $v \in \operatorname{Fix} y$, and for each i, let v_i be a non-zero vector in $H_i^\perp$. Then since $r_1 r_2 \cdots r_p v = v$, we have $r_2 \cdots r_p v = r_1^{-1} v$ and by Lemmas 3.17 and 1.6, it follows that $r_2 \cdots r_p v = r_1^{-1} v = v + \sum_{i=2}^{p} \lambda_i v_i = v + \lambda_1 v_1$ for certain scalars $\lambda_1, \ldots, \lambda_p \in \mathbb{C}$. By the linear independence of the v_i it follows that v is fixed by r_1^{-1} and by $r_2 \cdots r_p$. The statement (i) is now clear by induction.

For (ii), note that since g fixes $\operatorname{Fix} g$ pointwise, by Theorem 9.44 g is a product of reflections in hyperplanes H which contain $\operatorname{Fix} g$. Since each of these reflections fixes $\bigcap_{H \in \mathcal{A}(G),\ \operatorname{Fix} g \subseteq H} H$ pointwise, it follows that $\operatorname{Fix} g \supseteq \bigcap_{H \in \mathcal{A}(G),\ \operatorname{Fix} g \subseteq H} H$. The reverse inequality is trivial, and (ii) follows.

To see (iii), observe first that by (ii), each subspace $\operatorname{Fix} g$ is an intersection of hyperplanes in $\mathcal{A}(G)$ and hence lies in $\mathcal{L}_G$. The converse follows from Corollary 9.14, but may be seen directly as follows. Given $X \in \mathcal{L}_G$, it is clear (*cf.* Lemma 11.30) that X may be expressed as the intersection of an independent set $H_1, \ldots, H_p$ of hyperplanes in $\mathcal{A}$. But then (i) shows that $X = \operatorname{Fix} y$ for some element $y \in G$, which completes the proof of the Proposition. $\qquad\square$

5.2. Reflecting hyperplanes in eigenspaces. The following observation will be useful below.

Lemma 11.32. *Suppose $x \in G$ and that u, v are elements of V such that $xu = \mu u$ and $xv = \nu v$. Let F be a homogeneous polynomial of degree d in S^G. In the notation of Lemma 9.33, if $D_u F(v) \neq 0$, then $\mu \nu^{d-1} = 1$.*

Proof. By Lemma 9.33, we have $x(D_u F)(v) = D_{xu}(xF)(v) = \mu D_u F(v)$. On the other hand, we have also

$$x(D_u F)(v) = D_u F(x^{-1} v) = D_u F(\nu^{-1} v) = \nu^{-(d-1)} D_u F(v),$$

since $D_u F$ is homogeneous of degree $d - 1$. The lemma follows immediately. $\qquad\square$

The next result identifies the reflecting hyperplanes of the group $G(d) = N/C$. It was proved in the regular case in [146] and [77] and in the general case in [154].

Theorem 11.33. *Maintain the notation of Theorem 11.15. The reflecting hyperplanes of the group $\overline{N} := N/C$ on $E := V(g, \zeta)$ are the intersections with E of the reflecting hyperplanes of G which do not contain E.*

Proof. The element $x \in N$ acts on E as a reflection if and only if Fix $x \cap E$ has codimension 1 in E. If this is the case, then since, by Proposition 11.31 (*iii*), Fix x is an intersection of hyperplanes in $\mathcal{A}(G)$, it follows that Fix $x \cap E = H \cap E$, for some reflecting hyperplane H of G.

Conversely, suppose H is a reflecting hyperplane of G which does not contain E. Let $F_1, \ldots, F_a, F_{a+1}, \ldots, F_n$ be a set of basic invariants for G, ordered so that d divides d_i (the degree of F_i) if and only if $i \leq a$. We have seen (Corollary 11.17) that the restrictions to E of $F_1, \ldots, F_a$ form a set of homogeneous basic invariants for the reflection group $\overline{N}$ on E. Let $v_1, \ldots, v_a, v_{a+1}, \ldots, v_n$ be a basis of g-eigenvectors of G, such that $gv_i = \zeta_i v_i$ and $\zeta_i = \zeta$ if and only if $1 \leq i \leq a$. Let $X_1, \ldots, X_n$ be the dual basis of V^*. By Theorem 9.8 applied to $\overline{N}$, if $\Pi_N := \dfrac{\partial(F_1, \ldots, F_a)}{\partial(X_1, \ldots, X_a)}$, then the zero set of Π_N is the union of the reflecting hyperplanes of $\overline{N}$ acting on E. To show that $E \cap H$ is a reflecting hyperplane of $\overline{N}$, it therefore suffices to prove

(11.34)   *the polynomial Π_N vanishes on $E \cap H$,*

for if this is true, then $E \cap H$ is contained in the union of the reflecting hyperplanes of $\overline{N}$ on E, and hence must be one of them.

To prove (11.34), first observe that for $j > a$, $i \leq a$, and $v \in E$, we have $gv_j = \zeta_j v_j$, $gv = \zeta v$, and $\zeta_j \zeta^{d_i - 1} = \zeta_j \zeta^{-1} \neq 1$. It therefore follows from Lemma 11.32, taking $x = g$, that

(11.35) $D_{v_j} F_i(v) = 0.$

Next, take $u \in V$ such that $H^\perp = \mathbb{C}u$. Since $E \not\subseteq H$, $u \notin E^\perp$, whence $u = \sum_{i=1}^n \mu_i v_i$, and there is some $i \leq a$ such that $\mu_i \neq 0$. Moreover if r_H is a reflection in H, then $r_H u = \mu u$, with $\mu \neq 1$. Therefore, again applying Lemma 11.32 this time with $x = r_H$, we see that for any element $v \in H$, since $r_H v = v$, we have, for *any* i,

(11.36) $D_u F_i(v) = 0,$

which may be written in our chosen coordinates as

(11.37) $\displaystyle\sum_{j=1}^n \mu_j \frac{\partial F_i}{\partial X_j}(v) = 0.$

Now let v be any element of $E \cap H$. From (11.35), we see that for $j > a$ and $i \leq a$, $\frac{\partial F_i}{\partial X_j}(v) = 0$. Thus for such v, equation (11.37) may be thought of as an $a \times a$ system of linear equations for $\mu_1, \ldots, \mu_a$. Since this has a non-zero solution, it follows that $\frac{\partial(F_1, \ldots, F_a)}{\partial(X_1, \ldots, X_a)}(v) = 0$. This completes the proof of (11.34), and hence, of the theorem.   □

5.3. Irreducibility. Our objective is to prove the following result, which first appeared in [155].

Theorem 11.38. *Maintain the notation of Theorem 11.15. If G is an irreducible reflection group in V, then $\overline{N} := N/C$ is an irreducible reflection group on the eigenspace $E := V(g, \zeta)$.*

Proof. Let $\mathcal{A}(G)$ be the set of reflecting hyperplanes of G. If $\overline{N}$ does not act irreducibly on E, then $E = E_1 \oplus E_2$, where the E_i are non-zero subspaces of E, and any reflection in $\overline{N}$ contains either E_1 or E_2 in its fixed point set. By Theorem 11.33, the reflecting hyperplanes of $\overline{N}$ in E are precisely the intersections with E of those hyperplanes in $\mathcal{A}(G)$ which do not contain E. We therefore deduce that any hyperplane in $\mathcal{A}(G)$ contains E_1, E_2, or both. If every hyperplane in $\mathcal{A}(G)$ contains E, then G fixes the non-trivial subspace E pointwise, and hence by irreducibility, $G = 1$, and we have nothing to prove. Thus not every hyperplane in $\mathcal{A}(G)$ contains E.

Let $\mathcal{A}_1$ be the subset of $\mathcal{A}(G)$ consisting of those hyperplanes which contain E_1, but do not contain E_2, and let $\mathcal{A}_2 = \mathcal{A}(G) \setminus \mathcal{A}_1$. Let G_1 and G_2 respectively be the (reflection) groups generated by $\mathcal{A}_1$ and $\mathcal{A}_2$. Then we claim that G_1 is a normal subgroup of G. To see this, let $H \in \mathcal{A}_1$ and let r be a reflection with fixed point space H. It suffices to show that if s is a reflection of G, then $srs^{-1} \in G_1$. If $\mathrm{Fix}(s) = K \in \mathcal{A}_1$, then $s \in G_1$, and this is clear. Thus we may take $s = s_K$, with $\mathrm{Fix}(s) = K \in \mathcal{A}_2$. Then srs^{-1} is a reflection with hyperplane sH. But $sH \supseteq E_2$ if and only if $H \supseteq s^{-1}E_2 = E_2$, the latter equality holding because $K \supseteq E_2$. Hence $sH \in \mathcal{A}_1$, and we have proved that G_1 is a normal subgroup of G.

Next, observe that $G = G_1 G_2$, since the right side contains all the reflections which generate G. It follows that the fixed point space V^{G_1} is stabilised by G. By the irreducibility of G, this is impossible, whence $\overline{N}$ acts irreducibly on E, as asserted.   □

5.4. Codegrees. It was shown in Corollary 11.17 that the basic degrees of $G(d)$ are precisely those basic degrees of G which are divisible by d. The situation concerning the codegrees is not quite as straightforward, but the following result may be found in [155, Theorem C, p. 1176].

Theorem 11.39. *Let G be a reflection group, d a positive integer, and let $G(d)$ be the subquotient defined in Definition 11.19.*

(i)   *The codegrees of $G(d)$ are precisely those codegrees of G which are divisible by d if and only if d is a regular number for G.*

(ii)   *The codegrees of $G(d)$ form a sub-multiset of those codegrees of G which are divisible by d.*

We shall not give a detailed proof of Theorem 11.39, but here is a sketch. The codegrees of G may be conveniently described (see Definition 10.27) as follows. Let $u_1, \ldots, u_n$ be a homogeneous S^G-basis of $(S \otimes V)^G$. If $m_i^* = \deg u_i$, the codegrees are $\{d_1^*, \ldots, d_n^*\}$, where $d_i^* = m_i^* - 1$. Since $G(d)$ contains an element which operates on E (a maximal ζ_d-eigenspace) as scalar multiplication by $\exp\left(\frac{2\pi i}{d}\right)$, clearly d is regular for $G(d)$, and hence by Theorem 11.28 each codegree of $G(d)$ is divisible by d.

Define a map $\rho_0 : (S_V \otimes V)^G \to (S_E \otimes E)^N$ (where $S_V = S(V^*)$ etc.) as follows. Let $p_E : V \to E$ denote orthogonal projection to E and write Res_E^V for the restriction map on polynomial functions : $S_V \to S_E$. If $\rho = \mathrm{Res}_E^V \otimes p_E : S_V \otimes V \to S_E \otimes E)$, then clearly ρ is N-equivariant, and hence takes $(S_V \otimes V)^G$ into $(S_E \otimes E)^N$. Denote the restriction to $(S_V \otimes V)^G$ of ρ by ρ_0. Clearly ρ and ρ_0 are degree preserving. Hence, if we had

(11.40) $\rho_0 : (S_V \otimes V)^G \to (S_E \otimes E)^N$ is surjective,

the statement (ii) of the theorem would follow.

But (11.40) is proved when d is regular in [**155**, Theorem D, p. 1181]. Hence if d is regular, the coexponents of $G(d)$ form a sub-multiset of the set of codegrees of G which are divisible by d. But by Theorem 11.28, the number of these is just $a(d)$, whence the proof of the 'if' part of (i) is complete.

To prove the converse, we require (ii), which is proved in [**155**, §5] using some case by case arguments. Given (ii), if the codegrees of $G(d)$ are precisely those codegrees of G which are divisible by d, it follows that $b(d) = a(d)$, and hence by Theorem 11.28, that d is regular.

Remarks 11.41.

1. The proof of (11.40) for the regular case uses algebraic geometric arguments, and is case free. Hence from the above sketch, it is clear that the proof of the 'if' statement of Theorem 11.39 (i) is independent of any case by case analysis.

2. The statement (11.40) is unresolved in general. It is known (unpublished) for $G = \mathrm{Sym}(n)$ by some arguments involving Cremona varieties.

6. Eigenvalues of pseudoregular elements

In this section we derive information about the eigenvalues of pseudoregular elements on various spaces on which they act. The exposition in this section combines ideas from [**154**], [**148**] and [**28**].

6.1. The setup. Throughout this section, g will be a pseudoregular element of the unitary reflection group G, and $\zeta \in \mathbb{C}$ will be such that $E := V(g, \zeta)$ is maximal among the ζ-eigenspaces of elements of G. As usual, we denote by N and C respectively the setwise and pointwise stabilisers of E.

We fix a basis $v_1, v_2, \ldots, v_n$ of V with the following properties. First, the v_i are eigenvectors of g, so that $gv_i = \zeta_i v_i$ for $i = 1, \ldots, n$, and $\zeta_i = \zeta$ if and only if $1 \leq i \leq a(d) = a$. Thus $v_1, \ldots, v_a$ is a basis of E. Secondly, we note that C is the reflection (parabolic) subgroup of G generated by the reflections in hyperplanes which contain E. Moreover, C is normalised by g, for if $c \in C$ and $e \in E$, then $g^{-1}cge = g^{-1}c\zeta e = e$. The subspace V^C of C-fixed points of V contains E, and further is stable under g, for if $v \in V^C$ and $c \in C$, then $cgv = gg^{-1}cgv = gv$, so that $gv \in V^C$. Hence V^C is a sum of g-eigenspaces, and we may choose notation so that $v_1, \ldots, v_s$ is a basis of V^C; of course $s \geq a$.

Given the basis $v_1, \ldots, v_n$ of V described above, we denote by $X_1, \ldots, X_n$ the dual basis of V^*.

Note that when g is regular, which is an important special case of the results of this section, C is trivial, and so $V^C = V$.

We shall fix a set $\{F_1, F_2, \ldots, F_n\} \subset S$ of basic invariants for G, and similarly a set of basic invariants $\{P_1, P_2, \ldots, P_n\} \subset S$ for C. Note that g normalises the reflection group C since if $c \in C$ and $v \in E$, then $gcg^{-1}v = gc\zeta^{-1}v = g\zeta^{-1}v = v$, so that $gcg^{-1} \in C$. It follows from (9.46), which is proved in the course of the proof of Theorem 9.44, that the P_i may be chosen as eigenfunctions for g, i.e. so that

$$(11.42) \qquad gP_i = \varepsilon_i P_i \quad \text{for } i = 1, 2, \ldots, n.$$

Finally, we denote by $d_i(G)$ and $d_j(C)$ the degrees of F_i and P_j respectively. The unqualified degree d_i will always indicate $d_i(G)$.

6.2. Eigenvalues on V^C. We begin by proving

Proposition 11.43. *Let notation be as in §6.1 above. There is a permutation π of $\{1, 2, \ldots, n\}$ such that for $i = 1, \ldots, n$ we have*

$$\varepsilon_i = \zeta^{d_{\pi i}(G) - d_i(C)}.$$

Proof. This is very similar to that of Theorem 9.44. Consider the commutative diagram

$$\mathbb{C}^n \, (= V) \xrightarrow{\;\omega_C\;} \mathbb{C}^n \, (= V/C)$$

$$(11.44) \qquad\qquad {\scriptstyle \omega_G} \searrow \qquad \downarrow {\scriptstyle \omega_{G,C}}$$

$$\mathbb{C}^n \, (= V/G)$$

in which ω_G and ω_C are the orbit maps corresponding to the reflection groups G and C respectively, and $\omega_{G,C}$ may be thought of as follows. Since each F_i is

C-invariant, there are unique polynomials $Q_i(y_1, \ldots, y_n) \in \mathbb{C}[y_1, \ldots, y_n]$ such that $F_i = Q_i(P_1, \ldots, P_n)$. These polynomials are the coordinate functions of the map $\omega_{G,C}$.

As in the proof of Theorem 9.44, we apply the chain rule to deduce that

$$(11.45) \qquad \frac{\partial(F_1, \ldots, F_n)}{\partial(P_1, \ldots, P_n)}(X_1, \ldots, X_n) = \kappa \prod_{H \in \mathcal{A}(G),\, H \not\supseteq E} L_H^{e_H - 1},$$

where $\kappa \in \mathbb{C}$ is a non-zero constant. Hence this determinant does not vanish on E, and there is an element $v \in E$ and a permutation π of $\{1, 2, \ldots, n\}$ such that

$$\prod_{i=1}^{n} \frac{\partial F_{\pi i}}{\partial P_i}(v) \neq 0,$$

i.e.

$$(11.46) \qquad \frac{\partial F_{\pi i}}{\partial P_i}(v) \neq 0 \text{ for each } i.$$

Now let us evaluate $\left(g \dfrac{\partial F_j}{\partial P_i} \right)(v)$ in two different ways. On the one hand, as in the proof of Theorem 9.44, we have $g \dfrac{\partial F_j}{\partial P_i} = \varepsilon_i^{-1} \dfrac{\partial F_j}{\partial P_i}$, whence $g \left(\dfrac{\partial F_j}{\partial P_i} \right)(v) = \varepsilon_i^{-1} \dfrac{\partial F_j}{\partial P_i}(v)$. However we also have $\left(g \dfrac{\partial F_j}{\partial P_i} \right)(v) = \dfrac{\partial F_j}{\partial P_i}(g^{-1}v) = \dfrac{\partial F_j}{\partial P_i}(\zeta^{-1}v) = \zeta^{d_j(G) - d_i(C)} \dfrac{\partial F_j}{\partial P_i}(v)$, the last equality following because $\dfrac{\partial F_j}{\partial P_i}$ is homogeneous of degree $d_j(G) - d_i(C)$ in the X_k.

It follows that if $\dfrac{\partial F_j}{\partial P_i}(v) \neq 0$, then $\varepsilon_i = \zeta^{d_j(G) - d_i(C)}$. Combining this statement with (11.46), we obtain the Proposition. $\qquad \square$

Corollary 11.47. *There are integers $1 \leq i_1 < i_2 < \cdots < i_s \leq n$ such that the eigenvalues of g on V^C are $\{\, \zeta^{1-d_{i_j}} \mid j = 1, \ldots, s \,\}$.*

Proof. Since $X_1, \ldots, X_s$ are invariant under C, the basic invariant polynomials $P_1, \ldots, P_n$ for C may be chosen so that $P_i = X_i$ for $i = 1, \ldots, s$. Thus $\zeta_i = \varepsilon_i^{-1}$; writing $\pi j = i_j$ in the statement of Proposition 11.43, and taking into account that $d_i(C) = 1$ for $1 \leq i \leq s$, we obtain the result. $\qquad \square$

6.3. Eigenvalues on M-covariants. Let M be any G-module of finite dimension p. We have seen (Chapter 10 §1) that $(S \otimes_{\mathbb{C}} M^*)^G$ is free as an S^G-module, and that the M-exponents $q_j(M) = q_j^G(M)$, $j = 1, \ldots, p$ of G are the degrees of any set of S^G-basis elements of $((S \otimes_{\mathbb{C}} M^*)^G$. The same applies to the reflection subgroup C of G, whose M-exponents we write as $q_j^C(M)$.

Let $\psi \in (S \otimes_{\mathbb{C}} M^*)^C$. For any element $c \in C$, we have $cg\psi = g(g^{-1}cg)\psi = g\psi$ since $g^{-1}cg \in C$. Hence g acts on $(S \otimes_{\mathbb{C}} M^*)^C$. In analogy with the situation in (9.46), we show that an S^C-basis $\psi_1, \ldots, \psi_p$ of $(S \otimes_{\mathbb{C}} M^*)^C$ may be chosen so that

$$(11.48) \qquad g\psi_i = \eta_i \psi_i \text{ for } i = 1, \ldots, p.$$

To see this, let $\mathcal{H}_C$ be the space of C-harmonic polynomials in S. Then $h \in \mathcal{H}_C$ if and only if $D_P h = 0$ for all homogeneous elements $P \in S(V)^C$ of positive degree. Thus if $h \in \mathcal{H}_C$ and P is such an element of $S(V)^C$, $D_P(gh) = gD_{g^{-1}P}(h) = 0$, since $g^{-1}S(V)^C \subseteq S(V)^C$. Thus $g\mathcal{H}_C \subseteq \mathcal{H}_C$, so that g acts on $\mathcal{H}_C \otimes M^*$, and hence on $(\mathcal{H}_C \otimes M^*)^C$. Moreover g is evidently semisimple, and preserves degree, so that $(\mathcal{H}_C \otimes M^*)^C$ has a $\mathbb{C}$-basis of homogeneous g-eigenvectors. Such a basis is automatically a free S^C-basis of $(S \otimes_{\mathbb{C}} M^*)^C$.

Theorem 11.49. *Given a reflection group G in V, let M, g, ζ, C etc. be as above. There is a permutation π of $\{1, \ldots, p\}$ such that for $i = 1, \ldots, p$, we have*

$$\eta_i = \zeta^{q^G_{\pi i}(M) - q^C_i(M)}.$$

Proof. Let $y_1, \ldots, y_p$ be a basis of M^* and let $\psi_1, \ldots, \psi_p$ be an S^C-basis which satisfies (11.48). Then there are unique elements $k_{ij} \in S$ such that $\psi_i = \sum_{j=1}^p k_{ij} \otimes y_j$, and by Theorem 10.13,

$$(11.50) \qquad \det(k_{ij}) = \kappa \prod_{H \in \mathcal{A}(G),\, H \supseteq E} L_H^{C(H,M)},$$

where $\kappa \in \mathbb{C}$ is a non-zero constant, and $C(H, M)$ depends only on M as G_H-module, where G_H is the point stabiliser of H. Note also that $\deg k_{ij} = q_i^C(M)$ for each i, j such that $k_{ij} \neq 0$.

Now if $\varphi_1, \ldots, \varphi_p$ is any homogeneous S^G-basis of $(S \otimes_{\mathbb{C}} M^*)^C$, then writing $\varphi_i = \sum_{j=1}^p h_{ij} \otimes y_j$, we also have

$$(11.51) \qquad \det(h_{ij}) = \kappa' \prod_{H \in \mathcal{A}(G)} L_H^{C(H,M)},$$

where $\kappa' \in \mathbb{C}$ is a non-zero constant, and $C(H, M)$ is as above.

Moreover since the φ_i are C-invariant, there are *unique* polynomials $f_{ij} \in S^C$ $(1 \leq i, j \leq p)$ such that for each i, $\varphi_i = \sum_{j=1}^p f_{ij} \psi_j$, and from (11.50) and (11.51) it is clear that

$$(11.52) \qquad \det(f_{ij}) = \kappa'' \prod_{H \in \mathcal{A}(G),\, H \not\supseteq E} L_H^{C(H,M)},$$

for some non-zero constant κ''.

It follows that there is an element $v \in E$ such that $\det(f_{ij})(v) \neq 0$, and hence that for some permutation π of $\{1, \ldots, p\}$, we have for $i = 1, \ldots, p$,

$$(11.53) \qquad f_{\pi i, i}(v) \neq 0.$$

In analogy with the proof of Proposition 11.43, we next evaluate $(gf_{ij})(v)$ in two different ways. First observe that since $\varphi_i = \sum_{j=1}^{p} f_{ij}\psi_j$ is invariant under G, we have $g\varphi_i = \varphi_i = \sum_{j=1}^{p}(gf_{ij})(g\psi_j) = \sum_{j=1}^{p}\eta_j(gf_{ij})\psi_j$, whence by the uniqueness of the f_{ij}, we see that for $1 \le i, j \le p$, we have $gf_{ij} = \eta_j^{-1}f_{ij}$. Hence

$$(gf_{ij})(v) = \eta_j^{-1}f_{ij}(v).$$

But the relation $\varphi_i = \sum_{j=1}^{p} f_{ij}\psi_j$ implies that

$$h_{i\ell} = \sum_{j=1}^{p} f_{ij}k_{j\ell}.$$

Since $h_{i\ell}$ and $k_{j\ell}$ are homogeneous of degree $q_i^G(M)$ and $q_j^C(M)$ respectively, it follows, since $h_{i\ell}$ is non-zero for some ℓ, that f_{ij} is homogeneous of degree $q_i^G(M) - q_j^C(M)$.

Therefore $(gf_{ij})(v) = f_{ij}(g^{-1}v) = f_{ij}(\zeta^{-1}v) = \zeta^{-(q_i^G(M) - q_j^C(M))}f_{ij}(v)$, and hence if $f_{ij}(v) \ne 0$, we deduce that $\eta_j = \zeta^{q_i^G(M) - q_j^C(M)}$. Taking (11.53) into account, the proof is now complete.   □

6.4. The cases $M = V$ and $M = V^*$. We next apply Theorem 11.49 in the special cases when $M = V$ and $M = V^*$. Consider first the case $M = V$. In this case we may take the basis ψ_i of $(\mathcal{H}_C \otimes V^*)^C$ to be such that $\psi_i = 1 \otimes X_i$ for $i = 1, \ldots, s$. Then $g\psi_i = g(1 \otimes X_i) = \zeta_i^{-1}1 \otimes X_i$ for $1 \le i \le s$, since $gX_i = \zeta^{-1}X_i$, and $\deg \psi_i = q_i^C(V) = 0$ for $i = 1, \ldots, s$. The next statement is now immediate from Theorem 11.49.

Corollary 11.54. *Maintaining the above notation, there are integers $i_1, \ldots, i_s$, with $1 \le i_1 < \cdots < i_s \le n$ such that for $j = 1, \ldots, s$,*

$$\eta_j = \zeta^{-m_{i_j}},$$

where $m_1, \ldots, m_n$ are the exponents (see Definition 10.24) of G.

Next we turn to the case $M = V^*$. Here, dually to the previous case, we may take $\psi_i = 1 \otimes v_i$ for $i = 1, \ldots, s$, so that $g\psi_i = \eta_i\psi_i$ and $\deg \psi_i = 0$ for $i = 1, \ldots, s$. Again applying Theorem 11.49, the next statement is clear.

Corollary 11.55. *There are integers $k_1, \ldots, k_s$ with $1 \le k_1 < \cdots < k_s \le n$ such that for $i = 1, \ldots, s$,*

$$\eta_i = \zeta^{m_{k_i}^*},$$

where $m_1^*, \ldots, m_n^*$ are the coexponents (see Definition 10.27) of G.

6.5. The regular case. We have seen (Remark 11.27) that regular elements are pseudoregular. When the element g of this section is regular, the results above may be strengthened. In particular, since C is the trivial group in this case, $V^C = V$ and $s = n$. The results above may therefore be summarised as follows.

Theorem 11.56. *Let g be a regular element of the reflection group G, and let $V(g, \zeta)$ be a regular eigenspace.*

(i)   *The eigenvalues of g on V are $\{\, \zeta^{1-d_i} \mid i = 1, \dots, n \,\}$, where the d_i are the basic degrees of G.*

(ii)   *Let M be any G-module of finite dimension p. Then the eigenvalues of g on M are $\{\, \zeta^{-q_i(M)} \mid i = 1, \dots, p \,\}$, where the $q_i(M)$ are the M-exponents of G.*

(iii)   *The eigenvalues of g on V are $\{\, \zeta^{-m_i} \mid i = 1, \dots, n \,\}$, where the m_i are the exponents of G.*

(iv)   *The eigenvalues of g on V^* are $\{\, \zeta^{m_i^*} \mid i = 1, \dots, n \,\}$, where the m_i^* are the coexponents of G.*

Proof. Part (i) is immediate from Corollary 11.47, given that $V^C = V$. For (ii), note that since $C = 1$, we may take ψ_i in (11.48) to be $1 \otimes y_i$ ($i = 1, \dots, p$), where $y_1, \dots, y_p$ is a basis of Y^*, which may be assumed to consist of g-eigenvectors. Then the eigenvalues of g on M are $\eta_1^{-1}, \dots, \eta_p^{-1}$, and the assertion (ii) now follows from Theorem 11.49.

The statements (iii) and (iv) are immediate from Corollaries 11.54 and 11.55 respectively.   □

Remark 11.57. Note that comparing (i) and (iii) of the above theorem gives another proof of the fact (Proposition 10.23) that the exponents m_i of G satisfy $m_i = d_i - 1$, where the d_i are the basic degrees.

The above result also leads to the following remarkable numerical fact concerning exponents and coexponents.

Theorem 11.58. *Let G be a unitary reflection group in V, and let G have exponents $m_1, \dots, m_n$ and coexponents $m_1^*, \dots, m_n^*$. Let d be any integer. Then the following are equivalent.*

(i)   *The number of m_i congruent to -1 modulo d is equal to the number of m_i^* which are congruent to 1 modulo d.*

(ii)   *The sequences $-m_1, \dots, -m_n$ and $m_1^*, \dots, m_n^*$ are equal modulo d, up to order.*

(iii)   *d is a regular number for G.*

Proof. The equivalence of (i) and (iii) is the statement of Theorem 11.28. But (ii) trivially implies (i), and from Theorem 11.56 (iii) and (iv), we see that (iii) implies (ii).   □

CHAPTER 12

Reflection cosets and twisted invariant theory

Suppose G is a unitary reflection group in $V = \mathbb{C}^n$. Let γ be a linear transformation of finite order of V, such that $\gamma G = G\gamma$. The coset γG is then called a *reflection coset*, and many results of the last three chapters may be proved in the slightly more general context of reflection cosets, using the same arguments as those given. The advantage of this approach is that in the study of parabolic subgroups and reflection quotients, reflection cosets arise naturally even if one starts with a reflection group, so the context of reflection cosets is a natural one for the subject. Write $\langle G, \gamma \rangle$ for the group generated by G and γ. Since this group is finite, we may assume that $\langle G, \gamma \rangle$, and hence γG, leave invariant a positive definite hermitian form on V, i.e. that $\gamma G \subset U(V)$.

If $\langle G, \gamma \rangle$ acts on a complex vector space W, then since γ normalises G, γ stabilises the space W^G of G-invariant elements of W, and an element $w \in W^G$ is called a *(twisted) invariant* of the coset γG if $\gamma w = \varepsilon w$ for some $\varepsilon \in \mathbb{C}$. Much of the development of this chapter is taken from [155] and [28].

1. Reflection cosets

In this section, we shall show why it suffices to consider the case of reflection cosets γG where γ has finite order, and we shall state twisted versions of some of the key results concerning invariants and covariants. The proofs of the twisted versions are the same as those of the original results, and we generally leave it to the reader to provide details.

Proposition 12.1. *Let G be a unitary reflection group in $V = \mathbb{C}^n$. Let $GL(V)$ be the group of invertible linear transformations of V, and let ξ be an element of the normaliser $\mathcal{N} := N_{GL(V)}(G)$. Then there is an element $\gamma \in \mathcal{N}$ such that γ induces the same automorphism of G as ξ, and γ has finite order.*

Proof. It follows from Theorem 1.27 that there is a unique decomposition $V = V^G \oplus V_1 \oplus \cdots \oplus V_r$, where V^G denotes the space of G-fixed points of V, and each V_i is an irreducible G-submodule of V. Correspondingly, $G = G_1 \times \cdots \times G_r$, where G_i acts as an (irreducible) reflection group in V_i, and acts trivially on all other summands in the above decomposition of V. Of course distinct pairs (G_i, V_i) may be isomorphic as reflection groups.

We wish to show that for any element $\xi \in \mathcal{N}$, there is an element of finite order in $\xi C_{GL(V)}(G)$. We may assume that ξ acts trivially on V^G. Since $\mathcal{N}/C_{GL(V)}(G)$ is isomorphic to a subgroup of the (finite) automorphism group of G, it is finite. Hence there is a power ξ^m of ξ such that $\xi^m \in C_{GL(V)}(G)$. Since the V_i are the non-trivial irreducible G-submodules of V, ξ permutes the V_i; let the induced permutation of $\{1, \ldots, r\}$ be π. If $D(\lambda_1, \ldots, \lambda_r)$ denotes the element of $GL(V)$ which acts as λ_i on V_i and trivially on V^G, then $D(\lambda_1, \ldots, \lambda_r)$ commutes with ξ if and only if $\lambda_i = \lambda_{\pi i}$ for each i. By Schur's Lemma $\xi^m = D(\mu_1, \ldots, \mu_r)$, and since ξ commutes with ξ^m, $\mu_i = \mu_{\pi i}$ for each i, and there is an element $D(\lambda_1, \ldots, \lambda_r)$ which commutes with ξ such that $D(\lambda_1, \ldots, \lambda_r)^m = D(\mu_1, \ldots, \mu_r)^{-1}$. So $\xi D(\lambda_1, \ldots, \lambda_r)$ has finite order. $\qquad\square$

In view of the above proposition, we shall confine our attention to cosets γG where γ has finite order, and normalises G. The term 'reflection coset' will be reserved for this situation. Observe that one consequence is that γ is semisimple; that is, γ is diagonalisable in any representation of $\widetilde{G} := \langle G, \gamma \rangle$. Note that our term 'reflection coset' is synonymous with the term 'reflection datum' used by Broué et al. (see [37, Chapter IV]). Our terminology reflects the more general context in which we develop the subject. We shall see in Appendix C that every connected reductive algebraic group with an $\mathbb{F}_q$-structure determines a reflection coset. This gives rise to one set of applications of the material in this chapter.

We shall make repeated use of the argument used to prove (9.46), which relies on the semisimplicity of γ, to deduce that the space of G-invariants on various spaces has a basis of γ-eigenvectors. We begin with twisted analogues of some of the results of the previous three chapters.

2. Twisted invariant theory

Let γG be a reflection coset in V, as defined in the previous section, and write $\widetilde{G} := \langle G, \gamma \rangle$.

Proposition 12.2.

(i) The spaces J and $\mathcal{H}$ of G-invariant polynomials and of G-harmonic polynomials respectively are both stable under the normaliser $\mathcal{N} = N_{GL(V)}(G)$, and hence a fortiori under γ. In particular the Chevalley isomorphism of Corollary 9.39 is one of $\mathcal{N}$-modules.

(ii) Let M be any $\widetilde{G}$-module of dimension r. There is a homogeneous basis $u_1, \ldots, u_r$ of $(\mathcal{H} \otimes M^*)^G$ such that for each i, $\gamma u_i = \varepsilon_i(M) u_i$, for some root of unity $\varepsilon_i(M) \in \mathbb{C}$.

(iii)   *The (multi)set of pairs* $(\deg u_i, \varepsilon_i(M))$ *depends only on the coset* γG *and the module* M.

Proof. First observe that for any polynomial $F \in J = S^G$ and elements $g \in G$ and $\nu \in \mathcal{N}$, we have $g(\nu F) = (g\nu)F = (\nu(\nu^{-1}g\nu))F = \nu F$ since $\nu^{-1}g\nu \in G$. Thus $\nu F \in J$, whence the first statement. The same proof shows that dually, $S(V)^G \simeq \mathcal{D}_S^G$ (see Chapter 9 §5) is also stable under $\mathcal{N}$, whence so is the ideal $F(V)$ of $S(V)$ which is generated by $S(V)^G$. By Lemma 9.36, $\mathcal{H} = F(V)^\perp$, so that it follows from the $GL(V)$-invariance of the pairing $S(V) \otimes S \to \mathbb{C}$ (Corollary 9.34) that $\mathcal{H}$ is stable under $\mathcal{N}$. The fact that the map $J \otimes \mathcal{H} \to S$ respects the $\mathcal{N}$-action is evident from its definition. This proves (i).

To see (ii), we begin by observing that since γ acts on $\mathcal{H}$, so does $\widetilde{G} := \langle G, \gamma \rangle$, whence $\widetilde{G}$ acts on $\mathcal{H} \otimes M$. Moreover the same argument as above shows that the subspace $(\mathcal{H} \otimes M)^G$ is stable under γ. Since γ preserves degrees in this space, it acts on each homogeneous component $(\mathcal{H}_d \otimes M)^G$. Since γ has finite order, it is a semisimple transformation of this finite dimensional space, and therefore for each degree d, $(\mathcal{H}_d \otimes M)^G$ has a basis of eigenvectors for γ, which is the assertion (ii).

The last statement (iii) is immediate from (ii) since the relevant multiset is characterised as the set of eigenvalues of γ on the homogeneous components of $\mathcal{H} \otimes M$, labelled with the degree of the space in which the eigenvalue occurs. Since G acts trivially on these spaces, these eigenvalues depend only on the coset, and not on the representative γ. $\square$

We have already met the multiset $\{\deg u_i\}$ in Chapter 10; its elements are the M-exponents of G. The pairs above therefore generalise the case $\gamma = 1$.

With notation as in Proposition 12.2, the multiset $\{\varepsilon_i(M)\}$ will be called the *set of M-factors* of γG.

Example 12.3. Consider the special case $M = V$ above. By Proposition 10.22, if $F_1, \ldots, F_n$ is a set of basic invariants for G, then $dF_1, \ldots, dF_n$ is a basis of $(S \otimes V^*)^G$, where d is the derivative defined in Lemma 10.21. Since $gd = dg$ for any element $g \in GL(V)$ (Lemma 10.21 (iii)), it follows that the V-factors of γG are defined as follows. There is a set $\{F_i\}$ of basic invariants for G such that $\gamma F_i = \varepsilon_i F_i$ for each i. The ε_i are the V-factors of γG.

Dually, there is (§4.2) a homogeneous basis $u_1, \ldots, u_n$ of $(S \otimes V)^G$ such that $\gamma u_i = \varepsilon_i^* u_i$ for each i, and the ε_i^* are the V^*-factors of γG. We shall reserve the notation ε_i and ε_i^* for these special cases of the M-factors of γG.

Definition 12.4. The V-factors ε_i and V^*-factors ε_j^* will be referred to as simply *factors* and *cofactors* respectively.

Our next result includes a connection between the examples $M = V$ and $M = V^*$.

Let M be a $\widetilde{G}$-module, as above. Then M^* is similarly the contragredient $\widetilde{G}$-module, and we have an S-bilinear map $(S \otimes M) \times (S \otimes M^*) \to S$ given by

$$(12.5) \qquad (F \otimes y, F' \otimes y') \mapsto \langle y, y' \rangle FF'.$$

This map clearly respects the action of $\widetilde{G}$ on both sides, and it therefore restricts to a map

$$(12.6) \qquad (S \otimes M)^G \times (S \otimes M^*)^G \to S^G,$$

which we denote by $(-, -)_M$.

Let $u_1, \ldots, u_r$ and $w_1, \ldots, w_r$ denote S^G-bases of $(S \otimes M)^G$ and $(S \otimes M^*)^G$ respectively, such that for each i, $\gamma u_i = \varepsilon_i(M^*) u_i$ and $\gamma w_i = \varepsilon_i(M) w_i$.

Definition 12.7. The (M)-discriminant matrix Q_M is the matrix with (i, j)-entry $(u_i, w_j)_M \in S^G$. Its determinant $\Delta_M := \det Q_M$ is the 'M-discriminant' of G.

Recall (Proposition 10.7) that if $y_1, \ldots, y_r$ is any basis of M, and $u_i = \sum_j A_{ij} \otimes y_j$, then $\Pi_{M^*} := \det(A_{ij})$ is uniquely determined up to a non-zero constant multiple, and for $x \in G$, $x \Pi_{M^*} = \det_{M^*}(x) \Pi_{M^*}$.

Lemma 12.8. *Up to a non-zero constant multiple, we have*

$$\Pi_M \Pi_{M^*} = \Delta_M.$$

Proof. Let $y_1, \ldots, y_r$ be a basis of M, and let $y_1^*, \ldots, y_r^*$ be the dual basis of M^*. Then for $i = 1, \ldots, r$, we have $u_i = \sum_j A_{ij} \otimes y_j$ and $w_i = \sum_j A_{ij}^* \otimes y_j^*$ for unique homogeneous elements $A_{ij}, A_{ij}^* \in \mathcal{H} \subseteq S$. Hence $(u_i, w_j)_M = \sum_{k=1}^r A_{ik} A_{jk}^*$, and so $Q_M = (A_{ij})(A_{ij}^*)^t$, and taking determinants, the result follows. $\square$

In the special case where $M = V$, the polynomial $\Delta_V = \prod_{H \in \mathcal{A}(G)} L_H^{e_H}$ (see §4) is known as the *discriminant polynomial* of G. We shall denote it simply by Δ.

If Q is any matrix with entries in S and $g \in GL(V)$, then $\det gQ = g \det Q$. It follows that since

$$(12.9) \qquad \gamma(u_i, w_j)_M = \varepsilon_i(M^*) \varepsilon_j(M) (u_i, w_j)_M,$$

we have

$$\gamma \Delta_M = \det \gamma Q_M = \prod_i \varepsilon_i(M^*) \prod_j \varepsilon_j(M) \Delta_M,$$

so that all entries of Q_M are twisted invariants of γG, as is Δ_M.

3. Eigenspace theory for reflection cosets

In this section we shall state some twisted generalisations of the main theorems of Chapter 10, and show how they are used to prove twisted versions of the eigenspace theory developed in the last chapter. In particular we shall give a criterion for d to be a regular number for the reflection coset γG. We begin with a twisted version of Theorem 10.29.

3.1. Twisted Poincaré series.

Theorem 12.10. *Let γG be a unitary reflection coset in $V \simeq \mathbb{C}^n$ and let M be a $\langle G, \gamma \rangle$-module of dimension r, which is amenable (see also Definition 10.14) as a G-module. Let $\varepsilon_1(M), \ldots, \varepsilon_r(M)$ and $\varepsilon_1, \ldots, \varepsilon_n$ be the M-factors (see Proposition 12.2) and V-factors of γG respectively (recall that this means in particular that $\gamma F_i = \varepsilon_i F_i$ for some set $\{F_i\}$ of basic invariants of G). Let t and u be indeterminates. Then*

$$(12.11) \qquad |G|^{-1} \sum_{g \in G} \frac{\det_{M^*}(1 - ug\gamma)}{\det_{V^*}(1 - tg\gamma)} = \frac{\prod_{j=1}^r (1 - \varepsilon_j(M) u t^{q_j(M)})}{\prod_{i=1}^n (1 - \varepsilon_i t^{d_i})},$$

where the $q_j(M)$ are the M-exponents of G (see Definition 10.1), and the d_i are the basic degrees of G.

Proof. This is entirely analogous to that of Theorem 10.29. The crux is to compute the bigraded trace $\sum_{i,p \geq 0} \mathrm{Tr}(\gamma, (S \otimes \Lambda M^*)_{i,p}^G t^i u^p$ in two different ways. First, we apply a variant of Molien's Theorem (Theorem 4.13) which arises from the elementary observation that if W is a finite dimensional $\langle G, \gamma \rangle$-module, then γ stabilises the subspace W^G of G-fixed points of W, and

$$\mathrm{Tr}(\gamma, W^G) = |G|^{-1} \sum_{g \in G} \mathrm{Tr}(\gamma g, W).$$

The second evaluation simply uses the decomposition

$$(S \otimes \Lambda M^*)^G \simeq J \otimes (\mathcal{H} \otimes \Lambda M^*)^G,$$

already noted in the proof of Theorem 10.29, as well as the definition of the M-factors.

Because the details are very similar to the computation in the proof of Theorem 10.29, we leave the reader to fill them in. $\qquad \square$

Note that in view of Corollary 10.16, Theorem 12.10 applies in particular to the Galois twisted modules $M = V^\sigma$ and $M = (V^*)^\sigma$, where $\sigma \in \mathrm{Gal}(\overline{\mathbb{Q}}/\mathbb{Q})$ (see also Corollary 10.17).

For applications to eigenspace theory, we wish to prove a twisted analogue of Theorem 10.33. We shall need the following notation, which generalises that of Theorem 10.33. Let γG be a reflection coset in V. For $\zeta \in \mathbb{C}$ a root of unity, define

$$A(\zeta) = \{ i \mid \zeta^{d_i} \varepsilon_i = 1 \},$$

where d_i is the degree of the basic invariant corresponding to $\varepsilon_i (= \varepsilon_i(V))$.

If $\sigma \in \mathrm{Gal}(\overline{\mathbb{Q}}/\mathbb{Q})$ define

$$B_\sigma(\zeta) = \{\, j \mid \varepsilon_j(V^\sigma)\zeta^{\sigma + q_j(V^\sigma)} = 1\,\},$$

where the $q_j(M)$ are the M-exponents of G, and the $\varepsilon_j(M)$ are the M factors of γG. Denote by $a(\zeta)$ and $b_\sigma(\zeta)$ the respective cardinalities of these two sets.

Since all Galois twists of V are amenable (see Corollary 10.17), we may apply Theorem 10.33 when $M = V^\sigma$, for $\sigma \in \mathrm{Gal}(\overline{\mathbb{Q}}/\mathbb{Q})$. The result is as follows. Recall that for any linear transformation x of V, $V(x, \xi)$ denotes the ξ-eigenspace of x, and $d(x, \xi)$ denotes $\dim V(x, \xi)$.

Theorem 12.12. *Let γG be a reflection coset in V, and $\sigma \in \mathrm{Gal}(\overline{\mathbb{Q}}/\mathbb{Q})$. For any root of unity $\zeta \in \mathbb{C}$, let $A(\zeta)$ and $B_\sigma(\zeta)$ be the sets defined above. Then $a(\zeta) \leq b_\sigma(\zeta)$, and we have the following identity in the polynomial ring $\mathbb{C}[T]$. We write $\det'(\varphi)$ for the product of the non-zero eigenvalues of a linear transformation φ. If $a(\zeta) \neq b_\sigma(\zeta)$, then*

$$(12.13) \qquad \sum_{x \in \gamma G} T^{d(x,\zeta)} {\det'}_V (1 - \zeta^{-1} x)^{\sigma-1} = 0,$$

while if $a(\zeta) = b_\sigma(\zeta)$, then the left side is equal to

$$(12.14)$$

$$\prod_{j \in B_\sigma(\zeta)} (T + q_j(V^\sigma)) \prod_{j \notin B_\sigma(\zeta)} \left(1 - \varepsilon_j(V^\sigma)^{-1}\zeta^{-\sigma - q_j(V^\sigma)}\right) \prod_{i \notin A(\zeta)} \frac{d_i}{1 - \varepsilon_i^{-1}\zeta^{-d_i}}.$$

Proof. The proof is an exact analogue of the proof of Theorem 10.33. We begin with the complex conjugate of the formula (12.11) with $M = V^\sigma$, take ζ as above, and make precisely the same change of variable as in that proof, namely $T = \frac{1 - u\zeta^\sigma}{1 - t\zeta}$. The left side of equation (12.11) then has no pole at $t = \zeta^{-1}$. This yields the inequality $a(\zeta) \leq b_\sigma(\zeta)$ by inspection of the right side.

Evaluating the limit as $t \to \zeta^{-1}$ of both sides of (12.11), we obtain the stated formula. $\qquad\square$

Remark 12.15. Since γ may be replaced by $\zeta\gamma$ in the statement above (where ζ is any root of unity), it suffices to consider the case $\zeta = 1$ in both the statement and proof of Theorem 12.12. The coset γG is then replaced by $\zeta\gamma G$, the coexponents remain the same, but for any $\langle G, \gamma \rangle$-module M, given that ζ acts on M as scalar multiplication by $f(\zeta)$ (for some $f(\zeta) \in \mathbb{C}$), the $\zeta\gamma G$-exponents are just the $q_j(M)$, while the M-factors are $f(\zeta)^{-q_j(M)} f(\zeta)^{-1} \varepsilon_j(M)$, where $\varepsilon_j(M)$ is the relevant M-factor of γG.

It should also be noted that since we assume that all eigenvalues of elements of γG are roots of unity in $\mathbb{C}$, it is always the case that V^* is Galois conjugate to V as a representation of $\tilde{G}$.

Just as in Chapter 10, we have the following two special cases of Theorem 12.12.

Corollary 12.16. *Maintain the notation of Theorem 12.12. Then*

$$\sum_{x \in \gamma G} T^{d(x,\zeta)} = \prod_{j \in A(\zeta)} (T + m_j) \prod_{j \notin A(\zeta)} d_j,$$

where the $m_j = d_j - 1$ are the exponents (see (10.24)) of G.

This is simply the case $\sigma = \text{id}$ of Theorem 12.12. Note that when $\sigma = \text{id}$, $B_\sigma(\zeta) = \{ j \mid \varepsilon_j(V^\sigma)\zeta^{\sigma + q_j(V^\sigma)} = 1 \} = \{ j \mid \varepsilon_j \zeta^{1+m_j} = 1 \} = A(\zeta)$.

The case where σ is complex conjugation similarly leads to the exact analogue for cosets of Corollary 10.41. If σ is complex conjugation, $V^\sigma \simeq V^*$. In analogy with the notation of Definition 10.27, write $\varepsilon_i(V^*) = \varepsilon_i^*$, and $B_\sigma(\zeta) = B(\zeta) = \{ j \mid \varepsilon_j^* \zeta^{d_j^*} = 1 \}$. Denote the cardinalities of $A(\zeta)$ and $B(\zeta)$ by $a(\zeta)$ and $b(\zeta)$ respectively. The proof of the next result is precisely the same as that of Corollary 10.41.

Corollary 12.17. *With the above notation, we have $a(\zeta) \leq b(\zeta)$, and*

$$(-\zeta)^n \sum_{x \in \gamma G} \det(x^{-1})(-T)^{d(x,\zeta)} =$$

(12.18)

$$\begin{cases} 0 & \text{if } a(\zeta) < b(\zeta) \\ \displaystyle\prod_{j \in B(\zeta)} (T + m_j^*) \prod_{j \notin B(\zeta)} (1 - \varepsilon_j^{-1}\zeta^{-d_j^*}) \prod_{i \notin A(\zeta)} \frac{d_i}{1 - \varepsilon_i^{-1}\zeta^{-d_i}} & \text{if } a(\zeta) = b(\zeta), \end{cases}$$

where the $d_j^* = m_j^* - 1$ are the codegrees of G.

3.2. Twisted eigenspace theory. The results of Chapter 11 all have counterparts for reflection cosets. We state some of these here, leaving the reader to fill in details of the arguments in their proof, which are the same as those in Chapter 11, §§2, 3. We maintain the notation of the above discussion.

Theorem 12.19. *Let γG be a reflection coset in $V \simeq \mathbb{C}^n$. Let $F_1, \dots, F_n$ be a set of basic invariants for G such that $\gamma F_i = \varepsilon_i F_i$, and for a root of unity $\zeta \in \mathbb{C}$, let $A(\zeta), a(\zeta)$ be as above. Write $\mathcal{V}(F_i) := \{ v \in V \mid F_i(v) = 0 \}$, etc. Write $V(\zeta) = \bigcap_{i \notin A(\zeta)} \mathcal{V}(F_i)$. Then*

(i) $V(\zeta) = \bigcup_{x \in \gamma G} V(x, \zeta);$

(ii)   *the irreducible components of $V(\zeta)$ are the maximal eigenspaces among the $V(x, \zeta)$ (they all have dimension $a(\zeta) = |A(\zeta)|$);*

(iii)   *if E is one of the maximal eigenspaces of (ii), the restrictions to E of the F_i, $i \in A(\zeta)$, are algebraically independent;*

(iv)   *all the maximal eigenspaces of (ii) are conjugate under the action of G.*

Proof. Observe that $v \in V(x, \zeta)$ for some $x \in \gamma G$ if and only if, for some $g \in G$, $g\gamma v = \zeta v$, i.e. if and only if γv and ζv are in the same G-orbit. But this is equivalent to the condition $F_i(\gamma v) = F_i(\zeta v)$, i.e. that $\varepsilon_i^{-1} F_i(v) = \zeta^{-d_i} F_i(v)$ where $d_i =$

$\deg F_i$, or equivalently, that $F_i(v) = 0$ whenever $i \notin A(\zeta)$. This proves (i) and we leave the reader to use the arguments of Chapter 11 to complete the proof. □

In the same way as in Chapter 11 §3, this may be applied to prove the following twisted version of Theorem 11.15.

Theorem 12.20. *Let γG be a reflection coset in V, and for $\zeta \in \mathbb{C}$, let $E = V(x, \zeta)$ be a ζ-eigenspace, maximal among the ζ-eigenspaces of elements of γG. Let $N = \{ g \in G \mid gE = E \}$, and $C = \{ g \in G \mid ge = e \text{ for all } e \in E \}$. Let $F_1, \dots, F_n$ be a set of basic invariants for G such that for each i, $\gamma F_i = \varepsilon_i F_i$ and d_i is the degree of F_i. Then*

(i) $\overline{N} := N/C$ *acts as a unitary reflection group on E, whose hyperplanes are the intersections with E of the hyperplanes of G;*

(ii) *the restrictions to E of the F_i with $i \in A(\zeta)$ form a set of basic invariants of $\overline{N}$;*

(iii) *the coset xC is a reflection coset, and*

$$N = N_G(xC) := \{ g \in G \mid gxC = xC \}.$$

Proof. The proofs of (i) and (ii) are easily adapted from *loc. cit.*. As for (iii), to verify that xC is a reflection coset, it suffices to show that x normalises C. But both xC and Cx are characterised as the set of elements y of γG such that $E = V(y, \zeta)$; hence they are equal.

Now for $g \in G$, $gV(y, \zeta) = V(gyg^{-1}, \zeta)$. Hence by the paragraph above, $g \in G$ normalises E if and only if, for each element $y \in xC$, $gyg^{-1} \in xC$, which is the assertion (iii). □

3.3. Regular elements in reflection cosets.

In analogy with the untwisted case, we say that $x \in \gamma G$ is ζ-regular if $V(x, \zeta)$ contains a G-regular vector; x is regular if it is ζ-regular for some ζ.

Proposition 12.21. *Let γG be a reflection coset and let M be a $\widetilde{G}$-module ($\widetilde{G} = \langle G, \gamma \rangle$) of dimension r. If x is a ζ-regular element of γG, then the eigenvalues of x on M are $\{ \varepsilon_i(M^*)\zeta^{q_i(M^*)} \mid i = 1, \dots, r \}$.*

Proof. Let v be a regular element of $V(x, \zeta)$, and let $y_1, \dots, y_r$ be a basis of M such that for each i, $xy_i = \mu_i y_i$. Let $u_1, \dots, u_r$ be a homogeneous basis of $(\mathcal{H} \otimes M)^G$ such that $xu_i = \varepsilon_i(M^*)u_i$, and $\deg(u_i) = q_i(M^*)$ for each i. If we write $u_i = \sum_j A_{ij} \otimes y_j$, it then follows that for each i, j, $xA_{ij} = \varepsilon_i(M^*)\mu_j^{-1}A_{ij}$.

Since $xA_{ij}(v) = A_{ij}(x^{-1}v) = A_{ij}(\zeta^{-1}v) = \zeta^{-q_i(M^*)}A_{ij}(v)$, it follows that if $A_{ij}(v) \neq 0$, then $\mu_j = \varepsilon_i(M^*)\zeta^{q_i(M^*)}$. But by Theorem 10.13 $\det(A_{ij})$ is a product of linear forms corresponding to the reflecting hyperplanes of G. It follows from the regularity of v that $\det(A_{ij}(v)) \neq 0$, and hence that there is a permutation π of $\{1, \dots, r\}$ such that $A_{\pi i, i}(v) \neq 0$ for each i. The result follows. □

Corollary 12.22. *With the hypotheses of Proposition 12.21, the (multi)sets*

$$\{\, \varepsilon_i(M^*)\zeta^{q_i(M^*)} \mid i = 1, \ldots, r \,\} \quad and \quad \{\, \varepsilon_i(M)^{-1}\zeta^{-q_i(M)} \mid i = 1, \ldots, r \,\}$$

are equal.

Proof. Since the eigenvalues of x on M^* are the inverses of those of x on M, both sets are expressions for the set of eigenvalues of x on M. □

Theorem 12.23. *Let γG be a reflection coset in $V \simeq \mathbb{C}^n$. The following are equivalent.*

(i)   *There is a ζ-regular element $x \in \gamma G$.*
(ii)   *We have $a(\zeta) = b(\zeta)$, where $a(\zeta)$ and $b(\zeta)$ are the integers of Corollary 12.17.*
(iii) *The sets $\{\, \varepsilon_i^*\zeta^{m_i^*} \mid i = 1, \ldots, n \,\}$ and $\{\, \varepsilon_i^{-1}\zeta^{-m_i} \mid i = 1, \ldots, n \,\}$ are equal.*
(iv)   *For any $\widetilde{G}$-module M with $\dim M = r$, the sets*

$$\{\, \varepsilon_i(M^*)\zeta^{q_i(M^*)} \mid i = 1, \ldots, r \,\} \quad and \quad \{\, \varepsilon_i(M)^{-1}\zeta^{-q_i(M)} \mid i = 1, \ldots, r \,\}$$

are equal.

Proof. The implication $(i) \implies (iv)$ is Corollary 12.22. The implications $(iv) \implies (iii)$ and $(iii) \implies (ii)$ are trivial, since (iii) is the special case $M = V$ of (iv) and (ii) asserts that the number of occurrences of 1 in the two sets of (iii) is the same. It therefore remains only to prove that $(ii) \implies (i)$.

Now Corollary 12.16 shows that the maximal dimension of $V(x, \zeta)$ over $x \in \gamma G$ is $a(\zeta)$. By Corollary 12.17, it follows that if $a(\zeta) = b(\zeta)$, the coefficient of $T^{a(\zeta)}$ on the left side of (12.18) is non-zero. Let $E = V(x, \zeta)$ be such that $\dim E = a(\zeta)$. Then $\{\, y \in \gamma G \mid V(y, \zeta) = E \,\}$ is the reflection coset xC, where C is the pointwise stabiliser of E; i.e. $C = \{\, g \in G \mid ge = e \text{ for all } e \in E \,\}$. It follows that the sum of the terms on the left side of (12.18) which correspond to elements x with $V(x, \zeta) = E$ has a factor $\sum_{c \in C} \det_V(c)$, and since C is a reflection group, this is zero unless C is trivial. Consequently there is an element $x \in \gamma G$ such that $E = V(x, \zeta)$ is not contained in any reflecting hyperplane of G. This implies that E contains a G-regular vector, and the proof is complete. □

Remark 12.24. In a reflection group G it is trivial that $1 \in G$ is regular, with eigenvalue 1. However it is not obvious that any reflection coset γG contains a regular element. Moreover if G does not act irreducibly on V, it is clear that γ might be chosen so that there is no γG-regular vector in V. However, we do have (see also [**28**, Corollary 7.3] or [**161**]) the following.

Theorem 12.25. *If γG is a reflection coset and G acts irreducibly on V, then there is a ζ-regular element in γG, for some $\zeta \in \mathbb{C}$.*

The proof of this theorem uses the classification in a mild way.

4. Subquotients and centralisers

In this section, largely following [155, §3], we shall explore a connection between the subquotients arising from elements $x \in \gamma G$ such that $V(x, \zeta)$ is maximal, and centralisers of these elements. One of the main results is the following theorem, of which Theorem 11.24 (ii) is the special case $C = 1$ (i.e. the regular case).

Theorem 12.26. *Let γG be a reflection coset in V. Suppose $\zeta \in \mathbb{C}$ and that $x \in \gamma G$ is such that the eigenspace $E := V(x, \zeta)$ is maximal among ζ-eigenspaces of elements of γG. If we write N, C respectively for the stabiliser and pointwise stabiliser of E, there is an element $c \in C$ such that $N = C_G(xc).C$.*

For any element $x \in \gamma G$, the x-eigenspace decomposition of v may be written

$$(12.27) \qquad V = E_0 \oplus E_1 \oplus \cdots \oplus E_s,$$

where $E_i = V(x, \xi_i)$, the ξ_i are pairwise distinct, and $\xi_0 = \zeta$. For $i = 0, 1, \ldots, s$, define the parabolic subgroup C_i of G as the pointwise stabiliser of $E_0 \oplus E_1 \oplus \cdots \oplus E_i$, so that $C_0 = C$ and $C_i \supseteq C_{i+1}$ for each i; it will be convenient to use the notation $C_{-1} = G$. Note that x normalises each of the C_i, since x acts as a scalar on E_j for each j, whence if $c \in C_i$, xcx^{-1} acts trivially on each E_j for $j \leq i$. Thus xC_i is a reflection coset for each i.

Say that x is *(ζ-)quasi-regular* if there is a decomposition (12.27) such that for $i = 0, 1, \ldots, s$, E_i is a maximal ξ_i-eigenspace for the reflection coset xC_{i-1}. Thus in particular if x is quasi-regular, then x is pseudoregular (see Definition 11.26). We shall require the following result for the proof of Theorem 12.26.

Lemma 12.28. *If $x \in \gamma G$ is such that $E_0 := V(x, \zeta)$ is maximal among the ζ-eigenspaces of elements of γG, then there exists an element $c \in C = G_{E_0}$ such that xc is quasi-regular.*

Proof. Beginning with $(x = x_0, E_0)$, we shall recursively define pairs (x_i, E_i), $i = 0, 1, 2, \cdots$, satisfying, for each i, the following.

(i) $V \supseteq E_{i+1} \neq 0$, and E_{i+1} is orthogonal to $E_0 \oplus \cdots \oplus E_i$.

(ii) $x_{i+1} \in x_i C_i$.

(iii) For each pair i, k with $0 \leq k \leq i$, $E_k = V(x_i, \xi_k)$, and the ξ_k are pairwise distinct.

(iv) For each pair i, k with $0 \leq k \leq i$, E_k is a maximal ξ_k-eigenspace for the reflection coset $x_i C_{k-1}$.

Given such a sequence of pairs, By (i), there is an integer s such that $V = E_0 \oplus \cdots \oplus E_s$, and by (ii), $x_s \in xC_0 C_1 \cdots C_{s-1} \subseteq xC$. Moreover by (iii) and (iv), x_s is quasi-regular. Thus to complete the proof, it remains only to construct the sequence (x_i, E_i).

The first pair $(x_0, E_0) = (x, E)$ is given. Suppose therefore that we have (x_j, E_j) for $j = 0, \ldots, i$, and that $\bigoplus_{0 \leq j \leq i} E_j \subsetneq V$. We show how to construct (x_{i+1}, E_{i+1}) so that the conditions (i)–(iv) are satisfied.

Since $\bigoplus_{0 \leq j \leq i} E_j \neq V$, x_i has an eigenvalue ξ_{i+1} on $\bigoplus_{0 \leq j \leq i} E_j^\perp$, and by the maximality condition (iv), $\xi_{i+1} \neq \xi_j$ for $j = 0, 1, \ldots, i$. Let $x_{i+1} \in x_i C_i$ be such that $V(x_{i+1}, \xi_{i+1})$ is maximal among the ξ_{i+1}-eigenspaces of elements of $x_i C_i$.

Then properties (i) and (ii) are evident, and it remains only to prove (iii) and (iv). For (iii) we must show that for each k, $0 \leq k \leq i + 1$, we have $E_k = V(x_{i+1}, \xi_k)$. But by induction, $E_k = V(x_i, \xi_k)$, and since C_i acts trivially on E_k, $E_k \subseteq V(x_{i+1}, \xi_k)$. But by (iv), E_k is a maximal ξ_k-eigenspace for the reflection coset $x_i C_{k-1}$, whence we have equality.

Finally, for (iv), we must show that for each k with $0 \leq k \leq i + 1$, E_k is a maximal ξ_k-eigenspace for the reflection coset $x_{i+1} C_{k-1}$. If $k = i + 1$, this holds by construction. If $k \leq i$, then $x_{i+1} C_{k-1} = x_i c_i C_{k-1} = x_i C_{k-1}$, since $c_i \in C_i \subseteq C_{k-1}$. Thus (iv) follows from the induction hypothesis. This completes the proof of the lemma. □

We are now able to complete the

Proof of Theorem 12.26. In view of the previous lemma the theorem will follow from the following assertion:

(12.29)   *if $x \in \gamma G$ is quasi-regular, then $N = C_G(x).C$.*

To prove this, let $V = E_0 \oplus \cdots \oplus E_s$ be the x-eigenspace decomposition (12.27) of V which satisfies the conditions pertaining to the quasi-regularity of x (see above). Given an element $y \in N$, we shall construct a sequence $y_0, y_1, \ldots, y_s$ of 'approximations' to y with the following properties.

(i) $y_0 = y$ and $y_i \in N$ for each i.
(ii) $y_{i+1} \in C_i y_i$ for each i.
(iii) $y_i x y_i^{-1} x^{-1} \in C_i$ for each i.

Given such a sequence, we have $y_i \in C_{i-1} \cdots C_0 y = Cy$ by (ii) for each i, and from (iv), $y_s x = xy_s$ since $C_s = \{1\}$. Thus $y \in y_s C \subseteq C_G(x).C$. It follows that the existence of a sequence (y_i) as above implies (12.29), and hence Theorem 12.26.

It remains only to construct the sequence (y_i). Suppose we have $y_0, \ldots, y_k$. By (iv), $y_k x y_k^{-1} x^{-1}$ fixes $E_0 \oplus \cdots \oplus E_k$ pointwise, whence $y_k E_{k+1}$ is the maximal ξ_{k+1}-eigenspace $V(y_k x y_k^{-1}, \zeta_{k+1})$ for the coset $x C_k$. By Theorem 12.19 (iv), this implies that there is an element $c_k \in C_k$ such that $y_k E_{k+1} = c_k E_{k+1}$. Define $y_{k+1} := c_k^{-1} y_k \in C_k y_k$. We check the three conditions above; (i) and (ii) are trivial. As for (iii), we have $y_{k+1} x y_{k+1}^{-1} x^{-1} = c_k^{-1}(y_k x y_k^{-1} x^{-1}) x c_k x^{-1} \in C_k$. But since y_{k+1} fixes E_{k+1} setwise and x acts on E_{k+1} as a scalar, $y_{k+1} x y_{k+1}^{-1} x^{-1}$ fixes E_{k+1} pointwise, and hence lies in C_{k+1}. This completes the proof. □

5. Parabolic subgroups and the coinvariant algebra

In this section we explain a close relationship between the twisted invariants of a reflection coset and its parabolic subcosets. This will be exploited to prove an important result about the finer module structure of the coinvariant algebra $\mathcal{H}_G$ (here identified as the space of G-harmonic polynomials) and that of the coinvariant algebra of a parabolic subcoset.

Throughout this section we shall adopt the following notation: γG will be a reflection coset in $V \simeq \mathbb{C}^n$ and $v \in V$ is a ζ-eigenvector for γ, i.e. $\gamma(v) = \zeta v$. The pointwise stabiliser G_v of v in G is then a parabolic subgroup of G, which is normalised by γ, for if $g \in G_v$, then $\gamma g \gamma^{-1}(v) = \gamma g(\zeta^{-1}v) = \zeta^{-1}\gamma(v) = v$. Thus γG_v is a reflection subcoset of γG. In accord with the notation established above we write $\widetilde{G} = \langle G, \gamma \rangle$ and $\widetilde{G}_v = \langle G_v, \gamma \rangle$.

In general, if γG is a reflection coset and G' is a parabolic subgroup of G which is normalised by γ, we refer to the reflection coset $\gamma G'$ as a *parabolic subcoset* of γG.

5.1. Twisted invariants of parabolic subcosets.

Proposition 12.30. *Let γG_v be a reflection subcoset of γG as above. Let M be an r-dimensional $\widetilde{G}$-module, and write M_v for its restriction to G_v. Then we have an equality of (multi)sets*

$$\{\varepsilon_1(M)\zeta^{q_1(M)}, \ldots, \varepsilon_r(M)\zeta^{q_r(M)}\} = \{\varepsilon_1(M_v)\zeta^{q_1(M_v)}, \ldots, \varepsilon_r(M_v)\zeta^{q_r(M_v)}\},$$

where $\varepsilon_i(M_v)$ and $q_i(M_v)$ are the M_v-factor and the M_v-exponent of G_v.

Proof. Let $u_1, \ldots, u_r$ be an S^G-basis of $(S \otimes M^*)^G$ such that $\deg(u_i) = q_i(M)$ and $\gamma u_i = \varepsilon_i(M)u_i$ (see Proposition 12.2). If we fix a basis $y_1, \ldots, y_r$ of M^*, then we have, for each i, $u_i = \sum_j A_{ij} \otimes y_j$, where $A_{ij} \in S$ is either 0 or homogeneous of degree $q_i(M)$ for each j. Moreover by Theorem 10.13, we have, up to multiplication be a non-zero scalar,

$$(12.31) \qquad \det(A_{ij}) = \prod_{H \in \mathcal{A}(G)} L_H^{C(H,M)},$$

where $\mathcal{A}(G)$ is the set of reflecting hyperplanes of G, $L_H \in V^*$ is a linear form corresponding to $H \in \mathcal{A}(G)$, and $C(H, M)$ is an integer which depends only on the action of G_H on M, where G_H is the (cyclic) subgroup of G which fixes H pointwise.

Similarly, we have an S^G-basis $w_1, \ldots, w_r$ of $(S \otimes M^*)^{G_v}$ such that $\gamma w_i = \varepsilon_i(M_v)w_i$, and $\deg(w_i) = q_i(M_v)$. If we write $w_i = \sum_j B_{ij} \otimes y_j$, then in analogy with (12.31) we have, again up to a non-zero scalar,

$$(12.32) \qquad \det(B_{ij}) = \prod_{H \in \mathcal{A}(G_v)} L_H^{C(H,M_v)}.$$

Moreover since the elements u_i are *a fortiori* G-invariant, there are (unique) homogeneous elements $C_{ij} \in S^G$ such that for $1 = 1, \ldots, r$, we have

$$(12.33) \qquad\qquad u_i = \sum_{j=1}^{r} C_{ij} w_j.$$

Hence $\det(A_{ij}) = \det(C_{ij}) \det(B_{ij})$. Moreover from (12.31) and (12.32), we see that up to a non-zero scalar,

$$(12.34) \qquad\qquad \det(C_{ij}) = \prod_{H \in \mathcal{A}(G), \, H \ni v} L_H^{C(H,M)}.$$

This is because, by Theorem 9.44, $\mathcal{A}(G_v)$ consists precisely of those hyperplanes in $\mathcal{A}(G)$ which contain v, and clearly $C(H, M) = C(H, M_v)$ for $H \in \mathcal{A}(G_v)$.

Thus in particular, $\det(C_{ij})$ does not vanish at v.

Now let us evaluate $(\gamma(C_{ij}))(v)$ in two different ways. First, from (12.33), since $\gamma u_i = \varepsilon_i(M) u_i$ and $\gamma w_j = \varepsilon_j(M_v) w_j$, $\gamma C_{ij} = \varepsilon_i(M) \varepsilon_j^{-1}(M_v) C_{ij}$, whence $(\gamma(C_{ij}))(v) = \varepsilon_i(M) \varepsilon_j^{-1}(M_v) C_{ij}(v)$.

On the other hand, $(\gamma(C_{ij}))(v) = C_{ij}(\gamma^{-1} v) = \zeta^{-\deg(C_{ij})} C_{ij}(v)$. But

$$\deg(C_{ij}) = \deg(u_i) - \deg(w_j) = q_i(M) - q_j(M_v).$$

It follows that if $C_{ij}(v) \neq 0$, then $\varepsilon_i(M) \zeta^{q_i(M)} = \varepsilon_j(M_v) \zeta^{q_j(M_v)}$.

Since $\det(C_{ij}(v)) \neq 0$, there is a permutation π of $\{1, \ldots, r\}$ such that $C_{i,\pi i}(v) \neq 0$ for each i, and the result follows. $\qquad\square$

5.2. Application to the coinvariant algebra. We maintain the notation of §5.1. As usual, $\mathcal{H} = \bigoplus_i \mathcal{H}_i$ will denote the (graded) space of G-harmonic polynomials in S; similarly $\mathcal{H}_v = \bigoplus_j \mathcal{H}_{v,j}$ is the graded space of G_v-harmonic polynomials. By Proposition 12.2 (i), these spaces are graded representations of $\widetilde{G}$ and $\widetilde{G}_v$ respectively. We shall show how sums of certain graded components of $\mathcal{H}$, suitably twisted, are isomorphic as representations of $\widetilde{G}$, to representations induced from $\widetilde{G}_v$.

To state the result it is helpful to keep the following simple observation in mind. Let λ be a 1-dimensional representation of the cyclic group $\langle \gamma \rangle$. Since $\widetilde{G}/G \simeq \langle \gamma \rangle / \langle \gamma \rangle \cap G$, if the restriction of λ to $\langle \gamma \rangle \cap G$ is trivial, λ defines a (1-dimensional) representation of $\widetilde{G}$, which we also denote by λ. Now there is a natural 1-dimensional representation of $\langle \gamma \rangle$ on the line $\mathbb{C}v$, which we shall call θ_v, and we may also consider the powers θ_v^k, $k = 1, 2, \ldots$. First apply the above reasoning with G replaced by G_v. Since each element of G_v acts trivially on $\mathbb{C}v$, it is automatic that θ_v restricts trivially to $\langle \gamma \rangle \cap G_v$, and hence that we have a corresponding representation θ_v of G_v. However, the requirement that θ_v^k restrict trivially to $\langle \gamma \rangle \cap G$ is a condition on k. When it is satisfied, we have a corresponding representation θ_v^k of $\widetilde{G}$.

Theorem 12.35. *Let γG be a reflection coset in $V = \mathbb{C}^n$ and let $v \in V$ be an eigenvector of γ. Let G_v, $\mathcal{H}$, $\mathcal{H}_v$, etc., be as above, and let θ_v be the 1-dimensional representation of $\langle \gamma \rangle$ on $\mathbb{C}v$. Then in the notation of the previous paragraph, we have, for each integer $k \in \mathbb{Z}$,*

$$(12.36) \qquad \bigoplus_{\substack{i \geq 0 \\ \langle \gamma \rangle \cap G \subseteq \mathrm{Ker}\, \theta_v^{i+k}}} \mathcal{H}_i \otimes \theta_v^{k+i} \cong \mathrm{Ind}_{\widetilde{G_v}}^{\widetilde{G}} \left((\oplus_j \mathcal{H}_{v,j} \otimes \theta_v^j) \otimes \theta_v^k \right).$$

Proof. We show that the multiplicity of any $\widetilde{G}$-module M is the same in the left and right sides of (12.36). For any $\widetilde{G}$-module X, denote the multiplicity of M in X by $\langle X, M \rangle_{\widetilde{G}}$, and observe that if we denote the subspace of L-fixed points on a space Y by Y^L, we have $\langle X, M \rangle_{\widetilde{G}} = \dim(X \otimes M^*)^{\widetilde{G}} = \dim \left((X \otimes M^*)^G \right)^{\langle \gamma \rangle}$, since $\langle \gamma \rangle$ acts on $(X \otimes M^*)^G$.

Take X to be the left side of (12.36). Then, taking $\dim M = r$,

$$(12.37)$$
$$\langle X, M \rangle_{\widetilde{G}} = \sum_{\substack{i \geq 0 \\ \langle \gamma \rangle \cap G \subseteq \mathrm{Ker}\, \theta_v^{i+k}}} \dim(\mathcal{H}_i \otimes M^* \otimes \theta_v^{k+i})^{\widetilde{G}}$$

$$= \sum_{\substack{i \geq 0 \\ \langle \gamma \rangle \cap G \subseteq \mathrm{Ker}\, \theta_v^{i+k}}} \dim \left((\mathcal{H}_i \otimes M^* \otimes \theta_v^{k+i})^G \right)^{\langle \gamma \rangle}$$

$$= \sum_{\substack{i \geq 0 \\ \langle \gamma \rangle \cap G \subseteq \mathrm{Ker}\, \theta_v^{i+k}}} \dim \left((\mathcal{H}_i \otimes M^*)^G \otimes \theta_v^{k+i} \right)^{\langle \gamma \rangle}$$

$$= \sum_{\substack{\ell=0 \\ \langle \gamma \rangle \cap G \subseteq \mathrm{Ker}\, \theta_v^{q_\ell(M)+k}}}^{r} \langle \varepsilon_\ell(M), \theta_v^{-(q_\ell(M)+k)} \rangle_{\langle \gamma \rangle},$$

where, for any (suitable) element $\xi \in \mathbb{C}$, we write ξ for the character of $\langle \gamma \rangle$ which takes γ to ξ. Define $\zeta \in \mathbb{C}$ by $\theta_v(\gamma)v = \zeta v$. Then according to the notation just introduced, $\theta_v = \zeta$. Now evidently each of the characters $\varepsilon_\ell(M)$ of $\langle \gamma \rangle$ is trivial on $\langle \gamma \rangle \cap G$. Hence $\langle \varepsilon_\ell(M), \theta_v^{-(q_\ell(M)+k)} \rangle_{\langle \gamma \rangle} = 0$ if $\langle \gamma \rangle \cap G \not\subseteq \mathrm{Ker}\, \theta_v^{q_\ell(M)+k}$. It follows that

$$(12.38)$$
$$\langle X, M \rangle_{\widetilde{G}} = \sum_{\ell=0}^{r} \langle \varepsilon_\ell(M), \theta_v^{-(q_\ell(M)+k)} \rangle_{\langle \gamma \rangle}$$

$$= \sum_{\ell=0}^{r} \langle \varepsilon_\ell(M) \zeta^{q_\ell(M)}, \zeta^{-k} \rangle_{\langle \gamma \rangle}$$

$$= |\{\, \ell, 1 \leq \ell \leq r \mid \varepsilon_\ell(M) \zeta^{q_\ell(M)} = \zeta^{-k} \,\}|.$$

Let Y denote the right side of (12.36). Applying Frobenius reciprocity, a similar, but easier computation to that above shows that

$$(12.39) \qquad \langle Y, M \rangle_{\widetilde{G}} = |\{ \ell, 1 \le \ell \le r \mid \varepsilon_\ell(M_v) \zeta^{q_\ell(M_v)} = \zeta^{-k} \}|,$$

where M_v is the restriction to $\widetilde{G}_v$ of M.

Since M is arbitrary, the theorem is now an immediate consequence of Proposition 12.30. $\qquad\qquad\square$

Remark 12.40. The special case of Theorem 12.35 in which $\gamma \in G$ and v is regular (so that $G_v = 1$) is essentially equivalent to Springer's determination of the eigenvalues of a regular element (see also [213]).

In general if $\gamma \in G$ and $\zeta = \theta_v(\gamma)$ has order d, then the result may be written

$$\bigoplus_{i \equiv -k \pmod{d}} \mathcal{H}_i \cong \mathrm{Ind}_{\widetilde{G}_v}^{\widetilde{G}} \left((\oplus_j \mathcal{H}_{v,j} \otimes \theta_v^j) \otimes \theta_v^k \right).$$

In particular, for any integer d which divides a basic degree d_i for some i, the sum $\sum_{i \equiv -k \pmod{d}} \dim \mathcal{H}_i$ is independent of k.

6. Duality groups

In Chapter 6, §2 and §2.1, reference was made to 'Shephard groups', which are discussed in [63] in terms of generators and relations. As noted in *loc. cit.* we take a 'Shephard group' to be the symmetry group of a regular complex polytope (see [191]), in agreement with the definition in [181] (see also [180] and [182, 6.6.15]). Coxeter [62, 13.4] showed that these groups always have a presentation which generalises those considered in Chapter 6. For finite two generator groups, the concepts coincide.

It was proved by Orlik and Solomon that Shephard groups satisfy the condition of Definition 12.41 below. However the groups satisfying the condition form a larger class than the Shephard groups and, following Orlik and Terao, we shall call them 'duality groups'.

Cosets of duality groups always have regular eigenvalues of a particular type, which in the case of finite Coxeter groups (the 'real case') are the eigenvalues of the Coxeter elements. We explain this below. Some of the proofs make use of the exercises at the end of this chapter.

Definition 12.41. Let G be an irreducible reflection group in $V = \mathbb{C}^n$. Suppose the degrees d_i and codegrees d_i^* (see Definition 10.27) are ordered so that $d_1 \le d_2 \le \cdots \le d_n$ and $d_1^* \ge d_2^* \ge \cdots \ge d_n^*$. We say that G is a *duality group* if $d_i + d_i^* = d_n$ for all i.

Remark 12.42. It has been remarked by Orlik and Solomon [182, Appendix B.1] that the condition given is equivalent to the requirement that G be generated by

$n = \dim V$ reflections. This may be verified by inspection of the irreducible groups in the classification, but no conceptual proof is known.

The irreducible groups which are duality groups, but not Shephard groups, are $G_{24}, G_{27}, G_{29}, G_{33}$ and G_{34}.

We begin with the following result which may be found in [15] (see also [16] and [180, Theorem (5.2)]).

Theorem 12.43. *Let G be an irreducible unitary reflection group in $V = \mathbb{C}^n$ and assume that the degrees d_i and codegrees d_i^* are ordered as above. Let $N = \sum_i (d_i - 1)$ be the number of reflections of G (Theorem 4.14 (ii)) and write $N^* = \sum_i (d_i^* + 1)$. Suppose that G satisfies $d_i + d_i^* \le d_n$ for all i.*

(i)   *Every primitive d_n^{th} root of unity η is regular for G.*

(ii)   *For every pair i, j we have $d_i + d_j^* \le 2d_n - 1$.*

(iii)   *Let $F_1, \ldots, F_n$ be a set of basic invariants for G with $\deg F_i = d_i$. Let I be the ideal of S generated by $F_1, \ldots, F_{n-1}$. Then the discriminant polynomial $\Delta = \Delta_V$ (see Definition 12.7) is congruent to cF_n^k modulo I, where $c \in \mathbb{C}$, $c \ne 0$ and $k = (N + N^*)/d_n$.*

(iv)   *The discriminant matrix $Q = Q_V$ (see Definition 12.7) is congruent modulo I to a matrix of the form $F_n Q_0$, where Q_0 has entries in $\mathbb{C}$ and is non-singular.*

(v)   *G is a duality group; that is, $d_i + d_i^* = d_n$ for each i and the integer k of (iii) is equal to n.*

Proof. Observe first that by irreducibility, we have $d_n^* = 0$ and $d_i^* > 0$ for $1 \le i \le n - 1$. Hence by hypothesis, $1 \le d_i < d_n$ for $i < n$. Hence $d_i^* \le d_n - 1$ for $i < n$. The statement (ii) follows. Let η be a primitive d_n^{th} root of unity. Then in the notation of Theorem 12.23 (ii), we have $a(\eta) = b(\eta) = 1$, whence η is regular, i.e. there is an η-regular element of G.

Let $I = (F_1, \ldots, F_{n-1})$ as above. Then there is an η-regular element $v \in V$, and clearly $v \in \mathcal{V}(I)$. Since $\Delta(v) \ne 0$, this shows that $\Delta \notin I$. But since $\Delta \in S^G$, it is a polynomial in $F_1, \ldots, F_n$. Hence modulo I, Δ is of the form cF_n^k for some integer k. Since the degree of Δ is $N + N^*$, it follows that $k = (N + N^*)/d_n$.

Now the entries of Q are homogeneous elements of S^G, whence the argument of the last paragraph, together with (ii), shows that modulo I, each entry of Q is either constant or a scalar multiple of F_n. But the degree of the (i, j) entry Q_{ij} of Q is $d_i^* + d_j > 0$, and the first statement (iv) follows. The fact that Q_0 is non-singular follows from (iii).

Finally, note that $N + N^* = \sum_{i=1}^n (d_i + d_i^*) \le nd_n$. But by (iv), the degree of $\Delta = \det Q$ is nd_n, and (v) follows, completing the proof of the theorem. $\square$

The following consequence of Theorem 12.43 asserts that reflection cosets of duality groups contain 'Coxeter-like' elements.

Corollary 12.44. *Let γG be a reflection coset, where G is an irreducible duality group. As usual, denote the V-factors and V^*-factors of γG by $\varepsilon_1, \ldots, \varepsilon_n$ and $\varepsilon_1^*, \ldots, \varepsilon_n^*$ respectively. Then*

(i)   *there is a permutation π of $\{1, \ldots, n\}$ such that for each i, $d_i = d_{\pi(i)}$ and*
$$\varepsilon_{\pi(i)}^* = \varepsilon_i^{-1} \varepsilon_n;$$

(ii)   *if $\zeta \in \mathbb{C}$ satisfies $\zeta^{d_n} = \varepsilon_n^{-1}$, then ζ is regular for the coset γG.*

Proof. Write Q_{ij} for the (i, j) entry of the matrix Q of Theorem 12.43. It follows from Theorem 12.43 (*iv*) that if $(Q_0)_{ij} \neq 0$, then $\gamma(Q_{ij}) = \varepsilon_n Q_{ij}$, since in this case, $Q_{ij} = cF_n + F$, where $F \in \mathbb{C}[F_1, \ldots, F_{n-1}]$ and $\mathbb{C} \ni c \neq 0$. But by 12.9, $\gamma Q_{ij} = \varepsilon_i^* \varepsilon_j Q_{ij}$. Hence if $(Q_0)_{ij} \neq 0$, we have $\varepsilon_i^* = \varepsilon_j^{-1} \varepsilon_n$. But since Q_0 is non-singular, there is a permutation π of $\{1, \ldots, n\}$ such that for each i, $(Q_0)_{\pi(i),i} \neq 0$. Thus $\varepsilon_{\pi(i)}^* = \varepsilon_i^{-1} \varepsilon_n$. To see that $d_{\pi(i)} = d_i$ for all i, observe that it is evident from the proof of Theorem 12.43 (*iv*) that $(Q_0)_{ij} = 0$ if $d_i^* + d_j \neq d_n$. This shows that Q_0 is block diagonal, and the blocks are non-singular. Thus the permutation π may be assumed to preserve the blocks, which completes the proof of (*i*).

Let $\zeta \in \mathbb{C}$ be such that $\zeta^{d_n} = \varepsilon_n^{-1}$. We show ζ is regular for γG by checking the condition $a(\zeta) = b(\zeta)$ of Theorem 12.23 (*ii*). Now $a(\zeta)$ and $b(\zeta)$ are respectively the number of occurrences of 1 in the sets $\{\, \varepsilon_i \zeta^{d_i} \mid i = 1, \ldots, n \,\}$ and $\{\, \varepsilon_i^* \zeta^{d_i^*} \mid i = 1, \ldots, n \,\}$. But by (*i*), $\{\, \varepsilon_i^* \zeta^{d_i^*} \mid i = 1, \ldots, n \,\} = \{\, \varepsilon_i^{-1} \varepsilon_n \zeta^{d_n - d_i} \mid i = 1, \ldots, n \,\} = \{\, \varepsilon_i^{-1} \varepsilon_n \varepsilon_n^{-1} \zeta^{-d_i} \mid i = 1, \ldots, n \,\} = \{\, \varepsilon_i^{-1} \zeta^{-d_i} \mid i = 1, \ldots, n \,\}$, and (*ii*) is now immediate. $\qquad\square$

Note that Spaltenstein [200] has also defined 'Coxeter elements' of unitary reflection groups by regarding them as linear groups over a larger field with Galois action.

Exercises

1. Write out a complete proof of Theorem 12.10, including all details.

2. Write out a proof of Theorem 12.19, including the details omitted in the proof given in the text.

3. Write out a proof of Theorem 12.20, including all details omitted in the text above.

4. Let γG be a reflection coset in V and suppose that for $\zeta \in \mathbb{C}$, $x \in \gamma G$ is such that $V(x, \zeta)$ is a maximal ζ-eigenspace for γG. Show that in the usual notation (of §4) there is an element $c \in C$ such that $C_G(xc)/(C_G(xc) \cap C)$ acts faithfully on E as a reflection group.

5. Maintain the notation of the previous question. For $c \in C$ write $\varphi(c) = xcx^{-1}$ and say that $c, c' \in C$ are φ-conjugate if $c' = \varphi(d)cd^{-1}$ for some $d \in C$.

(i)   Show that for any element $g \in G$, the subspace $V(xg, \zeta)$ is a maximal
       ζ-eigenspace for γG if and only if xg is G-conjugate to an element of
       xC.

(ii)   Prove that the G-conjugacy classes of elements xg such that $V(xg, \zeta)$
       is maximal are in bijection with the equivalence classes $C/\sim$, where $\sim$
       is the equivalence relation on C generated by conjugacy and
       φ-conjugacy.

6. Let $g \in G$ and let H be a parabolic subgroup of G such that g normalises H.
   The reflection subcoset gH of G is said to be *special* if it consists precisely
   of the elements $x \in G$ such that $V(x, \zeta)$ is a maximal ζ-eigenspace for G.

   Prove that any power of a special parabolic coset contains a special parabolic coset.

7. Let γG be a reflection coset in $V = \mathbb{C}^n$. Recall that $V/G \cong \mathbb{C}^n$ and that
   V/G has coordinate ring $S^G = \mathbb{C}[F_1, \ldots, F_n]$, where the F_i are homogeneous, algebraically independent G-invariants, which may be chosen so that
   $\gamma F_i = \varepsilon_i F_i$ for each i. Let $I(\gamma)$ denote the ideal of S^G which is generated
   by $\{ F_i \mid \varepsilon_i = 1 \}$.

   (i)   Show that $\mathcal{V}(I(\gamma)) = (V/G)^\gamma$.

   (ii)   If $V^0 := V \setminus \bigcup_{H \in \mathcal{A}(G)} H$, show that

   $$V^0/G = V/G \setminus \mathcal{V}(\Delta),$$

   where $\Delta = \prod_{H \in \mathcal{A}(G)} L_H^{e_H}$ is the polynomial Δ_V of Lemma 12.8.

8. Continuing the notation of the previous question, prove the following.

   (i) ζ is regular for γG if and only if $\Delta \notin I(\zeta^{-1}\gamma)$.

   (ii)   If ζ is regular for γ and there is a unique index $i = i_0$ such that $\varepsilon_i = 1$,
          then Δ is a monic polynomial in F_{i_0}. (Note that Δ is uniquely express-
          ible as a polynomial in the F_i; the assertion is that this polynomial is
          monic in F_{i_0}.)

Some background in commutative algebra

In this appendix, we gather some elementary results from commutative algebra of which use is made in the text. We make no attempt to present results in the greatest generality applicable; on the contrary, our approach is concrete, and to the extent possible, directed towards the applications we require. Either sketch proofs or references to accessible texts are given for all statements. Our principal objective is the version of Krull's principal ideal theorem given in Theorem A.9 below. By 'ring' we shall always understand 'commutative ring with identity'.

Definition A.1. For any commutative ring R, the *dimension* $\dim R$ is the supremum of the set of integers r, such that there is a chain $P_0 \subsetneq P_1 \subsetneq \cdots \subsetneq P_r$ of prime ideals of R, of length r.

Recall that a ring R is *Artinian* if any descending chain of ideals of R stabilises, i.e. if $R \supseteq I_1 \supseteq I_2 \supseteq \ldots$ is a chain of ideals, then there is an integer n_0 such that for $n \geq n_0$, $I_n = I_{n_0}$. We shall require the following characterisation of Artinian rings, which may be found in [**8**, Theorem 8.5].

Theorem A.2. *The ring R is Artinian if and only if R is Noetherian, and $\dim R = 0$.*

The following elementary results from [**8**, Proposition 2.4] will also be useful.

Lemma A.3. *Let M be a finitely generated R-module, I an ideal of R, and $\varphi : M \to M$ an R-module endomorphism of M (this means that φ commutes with the elements of R as operators on M). Assume that $\varphi(M) \subseteq IM$. Then there are elements $a_1, \ldots, a_n \in I$ such that φ satisfies the equation*

$$\varphi^n + a_1 \varphi^{n-1} + \cdots + a_n = 0.$$

Proof. Suppose $t_1, \ldots, t_n$ generate M as an R-module. By hypothesis, there are elements $a_{ij} \in I$ such that for each i, $\varphi(t_i) = \sum_{j=1}^n a_{ij} t_j$. These relations may be written in matrix terms as $\sum_{j=1}^n p_{ij} t_j = 0$ for each i, where $p_{ij} = \delta_{ij}\varphi - a_{ij} \in I[\varphi]$. Multiplying by the adjoint of the matrix $P = (p_{ij})$, we see that $\det P$ is the zero endomorphism of M, and this is the desired relation. $\square$

Corollary A.4.

(*i*)   Let M be a finitely generated R-module, and suppose that $M = IM$ for some ideal I of R. Then there is an element $u = 1 + a \in R$, where $a \in I$, such that $uM = 0$.

(*ii*)   (*Nakayama's Lemma*) Suppose that M is a finitely generated R-module, and that I is an ideal of R which is contained in every maximal ideal of R. If $IM = M$, then $M = 0$.

Proof. The assertion (*i*) is immediate if we take φ to be the identity map in the lemma. For (*ii*), let $\mathcal{J}$ be the intersection of all maximal ideals of R (this is the *Jacobson radical* of R). By (*i*) there is an element $a \in \mathcal{J}$ such that $u = 1 + a$ acts as zero on M. But u is invertible, since the ideal it generates is R, whence $M = 0$.   □

We next point out some consequences which will be of interest to us.

Corollary A.5. Let R be a Noetherian ring, I an ideal of R, and M a finitely generated R-module. Then

$$\bigcap_{n=1}^{\infty} I^n M = \{\, x \in M \mid ux = 0 \text{ for some } u \in 1 + I \,\}.$$

Proof. Suppose $x \in M$ and $ux = 0$ for some $u = 1 - a \in 1 + I$. Then $x = ax = a^2 x = \cdots$ and thus $x \in \bigcap_{n=1}^{\infty} I^n M$.

Conversely, if $N = \bigcap_{n=1}^{\infty} I^n M$, then N is a submodule of the finitely generated R-module M and hence N is finitely generated, since R is Noetherian. Evidently $IN = N$, and so by Corollary A.4 (*i*) above, there is an element $u = 1 + a \in 1 + I$ such that $uN = 0$.   □

Corollary A.6. Let R be a Noetherian integral domain. Then $\bigcap_{n \geq 1} I^n = 0$ for any proper ideal I of R.

Proof. Taking $M = R$ in Corollary A.5, it follows that any element $x \in \bigcap_{n \geq 1} I^n$ satisfies $(1 + a)x = 0$ for some $a \in I$. If $x \neq 0$ this implies $a = -1 \in I$, and so $I = R$.   □

Definition A.7. Let R be an integral domain (i.e. R has no divisors of zero) and denote by F the quotient field of R. If P is a prime ideal of R, the *localisation* R_P of R at P is the subring of F consisting of the elements which may be expressed as $\frac{r}{s}$ where $r \in R$ and $s \in R \setminus P$.

This is a special case of a more general concept (see [**142**, p. 107]). The ring R_P has a unique maximal ideal, which consists of the elements $\frac{r}{s}$ where $r \in P$ and $s \in R \setminus P$.

The key facts we shall require concerning this construction are summarised as follows.

Proposition A.8. *Let* $\varphi : R \to R_P$ *be the inclusion map. The map* $Q \mapsto R_P\varphi(Q)$ *takes ideals of* R *to ideals of* R_P.

In particular, this map defines a bijection between the prime ideals of R *which are contained in* P, *and the set of all prime ideals of* R_P.

The proof is straightforward, and is left to the reader.

Effective use of the concept of dimension depends on Krull's principal ideal theorem (see [**8**, Corollary 11.17]), of which we shall prove the following version.

Theorem A.9. *Let* R *be a Noetherian integral domain, and let* f *be an element of* R *which is neither zero nor a unit. Let* P *be a prime ideal of* R, *minimal with respect to the requirement that* $P \supseteq (f)$, *where* (f) *is the (principal) ideal generated by* f.

Then P *is a minimal prime ideal of* R. *That is, if* Q *is a prime ideal of* R *such that* $Q \subsetneq P$, *then* $Q = 0$.

Proof. Since we are concerned with prime ideals $Q \subseteq P$, by Proposition A.8 we may replace R by R_P; that is, we may assume that R is a local ring with unique maximal ideal P. Since P is a minimal prime ideal containing (f), P is both a maximal and a minimal prime ideal of $R/(f)$, whence $\dim R/(f) = 0$. Since $R/(f)$ is Noetherian, it follows from Theorem A.2 that $R/(f)$ is Artinian.

Let $Q \subsetneq P$ be a prime ideal of R. By Corollary A.6, $\bigcap_{n \geq 0} Q^n = 0$. For any ideal I of R, define the ideal $\widetilde{I}$ by

$$\widetilde{I} = \{ r \in R \mid yr \in I \text{ for some } y \in R \setminus Q \}.$$

Clearly $I \subseteq \widetilde{I}$ and if $I \subseteq I'$, then $\widetilde{I} \subseteq \widetilde{I'}$. Thus the sequence $((\widetilde{Q^n} + (f))/(f))$, $n = 1, 2, \ldots$ is a descending sequence of ideals of the Artinian ring $R/(f)$, whence there is an integer m such that $\widetilde{Q^m} + (f) = \widetilde{Q^{m+1}} + (f)$. Hence for $x \in \widetilde{Q^m}$, we have $x = y + fz$ where $y \in \widetilde{Q^{m+1}}$. Now P is a minimal prime ideal containing f and therefore $f \notin Q$, whence $z \in \widetilde{Q^m}$. It follows that

$$\widetilde{Q^m} = \widetilde{Q^{m+1}} + f\widetilde{Q^m}$$

and thus for the R-module $M = \widetilde{Q^m}/\widetilde{Q^{m+1}}$ we have $M = PM$. Since R is Noetherian and M is finitely generated it follows from Corollary A.4 (*ii*) that $M = 0$, i.e. $\widetilde{Q^m} = \widetilde{Q^{m+1}} = \widetilde{Q^{m+2}} = \cdots$.

Now localise at Q. Since R_Q is a Noetherian integral domain and $R_Q\widetilde{Q^n} = R_Q Q^n = (R_Q Q)^n$, it follows from Corollary A.6 that $\bigcap_{n \geq 1} R_Q\widetilde{Q^n} = 0$. But $\bigcap_{n \geq 1} \widetilde{Q^n} \subseteq \bigcap_{n \geq 1} R_Q\widetilde{Q^n}$ and therefore $\bigcap_{n \geq 1} \widetilde{Q^n} = 0$.

It follows that for some integer n_0 we have $\widetilde{Q^{n_0}} = 0$ and hence $Q^{n_0} = 0$. If $x \in Q$, then $x^{n_0} = 0$, whence $x = 0$ since R is an integral domain. Thus $Q = 0$, as stated. $\qquad\square$

The next statement is an immediate consequence.

Corollary A.10. *With notation of Theorem A.9, we have* $\dim R/P = \dim R - 1$.

Remark A.11. Theorem A.9 remains true without the assumption that R is an integral domain, and has the same proof as that given. However in that case, there may be no prime $Q \subsetneq P$.

In our treatment of the dimension of an irreducible algebraic set, we shall require the Noether normalisation lemma, which we state in the following form [**142**, Theorem 2.1, p. 357].

Theorem A.12. *Let K be any field and let R an integral domain which is finitely generated as a K-algebra. If the transcendence degree of the quotient field of R over K is m, then there is a set of algebraically independent elements $Y_1, \ldots, Y_m \in R$, such that R is integral over the polynomial ring $K[Y_1, \ldots, Y_m]$.*

The context of its use is the following

Lemma A.13. *Let $R' \subseteq R$ be integral domains, and suppose that R' is integrally closed and that R is finitely generated as an R' module. Then for any maximal ideal M of R, $N = R' \cap M$ is a maximal ideal of R', and $\dim R_M = \dim R'_N$.*

Proof. N is evidently prime, and since the integral domain R/M is integral over R'/N, one is a field if and only if the other is, whence N is maximal.

Let $M \supsetneq P_1 \supsetneq \cdots \supsetneq P_m$ be a chain of prime ideals of R. Then by intersecting with R', we obtain a sequence $N \supseteq Q_1 \supseteq \cdots \supseteq Q_m$, where $Q_i = P_i \cap R'$. Moreover it may be easily verified that the inequalities are strict, so that $\dim N \geq \dim M$. Conversely, given a sequence $N \supsetneq Q_1 \supsetneq \cdots \supsetneq Q_m$ where the Q_i are prime ideals of R', it follows from the 'going-down theorem' (see, e.g. [**8**, Theorem (5.16)]) that $P_i = RQ_i$ is prime, and that we have a sequence of strict containments $M \supsetneq P_1 \supsetneq \cdots \supsetneq P_m$. Note that it is here that the condition that R_0 is integrally closed is used. $\square$

Forms over finite fields

In Chapter 8 we show that the finite primitive unitary reflection groups are closely related to symplectic, unitary and orthogonal groups over small finite fields; in many cases the reflection group can be identified as a subgroup of index 2 in an orthogonal group. This identification uses some basic facts about quadratic, alternating and hermitian forms and the corresponding finite orthogonal, symplectic and unitary groups. Most of this material can be found in Taylor [**216**] but for the convenience of the reader we summarise the necessary results in this appendix.

1. Basic definitions

Let V be a finite dimensional vector space over a field $\mathbb{F}$. A *quadratic form* on an V is a function $Q : V \to \mathbb{F}$ such that

$$Q(av) = a^2 Q(v) \quad \text{for all } a \in \mathbb{F}, v \in V, \quad \text{and}$$
$$\beta(u, v) := Q(u + v) - Q(u) - Q(v) \quad \text{is bilinear.}$$

The form β is called the *polar form* of Q. If the characteristic of $\mathbb{F}$ is not 2 and if β is a symmetric form defined on V, then $Q(v) := \frac{1}{2}\beta(v, v)$ is a quadratic form whose polar form is β.

A non-zero vector $u \in V$ is *singular* if $Q(u) = 0$; it is *non-singular* if $Q(u) \neq 0$. A subspace U of V is *totally singular* if $Q(u) = 0$ for all $u \in U$. The *radical* of β is the subspace $V^\perp := \{ v \in V \mid \beta(v, u) = 0 \text{ for all } u \in V \}$. The quadratic form Q is *non-degenerate* if every non-zero element of $V^\perp$ is non-singular.

Suppose that Q is non-degenerate. The *orthogonal group* $O(V, Q)$ is the group of linear transformations g of V that preserve Q; that is $Q(g(v)) = Q(v)$, for all $v \in V$.

A bilinear form β is *alternating* if $\beta(u, u) = 0$ for all u. The *symplectic group* $Sp(V)$ is the group of all linear transformations of V that preserve β.

Suppose that $\mathbb{F}$ is the finite field $\mathbb{F}_{q^2}$. For $x \in \mathbb{F}$ put $\bar{x} = x^q$. A form β in two variables is *hermitian* if it is linear in the first variable and if $\beta(v, u) = \overline{\beta(u, v)}$ for all u, v. The *unitary group* $U(\mathbb{F}_{q^2})$ is the group of all linear transformations of V that preserve β. Some references use the notation $U(n, q)$ to denote this group.

2. Witt's Theorem

Suppose that Q is a non-degenerate quadratic form on the vector space V and that β is the polar form of Q.

Theorem B.1 (Witt). *If U is a subspace of V and if $f : U \to V$ is an isometry, then there is an isometry $g : V \to V$ such that $g(u) = f(u)$ for all $u \in U$ if and only if $f(U \cap V^\perp) = f(U) \cap V^\perp$.*

Proof. See [**216**, Theorem 7.4]. $\qquad\qquad\qquad\qquad\qquad\qquad\qquad\qquad\qquad$ $\square$

It is a consequence of this theorem that any two maximal totally singular subspaces of V have the same dimension. This common dimension is called the *Witt index* of the form Q. If M is a totally singular subspace, then $M \subseteq M^\perp$ and the Witt index of Q is at most $\frac{1}{2} \dim V$.

Suppose that the polar from of Q is non-degenerate. If $\mathbb{F}$ is the finite field $\mathbb{F}_q$ and if V is a vector space of dimension n over $\mathbb{F}$, there are limited possibilities for the Witt index: if $n = 2m + 1$, the Witt index is m; if $n = 2m$ the Witt index is either m or $m - 1$. If n is odd, then up to isomorphism there is just one orthogonal group, which we denote by $O_n(\mathbb{F}_q)$. If $n = 2m$, there are two possibilities: the orthogonal group $O_{2m}^+(\mathbb{F}_q)$ of a form of Witt index m and the orthogonal group $O_{2m}^-(\mathbb{F}_q)$ of a form of Witt index $m - 1$.

3. The Wall form, the spinor norm and Dickson's invariant

Throughout this section we use $[V, f]$ to denote $\mathrm{Im}(1 - f)$.

3.1. The Wall form. For $f \in O(V, Q)$ and $u, v \in [V, f]$ define the *Wall form* of f to be

$$\chi_f(u, v) := \beta(w, v),$$

where w is a vector such that $u = w - f(w)$. (See Wall [**222**].)

Lemma B.2. *χ_f is a well-defined non-degenerate bilinear form on $[V, f]$ such that $\chi_f(u, u) = Q(u)$ for all $u \in [V, f]$.*

Proof. If $u = w - f(w) = w' - f(w')$, then $w - w' \in \mathrm{Ker}(1 - f) = [V, f]^\perp$. Thus for $v \in [V, f]$, $\beta(w, v) = \beta(w', v)$ and therefore $\chi_f(u, v)$ is well-defined.

If $\chi_f(u, v) = 0$ for all $v \in [V, f]$, then $\beta(w, v) = 0$ for all $w \in V$, and so $v = 0$. Thus χ_f is non-degenerate. On putting $f' := 1 - f$, the equation $Q(f(w)) = Q(w)$ becomes $Q(f'(w)) = \beta(w, f'(w))$ and therefore $\chi_f(u, u) = Q(u)$ for all $u \in [V, f]$. $\qquad\qquad$ $\square$

3.2. The spinor norm. Let $\mathbb{F}^\times$ be the multiplicative group of non-zero elements of $\mathbb{F}$ and let $\mathbb{F}^2$ be the subgroup of squares of elements of $\mathbb{F}^\times$.

For $f \in O(V, Q)$ the *spinor norm* $\nu(f)$ of f is the discriminant of χ_f. That is, if $e_1, e_2, \ldots, e_m$ is a basis for $[V, f]$, then $\nu(\chi_f)$ is the element $\det(\chi(e_i, e_j))\mathbb{F}^2$ in $\mathbb{F}^\times/\mathbb{F}^2$. The discriminant depends only on χ_f and not on the chosen basis.

A *reflection* in the group $O(V, Q)$ is a linear transformation defined by

$$r_u(v) = v - Q(u)^{-1}\beta(v, u)u,$$

where u is a non-singular vector of V that is not in the radical of the polar form β. Then $[V, r_u] = \langle u \rangle$ and hence $\nu(r_u) = Q(u)\mathbb{F}^2$. The spinor norm can be characterised as the homomorphism $\nu : O(V, Q) \to \mathbb{F}^\times/\mathbb{F}^2$ such that $\nu(r_u) = Q(u)\mathbb{F}^2$ for all reflections r_u (see [**216**, Theorem 11.50]).

3.3. The Dickson invariant. For $g \in O(V, Q)$ the *Dickson invariant* of g is

$$D(g) := \dim[V, g] \pmod 2$$

and the map

$$D : O(V, Q) \to \mathbb{Z}/2\mathbb{Z} : g \mapsto D(g)$$

is a homomorphism (see [**216**, Theorem 11.3]). If r is a reflection, then $D(r) = 1$. If the characteristic of $\mathbb{F}$ is not 2, then $D(g) = 0$ if $\det(g) = 1$ and $D(g) = 1$ if $\det(g) = -1$.

If $a \in \mathbb{F}^\times$, then $O(V, Q) = O(V, aQ)$. However, the spinor norm depends on Q: if $\tilde{\nu}$ is the spinor norm corresponding to aQ, then from [**216**, Lemma 11.49] $\tilde{\nu}(g) = a^{D(g)}\nu(g)$. The restrictions of ν and $\tilde{\nu}$ to Ker D coincide and therefore the kernel $\Omega(V, Q)$ of the restriction to Ker D of any spinor norm is well-defined.

4. Order formulae

Theorem B.3.

(i) $|Sp_{2m}(\mathbb{F}_q)| = q^{m^2} \prod_{i=1}^{m} (q^{2i} - 1).$

(ii) $|U_n(\mathbb{F}_{q^2})| = q^{\frac{1}{2}n(n-1)} \prod_{i=1}^{n} (q^i - (-1)^i).$

(iii) $|O_{2m}^\varepsilon(\mathbb{F}_q)| = 2q^{m(m-1)}(q^m - \varepsilon) \prod_{i=1}^{m-1} (q^{2i} - 1),$ *where* $\varepsilon = \pm 1.$

(iv) $|O_{2m+1}(\mathbb{F}_q)| = q^{m^2} \prod_{i=1}^{m} (q^{2i} - 1)$ *if q is even.*

(v) $|O_{2m+1}(\mathbb{F}_q)| = 2q^{m^2} \prod_{i=1}^{m} (q^{2i} - 1)$ *if q is odd.*

5. Reflections in finite orthogonal groups

It follows from Witt's Theorem that reflections r_u and r_v are conjugate in $O(V, Q)$ if and only if $Q(u) = Q(\lambda v)$ for some $\lambda \neq 0$.

If $\mathbb{F} = \mathbb{F}_q$ and q is a power of 2, then $\mathbb{F}^\times = \mathbb{F}^2$ and Dickson's map D is the only homomorphism from $O_n(\mathbb{F}_q)$ onto a group of order 2. In this case $O_n(\mathbb{F}_q)$ has a single conjugacy class of reflections.

On the other hand, if q is odd, then $\mathbb{F}^\times / \mathbb{F}^2$ is a group of order 2 and there are three homomorphisms from $O(V, Q)$ onto a group of order 2: the map D and the two spinor norm homomorphisms ν and $\tilde{\nu}$. The kernel of D is the *special orthogonal group* $SO(V, Q)$. There is no standard notation for the kernels of ν and $\tilde{\nu}$ – we refer to them as $\widehat{\Omega}(V, Q)$ and $\widetilde{\Omega}(V, Q)$. When q is odd, $O(V, Q)$ has two conjugacy classes of reflections r_u; they are distinguished by whether or not $Q(u)$ is a square and therefore one conjugacy class belongs to $\widehat{\Omega}(V, Q)$ and the other to $\widetilde{\Omega}(V, Q)$.

Suppose that $s \in \mathbb{F}_q$ is a non-square. If $n = 2m$, the group $O(V, Q)$ is $O_{2m}^\varepsilon(\mathbb{F}_q)$, where ε is 1 or -1, and we abbreviate ε to $+$ or $-$, where appropriate. We use a similar notation for $\widehat{\Omega}(V, Q)$, consistent with the notation for $O(V, Q)$. The forms Q and sQ define isometric geometries and therefore the groups $\widehat{\Omega}_{2m}^\varepsilon(\mathbb{F}_q)$ and $\widetilde{\Omega}_{2m}^\varepsilon(\mathbb{F}_q)$ and the two classes of reflections are interchanged by an outer automorphism of $O_n^\varepsilon(\mathbb{F}_q)$. In this case $-1 \in SO_{2m}^\varepsilon(\mathbb{F}_q)$ and from [**216**, p. 165], $-1 \in \Omega_{2m}^\varepsilon(\mathbb{F}_q)$ if and only if $q^m \equiv \varepsilon \pmod 4$. If $-1 \notin \Omega_{2m}^\varepsilon(\mathbb{F}_q)$, then $SO_{2m}^\varepsilon(\mathbb{F}_q) = \langle -1 \rangle \times \Omega_{2m}^\varepsilon(\mathbb{F}_q)$.

If both q and n are odd, we set $\varepsilon = 0$ and omit it from our symbols for the groups. The groups do not depend on the form but the geometries defined by Q and sQ are not isometric. In this case $-1 \notin SO_n(\mathbb{F}_q)$ and therefore we may choose the notation so that $\widetilde{\Omega}_n(\mathbb{F}_q) = \langle -1 \rangle \times \Omega_n(\mathbb{F}_q)$ and $\widehat{\Omega}_n(\mathbb{F}_q) \simeq SO_n(\mathbb{F}_q)$.

Lemma B.4. *The number of reflections in* $O_n^\varepsilon(\mathbb{F}_q)$ *is*

$$
\begin{cases}
q^{m-1}(q^m - \varepsilon) & \text{if } n = 2m, \\
q^{n-1} & \text{if } n \text{ and } q \text{ are odd}, \\
q^{n-1} - 1 & \text{if } n \text{ is odd and } q \text{ is even}.
\end{cases}
$$

Proof. If k is the Witt index of the quadratic form, then $\varepsilon = 2k - n + 1$ and from [**216**, Theorem 11.5] the number of singular vectors in V is $(q^{k-\varepsilon} + 1)(q^k - 1)$. Given non-singular vectors u and v, the reflections r_u and r_v are equal if and only if u is a scalar multiple of v. Therefore, except when n is odd and q is even, the number of reflections in $O_n^\varepsilon(\mathbb{F}_q)$ is

$$
(q^n - 1 - (q^{k-\varepsilon} + 1)(q^k - 1))/(q - 1),
$$

which, on substituting for ε, provides the first two formulae. In the remaining case, the radical of the polar form is spanned by a non-singular vector but does not correspond to a reflection. Hence the number of reflections in this case is $q^{n-1} - 1$. □

Theorem B.5. *Suppose that q is an odd prime power.*

(i)   *The number of reflections in $\widehat{\Omega}^{\varepsilon}_{2m}(\mathbb{F}_q)$ is $\frac{1}{2}q^{m-1}(q^m - \varepsilon)$.*

(ii)   *The number of reflections in $\widehat{\Omega}_{2m+1}(\mathbb{F}_q)$ is $\frac{1}{2}q^m(q^m - \varepsilon)$ and the number of reflections in $\widetilde{\Omega}_{2m+1}(\mathbb{F}_q) = \langle -1 \rangle \times \Omega_{2m+1}(\mathbb{F}_q)$ is $\frac{1}{2}q^m(q^m + \varepsilon)$, where $q^m \equiv \varepsilon \pmod 4$.*

Proof. (i) In this case $\widehat{\Omega}^{\varepsilon}_{2m}(\mathbb{F}_q)$ contains half the reflections of $O^{\varepsilon}_{2m}(\mathbb{F}_q)$ and the formula follows directly from the previous lemma.

(ii) If u is non-singular, the restriction of the quadratic form to the orthogonal complement of $\langle u \rangle$ has Witt index m or $m - 1$, depending on the value of $Q(u)$ (mod $\mathbb{F}^2_q$). If $n = 2m + 1$, it follows that the centraliser of the reflection r_u in $O_n(\mathbb{F}_q)$ is $\langle r_u \rangle \times O^{\varepsilon}_{2m}(\mathbb{F}_q)$. Thus, from Theorem B.3, the number of conjugates of r_u is $q^m(q^m + \varepsilon)$. We have $-1 \in \Omega^{\varepsilon}_{2m}(\mathbb{F}_q)$ if and only if $q^m \equiv \varepsilon \pmod 4$ and therefore $r_u \in \widehat{\Omega}_n(\mathbb{F}_q)$ if and only if r_u has $\frac{1}{2}q^m(q^m - \varepsilon)$ conjugates. $\square$

Theorem B.6.

(i)   *If q is a power of 2, then $O_3(\mathbb{F}_q) \simeq SL_2(\mathbb{F}_q)$;*

(ii)   *if q is odd, then $O_3(\mathbb{F}_q) \simeq C_2 \times PGL_2(\mathbb{F}_q)$.*

Proof. See Taylor [**216**, Theorem 11.6]. $\square$

A linear transformation $t \neq 1$ is a *transvection* if it fixes a hyperplane pointwise and if every eigenvalue of t is 1.

Theorem B.7.

(i)   *The group $Sp(V)$ is generated by its transvections.*

(ii)   *Except for $SU_3(\mathbb{F}_4)$, the group $SU_n(\mathbb{F}_{q^2})$ is generated by its transvections.*

(iii)   *Except for $O^+_4(\mathbb{F}_2)$, the group $O(V, Q)$ is generated by its reflections.*

(iv)   *If q is an odd prime power, each of the groups $\widehat{\Omega}^{\varepsilon}_{2m}(\mathbb{F}_q)$, $\widehat{\Omega}_{2m+1}(\mathbb{F}_q)$ and $\widetilde{\Omega}_{2m+1}(\mathbb{F}_q)$ has a single class of reflections and these reflections generate the group.*

Proof. For (i), (ii) and (iii) see [**216**, Theorem 8.5, Theorem 10.2, Corollary 11.42].
   $\square$

Applications and further reading

We shall indicate some areas of application for the basic material in this book, and provide direction and references for further study. Another useful source for some of the material below is [37]. We mention also the books [69] and [20], which focus on different aspects of the theory, principally in the context of real reflection groups.

1. The space of regular elements

Let G be a finite reflection group in $V = \mathbb{C}^n$.

Definition C.1. The space

$$M_G := V \setminus \bigcup_{H \in \mathcal{A}} H,$$

where $\mathcal{A}$ is the set of reflecting hyperplanes of G, is the space of $(G\text{-})$regular elements of V.

The topological space M_G, which is also known as the hyperplane complement associated to G, is of central importance in the study of presentations of G, and in linking the study of reflection groups to other areas of mathematics and physics. By Theorem 9.44, M_G may be characterised as the space of elements of V which are not fixed by any non-trivial element $g \in G$.

Example C.2. If $G = \mathrm{Sym}(n)$, acting on $\mathbb{C}^n$ by permuting coordinates, $M_G = M_n(\mathbb{C})$, the space of ordered configurations $(z_1, \ldots, z_n)$ of n distinct points in $\mathbb{C}$.

The space M_n arises in many different contexts, including low dimensional topology, configuration and moduli spaces and Hecke algebras.

In general, the spaces M_G and $X_G := M_G/G$ have been the subject of intense study in many different branches of mathematics and physics. We shall give a brief indication as to what these areas are, where the interested reader might start reading about them, and in some cases, open problems or suggested areas of further study.

1.1. Cohomology. The space $M_n(\mathbb{C}) = M_{\mathrm{Sym}(n)}$ was considered in [**4**], where Arnol'd gave a presentation of its cohomology ring. This work was generalised by Brieskorn [**36**] to the case where G is a finite real reflection group, i.e. a finite Coxeter group. Orlik and Solomon then showed [**177**] that the case of a general complex hyperplane complement could be dealt with similarly. Specifically they showed that the cohomology ring $H^*(M_G, \mathbb{C}) = \bigoplus_{i \geq 0} H^i(M_G, \mathbb{C})$ may be explicitly realised as a subcomplex of the de Rham complex of M_G. In [**144**] Lehrer showed that the space M_G, and in fact all hyperplane complements defined over number fields, are cohomologically pure; that is, the cohomology space $H_c^i(M_G, \mathbb{C})$ with compact supports consists entirely of classes of weight $i - n$. This is the basis of applications to Lie groups and Lie algebras over finite fields, because it connects the counting of rational points with the complex cohomology (see below).

Now it is clear that M_G has a G-action, which is transformed functorially to a linear action of G on $H^*(M_G, \mathbb{C})$. Moreover by the transfer theorem in cohomology, $H^*(X_G, \mathbb{C})$ is the 1_G-isotypic part of $H^*(M_G, \mathbb{C})$. This graded representation of G has been studied in many different contexts and for many special cases; see e.g. [**206, 10, 44, 153, 143**]. This problem is of fundamental importance in areas such as the study of conjugacy classes and adjoint orbits in reductive groups over finite fields and their Lie algebras, and in mathematical physics (see also [**188, 70**]) Three basic methods have been employed. One is the direct combinatorial algebraic manipulation of differential forms in the presentation of $H^*(X_G, \mathbb{C})$; secondly, there is the method of counting rational points, made available by [**144**], and thirdly, there is the method introduced in [**146**], which involves centralisers in unitary reflection groups. Because of its connection with results in this book, we state the basic theorem here (see [**146**, Theorem (2.1)]).

Theorem C.3. *Let G be a finite reflection group in $V = \mathbb{C}^n$, and let $g \in N_{GL(V)}(G)$. Then g acts on M_G; let t be an indeterminate, and write*

$$P_G(g,t) = \sum_{i \geq 0} \mathrm{Tr}(g, H^i(M_G, \mathbb{C}))t^i.$$

Suppose $f : C_G(g) \to \mathbb{C}$ satisfies the following two conditions.

(i) $f(1) = 1$.

(ii)   *For any parabolic subgroup $H \neq 1$ of G which is normalised by g, we have*

$$\sum_{x \in C_H(g)} f(x) = 0.$$

Then

$$P_G(g,t) = \sum_{x \in C_G(g)} f(x)(-t)^{n - \dim(\mathrm{Fix}\, x)}.$$

A function $f : C_G(g) \to \mathbb{C}$ which satisfies (i) and (ii) is called a Z-function. It is proved in [21] that for any G and $g \in N_{GL(V)}(G)$, there exists a Z-function $f : C_G(g) \to \mathbb{C}$, and formulae are given for such functions in all cases.

1.2. Eigenspaces and cohomology. We use the notation of Chapter 11, particularly Theorem 11.15.

If we take $g = 1$ in Theorem C.3, $C_G(g) = G$, and the function $f(x) = \det_V(x)$ is clearly a Z-function. It therefore follows immediately that

$$P_G(1,t) = \sum_{x \in G} \det_V(x)(-t)^{n - \dim(\text{Fix}\, x)}.$$

Applying Corollary 10.41, it follows directly that the Poincaré polynomial $P_G(t) := P_G(1,t)$ is given by

$$P_G(t) = \prod_{i=1}^{n}(1 + m_i^* t),$$

where the m_i^* are the coexponents of G. This formula was first proved in [178].

It therefore follows from Theorem 11.39 (ii) that if $E = V(g, \zeta)$ is the eigenspace on which N/C acts as a reflection group, and M_E is the corresponding hyperplane complement, then

(C.4) $P_E(t)$ divides $P_G(t)$.

Since divisibility of Poincaré polynomials often arises from fibrations or other topological sources, it is natural to pose the following problem.

Problem 1. Give a topological explanation of (C.4).

1.3. Local coefficients and the Milnor fibre. An important unsolved problem in this connection is the study of cohomology with local coefficients of the spaces X_G and M_G.

Problem 2. Let $\rho : \pi_1(X_G) \to GL_r(\mathbb{C})$ be a finite dimensional representation of the fundamental group $\pi_1(X_G)$. Let $\mathcal{L}_\rho$ be the corresponding local system on X_G. Compute $H^*(X_G, \mathcal{L}_\rho)$.

As an example of a partial solution in particular cases, see, e.g. [93, 187, 84, 70, 149]. For the context in mathematical physics, see [188].

One aspect of this problem relates to the Milnor fibre of the discriminant map. Let $\widetilde{M_G} = \{ (v, \xi) \in M_G \times \mathbb{C}^\times \mid \Delta(v) = \xi^m \}$, where $\Delta = \prod_{H \in \mathcal{A}(G)} L_H^{e_H}$ is thought of as a function $\Delta : V \to \mathbb{C}$, and $m = N + N^*$ is the degree of Δ. The group $\boldsymbol{\mu}_m$ of mth roots of unity in $\mathbb{C}$ acts on the second component, and the first projection

$p_1 : \widetilde{M}_G \to M_G$ is a $\boldsymbol{\mu}_m$-covering of M_G. If $F_G := \Delta^{-1}(1)$ is the Milnor fibre, then we have a commutative diagram

$$
\begin{array}{ccccc}
\widetilde{M}_G & \xrightarrow[(v,\xi)\mapsto v]{p_1} & M_G & \xrightarrow{/G} & X_G \\
{\scriptstyle (v,\xi)\mapsto \xi^{-1}v}\Big\downarrow & & \Big\downarrow{\scriptstyle /\mathbb{C}^\times} & & \Big\downarrow{\scriptstyle /\mathbb{C}^\times} \\
F_G & \xrightarrow{\boldsymbol{\mu}_m-\text{covering}} & \mathbb{P}(M_G) & \xrightarrow{/G} & \mathbb{P}(X_G),
\end{array}
$$

where $\mathbb{P}(M_G)$ is the projective hyperplane complement corresponding to M_G, and therefore $\mathbb{P}(M_G)$ and M_G have essentially the same cohomology, and similarly for $\mathbb{P}(X_G)$ and X_G.

Both rows of this diagram are coverings by the group $G \times \boldsymbol{\mu}_m$. It is therefore evident that among the local systems $\mathcal{L}_\rho$ are those which correspond to representations of $G \times \boldsymbol{\mu}_m$, which acts on the Milnor fibre in the obvious way. The following problem is therefore an important special case of Problem 2 (see [**70, 188, 78**]).

Problem 3. Determine the structure of $H^*(F_G, \mathbb{C})$ as a module for $G \times \boldsymbol{\mu}_m$.

1.4. Cohomology and counting rational points. As alluded to above, M_G may be thought of as the analytic space of $\mathbb{C}$-points of a scheme defined over the ring of integers of a number field. It is therefore possible to reduce M_G modulo prime ideals, and hence to speak of the variety $M_G(\overline{\mathbb{F}}_q)$ of its points over $\overline{\mathbb{F}}_q$, and the Frobenius morphism Frob_q associated to its $\mathbb{F}_q$-structure. This data may be used to connect the number of fixed points of (possibly twisted versions of) Frob_q with the equivariant structure of the cohomology of $M_G(\mathbb{C})$. See Kisin–Lehrer [**129, 128**] and Lehrer [**144, 149**] for details.

Among the applications of this point of view is an approach to counting rational points in the variety of regular and semisimple elements in a Lie group or Lie algebra over $\mathbb{F}_q$ (see [**147**]). Links with the homotopy theory of iterated loop spaces then emerge, since these provide an alternative way of approaching the regular semisimple varieties of classical (linear, orthogonal and symplectic) type [**152**]. This in turn leads to connections with mathematical physics.

2. Fundamental groups, braid groups, presentations

2.1. Fundamental groups. Since G acts regularly on M_G, the projection $p : M_G \to X_G$ ($= M_G/G$) is a regular covering, with group G. Since $\bigcup_{H \in \mathcal{A}(G)} H$ has (real) codimension 2 in V, M_G is path connected, and if x_0 is a point of M_G, we write $\pi_1(M_G, x_0)$ for the corresponding fundamental group. For $x_1 \in M_G$, there is a non-canonical isomorphism $\pi_1(M_G, x_0) \to \pi_1(M_G, x_1)$. The fundamental groups $\pi_1(X_G, p(x_0))$ and $\pi_1(M_G, x_0)$ are respectively known as the generalised braid group $B(G)$ and the 'pure braid group' $P(G)$ corresponding to G. Moreover we have a short exact sequence

$$(C.5) \qquad 1 \to \pi_1(M_G, x_0) \xrightarrow{p} \pi_1(X_G, p(x_0)) \xrightarrow{\eta} G \to 1.$$

The terminology comes from the ground breaking work of Artin [5] on the classical braid group, which in our context is the case $G = \mathrm{Sym}(n)$ of the above discussion. It is well known (see, e.g. [210, Appendix]) that the real unitary reflection groups are precisely the finite Coxeter groups; that is, they are characterised among all groups by the fact that they are finite groups, and have a presentation:

$$(C.6) \qquad \begin{aligned} G = \langle s_1, \dots, s_n \mid & (Q)\ s_i^2 = 1 \text{ for all } i, \\ & (BR)\ s_i s_j s_i \cdots = s_j s_i s_j \cdots \text{ for all } i, j \rangle, \end{aligned}$$

where the second relation (BR) has the same number of factors, m_{ij}, on both sides. The relations (BR) are usually referred to as the 'braid relations' and (Q) as the 'quadratic relations' for the s_i.

In this case the exact sequence has a particularly transparent form in terms of the above presentation. Let $A(G)$ be the group with generators $\sigma_1, \dots, \sigma_n$, and relations (BR) as above, with σ_i replacing s_i. Then $A(G) \cong B(G)$ in a way we shall shortly describe, and the map η of (C.5) simply maps σ_i to s_i. In this (real) case, the group $B(G)$ is called a *generalised Artin braid group*, and the kernel of this map is the pure (generalised) braid group. Artin's original work [5] shows that in the case $G = \mathrm{Sym}(n)$ the elements of $B(G)$ may be thought of as braids in the usual sense.

For a general unitary reflection group G, it has been proved in general by Bessis [14] that $B(G)$ has a 'Coxeter-like' presentation, such that the corresponding braid group $B(G)$ is obtained by omitting from this presentation the relations which stipulate the order of the relevant generators, just as in the case of the real groups, described above. Explicit presentations were given for the imprimitive groups by M. Broué, G. Malle and R. Rouquier [41], where they conjecture similar results for other groups. We shall describe the generators of $B(G)$ which are given in [41]. There will be one such generator σ_H for each hyperplane $H \in \mathcal{A}(G)$; it will be the projection under p of a path in M_G which we now describe. Intuitively, it is a loop with base x_0, around the hyperplane H.

Fix a hyperplane $H \in \mathcal{A}(G)$. Then we may choose $r > 0$ and $v_H \in H$ such that the closed ball $B(v_H, r) = \{ y \in V \mid |y - v_H| \le r \}$ intersects no other hyperplane $H'(\ne H) \in \mathcal{A}(G)$. Choose a point x_H on the boundary of $B(v_H, r)$ such that x_H is orthogonal to H, and take a path $\gamma : [0, 1] \to M_G$ beginning at x_0 and ending at x_H. Replacing γ with a shorter path if necessary, we may assume that $\gamma([0, 1]) \cap B(v_H, r) = \{ x_H \}$. Let δ be the path defined by $\delta(t) = \exp(\frac{2\pi i t}{e_H}) x_H$, where e_H is, as usual, the order of the pointwise stabiliser G_H of H. Then $\delta(0) = x_H$, while $\delta(1) = \exp(\frac{2\pi i}{e_H}) x_H$. But since $x_H \in H^\perp$, $s_H x_H = \exp(\frac{2\pi i}{e_H}) x_H$, for an appropriate reflection $s_H \in G$, which generates G_H. Hence $\delta(1) \in G x_H$, and so

$\delta \circ p$ is a loop on $p(x_H)$. Define $\sigma_{H,\gamma} := (\gamma \circ p)^{-1}(\delta \circ p)(\gamma \circ p)$. This is known as a 'braid reflection' or 'generator of the monodromy' for G. The following results may be found in [41].

Theorem C.7.

(i)   The monodromy generators $\sigma_{H,\gamma}$, over all $H \in \mathcal{A}(G)$, and paths γ, generate $\pi_1(X_G, p(x_0)) = B(G)$.

(ii) We have $\eta(\sigma_{H,\gamma}) = s_H$.

(iii) [41, Proposition 2.8] There is a choice of generators $\sigma_H = \sigma_{H,\gamma_0}$, one for each $H \in \mathcal{A}(G)$, such that $B(G)$ is generated by $\{\sigma_H \mid H \in \mathcal{A}(G)\}$.

2.2. Braid presentations. The proof of the above theorem is case free, but Broué, Malle and Rouquier go on to show that each of the groups $B(G)$, for G an irreducible unitary reflection group, has a 'Coxeter-like' presentation; this requires a good deal of case by case analysis. From Theorem C.7 it is not difficult to show that the generating set $\{s_H \mid H \in \mathcal{A}(G)\}$ may be cut down further, in analogy with the real case, when $\mathcal{A}(G)$ may be replaced by the subset of hyperplanes which form the walls of a chamber in a real form $V_{\mathbb{R}}$ of V. In [41] and subsequent work, building on the results of several others various authors produce, for each such G, a finite 'braid diagram' whose set $\mathcal{D} = \{s_i\}$ of nodes is in bijection with a generating set of braid generators of $B(G)$, which are referred to a 'distinguished braid generators'. They show that all relations in $B(G)$ are consequences of a finite set of 'homogeneous relations' of the form $s_{i_1} \cdots s_{i_k} = s_{j_1} \cdots s_{j_k}$ ($s_i \in \mathcal{D}$). These homogeneous relations may be read off from the braid diagram.

In [41], all but six of the exceptional groups are treated. The remaining ones are completed in [18] using methods developed in [14], but proofs for the final four cases are given in [16, Theorem 0.6].

In each case, the further relations which define G are then all of the form $s_i^{e_i} = 1$, so that the exact sequence (C.5) is again transparent. It would be desirable to have a case-free development of this material. A list of the braid diagrams may be found at http://people.math.jussieu.fr/~jmichel/table.pdf.

In addition to their use in relating the unitary reflection groups to the fundamental groups described above, these presentations are well adapted to the study of parabolic subgroups, and to the construction of Hecke algebras (see below).

2.3. The $K(\pi,1)$ question. The topological space X is a $K(\pi,1)$ space if the higher homotopy groups $\pi_i(X)$, $i \geq 2$ are all trivial. Such a space X is also then known as an Eilenberg–Mac Lane space, or classifying space for the group $\Gamma := \pi_1(X)$. It has the property that its (co)homology is the same as the group (co)homology of Γ, the latter being defined using a resolution of the trivial $\mathbb{Z}\Gamma$-module $\mathbb{Z}$.

The question as to whether the spaces M_G and X_G are $K(\pi, 1)$ spaces is classical. The case $G = \mathrm{Sym}(n)$ goes back to Fox and Neuwirth [99]; Brieskorn [35] established that M_G and X_G are $K(\pi, 1)$ spaces when G is real, with some exceptions, and Deligne [71] proved this result for a general real reflection group G. Orlik and Solomon [179] proved the corresponding result for all Shephard groups. Some special cases were also treated in [170].

The general question has recently been resolved by Bessis [16], who proved:

Theorem C.8. *For any finite unitary reflection group G, the spaces M_G and X_G are $K(\pi, 1)$.*

The proof uses the concept of 'dual braid monoid', which is beyond the scope of this book.

3. Hecke algebras

Hecke algebras are 'deformations' of reflection groups. They have applications in many areas, including reductive algebraic groups, various branches of representation theory, mathematical physics, statistical mechanics, low dimensional topology and knot invariants, and combinatorics, the latter particularly through affine Hecke algebras. We give here the merest hints as to how a study of these subjects might be approached, and how the material in this book relates to it.

3.1. The Coxeter case. Suppose G is the group with presentation (C.6). For the moment, G need not be finite. Let u be an indeterminate over $\mathbb{C}$ and write $R = \mathbb{C}[u, u^{-1}]$. Then the corresponding Hecke algebra $H_G(u)$ is the R-algebra defined by the presentation:

(C.9)
$$H_G(u) = \langle T_1, \ldots, T_n \mid (Q)\ (T_i - u)(T_i + u^{-1}) = 0 \text{ for all } i,$$
$$(BR)\ T_i T_j T_i \cdots = T_j T_i T_j \cdots \text{ for all } i, j \rangle,$$

where there are m_{ij} factors on each side of the braid relations (BR) (*cf.* (C.6)).

These algebras were first introduced by Iwahori [120] when $G = W$, the Weyl group of a finite dimensional simple complex Lie algebra $\mathfrak{g}$, with u replaced by a prime power q. He showed that $H_W(q)$ is isomorphic to the endomorphism algebra of the permutation representation of the finite reductive group $\overline{G}(\mathbb{F}_q)$ over the Galois field $\mathbb{F}_q$ of q elements on its flag variety. Here $\overline{G}$ is the group scheme (see [121]) with Weyl group W.

If $\varphi : R \to A$ is a homomorphism of commutative rings, the *specialisation* of $H_G(u)$ at φ, denoted $H_G(\varphi)$, is the A-algebra $A \otimes_R H_G(u)$. This has a presentation like (C.9), with u replaced by $\varphi(u) \in A$. Thus $H_W(q)$ is $H_W(\varphi_q)$, where $\varphi_q : R \to \mathbb{C}$ is the algebra homomorphism defined by $u \mapsto q$.

It is known that in general $H_G(u)$ has an R-basis $\{T_g \mid g \in G\}$, where $T_g = T_{i_1} \cdots T_{i_\ell}$ if $g = s_{i_1} \cdots s_{i_\ell}$ is any 'reduced expression' for $g \in G$, i.e. an expression of minimal length in the generators $s_1, \ldots, s_n$. This is one of its key properties. Clearly $H_G(\varphi_1) \cong \mathbb{C}G$, where $\varphi_1 : R \to \mathbb{C}$ is defined by $u \mapsto 1$. Thus $H_G(u)$ is indeed a deformation of G.

When $G = \mathrm{Sym}(n)$, $H_G(u)$ is intimately related to Jones' invariant of oriented links (see [122]). The case when G is a Weyl group of type B_n also has topological relevance through its connection with Kaufmann's ambient isotopy invariant [19]. The study of $H_G(u)$ initiated by Kazhdan and Lusztig in [126] has led to numerous connections with representation theory and geometry (see [127]). For an indication of some of these connections, see [49].

The decomposition of the permutation representation of the finite reductive group $\overline{G}(\mathbb{F}_q)$ on its flag variety is a special case of the decomposition of an induced cuspidal representation (see [48, 81]). It was proved by Howlett and Lehrer [114] that in the general case, the endomorphism algebras are still essentially Hecke algebras, but with extra parameters.

Let $u_1, \ldots, u_c$ be indeterminates which are in bijection with the set $\mathcal{A}(G)/G$ of G-orbits of reflecting hyperplanes of G. Define $A = \mathbb{C}[u_1^{\pm 1}, \ldots, u_c^{\pm 1}]$. The more general version $H_G(u_1, \ldots, u_c)$ of $H_G(u)$ is the A-algebra defined by the presentation

(C.10)
$$H_G(u_1, \ldots, u_c) = \langle T_1, \ldots, T_n \mid (Q) \ (T_i - u_i)(T_i + u_i^{-1}) = 0 \text{ for all } i,$$
$$(BR) \ T_i T_j T_i \cdots = T_j T_i T_j \cdots \text{ for all } i, j \rangle,$$

where u_i is the parameter corresponding to the G-orbit of $\mathcal{A}(G)$ in which $H = \mathrm{Fix}\, s_i$ lies (cf. (C.6)).

This more general 'multi-parameter' Hecke algebra enjoys many of the properties of $H_G(u)$. We shall see below how its construction generalises to arbitrary unitary reflection groups.

3.2. Hecke algebras for finite unitary reflection groups.

For any finite reflection group G in $V = \mathbb{C}^n$ denote by $\mathcal{O}_1, \ldots, \mathcal{O}_\ell$ the orbits of G on $\mathcal{A}(G)$, and for $1 \le i \le \ell$, let e_i be the common value of e_H for $H \in \mathcal{O}_i$ (see also Theorem 9.19). Let $\mathbf{u} := \{ u_{i,j} \mid 1 \le i \le c; \ 0 \le j \le e_i - 1 \}$ be a set of indeterminates over $\mathbb{C}$. With notation as in §2.2, define the (generic) Hecke algebra of G by

(C.11)
$$H_G(\mathbf{u}) = \langle T_i \, ; s_i \in \mathcal{D} \mid (T_i - u_{i,0})(T_i - u_{i,1}) \cdots (T_i - u_{i,e_i - 1}) = 0 \text{ for all } i,$$
$$(H) \ T_{i_1} T_{i_2} \cdots T_{i_k} = T_{j_1} T_{j_2} \cdots T_{j_k} \rangle,$$

where we have a relation (H) whenever the corresponding homogeneous relation $s_{i_1} \cdots s_{i_k} = s_{j_1} \cdots s_{j_k}$ holds in G. Note that the rôle of the diagram $\mathcal{D}$ in the definition of $H_G(\mathbf{u})$ is only to make the presentation explicit. In fact $H_G(\mathbf{u})$ is defined intrinsically as the quotient of the group ring of $B(G)$ by the ideal generated by appropriate polynomials in the braid generators (see §2.2).

Note that if $\zeta_k = \exp(\frac{2\pi i}{k})$, the specialisation defined by $u_{i,j} \mapsto \zeta_{e_i}^j$ makes (C.11) into a presentation of $\mathbb{C}G$, so that we do have a deformation of $\mathbb{C}G$, as in the Coxeter case.

There is a closer analogy with the real case. Observe that since the T_i are invertible and satisfy all the relations which define the presentation of $B(G)$ in terms of the generators $s_i \in \mathcal{D}$, we have a group homomorphism

$$\beta : B(G) \to H_G(\mathbf{u}),$$

defined by $\beta(s_i) = T_i$ for $s_i \in \mathcal{D}$, whose image lies in the group of units of $H_G(\mathbf{u})$.

The next proposition depends on case by case analysis, and is not known in all cases. It known of course when G is real, when $G = G(m,p,n)$ and for some other low dimensional groups G. For the two-dimensional case, see [94].

Proposition C.12. *([37, Theorem-Assumption 3.16]) For each element $g \in G$ there is an element $\tilde{g} \in B(G)$ such that $\eta(\tilde{g}) = g$ (see (C.5)), and such that the set $\{\beta(\tilde{g}) \mid g \in G\}$ is an A-basis of $H_G(\mathbf{u})$. Moreover $\tilde{\eta}(s_i) = s_i$ for $s_i \in \mathcal{D}$.*

One consequence of this proposition is that $H_G(\mathbf{u})$ is a quotient of the group ring $AB(G)$. We take the opportunity to mention a problem which remains unsolved even for the case of $G = \mathrm{Sym}(n)$, and whose solution would have important algebraic and topological consequences.

Problem 4. Is the group homomorphism $\beta : B(G) \to H_G(\mathbf{u})$ injective? If not give sufficient conditions for its injectivity. In general, determine its kernel.

The interested reader may find further information in [156].

Given Proposition C.12, one might expect that a generic specialisation of $H_G(\mathbf{u})$, i.e. one where the specialised values $\varphi(u_i)$ are 'generic', is isomorphic to $\mathbb{C}G$. In point of fact, the proof of the basic properties of $H_G(\mathbf{u})$ involve the realisation of homomorphisms $\tau : H_G(\mathbf{u}, \varphi) \to \mathbb{C}G$ from certain specialisations $H_G(\mathbf{u}, \varphi)$ of $H_G(\mathbf{u})$ to $\mathbb{C}G$ in terms of the monodromy action of $\pi_1(M_G, x_0)(= B(G))$ on solutions of certain differential equations on M_G (see [41, §4]). We shall give a rough intuitive idea of the arguments used to prove this here. For an accurate exposition the reader is referred to [41, §4].

Let $\mathcal{F} = \mathcal{F}(M_G, \mathbb{C}G)$ be the space of meromorphic functions $F : M_G \to \mathbb{C}G$. Since $M_G \times V$ may be identified with the tangent manifold of M_G, the space of $\mathbb{C}G$-valued differential forms on M_G may be thought of as the algebra $\mathcal{F} \otimes \Lambda V^*$. Now let ω be a 1-form, and consider the differential equation

(C.13) $$dF = \omega F,$$

for $F \in \mathcal{F}$, where the product on the right is in the algebra $\mathcal{F} \otimes \Lambda V^*$. Then for each point $y \in M_G$, there is a unique solution $x \mapsto F(x,y)$ of (C.13) defined for x in a neighbourhood U_y of y, such that $F(y, y) = 1 \in \mathbb{C}G$. If ω is G-stable, then for $g \in G$, $gF(x, y)g^{-1} = F(gx, gy)$.

If $\gamma : [0, 1] \to M_G$ is a path, then the solution $x \mapsto F(x, \gamma(0))$ may be extended analytically to solutions $x \mapsto F(x, \gamma(t))$ for $t \in [0, 1]$. Hence by a standard patching argument, for a suitable partition $1 = t_k > t_{k-1} > \cdots > t_1 > t_0 = 0$ of $[0, 1]$, we may unambiguously define

$$S_\omega(\gamma) := F(\gamma(t_k), \gamma(t_{k-1}))F(\gamma(t_{k-1}), \gamma(t_{k-2})) \cdots F(\gamma(t_1), \gamma(t_0)) \in \mathbb{C}G.$$

Moreover if ω is integrable (meaning that $d\omega + \omega \wedge \omega = 0$), $S_\omega(\gamma)$ depends only on the homotopy class of γ. In particular, we obtain algebra homomorphisms $S_\omega :$ $\mathbb{C}B(G) \to \mathbb{C}G$. The key is to select ω so that S_ω vanishes on the kernel of the surjection $\mathbb{C}B(G) \to H_G(\mathbf{u}, \varphi)$. This selection is achieved as follows.

The form ω may be written $\omega = \sum_{H \in \mathcal{A}(G)} f_H \frac{dL_H}{L_H}$, where the L_H are the usual linear forms. Choose elements $\{ a_{i,j} \in \mathbb{C}) \mid 1 \le i \le c; \ 0 \le j \le e_i - 1 \}$ and define $q_{i,j} = \exp(\frac{-2\pi i a_{i,j}}{e_i})$. As above, we write $a_{H,j} = a_{i,j}$, $q_{H,j} = q_{i,j}$ and $e_i = e_H$ if $H \in \mathcal{O}_i$. For each hyperplane $H \in \mathcal{A}(G)$, define

$$\varepsilon_j(H) = e_H^{-1} \sum_{k=0}^{e_H - 1} \zeta_{e_H}^{jk} s_H^k \in \mathbb{C}G \text{ for } j = 0, 1, \dots, e_H - 1.$$

Finally, take $f_H = \sum_{j=0}^{e_H - 1} a_{H,j} \varepsilon_j(H)$.

Then $\omega = \sum_{H \in \mathcal{A}(G)} f_H \frac{dL_H}{L_H}$ is G-stable and integrable. It is shown in [41, Theorem 4.12] that the homomorphism S_ω factors through $H_G(\mathbf{u}, \varphi)$, where $\varphi(u_{i,j}) = q_{i,j} \zeta_{e_i}^j$.

This construction provides homomorphisms

$$H_G(\mathbf{u}, \varphi) \xrightarrow{F_\varphi} \mathbb{C}G$$

from the specialisations $H_G(\mathbf{u}, \varphi)$ of $H_G(\mathbf{u})$ to $\mathbb{C}G$. However not all aspects of the construction and structure of the generic Hecke algebras $H_G(\mathbf{u})$ have yet been successfully addressed. In particular, Proposition C.12 has not yet been verified for all the irreducible groups G. Much of the theory of parabolic subalgebras, as well as generic isomorphism with $\mathbb{C}G$, is predicated upon it, so a unified approach to this question is very desirable.

The differential equation (C.13) is a generalisation of the celebrated Knizhnik–Zamolodchikov (KZ) equation [133]; the interested reader may find more about this context in [134, 135].

In addition to the ideas touched on above, there are significant interactions between unitary reflection groups, the theory of Painlevé equations, and singularity theory, in which reflection groups arise through monodromy actions on solutions of the equations. An account of this may be found in [27].

3.3. The nil Hecke algebra. We have stipulated above that u should be invertible in the Hecke algebra $H_G(u)$ when G is a real group, but if we replace the quadratic relation (Q) in (C.9) by $T_i^2 = 0$, we obtain the 'nil Hecke algebra' $H_G(0)$, which has well known connections with geometry (see, for example, [138, 139]) and representation theory (see [115, 116]). Moreover the Demazure operators δ_w (see Chapter 9, exercises) span an algebra isomorphic to $H_G(0)$ in this case. In analogy with the real case, one might ask about the algebra obtained by setting $u_{i,j} = 0$ for all i, j in the relation (C.11). Moreover from Exercise 5 in Chapter 9 one sees that the 'Demazure operators' δ_r satisfy $\delta_r^e = 0$, where e is the order of the reflection $r \in G$. It is therefore natural to pose the following problem.

Problem 5. Determine whether the Demazure operators δ_r of the exercises in Chapter 9 satisfy the homogeneous relations of (C.11) above. More generally, does the 'nil Hecke' algebra generated by the δ_r have representation theoretical significance?

3.4. Cellular algebras and Hecke algebras. The representation theory of the Hecke algebras discussed above enters almost every application of the theory. We have seen in §3.2 that generically, the Hecke algebra is isomorphic to $\mathbb{C}G$. However, important applications (for example, see [157, 158, 3]) relate to specialisations which are not semisimple. Cellular algebras are a class of algebras which were specifically introduced Graham and Lehrer in [104] for the analysis of algebras which are generically semisimple, but which have significant non-semisimple deformations. The cellular structure permits the analysis of these non-semisimple algebras to be reduced to certain problems in linear algebra, specifically to the study of certain bilinear forms.

Many algebras of geometric and topological significance have been shown to have a cellular structure (see [105]). In the context of the present work, the most significant result is the following theorem of Geck [100].

Theorem C.14. *The algebras $H_G(u)$ have a cellular structure when G is a finite Coxeter group.*

In the original paper [104] it was also shown that $H_G(\mathbf{u})$ has a cellular structure when G is the imprimitive group $G(m, 1, n)$. The algebra $H_G(\mathbf{u})$ is known as the Ariki–Koike algebra in this case. Its study is closely related to that of the affine Hecke algebra of type A (see also [106]).

In [100] Geck indicates how it might be proved $H_G(\mathbf{u})$ has a cellular structure for any real G. Some partial results along these lines are known. However, the following problem remains open.

Problem 6. Does the Hecke algebra $H_G(\mathbf{u})$ have a cellular structure for an arbitrary unitary reflection group G? If not, determine for which groups this is true. For these groups, develop the cellular theory explicitly, particularly in the singular case.

4. Reductive groups over finite fields

There are many ways in which unitary reflection groups enter the structure and representation theory of reductive algebraic groups, particularly those over finite fields. Some of these are still conjectural. It is well beyond the scope of this book to give a comprehensive treatment of the current state of these theories. In this section we shall describe the general areas of application, the direction of current trends, some links with the theory presented in this book, and indicate sources for further reading. In this section we shall use the following notation: $\mathbf{G}$ will denote a connected reductive algebraic group defined over the finite field $\mathbb{F}_q$ and $\mathbf{F} : \mathbf{G} \to \mathbf{G}$ will denote the Frobenius endomorphism corresponding to the $\mathbb{F}_q$-structure (see, e.g. [81]); $\mathbf{T}$ is an $\mathbf{F}$-stable maximal torus of $\mathbf{G}$, which we take to be contained in an $\mathbf{F}$-stable Borel subgroup $\mathbf{B}$. This determines the pair $\mathbf{T} \subseteq \mathbf{B}$ up to conjugacy by the finite group $\mathbf{G}^{\mathbf{F}}$ of $\mathbf{F}$-stable points of $\mathbf{G}$. Denote the real reflection group $N_{\mathbf{G}}(\mathbf{T})/\mathbf{T}$ by W. This is the Weyl group of $\mathbf{G}$ with respect to $\mathbf{T}$. If $\mathbf{H}$ is an $\mathbf{F}$-stable closed subgroup of $\mathbf{G}$, $\mathbf{H}^{\mathbf{F}}$ denotes its group of $\mathbf{F}$-stable (or 'rational') points.

Note that since $\mathbf{T}$ is $\mathbf{F}$-stable, $\mathbf{F}$ acts on W. The reflection group W may be realised in the space $V = Y(\mathbf{T}) \otimes_{\mathbb{Z}} \mathbb{C}$, where $Y(\mathbf{T})$ is the cocharacter group of $\mathbf{T}$. The $\mathbf{F}$-action on W is induced by an element $\gamma \in N_{GL(V)}(W)$ as in Chapter 12, so that $\mathbf{G}$ together with its $\mathbf{F}$-structure give rise to a reflection coset γW.

4.1. Twisting of tori and reflection groups.

All maximal tori of $\mathbf{G}$, and in particular all $\mathbf{F}$-stable maximal tori, are conjugate under $\mathbf{G}$. However, the $\mathbf{F}$-stable maximal tori are not necessarily conjugate under $\mathbf{G}^{\mathbf{F}}$. In fact the $\mathbf{G}^{\mathbf{F}}$-conjugacy classes of $\mathbf{F}$-stable maximal tori are in bijection with the $\mathbf{F}$-conjugacy classes in W (see also [205, 209, 144]). This is seen as follows. Start with the $\mathbf{F}$-stable torus $\mathbf{T}$. If $\mathbf{T}'$ is another $\mathbf{F}$-stable maximal torus, there is an element $g \in \mathbf{G}$ such that $\mathbf{T}' = g\mathbf{T}g^{-1}$. The equation $\mathbf{F}(\mathbf{T}') = \mathbf{T}'$ yields $g^{-1}\mathbf{F}(g) \in N_{\mathbf{G}}(\mathbf{T})$; taking $g^{-1}\mathbf{F}(g)$ modulo $\mathbf{T}$ defines a map $\mathbf{T}' \mapsto w \in W$, which is easily shown to be well defined up to $\mathbf{F}$-conjugacy, where $w' \sim_{\mathbf{F}} w$ if $w' = vw\mathbf{F}(v)^{-1}$ for some $v \in W$. Furthermore using Lang's Theorem ([209, 205]) it is easily seen that this defines a bijection between the $\mathbf{G}^{\mathbf{F}}$-orbits of $\mathbf{F}$-stable maximal tori, and the $\mathbf{F}$-conjugacy classes in W.

If $\mathbf{T}' = g\mathbf{T}g^{-1}$ with $g^{-1}\mathbf{F}(g) = \dot{w}$, where $\dot{w} \in N_{\mathbf{G}}(\mathbf{T})$ maps to $w \in W$ under the natural map, we say that $\mathbf{T}'$ is w-twisted. It is easily seen that if $\mathbf{T}'$ is w-twisted, then the following diagram, in which we use the notation $\operatorname{int} x$ for the map $y \mapsto xyx^{-1}$, commutes,

$$
\begin{array}{ccc}
\mathbf{T} & \xrightarrow{\;w\mathbf{F}\;} & \mathbf{T} \\
{\scriptstyle \operatorname{int} g} \downarrow & & \downarrow {\scriptstyle \operatorname{int} g} \\
\mathbf{T}' & \xrightarrow{\;\mathbf{F}\;} & \mathbf{T}'
\end{array}
$$

where $w\mathbf{F} = \operatorname{int} w \circ \mathbf{F}$.

Using this diagram, it is easy to show the following.

Lemma C.15. *Let* $\mathbf{T}'$ *be a* w-*twisted rational maximal torus of* $\mathbf{G}$. *Then writing* $W' = N_{\mathbf{G}}(\mathbf{T}')/\mathbf{T}'$, *we have* $W'^{\mathbf{F}} \simeq C_W(w\gamma)$.

Thus we are naturally led to a consideration of centralisers in reflection cosets, which figure prominently in our development (see §4 and Theorems 11.24 and C.3). The reader is referred to [144] for further developments. There is a similar 'relative' twisting theory for Levi subgroups of $\mathbf{G}$ (see [87, 80]).

4.2. Deligne–Lusztig varieties, Green functions and generalised Hecke algebras.

Let $\mathbf{G}$ and $\mathbf{F}$ be as above. In the celebrated work [72], Deligne and Lusztig defined algebraic varieties $X(w)$ over $\overline{\mathbb{F}}_q$ for each $w \in W$, whose ℓ-adic cohomology is instrumental in the study of the ordinary characters of the finite groups $\mathbf{G}^{\mathbf{F}}$. The basic construction of generalised characters arises from the observation that if $\mathbf{T}_w$ is a w-twisted maximal torus of $\mathbf{G}$ (see §4.1 above), then $\mathbf{G}^{\mathbf{F}} \times \mathbf{T}_w^{\mathbf{F}}$ acts on $X(w)$. Thus for any character θ_w of $\mathbf{T}_w^{\mathbf{F}}$, $\mathbf{G}^{\mathbf{F}}$ acts on the θ_w-isotypic component of $H_c^*(X(w), \overline{\mathbb{Q}}_\ell)$. Then the generalised characters

$$R_{\mathbf{T}_w}^{\mathbf{G}}(\theta_w)(-) = \sum_i (-1)^i \operatorname{trace}(-, H_c^i(X(w), \overline{\mathbb{Q}}_\ell))$$

of $\mathbf{G}^{\mathbf{F}}$ have been the major object of study in the area. Their decomposition into irreducible characters has involved the theory of character sheaves, and generally, intersection cohomological methods (see [196, 80]).

An issue central to this study has been the determination of the values of the $R_{\mathbf{T}_w}^{\mathbf{G}}(\theta_w)$ and of their irreducible constituents on unipotent elements of $\mathbf{G}^{\mathbf{F}}$. These are the 'Green functions', and their study involves the 'Springer representations' (see, e.g. [198, 160]). This subject has close links with Hecke algebras $H_W(\mathbf{u})$, generalising the notion of fake degrees, discussed in §1 above (see Definition 10.1). This theory has been developed further in a context where its implications are not yet fully understood; that is, 'Green functions' have been defined for some of the more general Hecke algebras $H_G(\mathbf{u})$ (see [197]), which potentially have connections to the representation theory of reductive groups over finite fields.

4.3. Braid groups, the Deligne–Lusztig complex and unipotent characters.

An irreducible character of $\mathbf{G}^{\mathbf{F}}$ is unipotent if it occurs with non-zero multiplicity in a generalised character of the form $R_{\mathbf{T}_w}^{\mathbf{G}}(1_{\mathbf{T}_w^{\mathbf{F}}})$. The determination of the unipotent character values is a central part of the character theory of the groups $\mathbf{G}^{\mathbf{F}}$. In [82, 83] Digne, Michel and Rouquier generalise the construction of Deligne–Lusztig varieties in a way which links their study directly with unitary reflection groups, the braid groups $B(G)$ mentioned above, and their corresponding Hecke algebras. It is beyond our scope to describe the construction; our purpose is only to give some sense as to how some of the connections arise. We maintain the notation introduced earlier in this section, and for simplicity assume that $\mathbf{G}$ is $\mathbf{F}$-split, which means that $\mathbf{F}$ acts trivially on W. Our treatment is necessarily imprecise, and the reader is referred to [82, 83] for details.

Let $B(W) \simeq A(W)$ be the generalised braid group associated with the Weyl group W (§2.1). It has a distinguished set $\mathcal{D}$ of generators, which are in bijection with the (Coxeter) generators of W. Denote by $B(W)^+$ the monoid generated by $\mathcal{D}$; this is usually referred to as the 'braid monoid', and generalises in the obvious way to an arbitrary unitary braid group, given the results outlined in §2. In this (Coxeter) case, the surjection $\eta : B(W) \to W$ (cf. C.5) has a distinguished section, which we shall denote by $w \mapsto \mathbf{w}$, such that for $w \in W$, $\mathbf{w} \in B(W)^+$. For each $\mathbf{b} \in B(W)^+$, one constructs an analogue $X(\mathbf{b})$ of $X(w)$; when $\mathbf{b} = \mathbf{w}$ for some $w \in W$, $X(\mathbf{w}) \simeq X(w)$.

In general, for any element $\mathbf{b} \in B(W)^+$, if $\mathbf{c} \in B(W)^+$ 'divides' $\mathbf{b}$ on the left (i.e. $\mathbf{c}^{-1}\mathbf{b} \in B(W)^+$), then there is a morphism $e_{\mathbf{c}} : \mathbf{X}(\mathbf{b}) \to \mathbf{X}(\mathbf{c}^{-1}\mathbf{bc})$. Fix $e_{\mathbf{b}}$ and consider the category whose morphisms are compositions of such $e_{\mathbf{c}}$.

By studying the endomorphism algebras in this category this leads, under some conditions, and for some $\mathbf{b} \in B^+$, to an action of the centraliser $C_{B(W)}(\mathbf{b})$ on the variety $\mathbf{X}(\mathbf{b})$, and this action commutes with the action of $\mathbf{G}^{\mathbf{F}}$, so that given an identification $\overline{\mathbb{Q}}_\ell \xrightarrow{\sim} \mathbb{C}$, we have a homomorphism

(C.16) $h : \mathbb{C}C_{B(W)}(\mathbf{b}) \to \mathrm{End}_{\mathbf{G}^{\mathbf{F}}} H_c^*(X(\mathbf{b}), \overline{\mathbb{Q}}_\ell),$

where $H_c^*(X(\mathbf{b}), \overline{\mathbb{Q}}_\ell) = \bigoplus_{i \geq 0} H_c^i(X(\mathbf{b}), \overline{\mathbb{Q}}_\ell)$.

Let us consider the following 'wish list' of properties of the (conjectured) map h (C.16), when $\mathbf{b}$ is a dth root of π. These have been proved for some special cases, but in general, still present formidable challenges.

1. The map h may be defined for all (many?) $\mathbf{b}$ such that $\eta(\mathbf{b})$ is a regular element of W.

2. For such elements $\mathbf{b}$, $C_W(\eta(\mathbf{b}))$ is a unitary reflection group, and we have

$$C_{B(W)}(\mathbf{b}) \simeq B(C_W(\eta(\mathbf{b})).$$

Thus $C_{B(W)}(\mathbf{b})$ is the braid group of a (unitary) reflection group in this case.

3. The action (C.16) of $\mathbb{C}C_{B(W)}(\mathbf{b})$ on $H_c^*(X(\mathbf{b}), \overline{\mathbb{Q}}_\ell)$ factors through a specialisation $H_{C_{B(W)}(\mathbf{b})}(\mathbf{u}, \varphi)$ of its corresponding Hecke algebra (cf. §3.2).

4. Each of the algebras $\mathbb{C}\mathbf{G}^{\mathbf{F}}$ and $h(\mathbb{C}C_{B(W)}(\mathbf{b}))$ is the full centraliser of the other in $\mathrm{End}_{\mathbb{C}}(H_c^*(X(\mathbf{b}), \overline{\mathbb{Q}}_\ell))$.

The study of property 4 is linked to the question of whether the $\mathbf{G}^{\mathbf{F}}$ modules $H_c^i(X(\mathbf{b}), \overline{\mathbb{Q}}_\ell))$ have a common constituent for distinct i.

The properties in the wish list are known to hold in some cases. Most is known about the case where $\mathbf{b}$ is a 'root of π', which we now explain. When W is irreducible, the centre $Z(B(W))$ is cyclic, generated by an element π, which may be taken to be $\mathbf{w_0}^2$, where w_0 is the 'longest element' of W. The element $\mathbf{b} \in B(W)$ is a dth root of $\boldsymbol{\pi}$ if $\mathbf{b}^d = \boldsymbol{\pi}$. It is known [42] that for such an element $\mathbf{b}$, $\eta(\mathbf{b})$ is a regular element of W (see Definition 11.21) for the eigenvalue $\exp\left(\frac{2\pi i}{d}\right)$. In particular, the centraliser $C_W(\eta(\mathbf{b}))$ is a reflection group (Theorem 11.24 (iii)), which of course need not be real. The property (2) has been investigated in the more general setting where W is replaced by an arbitrary unitary reflection group G. It has been proved in several significant cases in [17].

4.4. Speculation, Spetses and dreaming. In this book, we have not presented much of the available material on the representation theory of reflection groups and their Hecke algebras, although we have tried to indicate where the interested reader may find such material. It should be clear, however, from this appendix, that there is a close relationship between the representation theory of the Hecke algebras $H_W(u)$ and the unipotent characters of $\mathbf{G}^F$. Many of the aspects of the representation theory of $H_W(u)$ which are involved in this relationship (such as Green functions, Schur elements, fake degrees) may be studied in a context much more general than simply for Weyl groups. This has led to the question as to whether, given a unitary reflection group G, there is some class $\mathcal{C}_G$ of objects for which G plays a role analogous to that of a 'Weyl group'. These hypothetical objects have been called 'Spetses', after the place where they were first considered. One suggestion for such objects has come from the theory of p-completions of groups and p-compact groups (see [2]). The reader is also referred to [40] for the spirit of these speculations.

As a specific instance of the type of question which might be considered in this connection is the following. The results of [145] show that many structures associated with reductive groups over the finite field $\mathbb{F}_q$, such as conjugacy classes, tori, parabolic quotients, regular elements, etc., are counted by evaluating equivariant Poincaré polynomials for Weyl groups at q. This makes sense for more general unitary reflection groups, but as yet, there is no similar counting interpretation in the general case.

We have barely mentioned the φ_d-Sylow theory [37, Chapter IV] of Broué, Malle and Michel, in which specialisations of certain Hecke algebras of unitary reflection groups play a central role in studying 'd-Harish–Chandra series' of representations of the groups $\mathbf{G}^F$, which were introduced in [39]. This is close to the origins of the material linking the more general Hecke algebras with reductive groups over finite fields. The subject received its initial impetus from Broué's 'abelian defect conjecture', which is beyond the scope of this book, but which arises from the modular representation theory of finite groups. In φ_d-Sylow theory, the usual p-subgroups of a finite group, for p a prime, are replaced by φ_d-subgroups of $\mathbf{G}^F$, where these are defined as those whose 'generic order', which is a polynomial in the cardinality q of $\mathbb{F}_q$, is a power of the dth cyclotomic polynomial $\varphi_d(q)$. In the split case, the φ_d-subgroups correspond precisely to ζ_d eigenspaces of elements of the Weyl group W, where ζ_d is a primitive d^{th} root of unity. In general, one needs a reflection coset γW. The statement that the maximal ones among these, the 'φ_d-Sylow subgroups', are all conjugate, is equivalent to the statement that the maximal ζ_d eigenspaces are conjugate under W; this is Proposition 11.14 (*iii*) or Theorem 12.20 (*iv*) in the twisted case.

The φ_d-Sylow subgroups are closely connected to the reflection subquotients $G(d)$ (see Definition 11.19). Specifically, the normaliser of a φ_d-Sylow subgroup modulo its centraliser is isomorphic to $W(d)$, where W is the Weyl group. The connection between the 'p' and 'd' here is that d is the least integer such that p divides $q^d - 1$.

The eigenspace theory of Chapter 11 and its twisted generalisation (Chapter 12 §3) suggest that there are credible analogues for reflection groups (in fact for all linear groups) of the Quillen complex of [**185**]. Empirical evidence suggests that the following simplicial complex is worthy of serious study, possibly in a context wider than just the unitary reflection groups.

Definition C.17. Let $V = \mathbb{C}^n$ and $G \subseteq GL(V)$ be a finite group. For $\zeta \in \mathbb{C}$ define $\mathcal{S}_\zeta(G)$ as the set $\{ V(g, \zeta) \mid g \in G \}$, partially ordered by the reverse of inclusion.

Now there is an abstract simplicial complex associated with any partially ordered set (see [**67**]), and we may therefore speak of its homology and cohomology. Clearly G acts on $\mathcal{S}_\zeta(G)$, and hence on its homology. One therefore obtains representations of G in this way.

Problem 7. Study connections between the structure and representations of G and the topology of the posets $\mathcal{S}_\zeta(G)$.

When G is a unitary reflection group and $\zeta = 1$, one obtains $\mathcal{S}_1(G) = \mathcal{L}(G)$, the lattice of intersections of the reflecting hyperplanes of G. In this case it is known that $\mathcal{S}_1(G)$ is spherical, so that it has non-zero cohomology only in degrees 0 and $n - 1$, and there is an isomorphism of G-modules $H^{n-1}(\mathcal{S}_1(G), \mathbb{Z}) \simeq H^{n-1}(M_G, \mathbb{Z})$, where M_G is the hyperplane complement defined by G. Thus, already in this rather straightforward case, the space $\mathcal{S}_\zeta(G)$ is interesting.

Of course the same questions may be asked when G is replaced by a reflection coset. The paper [**21**] is essentially a study of $\mathcal{S}_1(C_G(\gamma))$, for G a reflection group, and $\gamma \in N_{GL(V)}(G)$.

APPENDIX D

Tables

In this appendix we provide tables of various properties and invariants associated with the irreducible finite unitary reflection groups.

In the Shephard and Todd notation (from [193]) there are three infinite families of reflection groups and they are numbered 1, 2 and 3: the symmetric groups $\mathrm{Sym}(n)$, the imprimitive groups $G(m, p, n)$, for $n \geq 2$, and the cyclic groups $\mathcal{C}_n$, respectively.

In addition there are 34 primitive groups of rank > 1, numbered from 4 to 37 and denoted by the symbols G_4, $G_5, \ldots, G_{37}$. There is a small amount of overlap between these families: $\mathrm{Sym}(2)$ is the cyclic group $\mathcal{C}_2$, and the groups $\mathrm{Sym}(3)$ and $\mathrm{Sym}(4)$ are imprimitive and isomorphic (as reflection groups) to $G(3,3,2)$ and $G(2,2,3)$, respectively.

The maximal reflection subgroups. In Table D.4 we provide a list of all maximal line subsystems of the irreducible line systems of rank at least three. This is equivalent to a list of the maximal reflection subgroups of all irreducible groups of rank at least three, which of course provides sufficient information for the determination of all reflection subgroups of the irreducible groups of rank at least three. The corresponding information for the groups of rank two may be found in Chapter 6.

The table of reflection cosets. A list of reflection cosets of the irreducible unitary reflection groups is provided in Table D.5, which is complete in the following sense.

Given an irreducible reflection group G in V, let $N := N_{GL(V)}(G)$ be its normaliser and let $Z := C_{GL(V)}(G)$ be its centraliser in $GL(V)$. Since G is irreducible, Z is $\mathbb{C}^\times$, acting via scalar multiplication on V, and we are interested in $N/\widehat{G}$, where $\widehat{G} = ZG$. It is easy to see that $N/\widehat{G}$ is finite: every element $g \in N$ defines an automorphism of G via conjugation and so there is a homomorphism $N \to \mathrm{Aut}(G)$, whose kernel is Z. Thus N/Z is isomorphic to a subgroup of the finite group $\mathrm{Aut}(G)$

TABLE D.1. The finite primitive reflection groups of rank 2

#	G	$\lvert G\rvert$	2	3	4	5	$\mathbb{Z}(G)$
4	$SL_2(\mathbb{F}_3)$	$2^3\cdot 3$		4			$\mathbb{Z}[\omega]$
5	$\mathcal{C}_3\times\mathcal{T}$	$2^3\cdot 3^2$		$4+4$			$\mathbb{Z}[\omega]$
6	$\mathcal{C}_4\circ SL_2(\mathbb{F}_3)$	$2^4\cdot 3$	6	4			$\mathbb{Z}[i,\omega]$
7	$\mathcal{C}_3\times(\mathcal{C}_4\circ\mathcal{T})$	$2^4\cdot 3^2$	6	$4+4$			$\mathbb{Z}[i,\omega]$
8	$\mathcal{T}\mathcal{C}_4$	$2^5\cdot 3$	6		6		$\mathbb{Z}[i]$
9	$\mathcal{C}_8\circ\mathcal{O}$	$2^6\cdot 3$	$12+6$		6		$\mathbb{Z}[\zeta_8]$
10	$\mathcal{C}_3\times\mathcal{T}\mathcal{C}_4$	$2^5\cdot 3^2$	6	8	6		$\mathbb{Z}[i,\omega]$
11	$\mathcal{C}_3\times(\mathcal{C}_8\circ\mathcal{O})$	$2^6\cdot 3^2$	$12+6$	8	6		$\mathbb{Z}[\zeta_8,\omega]$
12	$GL_2(\mathbb{F}_3)$	$2^4\cdot 3$	12				$\mathbb{Z}[i\sqrt{2}]$
13	$\mathcal{C}_4\circ\mathcal{O}$	$2^5\cdot 3$	$12+6$				$\mathbb{Z}[\zeta_8]$
14	$\mathcal{C}_3\times GL_2(\mathbb{F}_3)$	$2^4\cdot 3^2$	12	8			$\mathbb{Z}[\omega,i\sqrt{2}]$
15	$\mathcal{C}_3\times(\mathcal{C}_4\circ\mathcal{O})$	$2^5\cdot 3^2$	$12+6$	8			$\mathbb{Z}[\zeta_8,\omega]$
16	$\mathcal{C}_5\times\mathcal{I}$	$2^3\cdot 3\cdot 5^2$				12	$\mathbb{Z}[\zeta_5]$
17	$\mathcal{C}_5\times(\mathcal{C}_4\circ\mathcal{I})$	$2^4\cdot 3\cdot 5^2$	30			12	$\mathbb{Z}[i,\zeta_5]$
18	$\mathcal{C}_{15}\times\mathcal{I}$	$2^3\cdot 3^2\cdot 5^2$		20		12	$\mathbb{Z}[\omega,\zeta_5]$
19	$\mathcal{C}_{15}\times(\mathcal{C}_4\circ\mathcal{I})$	$2^4\cdot 3^2\cdot 5^2$	30	20		12	$\mathbb{Z}[i,\omega,\zeta_5]$
20	$\mathcal{C}_3\times\mathcal{I}$	$2^3\cdot 3^2\cdot 5$		20			$\mathbb{Z}[\omega,\tau]$
21	$\mathcal{C}_3\times(\mathcal{C}_4\circ\mathcal{I})$	$2^4\cdot 3^2\cdot 5$	30	20			$\mathbb{Z}[i,\omega,\tau]$
22	$\mathcal{C}_4\circ\mathcal{I}$	$2^4\cdot 3\cdot 5$	30				$\mathbb{Z}[i,\tau]$

$$\omega = \tfrac{1}{2}(-1+i\sqrt{3}), \quad \tau = \tfrac{1}{2}(1+\sqrt{5}), \quad \zeta_k = \exp(2\pi i/k)$$

and, *a fortiori*, $N/\widehat{G}$ is finite. In the table, we provide a complete list of representatives for the cosets γG, where γ is taken modulo $\widehat{G}$, and is chosen so that if the order of $\gamma\widehat{G}$ is r, then $\gamma^r \in G$.

1. The primitive unitary reflection groups

Tables D.1 and D.2 list every finite primitive unitary reflection group of rank at least 2. The column headed # gives the Shephard and Todd numbering. The columns

TABLE D.2. The finite primitive reflection groups of rank ≥ 3

#	G	$|G|$	2	3	$\mathbb{Z}(G)$
1	$W(\mathcal{A}_n)$	$(n+1)!$	$\frac{1}{2}n(n+1)$		$\mathbb{Z}$
23	$W(\mathcal{H}_3)$	$2^3 \cdot 3 \cdot 5$	15		$\mathbb{Z}[\tau]$
24	$W(\mathcal{J}_3^{(4)})$	$2^4 \cdot 3 \cdot 7$	21		$\mathbb{Z}[\lambda]$
25	$W(\mathcal{L}_3)$	$2^3 \cdot 3^4$		12	$\mathbb{Z}[\omega]$
26	$W(\mathcal{M}_3)$	$2^4 \cdot 3^4$	9	12	$\mathbb{Z}[\omega]$
27	$W(\mathcal{J}_3^{(5)})$	$2^4 \cdot 3^3 \cdot 5$	45		$\mathbb{Z}[\omega, \tau]$
28	$W(\mathcal{F}_4)$	$2^7 \cdot 3^2$	$12+12$		$\mathbb{Z}$
29	$W(\mathcal{N}_4)$	$2^9 \cdot 3 \cdot 5$	40		$\mathbb{Z}[i]$
30	$W(\mathcal{H}_4)$	$2^6 \cdot 3^2 \cdot 5^2$	60		$\mathbb{Z}[\tau]$
31	$W(\mathcal{O}_4)$	$2^{10} \cdot 3^2 \cdot 5$	60		$\mathbb{Z}[i]$
32	$W(\mathcal{L}_4)$	$2^7 \cdot 3^5 \cdot 5$		40	$\mathbb{Z}[\omega]$
33	$W(\mathcal{K}_5)$	$2^7 \cdot 3^4 \cdot 5$	45		$\mathbb{Z}[\omega]$
34	$W(\mathcal{K}_6)$	$2^9 \cdot 3^7 \cdot 5 \cdot 7$	126		$\mathbb{Z}[\omega]$
35	$W(\mathcal{E}_6)$	$2^7 \cdot 3^4 \cdot 5$	36		$\mathbb{Z}$
36	$W(\mathcal{E}_7)$	$2^{10} \cdot 3^4 \cdot 5 \cdot 7$	63		$\mathbb{Z}$
37	$W(\mathcal{E}_8)$	$2^{14} \cdot 3^5 \cdot 5^2 \cdot 7$	120		$\mathbb{Z}$

$$\omega = \tfrac{1}{2}(-1 + i\sqrt{3}), \quad \tau = \tfrac{1}{2}(1 + \sqrt{5}), \quad \lambda = -\tfrac{1}{2}(1 + i\sqrt{7})$$

headed 2, 3, 4 and 5 give the number of cyclic subgroups of reflections of the given order. Where there is more than one conjugacy class this is indicated by giving the size of each class; for example, there are two classes of cyclic subgroups of reflections of order 3 in $G_5 \simeq \mathcal{C}_3 \times \mathcal{T}$, each of size 4 and this is indicated by the notation $4+4$. The final column gives the ring of definition of G; in each case it is a principal ideal domain.

Except for $W(\mathcal{M}_3)$ and $W(\mathcal{F}_4)$, each primitive unitary reflection group of rank at least three has a single orbit on its line system. The group $W(\mathcal{M}_3)$ has two orbits: the 9 lines spanned by the roots of the reflections of order two and the 12 lines spanned by the roots of the reflections of order three. In the case of $W(\mathcal{F}_4)$ all reflections have order two and $W(\mathcal{F}_4)$ has two orbits on lines.

2. Degrees and codegrees

The following table lists the degrees and codegrees of the finite primitive reflection groups G_4, $G_5, \ldots, G_{37}$ and for convenience, for each group G of rank at least 3, we give the name of the corresponding line system $\mathfrak{L}(G)$.

For the infinite families (Shephard and Todd numbers 1, 2 and 3) we have the following data.

1. The order of $W(\mathcal{A}_n) \simeq \mathrm{Sym}(n+1)$ is $(n+1)!$, its degrees are $2, 3, \ldots, n+1$ and its codegrees are $0, 1, \ldots, n-1$.

2. The order of $G(m, p, n)$ is $m^n \, n!/p$; its degrees are $m, 2m, \ldots, (n-1)m$, and nm/p. If $p \neq m$, its codegrees are $0, m, 2m, \ldots, (n-1)m$. If $p = m$, its codegrees are $0, m, 2m, \ldots, (n-2)m$ and $(n-1)m - n$.

3. The order of the cyclic group $\mathcal{C}_m$ is m. Its degree is m and its codegree is 0.

Let $d_1, \ldots, d_n$ and $d_1^*, \ldots, d_n^*$ respectively be the degrees and codegrees of the primitive unitary reflection group G. The case $\zeta = 1$ of Corollary 10.39 is the polynomial identity

$$\sum_{g \in G} T^{\dim(\mathrm{Fix}\, g)} = \prod_{i=1}^{n}(T + m_i),$$

where the $m_i = d_i - 1$ are the exponents of G. Replacing T by T^{-1} and multiplying by T^n, the identity becomes

$$\sum_{k=0}^{n} b_k T^k = \prod_{i=1}^{n}(1 + m_i T),$$

where b_k is the number of elements of G whose fixed point space has dimension $n - k$. The identity in this form was obtained by Shephard and Todd [193] as a consequence of their classification of the finite unitary reflection groups. A case-free proof was first given by Solomon [199].

Similarly the case $\zeta = 1$ of Corollary 10.41 yields the polynomial identity

$$\sum_{g \in G} \det(g)\, T^{\dim(\mathrm{Fix}\, g)} = \prod_{i=1}^{n}(T - m_i^*),$$

where the $m_i^* = d_i^* + 1$ are the coexponents of G. The latter identity may also be deduced from Exercise 8 of Chapter 10 as the case $\zeta = 1$ and $\lambda = \det^{-1}$.

Using these identities the degrees and codegrees of G may easily be computed using a computer algebra system, such as Magma [32], which contains matrix representations for the unitary reflection groups and the ability to factorise polynomials.

TABLE D.3. Degrees and codegrees of the primitive groups

#	$\mathfrak{L}(G)$	$\lvert G\rvert$	N	degrees	codegrees
4		$2^3\,3$	8	4, 6	0, 2
5		$2^3\,3^2$	16	6, 12	0, 6
6		$2^4\,3$	14	4, 12	0, 8
7		$2^4\,3^2$	22	12, 12	0, 12
8		$2^5\,3$	18	8, 12	0, 4
9		$2^6\,3$	30	8, 24	0, 16
10		$2^5\,3^2$	34	12, 24	0, 12
11		$2^6\,3^2$	46	24, 24	0, 24
12		$2^4\,3$	12	6, 8	0, 10
13		$2^5\,3$	18	8, 12	0, 16
14		$2^4\,3^2$	28	6, 24	0, 18
15		$2^5\,3^2$	34	12, 24	0, 24
16		$2^3\,3\,5^2$	48	20, 30	0, 10
17		$2^4\,3\,5^2$	78	20, 60	0, 40
18		$2^3\,3^2\,5^2$	88	30, 60	0, 30
19		$2^4\,3^2\,5^2$	118	60, 60	0, 60
20		$2^3\,3^2\,5$	40	12, 30	0, 18
21		$2^4\,3^2\,5$	70	12, 60	0, 48
22		$2^4\,3\,5$	30	12, 20	0, 28
23	$\mathcal{H}_3$	$2^3\,3\,5$	15	2, 6, 10	0, 4, 8
24	$\mathcal{J}_3^{(4)}$	$2^4\,3\,7$	21	4, 6, 14	0, 8, 10
25	$\mathcal{L}_3$	$2^3\,3^4$	24	6, 9, 12	0, 3, 6
26	$\mathcal{M}_3$	$2^4\,3^4$	33	6, 12, 18	0, 6, 12
27	$\mathcal{J}_3^{(5)}$	$2^4\,3^3\,5$	45	6, 12, 30	0, 18, 24
28	$\mathcal{F}_4$	$2^7\,3^2$	24	2, 6, 8, 12	0, 4, 6, 10
29	$\mathcal{N}_4$	$2^9\,3\,5$	40	4, 8, 12, 20	0, 8, 12, 16
30	$\mathcal{H}_4$	$2^6\,3^2\,5^2$	60	2, 12, 20, 30	0, 10, 18, 28
31	$\mathcal{O}_4$	$2^{10}\,3^2\,5$	60	8, 12, 20, 24	0, 12, 16, 28
32	$\mathcal{L}_4$	$2^7\,3^5\,5$	80	12, 18, 24, 30	0, 6, 12, 18
33	$\mathcal{K}_5$	$2^7\,3^4\,5$	45	4, 6, 10, 12, 18	0, 6, 8, 12, 14
34	$\mathcal{K}_6$	$2^9\,3^7\,5\,7$	126	6, 12, 18, 24, 30, 42	0, 12, 18, 24, 30, 36
35	$\mathcal{E}_6$	$2^7\,3^4\,5$	36	2, 5, 6, 8, 9, 12	0, 3, 4, 6, 7, 10
36	$\mathcal{E}_7$	$2^{10}\,3^4\,5\,7$	63	2, 6, 8, 10, 12, 14, 18	0, 4, 6, 8, 10, 12, 16
37	$\mathcal{E}_8$	$2^{14}\,3^5\,5^2\,7$	120	2, 8, 12, 14, 18, 20, 24, 30	0, 6, 10, 12, 16, 18, 22, 28

3. Cartan matrices

$$\mathcal{H}_3 : \begin{bmatrix} 2 & -\tau & 0 \\ -\tau & 2 & -1 \\ 0 & -1 & 2 \end{bmatrix} \quad \mathcal{J}_3^{(4)} : \begin{bmatrix} 2 & -1 & -\lambda \\ -1 & 2 & -1 \\ 1+\lambda & -1 & 2 \end{bmatrix} \quad \mathcal{J}_3^{(5)} : \begin{bmatrix} 2 & -\tau & -\omega \\ -\tau & 2 & -\omega^2 \\ -\omega^2 & -\omega & 2 \end{bmatrix}$$

$$\mathcal{L}_3 : \begin{bmatrix} 1-\omega & \omega^2 & 0 \\ -\omega^2 & 1-\omega & -\omega^2 \\ 0 & \omega^2 & 1-\omega \end{bmatrix} \quad \mathcal{M}_3 : \begin{bmatrix} 1-\omega & -\omega^2 & 0 \\ \omega^2 & 1-\omega & -1 \\ 0 & -1+\omega & 2 \end{bmatrix}$$

$$\mathcal{F}_4 : \begin{bmatrix} 2 & -1 & 0 & 0 \\ -1 & 2 & -2 & 0 \\ 0 & -1 & 2 & -1 \\ 0 & 0 & -1 & 2 \end{bmatrix} \quad \mathcal{H}_4 : \begin{bmatrix} 2 & -\tau & 0 & 0 \\ -\tau & 2 & -1 & 0 \\ 0 & -1 & 2 & -1 \\ 0 & 0 & -1 & 2 \end{bmatrix}$$

$$\mathcal{L}_4 : \begin{bmatrix} 1-\omega & \omega^2 & 0 & 0 \\ -\omega^2 & 1-\omega & -\omega^2 & 0 \\ 0 & \omega^2 & 1-\omega & \omega^2 \\ 0 & 0 & -\omega^2 & 1-\omega \end{bmatrix} \quad \mathcal{N}_4 : \begin{bmatrix} 2 & -1 & i+1 & 0 \\ -1 & 2 & -i & 0 \\ -i+1 & i & 2 & -1 \\ 0 & 0 & -1 & 2 \end{bmatrix}$$

$$\mathcal{O}_4 : \begin{bmatrix} 2 & -1 & i+1 & 0 & -i+1 \\ -1 & 2 & -i & 0 & 0 \\ -i+1 & i & 2 & -1 & -i+1 \\ 0 & 0 & -1 & 2 & -1 \\ i+1 & 0 & i+1 & -1 & 2 \end{bmatrix} \quad \mathcal{K}_5 : \begin{bmatrix} 2 & -1 & 0 & 0 & 0 \\ -1 & 2 & -1 & -1 & 0 \\ 0 & -1 & 2 & -\omega & 0 \\ 0 & -1 & -\omega^2 & 2 & -\omega^2 \\ 0 & 0 & 0 & -\omega & 2 \end{bmatrix}$$

$$\mathcal{K}_6 : \begin{bmatrix} 2 & -1 & 0 & 0 & 0 & 0 \\ -1 & 2 & -1 & 0 & 0 & 0 \\ 0 & -1 & 2 & -1 & -1 & 0 \\ 0 & 0 & -1 & 2 & -\omega & 0 \\ 0 & 0 & -1 & -\omega^2 & 2 & -\omega^2 \\ 0 & 0 & 0 & 0 & -\omega & 2 \end{bmatrix} \quad \mathcal{E}_6 : \begin{bmatrix} 2 & -1 & 0 & 0 & 0 & 0 \\ -1 & 2 & -1 & 0 & 0 & 0 \\ 0 & -1 & 2 & -1 & -1 & 0 \\ 0 & 0 & -1 & 2 & 0 & 0 \\ 0 & 0 & -1 & 0 & 2 & -1 \\ 0 & 0 & 0 & 0 & -1 & 2 \end{bmatrix}$$

$$\mathcal{E}_7 : \begin{bmatrix} 2 & -1 & 0 & 0 & 0 & 0 & 0 \\ -1 & 2 & -1 & 0 & 0 & 0 & 0 \\ 0 & -1 & 2 & -1 & 0 & 0 & 0 \\ 0 & 0 & -1 & 2 & -1 & -1 & 0 \\ 0 & 0 & 0 & -1 & 2 & 0 & 0 \\ 0 & 0 & 0 & -1 & 0 & 2 & -1 \\ 0 & 0 & 0 & 0 & 0 & -1 & 2 \end{bmatrix}$$

$$\mathcal{E}_8 : \begin{bmatrix} 2 & -1 & 0 & 0 & 0 & 0 & 0 & 0 \\ -1 & 2 & -1 & 0 & 0 & 0 & 0 & 0 \\ 0 & -1 & 2 & -1 & 0 & 0 & 0 & 0 \\ 0 & 0 & -1 & 2 & -1 & 0 & 0 & 0 \\ 0 & 0 & 0 & -1 & 2 & -1 & -1 & 0 \\ 0 & 0 & 0 & 0 & -1 & 2 & 0 & 0 \\ 0 & 0 & 0 & 0 & -1 & 0 & 2 & -1 \\ 0 & 0 & 0 & 0 & 0 & 0 & -1 & 2 \end{bmatrix}$$

$$\omega = \tfrac{1}{2}(-1+i\sqrt{3}), \quad \tau = \tfrac{1}{2}(1+\sqrt{5}), \quad \lambda = -\tfrac{1}{2}(1+i\sqrt{7})$$

TABLE D.4. Maximal subsystems of primitive line systems

$\mathfrak{M}$	$\dim \mathfrak{L} = \dim \mathfrak{M}$	$\dim \mathfrak{M} - 1$
$\mathcal{H}_3$	$3\mathcal{A}_1$	$\mathcal{A}_2,\ \mathcal{D}_2^{(5)}$
$\mathcal{J}_3^{(4)}$	$\mathcal{B}_3^{(2)}$ (two classes)	
$\mathcal{J}_3^{(5)}$	$\mathcal{H}_3$ (two classes), $\mathcal{B}_3^{(2)}$ (two classes), $\mathcal{D}_3^{(3)}$	
$\mathcal{L}_3$	$3\mathcal{L}_1$	$\mathcal{L}_2$
$\mathcal{M}_3$	$\mathcal{L}_3,\ \mathcal{B}_3^{(3)},\ \mathcal{L}_2 \perp \mathcal{A}_1$	
$\mathcal{F}_4$	$\mathcal{B}_4^{(2)}$ (two classes), $2\mathcal{A}_2$	
$\mathcal{H}_4$	$\mathcal{A}_4,\ \mathcal{D}_4^{(2)},\ 2\mathcal{D}_2^{(5)},\ \mathcal{H}_3 \perp \mathcal{A}_1,\ 2\mathcal{A}_2$	
$\mathcal{L}_4$	$\mathcal{L}_3 \perp \mathcal{L}_1,\ 2\mathcal{L}_2$	
$\mathcal{N}_4$	$\mathcal{A}_4$ (two classes), $\mathcal{B}_4^{(2)},\ \mathcal{D}_4^{(4)}$	
$\mathcal{O}_4$	$\mathcal{B}_4^{(4)},\ \mathcal{F}_4,\ \mathcal{N}_4$	
$\mathcal{K}_5$	$\mathcal{A}_5,\ \mathcal{D}_4^{(2)} \perp \mathcal{A}_1,\ \mathcal{D}_3^{(3)} \perp \mathcal{A}_2$	$\mathcal{D}_4^{(3)}$
$\mathcal{K}_6$	$\mathcal{A}_6$ (two classes), $\mathcal{D}_6^{(2)},\ \mathcal{D}_6^{(3)},\ \mathcal{E}_6,\ \mathcal{K}_5 \perp \mathcal{A}_1$	
$\mathcal{E}_6$	$\mathcal{A}_5 \perp \mathcal{A}_1,\ 3\mathcal{A}_2$	$\mathcal{D}_5^{(2)}$
$\mathcal{E}_7$	$\mathcal{A}_7,\ \mathcal{D}_6^{(2)} \perp \mathcal{A}_1,\ \mathcal{A}_5 \perp \mathcal{A}_2$	$\mathcal{E}_6$
$\mathcal{E}_8$	$\mathcal{A}_8,\ \mathcal{D}_8^{(2)},\ \mathcal{E}_7 \perp \mathcal{A}_1,\ \mathcal{E}_6 \perp \mathcal{A}_2,\ 2\mathcal{A}_4$	

4. Maximal subsystems

Let $\mathfrak{M}$ be a star-closed line system whose reflection group $W(\mathfrak{M})$ is primitive. A star-closed subsystem $\mathfrak{L}$ of $\mathfrak{M}$ is maximal if and only if the extension $\mathfrak{L} \subset \mathfrak{M}$ is minimal. Therefore the maximal subsystems $\mathfrak{L}$ of $\mathfrak{M}$ such that $\mathfrak{L}$ is indecomposable can be read off from the results of Chapters 7 and 8.

With a little more work, using the descriptions given in Chapter 7 § 6, it is possible to determine all decomposable maximal subsystems. If the rank of $W(\mathfrak{M})$ is two, the subsystems can be found in the tables and diagrams of Chapter 6 and will not be reproduced here.

5. Reflection cosets

In the table of reflection cosets (Table D.5), the matrices for γ have been chosen so that $\gamma^k \in G$, where k is the order of the coset γG, modulo scalars.

TABLE D.5. Reflection cosets

γ	G	$d_1, d_2, \ldots, d_n$ $\varepsilon_1, \varepsilon_2, \ldots, \varepsilon_n$	$d_1^*, d_2^*, \ldots, d_n^*$ $\varepsilon_1^*, \varepsilon_2^*, \ldots, \varepsilon_n^*$	regular ζ
$\operatorname{diag}(\zeta_{em/p}, 1, \ldots, 1)$	$G(m,p,n)$ $p \neq m$	$m, 2m, \ldots, (n-1)m, nm/p$ $1, 1, \ldots, 1, \zeta_e^{-1}$	$0, m, 2m, \ldots, (n-1)m$ $1, 1, 1, \ldots, 1$	$\zeta^{nm/p} = \zeta_e$
$\operatorname{diag}(\zeta_{em/p}, 1, \ldots, 1)$	$G(m,m,n)$	$m, 2m, \ldots, (n-1)e, n$ $1, 1, \ldots, 1, \zeta_e^{-1}$	$0, m, \ldots, (n-2)e, (n-1)m - n$ $1, 1, \ldots, 1, \zeta_e$	$\zeta^{nm/p} = \zeta_e$ or $\zeta^{(n-1)m} = 1$
$\frac{1}{2}\begin{bmatrix} i-1 & i-1 \\ i+1 & -i-1 \end{bmatrix}$	$G(4,2,2)$	$4, 4$ ω, ω^2	$0, 4$ $1, 1$	$\zeta^4 = 1$
$\frac{1}{\sqrt{2}}\begin{bmatrix} 1 & 1 \\ 1 & -1 \end{bmatrix}$	G_5	$6, 12$ $1, -1$	$0, 6$ $1, -1$	$|\zeta| \in$ $\{1, 2, 3, 6, 8, 24\}$
$\frac{1}{2}\begin{bmatrix} i+1 & i-1 \\ i-1 & i+1 \end{bmatrix}$	G_7	$12, 12$ $1, -1$	$0, 12$ $1, -1$	$\zeta^{12} = 1$
$\frac{i}{1-\omega}\begin{bmatrix} 1 & \omega^2 \\ 1 & \omega \end{bmatrix}$	$G(3,3,3)$	$3, 3, 6$ $-1, -1, 1$	$0, 3, 3$ $1, -1, -1$	$\zeta^6 = 1$
$\frac{1}{2}\begin{bmatrix} 1 & 1 & 1 & 1 \\ 1 & 1 & -1 & -1 \\ 1 & -1 & -1 & 1 \\ 1 & -1 & 1 & -1 \end{bmatrix}$	$G(2,2,4)$	$2, 4, 4, 6$ $1, \omega, \omega^2, 1$	$0, 2, 2, 4$ $1, \omega, \omega^2, 1$	$|\zeta| \in$ $\{1, 2, 3, 6, 12\}$
$\frac{1}{\sqrt{2}}\begin{bmatrix} 0 & 0 & 1 & -1 \\ 0 & 1 & -1 & 0 \\ 1 & 0 & 0 & 0 \\ 1 & -1 & 0 & 0 \end{bmatrix}$	$W(\mathcal{F}_4) = G_{28}$	$2, 6, 8, 12$ $1, -1, 1, -1$	$0, 4, 6, 10$ $1, -1, 1, -1$	$|\zeta| \in$ $\{1, 2, 4, 8, 12, 24\}$

Bibliography

[1] P. N. Achar and A.-M. Aubert. On rank 2 complex reflection groups. *Comm. Algebra*, 36 (6): 2092–2132, 2008.

[2] K. K. S. Andersen, J. Grodal, J. M. Møller, and A. Viruel. The classification of p-compact groups for p odd. *Ann. of Math. (2)*, 167(1):95–210, 2008.

[3] S. Ariki. Modular representation theory of Hecke algebras, a survey. In *Infinite-dimensional aspects of representation theory and applications*, volume 392 of *Contemp. Math.*, pages 1–14. Amer. Math. Soc., Providence, RI, 2005.

[4] V. I. Arnol′d. The cohomology ring of the group of dyed braids. *Mat. Zametki*, 5:227–231, 1969.

[5] E. Artin. Theory of braids. *Ann. of Math. (2)*, 48:101–126, 1947.

[6] M. Aschbacher. *Finite Group Theory*, volume 10 of *Cambridge Studies in Advanced Mathematics*. Cambridge University Press, Cambridge, 1986.

[7] M. Aschbacher. *3-Transposition Groups*, volume 124 of *Cambridge Tracts in Mathematics*. Cambridge University Press, Cambridge, 1997.

[8] M. F. Atiyah and I. G. Macdonald. *Introduction to Commutative Algebra*. Addison-Wesley, Reading, Massachusetts, 1969.

[9] G. Bagnera. I gruppi finiti di trasformazioni lineari dello spazio che contengono omologie. *Rend. Circ. Mat. Palermo (2)*, 19:1–56, 1905.

[10] H. Barcelo and A. Goupil. Combinatorial aspects of the poincaré polynomial associated with a reflection group. In H. Barcelo and G. Kalai, editors, *Jerusalem Combinatorics '93*, volume 178 of *Contemporary Mathematics*. Amer. Math. Soc., Providence RI, 1994.

[11] M. Benard. Schur indices and splitting fields of the unitary reflection groups. *J. Algebra*, 38:318–342, 1976.

[12] D. J. Benson. *Polynomial Invariants of Finite Groups*, volume 190 of *London Mathematical Society Lecture Notes Series*. Cambridge University Press, Cambridge, 1993.

[13] D. Bessis. Sur le corps de définition d'une groupe de réflexions complexe. *Comm. Algebra*, 25:2703–2716, 1997.

[14] D. Bessis. Zariski theorems and diagrams for braid groups. *Invent. Math.*, 145:487–507, 2001.

[15] D. Bessis. Topology of complex reflection arrangements. *arXiv.org/abs/math/0411645v1*, 2004.

[16] D. Bessis. Finite complex reflection arrangements are $K(\pi, 1)$. *arXiv.org/abs/math/0610777*, 2007.

[17] D. Bessis, F. Digne, and J. Michel. Springer theory in braid groups and the Birman–Ko–Lee monoid. *Pacific J. Math.*, 205:287–310, 2002.

[18] D. Bessis and J. Michel. Explicit presentations for exceptional braid groups. *Experiment. Math.*, 13(3):257–266, 2004.

[19] J. S. Birman and H. Wenzl. Braids, link polynomials and a new algebra. *Trans. Amer. Math. Soc.*, 313(1):249–273, 1989.

[20] A. Björner and F. Brenti. *Combinatorics of Coxeter groups*, volume 231 of *Graduate Texts in Mathematics*. Springer, New York, 2005.

[21] J. Blair and G. I. Lehrer. Cohomology actions and centralisers in unitary reflection groups. *Proc. London Math. Soc. (3)*, 83:582–604, 2001.

[22] H. F. Blichfeldt. On the order of linear homogeneous groups. *Trans. Amer. Math. Soc.*, 4:387–397, 1903.

[23] H. F. Blichfeldt. On the order of linear homogeneous groups (second paper). *Trans. Amer. Math. Soc.*, 5:310–325, 1904.

[24] H. F. Blichfeldt. The finite discontinuous primitive groups of collineations in four variables. *Math. Ann.*, 60:204–231, 1905.

[25] H. F. Blichfeldt. The finite, discontinuous primitive groups of collineations in three variables. *Math. Ann.*, 63:552–572, 1907.

[26] H. F. Blichfeldt. *Finite Collineation Groups*. University of Chicago Press, Chicago, 1917.

[27] P. Boalch. Painlevé equations and complex reflections. In *Proceedings of the International Conference in Honor of Frédéric Pham (Nice, 2002)*, volume 53, pages 1009–1022, 2003.

[28] C. Bonnafé, G. I. Lehrer, and J. Michel. Twisted invariant theory for reflection groups. *Nagoya Math. J.*, 182:135–170, 2006.

[29] A. Borel. The work of Chevalley in Lie groups and algebraic groups. In S. Ramanan, editor, *Proceedings of the Hyderabad Conference on algebraic groups*, pages 1–22, Bombay, 1991. National Board for Higher Mathematics.

[30] A. Borel. *Essays in the History of Lie Groups and Algebraic Groups*, volume 21 of *History of Mathematics*. Amer. Math. Soc., Providence RI, 2001.

[31] A. Borel and J. De Siebenthal. Les sous-groupes fermés de rang maximum des groupes de Lie clos. *Comment. Math. Helv.*, 23:200–221, 1949.

[32] W. Bosma, J. Cannon, and C. Playoust. The Magma algebra system. I. The user language. *J. Symbolic Comput.*, 24(3-4):235–265, 1997. Computational algebra and number theory (London, 1993).

[33] N. Bourbaki. *Groupes et algèbres de Lie*. Chapitres 4, 5 et 6. Hermann, Paris, 1968.

[34] R. Brauer. Über endliche lineare Gruppen von Primzahlgrad. *Math. Ann.*, 169:73–96, 1967.

[35] E. Brieskorn. Die Fundamentalgruppe des Raumes der regulären Orbits einer endlichen komplexen Spiegelungsgruppe. *Invent. Math.*, 12:37–61, 1971.

[36] E. Brieskorn. Sur les groupes de tresses (d'après V. I. Arnol'd). In *Séminaire Bourbaki 24ᵉ année 1971/2*, volume 317 of *Lecture Notes in Mathematics*. Springer-Verlag, Berlin, 1973.

[37] M. Broué. Reflection groups, braid groups, Hecke algebras, finite reductive groups. In *Current developments in mathematics, 2000*, pages 1–107. Int. Press, Somerville, MA, 2001.

[38] M. Broué and G. Malle. Théorèmes de Sylow génériques pour les groupes réductifs sure les corps finis. *Math. Ann.*, 292:241–262, 1992.

[39] M. Broué, G. Malle, and J. Michel. Generic blocks of finite reductive groups. *Astérisque*, 212:7–92, 1993.

[40] M. Broué, G. Malle, and J. Michel. Towards Spetses. I. *Transform. Groups*, 4:157–218, 1999.

[41] M. Broué, G. Malle, and R. Rouquier. Complex reflection groups, braid groups, Hecke algebras. *J. Reine Angew. Math.*, 500:127–190, 1998.

[42] M. Broué and J. Michel. Sur certains éléments réguliers des groupes de Weyl et les variétés de Deligne–Lusztig associés. In M. Cabanes, editor, *Finite Reductive Groups: Related Structures and Representations*, volume 141 of *Progress in Mathematics*, pages 73–140. Birkhaüser, 1997.

[43] A. E. Brouwer, A. M. Cohen, and A. Neumaier. *Distance-regular graphs*, volume 18 of *Ergebnisse der Mathematik und ihrer Grenzgebiete (3) [Results in Mathematics and Related Areas (3)]*. Springer-Verlag, Berlin, 1989.

[44] A. R. Calderbank, P. Hanlon, and S. Sundaram. Representations of the symmetric group in deformations of the free Lie algebra. *Trans. Amer. Math. Soc.*, 341(1):315–333, 1994.

[45] P. J. Cameron, J. M. Goethals, J. J. Seidel, and E. E. Shult. Line graphs, root systems and elliptic geometry. *J. Algebra*, 43:305–327, 1976.

[46] P. J. Cameron and J. H. van Lint. *Designs, graphs, codes and their links*, volume 22 of *London Mathematical Society Student Texts*. Cambridge University Press, Cambridge, 1991.

[47] H. Can. Some combinatorial results for complex reflection groups. *Europ. J. Combinatorics*, 19:901–909, 1998.

[48] R. W. Carter. *Finite Groups of Lie Type – Conjugacy Classes and Complex Characters*. John Wiley and Sons, Chichester, New York, 1985.

[49] I. Cherednik, Y. Markov, R. Howe, and G. Lusztig. *Iwahori-Hecke algebras and their representation theory*, volume 1804 of *Lecture Notes in Mathematics*. Springer-Verlag, Berlin, 2002. Lectures from the C.I.M.E. Summer School held in Martina-Franca, June 28–July 6, 1999, edited by M. Welleda Baldoni and Dan Barbasch.

[50] C. Chevalley. The Betti numbers of the exceptional simple Lie groups. In *Proc. Internat. Congr. Math. (Cambridge, Mass., 1950)*, volume 2, pages 21–24. Amer. Math. Soc., 1952.

[51] C. Chevalley. Invariants of finite groups generated by reflections. *Amer. J. Math.*, 77:778–782, 1955.

[52] A. Clark and J. Ewing. The realization of polynomial algebras as cohomology rings. *Pacific J. Math.*, 50:425–434, 1974.

[53] G. Cliff, J. Ritter, and A. Weiss. Group representations and integrality. *J. Reine Angew. Math.*, 426:193–202, 1992.

[54] A. M. Cohen. Finite complex reflection groups. *Ann. Sci. École Norm. Sup. (4)*, 9:379–436, 1976.

[55] A. M. Cohen. Erratum: "Finite complex reflection groups". *Ann. Sci. École Norm. Sup. (4)*, 11:613, 1978.

[56] J. H. Conway, R. T. Curtis, S. P. Norton, R. A. Parker, and R. A. Wilson. *Atlas of Finite Groups*. Clarendon Press, Oxford, 1985.

[57] J. H. Conway and N. J. A. Sloane. The Coxeter–Todd lattice, the Mitchell group and related sphere packings. *Math. Proc. Cambridge Philos. Soc.*, 93:421–440, 1983.

[58] J. H. Conway and N. J. A. Sloane. *Sphere packings, lattices and groups*, volume 290 of *Grundlehren der Mathematischen Wissenschaften [Fundamental Principles of Mathematical Sciences]*. Springer-Verlag, New York, third edition, 1999. With additional contributions by E. Bannai, R. E. Borcherds, J. Leech, S. P. Norton, A. M. Odlyzko, R. A. Parker, L. Queen and B. B. Venkov.

[59] J. H. Conway and D. A. Smith. *On Quaternions and Octonions*. A K Peters, Natick, Massachusetts, 1985.

[60] H. S. M. Coxeter. Quaternions and reflections. *Amer. Math. Monthly*, 53:136–146, 1946.

[61] H. S. M. Coxeter. The symmetry groups of the regular complex polygons. *Arch. Math.*, 13:86–97, 1962.

[62] H. S. M. Coxeter. *Regular Complex Polytopes*. Cambridge University Press, Cambridge, second edition, 1991.

[63] H. S. M. Coxeter and W. O. J. Moser. *Generators and Relations for Discrete Groups*. Springer-Verlag, Berlin, 4th edition, 1980.

[64] H. S. M. Coxeter and J. A. Todd. An extreme duodenary form. *Canadian J. Math.*, 5:384–392, 1953.

[65] D. W. Crowe. The groups of the regular complex polygons. *Canad. J. Math.*, 13:149–156, 1961.

[66] D. W. Crowe. Some two-dimensional unitary groups generated by three reflections. *Canad. J. Math.*, 13:418–426, 1961.

[67] C. W. Curtis and G. I. Lehrer. Homology representations of finite groups of Lie type. In *Papers in algebra, analysis and statistics (Hobart, 1981)*, volume 9 of *Contemp. Math.*, pages 1–28. Amer. Math. Soc., Providence, R.I., 1981.

[68] D. Cvetković, P. Rowlinson, and S. Simić. *Spectral Generalizations of Line Graphs. On graphs with least eigenvalue −2*, volume 314 of *London Mathematical Society Lecture Note Series*. Cambridge University Press, Cambridge, 2004.

[69] M. W. Davis. *The geometry and topology of Coxeter groups*, volume 32 of *London Mathematical Society Monographs Series*. Princeton University Press, Princeton, NJ, 2008.

[70] C. De Concini, C. Procesi, and M. Salvetti. Arithmetic properties of the cohomology of braid groups. *Topology*, 40(4):739–751, 2001.

[71] P. Deligne. Les immeubles des groupes de tresses généralisés. *Invent. Math.*, 17:273–302, 1972.

[72] P. Deligne and G. Lusztig. Representations of reductive groups over finite fields. *Ann. of Math. (2)*, 103(1):103–161, 1976.

[73] P. Deligne and G. D. Mostow. Monodromy of hypergeometric functions and nonlattice integral monodromy. *Inst. Hautes Études Sci. Publ. Math.*, 63:5–89, 1986.

[74] P. Deligne and G. D. Mostow. *Commensurabilities among lattices in $PU(1, n)$*, volume 132 of *Annals of Mathematics Studies*. Princeton University Press, Princeton, 1993.

[75] P. Delsarte, J. M. Goethals, and J. J. Seidel. Bounds for systems of lines, and Jacobi polynomials. *Philips Research Reports*, 30:91*–105*, 1975.

[76] M. Demazure. Invariants symétriques entiers des groupes de Weyl et torsion. *Invent. Math.*, 21:287–301, 1973.

[77] J. Denef and F. Loeser. Regular elements and monodromy of discriminants of finite reflection groups. *Indag. Math. (N.S.)*, 6:129–143, 1995.

[78] G. Denham and N. Lemire. Equivariant Euler characteristics of discriminants of reflection groups. *Indag. Math. (N.S.)*, 13(4):441–458, 2002.

[79] J. A. Dieudonné and J. B. Carrell. Invariant theory, old and new. *Adv. Math.*, 4:1–80, 1970.

[80] F. Digne, G. Lehrer, and J. Michel. The space of unipotently supported class functions on a finite reductive group. *J. Algebra*, 260(1):111–137, 2003. Special issue celebrating the 80th birthday of Robert Steinberg.

[81] F. Digne and J. Michel. *Representations of finite groups of Lie type*, volume 21 of *London Mathematical Society Student Texts*. Cambridge University Press, Cambridge, 1991.

[82] F. Digne and J. Michel. Endomorphisms of Deligne–Lusztig varieties. *Nagoya Math. J.*, 183:35–103, 2006.

[83] F. Digne, J. Michel, and R. Rouquier. Cohomologie des variétés de Deligne–Lusztig. *Adv. Math.*, 209(2):749–822, 2007.

[84] A. Dimca and A. Libgober. Local topology of reducible divisors. In *Real and complex singularities*, Trends Math., pages 99–111. Birkhäuser, Basel, 2007.

[85] I. V. Dolgachev. Reflection groups in algebraic geometry. *Bull. Amer. Math. Soc. (N.S.)*, 45(1):1–60 (electronic), 2008.

[86] L. Dornhoff. *Group Representation Theory, Part A. Ordinary Representation Theory*. Marcel Dekker, New York, 1971.

[87] J. M. Douglass. A formula for the number of F-stable Levi factors in a finite reductive group. *Comm. Algebra*, 22(13):5447–5455, 1994.

[88] P. Du Val. *Homographies Quaternions and Rotations*. Oxford Mathematical Monographs. Oxford University Press, Clarendon Press, Oxford, 1964.

[89] C. F. Dunkl and E. M. Opdam. Dunkl operators for complex reflection groups. *Proc. London Math. Soc. (3)*, 86(1):70–108, 2003.

[90] M. Dyer. Reflection subgroups of Coxeter systems. *J. Algebra*, 135:57–73, 1990.

[91] M. Dyer. Embeddings of root systems I: Root systems over commutative rings. *To appear*.

[92] E. B. Dynkin. Semisimple subalgebras of semisimple Lie algebras. *Mat. Sbornik N.S.*, 30(72):349–462 (3 plates), 1952.

[93] H. Esnault, V. Schechtman, and E. Viehweg. Cohomology of local systems on the complement of hyperplanes. *Invent. Math.*, 109(3):557–561, 1992.

[94] P. Etingof and E. Rains. New deformations of group algebras of Coxeter groups. II. *Geom. Funct. Anal.*, 17(6):1851–1871, 2008.

[95] W. Feit. On integral representations of finite groups. *Proc. London Math. Soc. (3)*, 29:633–683, 1974.

[96] W. Feit. Some integral representations of complex reflection groups. *J. Algebra*, 260:138–153, 2003.

[97] B. Fischer. Finite groups generated by 3-transpositions. *Invent. Math.*, 13:232–246, 1971.

[98] L. Flatto. Invariants of finite reflection groups. *Enseign. Math. (2)*, 24:237–292, 1978.

[99] R. Fox and L. Neuwirth. The braid groups. *Math. Scand.*, 10:119–126, 1962.

[100] M. Geck. Hecke algebras of finite type are cellular. *Invent. Math.*, 169(3):501–517, 2007.

[101] G. Gonzalez-Sprinberg and J. L. Verdier. Construction géométrique de la correspondence de McKay. *Ann. Sci. École Norm. Sup. (4)*, 16:409–449, 1983.

[102] P. Gordon. Ueber endliche Gruppen linearen Transformationen einer Veränderlichen. *Math. Ann.*, 12:23–46, 1877.

[103] M. E. Goursat. Sur les substitutions orthogonales et les divisions régulières de l'espace. *Ann. Sci. École Norm. Sup.*, 6:1–102, 1889.

[104] J. J. Graham and G. I. Lehrer. Cellular algebras. *Invent. Math.*, 123(1):1–34, 1996.

[105] J. J. Graham and G. I. Lehrer. Cellular algebras and diagram algebras in representation theory. In *Representation theory of algebraic groups and quantum groups*, volume 40 of *Adv. Stud. Pure Math.*, pages 141–173. Math. Soc. Japan, Tokyo, 2004.

[106] I. Grojnowski and M. Vazirani. Strong multiplicity one theorems for affine Hecke algebras of type A. *Transform. Groups*, 6(2):143–155, 2001.

[107] E. A. Gutkin. Matrices connected with groups generated by mappings. *Funct. Anal. Appl.*, 7:153–154, 1973. Translated from *Funktsional. Anal. i Prilozhen.* 7, 81–82 (1973).

[108] C. M. Hamill. On a finite group of order 6,531,840. *Proc. London Math. Soc. (3)*, 52:401–454, 1951.

[109] E. M. Hartley. A sextic primal in five dimensions. *Math. Proc. Cambridge Philos. Soc.*, 46:91–105, 1950.

[110] M. Hazewinkel, W. Hesselink, D. Siersma, and F. D. Veldkamp. The ubiquity of the Coxeter–Dynkin diagrams. *Nieuw Arch. Wisk. (4)*, 25:257–307, 1977.

[111] J. F. C. Hessel. Krystallometrie oder Krystallonomie und Krystallographie. In *Physikalische Wörterbuch*. Gehler, Leipzig, 1830. republished in Oswald's "Klassiker der exacten Wissenschaften", No. 88, 89. Leipzig: W. Engelmann, 1897.

[112] D. Hilbert. Über die Theorie der algebraischen Formen. *Math. Ann.*, 36:473–534, 1890.

[113] H. Hiller. *Geometry of Coxeter Groups*, volume 54 of *Research Notes in Mathematics*. Pitman, Boston–London–Melbourne, 1982.

[114] R. B. Howlett and G. I. Lehrer. Induced cuspidal representations and generalised Hecke rings. *Invent. Math.*, 58(1):37–64, 1980.

[115] R. B. Howlett and G. I. Lehrer. Embeddings of Hecke algebras in group algebras. *J. Algebra*, 105(1):159–174, 1987.

[116] R. B. Howlett and G. I. Lehrer. On the integral group algebra of a finite algebraic group. *Astérisque*, 168:9–10, 141–155, 1988. Orbites unipotentes et représentations, I.

[117] R. B. Howlett and J.-y. Shi. On regularity of finite reflection groups. *Manuscripta Math.*, 102(3):325–333, 2000.

[118] M. C. Hughes and A. O. Morris. Root systems for two dimensional complex reflection groups. *Sém. Lothar. Combin.*, 45:Art. B45e, 18 pp. (electronic), 2000/01.

[119] J. E. Humphreys. *Reflection Groups and Coxeter Groups*, volume 29 of *Cambridge Studies in Advanced Mathematics*. Cambridge University Press, 1990.

[120] N. Iwahori. On the structure of a Hecke ring of a Chevalley group over a finite field. *J. Fac. Sci. Univ. Tokyo Sect. I*, 10:215–236 (1964), 1964.

[121] J. C. Jantzen. *Representations of algebraic groups*, volume 107 of *Mathematical Surveys and Monographs*. American Mathematical Society, Providence, RI, second edition, 2003.

[122] V. F. R. Jones. Hecke algebra representations for braid groups and link polynomials. *Ann. of Math. (2)*, 126:335–388, 1987.

[123] C. Jordan. Mémoire sur les équations differentielles linéaires à intégrale algébrique. *J. Reine Angew. Math.*, 84:89–214, 1878.

[124] R. Kane. *Reflection groups and invariant theory*. CMS Books in Mathematics/Ouvrages de Mathématiques de la SMC, 5. Springer-Verlag, New York, 2001.

[125] W. M. Kantor. Generation of linear groups. In *The Geometric Vein*, pages 497–509. Springer, New York, 1981.

[126] D. Kazhdan and G. Lusztig. Representations of Coxeter groups and Hecke algebras. *Invent. Math.*, 53(2):165–184, 1979.

[127] D. Kazhdan and G. Lusztig. Schubert varieties and Poincaré duality. In *Geometry of the Laplace operator (Proc. Sympos. Pure Math., Univ. Hawaii, Honolulu, Hawaii, 1979)*, Proc. Sympos. Pure Math., XXXVI, pages 185–203. Amer. Math. Soc., Providence, R.I., 1980.

[128] M. Kisin and G. Lehrer. Eigenvalues of Frobenius and Hodge numbers. *Pure Appl. Math. Q.*, 2(2):497–518, 2006.

[129] M. Kisin and G. I. Lehrer. Equivariant Poincaré polynomials and counting points over finite fields. *J. Algebra*, 247(2):435–451, 2002.

[130] F. Klein. Ueber die Transformationen siebenter Ordnung der elliptischen Funktionen. *Math. Ann.*, 14:428–471, 1879.

[131] F. Klein. *Vorlesungen über das Ikosaeder und die Auflösung der Gleichung vom fünften Grade*. Teubner, Leipzig, 1884.

[132] M. Kneser. Über die Ausnahme-Isomorphismen zwischen endlichen klassischen Gruppen. *Abh. Math. Sem. Univ. Hamburg*, 31:136–140, 1967.

[133] V. G. Knizhnik and A. B. Zamolodchikov. Current algebra and Wess–Zumino model in two dimensions. *Nuclear Phys. B*, 247(1):83–103, 1984.

[134] T. Kohno. Hecke algebra representations of braid groups and classical Yang–Baxter equations. In *Conformal field theory and solvable lattice models (Kyoto, 1986)*, volume 16 of *Adv. Stud. Pure Math.*, pages 255–269. Academic Press, Boston, MA, 1988.

[135] T. Kohno. Elliptic KZ system, braid group of the torus and Vassiliev invariants. *Topology Appl.*, 78(1–2):79–94, 1997. Special issue on braid groups and related topics (Jerusalem, 1995).

[136] T. H. Koornwinder. A note on the absolute bound for systems of lines. *Nederl. Akad. Wetensch. Proc. Ser. A* **79**=*Indag. Math.*, 38(2):152–153, 1976.

[137] B. Kostant. On finite subgroups of $SU(2)$, simple Lie algebras, and the McKay correspondence. *Proc. Nat. Acad. Sci. U.S.A.*, 81:5275–5277, 1984.

[138] B. Kostant and S. Kumar. The nil Hecke ring and cohomology of G/P for a Kac–Moody group G. *Adv. in Math.*, 62(3):187–237, 1986.

[139] B. Kostant and S. Kumar. T-equivariant K-theory of generalized flag varieties. *J. Differential Geom.*, 32(2):549–603, 1990.

[140] M. Krishnasamy and D. E. Taylor. Embeddings of complex line systems and finite reflection groups. *J. Aust. Math. Soc.*, 85:211–228, 2008.

[141] L. Kronecker. Les facteurs irréductibles de l'expression $x^n - 1$. *J. Math. Pures Appl. (9)*, 19:177–192, 1854.

[142] S. Lang. *Algebra*, volume 211 of *Graduate Texts in Mathematics*. Springer, New York, Berlin, revised third edition, 2002.

[143] G. I. Lehrer. On the Poincaré series associated with coxeter group actions on complements of hyperplanes. *J. London Math. Soc. (2)*, 36:275–294, 1987.

[144] G. I. Lehrer. The *l*-adic cohomology of hyperplane complements. *Bull. London Math. Soc.*, 24(1):76–82, 1992.

[145] G. I. Lehrer. Rational tori, semisimple orbits and the topology of hyperplane complements. *Comment. Math. Helv.*, 67(2):226–251, 1992.

[146] G. I. Lehrer. Poincaré polynomials for unitary reflection groups. *Invent. Math.*, 120:411–425, 1995.

[147] G. I. Lehrer. The cohomology of the regular semisimple variety. *J. Algebra*, 199(2):666–689, 1998.

[148] G. I. Lehrer. A new proof of Steinberg's fixed-point theorem. *Int. Math. Research Notes*, 28:1407–1411, 2004.

[149] G. I. Lehrer. Rational points and cohomology of discriminant varieties. *Adv. Math.*, 186(1):229–250, 2004.

[150] G. I. Lehrer. Remarks concerning linear characters of reflection groups. *Proc. Amer. Math. Soc.*, 133(11):3163–3169 (electronic), 2005.

[151] G. I. Lehrer and J. Michel. Invariant theory and eigenspaces for unitary reflection groups. *C. R. Math. Acad. Sci. Paris*, 336(10):795–800, 2003.

[152] G. I. Lehrer and G. B. Segal. Homology stability for classical regular semisimple varieties. *Math. Z.*, 236(2):251–290, 2001.

[153] G. I. Lehrer and L. Solomon. On the action of the symmetric group on the cohomology of the complement of its reflecting hyperplanes. *J. Algebra*, 44:225–228, 1986.

[154] G. I. Lehrer and T. A. Springer. Intersection multiplicities and reflection subquotients of unitary reflection groups. I. In *Geometric Group Theory Down Under (Canberra 1996)*, pages 181–193. de Gruyter, Berlin, 1999.

[155] G. I. Lehrer and T. A. Springer. Reflection subquotients of unitary reflection groups. *Canad. J. Math.*, 51:1175–1193, 1999.

[156] G. I. Lehrer and N. Xi. On the injectivity of the braid group in the Hecke algebra. *Bull. Austral. Math. Soc.*, 64(3):487–493, 2001.

[157] G. I. Lehrer and R. B. Zhang. Strongly multiplicity free modules for Lie algebras and quantum groups. *J. Algebra*, 306(1):138–174, 2006.

[158] G. I. Lehrer and R. B. Zhang. A Temperley–Lieb analogue for the BMW algebra. *To appear*.

[159] G. Lusztig. Some examples of square integrable representations of semisimple *p*-adic groups. *Trans. Amer. Math. Soc.*, 277(2):623–653, 1983.

[160] G. Lusztig. Green functions and character sheaves. *Ann. of Math. (2)*, 131(2):355–408, 1990.

[161] G. Malle. Splitting fields for extended complex reflection groups and Hecke algebras. *Transform. Groups*, 11(2):195–216, 2006.

[162] J. McKay. Graphs, singularities and finite groups. In *Proc. Symp. in Pure Math.*, volume 37, pages 183–186, Providence RI, 1980. Amer. Math. Soc.

[163] G. A. Miller, H. F. Blichfeldt, and L. E. Dickson. *Theory and Applications of Finite Groups*. G. E. Stechert & Co., New York, 1938. Reprint of the 1916 edition of John Wiley & Sons.

[164] H. H. Mitchell. Determination of the ordinary and modular ternary linear groups. *Trans. Amer. Math. Soc.*, 12:207–242, 1911.

[165] H. H. Mitchell. Determination of the finite quaternary linear groups. *Trans. Amer. Math. Soc.*, 14:123–142, 1913.

[166] H. H. Mitchell. Determination of all primitive collineation groups in more than four variables which contain homologies. *Amer. J. Math.*, 36:1–12, 1914.

[167] T. Molien. Über die Invarianten der linearen Substitutionsgruppe. *Sitzungsber. König. Preuss. Akad. Wiss.*, pages 1152–1156, 1897.

[168] D. Mumford. Hilbert's fourteenth problem – the finite generation of subrings as rings of invariants. In F. E. Bowder, editor, *Mathematical Developments Arising from Hilbert's Problems*, volume 28 of *Proc. Symp. in Pure Math.* Amer. Math. Soc., Providence RI, 1976.

[169] M. Nagata. On the 14th problem of Hilbert. *Amer. J. Math.*, 81:766–772, 1959.

[170] T. Nakamura. A note on the $K(\pi, 1)$-property of the orbit space of the unitary reflection group $G(m, \ell, n)$. *Sci. Papers College Arts Sci. Univ. Tokyo*, 33:1–6, 1983.

[171] G. Nebe. The root lattices of the complex reflection groups. *J. Group Theory*, 2:15–38, 1999.

[172] M. D. Neusel and L. Smith. *Invariant Theory of Finite Groups*, volume 94 of *Mathematical Surveys and Monographs*. Amer. Math. Soc., Providence RI, 2002.

[173] E. Noether. Der Endlichkeitssatz der Invarianten endlicher Gruppen. *Math. Ann.*, 77:89–93, 1916.

[174] E. M. Opdam. A remark on the irreducible characters and fake degrees of finite real reflection groups. *Invent. Math.*, 120:447–454, 1995.

[175] E. M. Opdam. *Lecture notes on Dunkl operators for real and complex reflection groups*, volume 8 of *MSJ Memoirs*. Mathematical Society of Japan, Tokyo, 2000. With a preface by Toshio Oshima.

[176] P. Orlik, V. Reiner, and A. V. Shepler. The sign representation for Shephard groups. *Math. Ann.*, 322(3):477–492, 2002.

[177] P. Orlik and L. Solomon. Combinatorics and topology of complements of hyperplanes. *Invent. Math.*, 56:167–189, 1980.

[178] P. Orlik and L. Solomon. Unitary reflection groups and cohomology. *Invent. Math.*, 59:77–94, 1980.

[179] P. Orlik and L. Solomon. Braids and discriminants. In J. S. Birman and A. Lebgober, editors, *Braids*, volume 78 of *Contemporary Mathematics*, pages 605–613. Amer. Math. Soc., Providence RI, 1988.

[180] P. Orlik and L. Solomon. Discriminants in the invariant theory of reflection groups. *Nagoya Math. J.*, 109:23–45, 1988.

[181] P. Orlik and L. Solomon. The Hessian map in the invariant theory of reflection groups. *Nagoya Math. J.*, 109:1–21, 1988.

[182] P. Orlik and H. Terao. *Arrangements of hyperplanes*, volume 300 of *Grundlehren der Mathematischen Wissenschaften*. Springer-Verlag, Berlin, 1992.

[183] A. Pianzola and A. Weiss. Monstrous E_{10}'s and a generalization of a theorem of L. Solomon. *C. R. Math. Rep. Acad. Sci. Canada*, 11(5):189–194, 1989.

[184] V. L. Popov. *Discrete Complex Reflection Groups*. Communications of the Mathematical Institute. Rijksuniversiteit Utrecht, Mathematical Institute, Utrecht, 1982.

[185] D. Quillen. Homotopy properties of the poset of nontrivial p-subgroups of a group. *Adv. in Math.*, 28(2):101–128, 1978.

[186] G. R. Robinson. On linear groups. *J. Algebra*, 131:527–534, 1990.

[187] V. Schechtman, H. Terao, and A. Varchenko. Local systems over complements of hyperplanes and the Kac-Kazhdan conditions for singular vectors. *J. Pure Appl. Algebra*, 100(1–3):93–102, 1995.

[188] G. Segal and A. Selby. The cohomology of the space of magnetic monopoles. *Comm. Math. Phys.*, 177(3):775–787, 1996.

[189] J.-P. Serre. Groupes finis d'automorphismes d'anneaux locaux réguliers. *Colloque d'Algèbre EN-SJF, Paris*, 8:1–11, 1967.

[190] O. P. Shcherbak. Wave fronts and reflection groups. *Uspekhi Mat. Nauk*, 43(3(261)):125–160, 271, 272, 1988.

[191] G. C. Shephard. Regular complex polytopes. *Proc. London Math. Soc. (3)*, 2:82–97, 1952.

[192] G. C. Shephard. Unitary groups generated by reflections. *Canad. J. Math.*, 5:364–383, 1953.

[193] G. C. Shephard and J. A. Todd. Finite unitary reflection groups. *Canad. J. Math.*, 6:274–304, 1954.

[194] A. V. Shepler. Semi-invariants of finite reflection groups. *J. Algebra*, 220(1):314–326, 1999.

[195] J.-y. Shi. Simple root systems and presentations for certain complex reflection groups. *Comm. Algebra*, 33(6):1765–1783, 2005.

[196] T. Shoji. Representations of finite Chevalley groups. In *Groups and combinatorics – in memory of Michio Suzuki*, volume 32 of *Adv. Stud. Pure Math.*, pages 369–378. Math. Soc. Japan, Tokyo, 2001.

[197] T. Shoji. On Green functions associated to complex reflection groups [translation of sugaku 54 (2002), no. 1, 69–85]. *Sugaku Expositions*, 18(2):123–141, 2005. Sugaku Expositions.

[198] T. Shoji. Generalized Green functions and unipotent classes for finite reductive groups. I. *Nagoya Math. J.*, 184:155–198, 2006.

[199] L. Solomon. Invariants of finite reflection groups. *Nagoya Math. J.*, 22:57–64, 1963.

[200] N. Spaltenstein. Coxeter classes of unitary reflection groups. *Invent. Math.*, 119:297–316, 1995.

[201] T. A. Springer. Regular elements of reflection groups. *Invent. Math.*, 25:159–198, 1974.

[202] T. A. Springer. *Invariant Theory*, volume 585 of *Lecture Notes in Mathematics*. Springer-Verlag, Berlin–Heidelberg–New York, 1977.

[203] T. A. Springer. Poincaré series of binary polyhedral groups and McKay's correspondence. *Math. Ann.*, 278:99–116, 1987.

[204] T. A. Springer. Some remarks on characters of binary polyhedral groups. *J. Algebra*, 131:641–647, 1990.

[205] T. A. Springer and R. Steinberg. Conjugacy classes. In *Seminar on Algebraic Groups and Related Finite Groups (The Institute for Advanced Study, Princeton, N.J., 1968/69)*, Lecture Notes in Mathematics, Vol. 131, pages 167–266. Springer, Berlin, 1970.

[206] R. P. Stanley. Some aspects of groups acting on finite posets. *J. Combin. Theory Ser. A*, 32(2):132–161, 1982.

[207] R. Steinberg. Invariants of finite reflection groups. *Canad. J. Math.*, 12:616–618, 1960.

[208] R. Steinberg. Differential equations invariant under finite reflection groups. *Trans. Amer. Math. Soc.*, 112:392–400, 1964.

[209] R. Steinberg. *Endomorphisms of Linear Algebraic Groups*, volume 80 of *Mem. Amer. Math. Soc.* Amer. Math. Soc., Providence RI, 1968.

[210] R. Steinberg. *Lectures on Chevalley groups.* Yale University, New Haven, Conn., 1968. Notes prepared by John Faulkner and Robert Wilson.

[211] R. Steinberg. Finite subgroups of SU_2, Dynkin diagrams and affine Coxeter elements. *Pacific J. Math.*, 118:587–598, 1985.

[212] R. Stekolshchik. *Notes on Coxeter transformations and the McKay correspondence.* Springer-Verlag, Berlin, Heidelberg, 2008.

[213] J. R. Stembridge. On the eigenvalues of representations of reflection groups and wreath products. *Pacific J. Math.*, 140(2):353–396, 1989.

[214] W. I. Stringham. Determination of the finite quaternion groups. *Amer. J. Math.*, 4:345–357, 1881.

[215] D. E. Taylor. Some classical theorems on division rings. *Enseign. Math. (2)*, 20:293–298, 1974.

[216] D. E. Taylor. *The Geometry of the Classical Groups*, volume 9 of *Sigma Series in Pure Mathematics*. Heldermann Verlag, Berlin, 1992.

[217] W. Threlfall and H. Seifert. Topologische Untersuchung der Diskontinuitätsbereiche endlicher Bewegungsgruppen der dreidimensionalen sphärischen Raumes. *Math. Ann.*, 104:1–70, 1931.

[218] J. A. Todd. The invariants of a finite collineation group in five dimensions. *Math. Proc. Cambridge Philos. Soc.*, 46:73–90, 1950.

[219] B. Totaro. Towards a Schubert calculus for complex reflection groups. *Math. Proc. Cambridge Philos. Soc.*, 134(1):83–93, 2003.

[220] H. Valentiner. De endelige Transformations-Grupper Theori. Avec un résumé en français. *K. danske vidensk. selsk. (Copenhagen)*, 5(6):64–235, 1889.

[221] E. B. Vinberg. Discrete linear groups generated by reflections. *Math. USSR-Izv.*, 5:1083–1119, 1971.

[222] G. E. Wall. On the conjugacy classes in the unitary, symplectic and orthogonal groups. *J. Austral. Math. Soc.*, 3:1–62, 1963.

[223] L. C. Washington. *Introduction to cyclotomic fields*, volume 83 of *Graduate Texts in Mathematics*. Springer-Verlag, New York, second edition, 1997.

[224] A. Wiman. Ueber eine einfache gruppe von 360 ebenen collineationen. *Math. Annalen*, 47:531–556, 1896.

[225] A. Wiman. Endliche Gruppen linearer Substitutionen. In W. F. Meyer, editor, *Encyklopädie der Mathematischen Wissenschaften*, volume I of *part I*, pages 523–554. B. G. Teubner, Leipzig, 1899.

[226] E. Witt. Spiegelungsgruppen und Aufzählung halbeinfacher Liescher Ringe. *Abh. Math. Sem. Univ. Hamburg*, 14:289–322, 1941.

Index of notation

$\langle a \mid b \rangle$ Cartan coefficient, 17
γG reflection coset, 228
$\perp$ orthogonal, 9
$[V, g]$ image of $1 - g$, 9
$[x, y]$ commutator, 13

$a(d)$ number of degrees divisible by d, 203
$\mathcal{A}(G)$ reflecting hyperplanes of G, 176
$A[\![t]\!]$ formal power series, 54
$\mathrm{Av}(P)$ averaging operator, 42

$\mathcal{B}_n^{(3)}$ mixed $(3, 6)$-system, 148
$\mathcal{B}_n^{(k)}$ line system, 104
$b(d)$ number of codegrees divisible by d, 205

$C_G(X)$ centraliser, 13
$C(G, M)$ sum of the M-exponents of G, 195
$\mathcal{C}_n$ cyclic group, 13, 73

$\mathcal{D}_n^{(k)}$ line system, 104
$[f, P]$ non-degenerate pairing, 181
$D(g)$ Dickson's invariant, 252
$d(g, \zeta)$ dimension of $V(g, \zeta)$, 203
$\dim A$, 211
$\Delta = \Delta_V$ discriminant polynomial, 231
$\mathcal{D}_m$ binary dihedral group, 73
δ_{r_i} Demazure operator, 188
$\mathcal{D}_S$ algebra of differential operators, 179
d-regular element, 217

$\mathcal{E}_6, \mathcal{E}_7, \mathcal{E}_8$ 3-systems, 106–107
$\mathrm{End}_G(V)$ endomorphism ring, 14

$F = SJ^+$ ideal generated by invariants, 42
$\mathcal{F}_4$ 4-system, 109
$f_\chi(t)$ fake degree, 63
$\mathrm{Fix}\, g$ fixed points of g, 9
$f_M(t)$ fake degree, 192

$G', [G, G]$ derived group, 13
G_A pointwise stabiliser, 12
$G \circ H$ central product, 13
$G(d)$ reflection subquotient, 215
$GL(V)$ general linear group, 8
$G(m, p, n)$ imprimitive reflection group, 25
$\Gamma_a, \Gamma_b, \Gamma_c, \Delta, \Lambda$ Goethals–Seidel decomposition, 111

$H \leq G$ subgroup, 12
$H \trianglelefteq G$ normal subgroup, 12
$\mathbb{H}$ quaternions, 67
$\mathcal{H}_3, \mathcal{H}_4$ 5-systems, 110
$\mathcal{H}$ space of G-harmonic polynomials, 183
$\mathrm{Hom}_G(V, W)$ linear transformations, 14
$H \wr G$ wreath product, 24

$\mathcal{I}$ binary icosahedral group, 73
$\mathcal{I}(A)$ ideal of A, 208

$J = S^G$ algebra of invariants, 41
$\mathcal{J}_3^{(4)}$ 4-system, 108
$\mathcal{J}_3^{(5)}$ 5-system, 110
$\mathrm{Jac}(\omega_G)$ Jacobian matrix, 172

$\mathcal{K}_5, \mathcal{K}_6$ 3-systems, 107–108

$\mathcal{L}_2, \mathcal{L}_3, \mathcal{L}_4$ ternary 6-systems, 148
$\Lambda(V)$ exterior algebra, 57
λ root of $\lambda^2 + \lambda + 2 = 0$, 108, 153
L_H, 46
$L(q), R(q)$ left and right multiplication, 68

$\mathcal{M}_3$ mixed $(3, 6)$-system, 149
Δ_M M-discriminant, 231
$\varepsilon_i(M)$ M-factors, 230
M_G hyperplane complement, 255

$\mathcal{N}_4$ 4-system, 109
$N_G(X)$ normaliser, 13
N/C reflection subquotient, 214
$\mathrm{N}(q)$ quaternion norm, 67

$\mathcal{O}$ binary octahedral group, 73
$\mathcal{O}_4$ 4-system, 109
ω cube root of unity, 10, 103
$\widehat{\Omega}(V,Q), \widehat{\Omega}_{2m}^{\varepsilon}(\mathbb{F}_q), \widehat{\Omega}_n(\mathbb{F}_q)$ kernel of the
    spinor norm, 253
$O_n(\mathbb{F}_q), O_{2m}^+(\mathbb{F}_q), O_{2m}^-(\mathbb{F}_q)$ finite
    orthogonal groups, 251
$O_p(G)$ largest normal p-subgroup, 13, 156

$[f, P]$ pairing of $S(V) \times S(V^*)$, 181
$\Phi(P)$ Frattini subgroup, 13
φ_d-Sylow theory, 269
Π skew invariant, 172
Π_M M-skew polynomial, 194
$P_M^G(t)$ Poincaré polynomial, 55
$P_{(S \otimes \Lambda M^*)^G}(t, u)$ Poincaré series, 201

$\mathcal{Q}$ quaternion group, 73
$\mathbb{Q}(G)$ field of definition, 19
$q_i(M)$ M-exponent, 192
$q \cdot r$ inner product of quaternions, 69

r_a reflection of order two, 101
$R_{\mathbf{T}_w}^{\mathbf{G}}(\theta_w)$ Deligne–Lusztig character, 267

S^3 unit sphere, 69
S/F coinvariant algebra, 51
$Sp(V)$ symplectic group, 250
$(S \otimes M)^G$ module of M-covariants, 192
$SU_n(\mathbb{C})$ special unitary group, 8
$S(V)$ symmetric algebra, 40, 56
$S(V^*)$ coordinate ring, 40
$\mathrm{Sym}(n)$ symmetric group, 11
$\mathcal{S}_\zeta(G)$ poset of eigenspaces, 270

$\mathcal{T}$ binary tetrahedral group, 73
t_a reflection of order three, 147
τ golden section: $\tau^2 = \tau + 1$, 73, 91
$\mathrm{Tr}(q)$ quaternion trace, 67
$T^r(V)$ tensor power, 39
$T(V)$ tensor algebra, 39

$U(V), U_n(\mathbb{C})$ unitary group, 8
$U_n(\mathbb{F}_{q^2})$ finite unitary group, 250

$V(d)$ union of eigenspaces, 212
$V(g, \zeta)$ ζ-eigenspace of g, 203
V^σ Galois twist, 198
$\mathcal{V}(T)$ variety of T, 208

$W(A_m)$, 29
$W(\mathcal{A}_n) = \mathrm{Sym}(n+1)$, structure of, 157
$W(C)$ Weyl group of a Cartan matrix, 18
$W(\mathcal{E}_6) = G_{35}$, structure of, 167
$W(\mathcal{E}_7) = G_{36}$, structure of, 167
$W(\mathcal{E}_8) = G_{37}$, structure of, 167
$W(\mathcal{F}_4) = G_{28}$, structure of, 165
$W(\mathcal{H}_3) = G_{23}$, structure of, 159
$W(\mathcal{H}_4) = G_{30}$, structure of, 158
$W(\mathcal{J}_3^{(4)}) = G_{24}$, structure of, 160
$W(\mathcal{J}_3^{(5)}) = G_{27}$, structure of, 160
$W(\mathcal{K}_5) = G_{33}$, structure of, 166
$W(\mathcal{K}_6) = G_{34}$, structure of, 166
$W(\mathfrak{L})$ Weyl group of a line system, 102
$W(\mathcal{L}_3) = G_{25}$, structure of, 161
$W(\mathcal{L}_4) = G_{32}$, structure of, 162
$W(\mathcal{M}_3) = G_{26}$, structure of, 161
$W(\mathcal{N}_4) = G_{29}$, structure of, 165
$W(\mathcal{O}_4) = G_{31}$, structure of, 165

X/G orbits of G on X, 12

$Z(G)$ centre, 13
$\mathbb{Z}(G)$ ring of definition, 20

Index

algebra
  coinvariant, *see* coinvariant algebra
  division algebra, 67
  Hecke, 261
  of differential operators, 179
  of invariants, 41
  symmetric, 40
  tensor, 39
algebraic independence, 36
algebraic set
  reducible, 209
amenable module, 197
Ariki, S., 265
Arnol'd, V. I., 256
Artinian ring, 246
averaging operator, 42

basic invariants, 49
basis
  orthogonal, 8
bigraded modules, 56
bigrading, 201
bilinear form, *see* form
bilinear pairing, 180
Bott–Solomon–Tits formula, 190
braid relation, 259
Brieskorn, E., 256
Broué, M., 255

Cartan coefficient, 17, 154
Cartan matrix, 18, 35, 91
cellular structure
  of a Hecke algebra, 265
centralisers of elements
  and reflection subquotients, 237
chain rule, 187
character, 19, 24

classification
  imprimitive reflection groups, 28
  primitive reflection groups, 151
  rank two reflection groups, 85–90
codegrees, 201
  of a reflection subquotient, 221
coexponents, 201
cofactor, 230
cohomology, 256
  of hyperplane complement, 257
coinvariant algebra, 51
  and parabolic subgroups, 240, 241
  as graded G-module, 52, 53
commutative algebra
  dimension, 246
  finitely generated, 43
  integral extension, 43
  localisation, 247
  of finite type, 43
coordinate ring, 40, 209
covariant, 191
Coxeter elements
  in duality groups, 244
Coxeter group, 2, 27, 105
Cremona variety, 222

degree, of a monomial, 40
degrees, 51
  codegrees, 201
Deligne–Lusztig character, 267
Demazure operators, 188, 265
derivation, 179
Dickson invariant, 252
differential operator, 179
dimension
  and transcendence degree, 212
  of a ring, 211

directional derivative, 179
dual space, 10
duality group, 242

eigenspace theory
    twisted, 234
eigenspaces, 203
    and invariant hypersurfaces, 212
exponents, 51, 200
    M-exponents, 192, 195
    coexponents, 201

factor, 230
faithful representation, 14
fake degree, 63, 191, 192
    of an amenable module, 199
5-systems
    classification of, 146
fixed point space, 9
form
    alternating, 158, 162, 250
    hermitian, 7, 157, 250
    polar, 158, 250
    quadratic, 158, 250
formal power series, 54
4-systems
    classification of, 133
frame, 160
    special frame, 163
free
    associative algebra, 39
    commutative algebra, 40

Galois twist, 198
grading
    polynomial ring, 40
graph
    of a set of reflections, 16
Grothendieck ring, 54
group
    Artin braid group, 259
    binary
        dihedral, 73
        icosahedral, 73, 84, 158
        octahedral, 73, 84, 165
        tetrahedral, 73, 84
    central product, 13
    centraliser, 13
    centre, 13
    commutator, 13

dihedral, 26
duality, 242
elementary abelian, 13
extension, 13
extraspecial, 13, 31, 161, 164
Frattini subgroup, 13
fundamental, 258
imprimitive, 23
normal closure, 13, 85, 139
normaliser, 13
orthogonal, 158, 250
primitive, 23
quaternion, 27
reflection subgroup, 16
representation, 14
semidirect product, 13
Shephard, 87, 242
Sylow subgroup, 13
symmetric group, 11, 25, 104
unitary, 8
unitary reflection group, 10
Weyl group, 18
wreath product, 24, 110, 157

harmonic function, 183
    as a derivative, 183
harmonic polynomial, 183
Hessian, 94
hyperplane
    reflecting, 9

imprimitive reflection group, 25–37
inner product
    hermitian, 7, 71
    of quaternions, 69
integer
    Eisenstein, 153
    Gaussian, 153
    golden, 153
    Kleinian, 153
integral dependence
    and surjectivity, 45
intertwining number, 191
invariant
    basic, 49
    differential operator, 183
    polynomial, 36
    semi-invariant, 93
invariant pairing, 182

invariants
  finite generation, 43
irreducible algebraic set, 210
irreducible component
  of an algebraic set, 211

Jacobian, 94, 172
  rank of, 174
Jacobian matrix, 172

$K(\pi, 1)$, 260
Knizhnik, V. G., 264
Krull's Hauptidealsatz, 248

Lagrange interpolation, 40
Lehrer, G.I., 2, 5, 171, 186, 213, 256, 262, 265
line system, 99
  k-system, 101
  dimension, 104
  Euclidean, 101
  indecomposable, 102
  minimal extension, 104
  simple extension, 104
  star-closed, 101
  star-closure, 103
  ternary k-system, 148
linear character, 93, 176
Lusztig, G., 262, 267

Maschke's Theorem, 14
matrix
  of a linear transformation, 14
  orthogonal, 106, 143
  unitary, 8, 107
Milnor fibre, 257
M-discriminant, 231
module, 14
  amenable, 197
  extension, 14
  graded, 52, 54
  imprimitive, 23
  irreducible, 14
  isotypic component, 23
  module homomorphism, 14
  primitive, 23
  submodule, 14
module of covariants, 192
monodromy, 260

morphism
  surjective, 45

Nakayama's Lemma, 247
nil Hecke algebra, 265
Noetherian
  module, 42
  ring, 42

orbit
  and invariants, 41
  group action, 12
orbit map, 171
  Jacobian of, 173
Orlik and Terao, 218
Orlik–Solomon, 201
orthogonal complement, 9, 69

Painlevé equations, 264
parabolic subgroup, 171
  twisted invariants of, 239
partially ordered set, 270
Poincaré polynomial
  of a hyperplane complement, 257
Poincaré series, 55
  twisted two-variable, 232
polynomial
  elementary symmetric, 36
polynomial map
  surjective, 210
positive definite hermitian form, 7
prime ideal, 155
principal ideal domain, 155
pseudo-reflection, 9
pseudoregular, 217
pseudoregular elements
  eigenvalues on covariants, 225
  eigenvalues on invariants, 223

quasi-regular element, 237
quaternion group, 27, 73
quaternions, 67–77
  pure, 70, 159

rational points
  over finite fields, 258
real structure, 180
reflecting hyperplanes, 173
  of a reflection subquotient, 220

reflection, 9, 11
  unitary, 9
reflection coset, 228
reflection group, 10
  centre, 50
  construction from a Cartan matrix,
    18
  decomposition into irreducibles, 16
  field of definition, 19
  imprimitive, 25
  rank, 16
  reflection representation, 15
  ring of definition, 20
  root system, 21
  Shephard subgroup, 17
  support, 16
reflection subquotient, 214, 215
  irreducibility, 221
regular
  eigenspace, 216
  integer, 216
  vector, 216
regular element
  of a reflection coset, 236
regular vector, 185
regularity
  criteria for, 227
representation
  contragredient, 14
  definable over a subring, 14
  group representation, 14
  monomial, 25
  natural representation, 15
  reflection representation, 15
Reynolds operator, *see* averaging operator
ring of definition, 20, 153
root, 10
  long, 10, 101, 147
  short, 10, 157
  tall, 10, 147
root systems, 21
  for $G(m, p, n)$, 34
  for primitive groups, 153–155
  for rank 2 groups, 93

Schur's Lemma, 15
semi-invariant polynomial, 176
  corresponding to character, 178
Shephard and Todd theorem, 152
Shephard group, 87, 242
Shephard–Todd–Chevalley theorem, 48
simplicial complex
  of eigenspaces, 270
simply laced, 102
skew invariant, 172
skew polynomial, 173
Solomon, L., 201
space of regular elements, 255
spinor norm, 158, 252
Springer, T.A., 2, 5, 39, 44, 61, 63, 213, 215,
    242
star, k-star, 101
star-closed, *see* line system
Steinberg, R., 2, 4, 171, 174, 186
system of imprimitivity, 23

Taylor, D.E., 68, 250
tensor
  decomposition of S, 184
tensor power, 39
3-systems
  classification of, 127
transformation
  fixed points of, 9
  unitary, 8
twisted invariant, 228
twisting of tori, 266

unitary reflection, 10

vector space
  graded, 54

Weyl group, 18
Weyl group of a reflection, 18
Witt index, 159, 251

Z-function, 257
Zamolodchikov, A. B., 264
Zariski topology, 209

Printed in the United States
by Baker & Taylor Publisher Services